RAPID APPLICATION DEVELOPMENT USING VISUAL BASIC AND VISUAL C++

Pankaj Sharma
(Astt. Prof.)
Department of Information Technology
ABES, Ghaziabad

INNOVA PRESS

An Imprint of

S.K. KATARIA & SONS®

Publisher of Engineering Books
4885/109, Prakash Mahal, Dr. Subhash Bhargav Lane,
Opposite Delhi Medical Association, Daryaganj, New Delhi–110002 (INDIA)
Phone: +91-11-23243489, 43551243; *Mobile:* +91-9871775858
e-mail: katariabook@gmail.com, katariabooks@yahoo.com
Website: www.skkatariaandsons.com
(e-books, subscriptions and bulk buying also available on our website)

Published by:

S.K. Kataria & Sons®

4885/109, Prakash Mahal, Dr. Subhash Bhargav Lane,
Opposite Delhi Medical Association, Daryaganj, New Delhi–110002
Phone: +91-11-23243489, 43551243; Mobile: +91-9871775858
e-mail: katariabook@gmail.com, katariabooks@yahoo.com

First Edition : August, 2010; Second Edition: August, 2011, Reprint: 2023

Price: ₹ 550/-

Laser Typesetting by : **PLUS COMPUTERS,** Shahdara, Delhi-110093.

Printed at : Arora Printing Press, Delhi–110051

Preface.....

Rapid Application Development using Visual Basic and Visual C++ is a core subject for undergradute and Post graduate students in Computer Science Engineering (CSE), Information Technology Streams and Master of Computer Applications (MCA). This book is intended to serve as a textbook in accordance with the syllabus of Rapid Application Development using Visual Basic and Visual C++ offered by various universities in India as well as abroad.

In this book, a significant effort has been made to find simple ways to develop theoretical as well as Practical concepts of Rapid Application Development. A very strong emphasis, has been given on Practical examples. Neat and clean diagrams have been used for explanation.

Every effort has been made to make the concepts simple and comprehensive. This book is divided into eight chapters. The organisation of these chapters are :

Chapter 1	Gives the introduction to Windows programming.
Chapter 2	Describes the various parts of Visual C++ and Visual Programs.
Chapter 3	Introduces the concept of event oriented windows programming and device context.
Chapter 4	Explains how to work with mouse and keyboard in VC++
Chapter 5	Describes various controls and dialogues in Visual Basic and Visual C++
Chapter 6	Introduces the concept of serialisation and file handling.
Chapter 7	Gives the overview of win32 static DLL
Chapter 8	Explains the concept of Database connectivity using VB and CV++.

Pankaj Sharma
ABES, Ghaziabad

Acknowledgement.....

I get this opportunity to express my gratitude and thanks to the respected Director (administration) of ABES Engg. College, **Dr. S.K. Kansal** for his valuable technical suggestion and constant encouragement, without which this book would not have to come into existence.

I express kind gratitude of **Prof. R.G. Mendiratta**, Director (Academic) ABES Engg. College, Ghaziabad.

I wish to thank **Prof. C.L. Mittal** (Dean UG) ABES Engg. College, Gzb. I also like to remember the motivation initiated by **Shri M.D Punia** my father-in-Law, (Lead District Manager, PNB, Karnal).

I am grateful to **Prof. V.K. Srivastava,** Director of ABES, IT, Ghaziabad, for helping me through his insightful suggestions and construction inputs.

I wish to thank **Prof. Pradeep Sharma**, Head of Department of Information Technology, Department of ABES Engg. College, Ghaziabad, for their constant co operation, directly inspiration and encouragement.

I like to remember the motivation initiated by **Prof. D. Pandey** (HOD CSE) and Prof. Sunil Sharma (AHODCSE), ABES, Engg. College. I am also thankful to Prof. K.P. Jayant (HODMCA) for his inspiration.

I thank to my elder brother **Mr. Sanjay Sharma** (KIET), Ghaziabad for inspiration and encouragement.

I thank to **Mr. Aswin Perti,** Lecturer in IT Department for his technical suggestions.

I am grateful to my younger Brother **Sandeep** and daughter **Prisha** for inspiring me for this project.

I deeply express my heartful thanks to the **Mr. Sanjeev Kataria** for **S.K. Kataria & Sons** for publishing this book in such a beautiful getup and well in time.

Lastly I am grateful to Mr. Manoj Tyagi and his staff from Plus Computer's for thier hardwork.

Pankaj Sharma

Dedicated This book to

My Parents, Wife Priyanka, Daughter Prisha

and

Readers of this book

Contents.....

Chapter-6

Chapter-7

Chapter-8

1

Introduction To Windows Programming

1.0 INTRODUCTION

C++ is an object-oriented version of the good old C language. Visual C++ is further extension to C++ that includes all the necessary supports for Windows programming in 32-bit format on both Windows 9X and NT platforms.

Before one attempts to build VC++ applications, it is desirable as well as advisable to have a basic understanding of the fundamentals of the Windows programming. Some of the most important concepts are dealt with here under. Windows (with capital 'W' and 's' in the end) is used to refer to Microsoft's operating system whereas window (singular form with small 'w') refers to a rectangular area on the computer's screen, in which an application executes. An application window usually has tools and means, visual or otherwise, to perform window-related actions - *viz.* minimising, maximising, closing the window, moving the window and the like. Visual C++, an extension of C++ programming language, is one of the many programming languages employed to develop Windows programs. It has rich features to support all the aspects of Windows programming. Among others such languages are Visual Basic, VJ++ etc. But the power and flexibility that Visual C++ provides to the programmer is unmatched.

It utilises pre-designed classes and resources for visual components, which can be easily incorporated in an application. Some of these visual components are: application window, icons, menu, toolbars, combo box, edit box, sliders etc. Using these standard resources and components gives a consistent look and feel to an application.

Chapter Outline

- Introduction
- Event Driven Programming Model
- Visual Basic Development Environment
- The Visual C++ Development Environment
- Components of Visual C++ and Visual Basic

1.1 EVENT DRIVEN PROGRAMMING MODEL **(MDU 2007)**

In the traditional programming model, the flow of program execution, follows a certain path set down by the programmer. This type of programming is known as sequential programming.

In the event driven programming model, application on execution, sets up variables and structures, and perform other initialisations, just as a typical procedural program does. The windows application now sits there, waiting for user input of some form or the other. This input can be in the form of a mouse click or a keystroke. As soon as user provides this input, a series of events follows and the application responds to this events.

Windows programming is different in the sequence driven programming model, the operating system simply executes the program and then waits for it to finish while programming for windows, the OS not only executes the program, it talks to it! And the Application talks back *i.e.,* They communicate. Windows programming is based an event driven programming model.

1.2 VISUAL BASIC DEVELOPMENT ENVIRONMENT **(V.Imp)**

In design-time environment you create and modify objects, design screens (forms) and enter code. The run-time environment simulates running the program that you created in the designing environment.

The run-time simulation has several advantages. One, it makes the development process faster than it would be if you had to stop to compile a program, and then run it, to determine the impact of the changes you have done in the source program. Another, the VB run-time environment provides tools that help you understand exactly what goes on when your application runs.

1.2.1 Projects and Forms

Starting a New Project

Visual Basic programs are organised into projects. It means that the code written as procedures (events) will become program when the project is compiled. The forms contained in the project become windows.

Only one project can be opened at a time. Of course, you can open more than one instance of VB and you can open a project in each instance. In an instance, you can add more than one project to create a project group. But in that also only one project can act as the active project.

You can start a new project by using the project dialog displayed when you start the VB.

To start a new project :

1. Start Visual Basic

   The New Project dialog is displayed (See Fig. 1.1)

2. Select Standard Exe project type
3. Press Open

The large dialog box in the middle of the screen contains three tabs below the colourful banner at the top of the dialog. Use the default New tab. Leave the default standard EXE icon highlighted as you click the Open button in the lower-right corner of this dialog box.

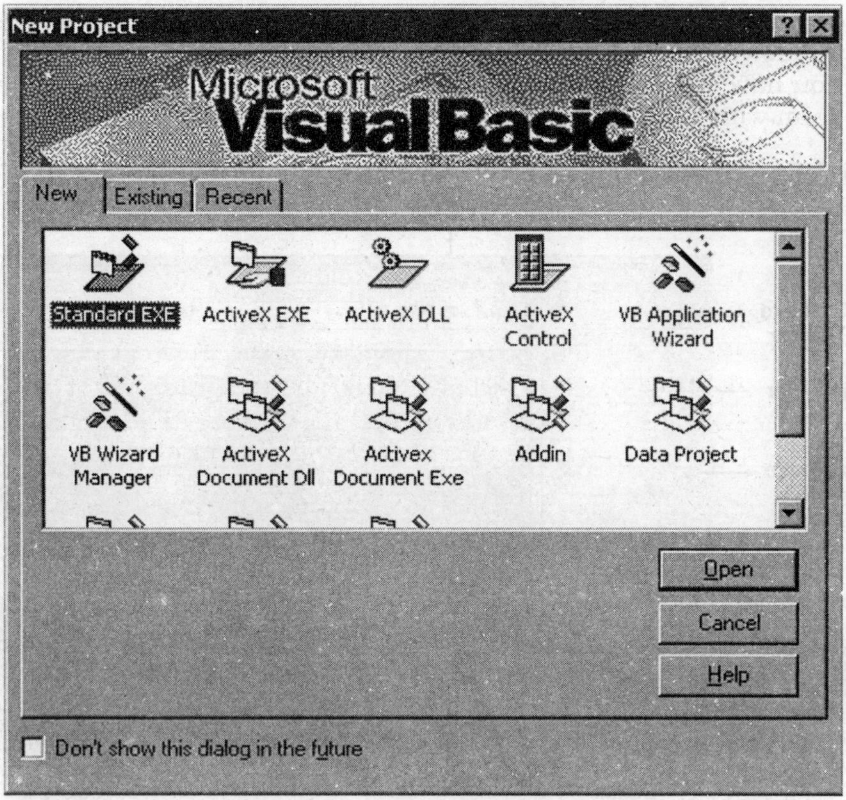

Fig. 1.1. Starting New Project

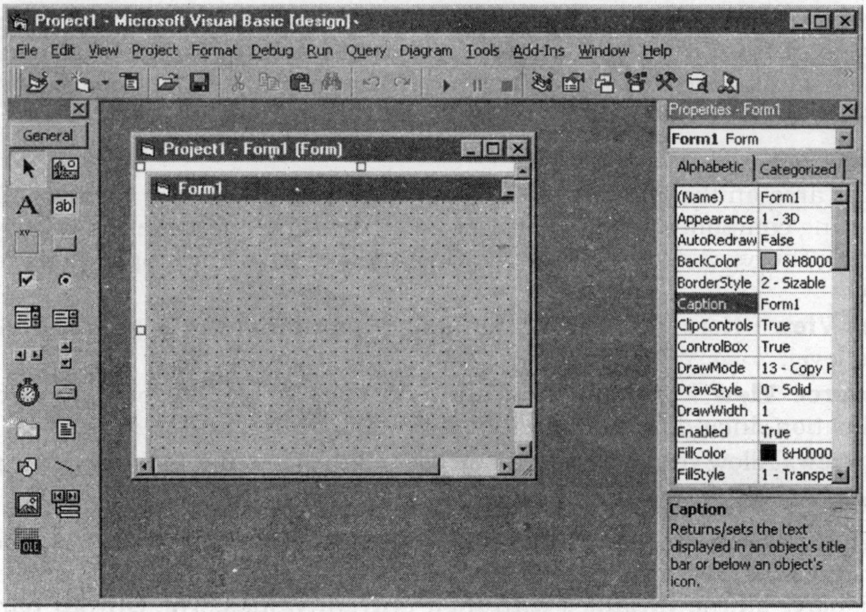

Fig. 1.2 The featureless Visual Basic Environment

In the above Fig. 1.2, the form window has been moved and resized to reveal the other Visual Basic windows under it. Visual Basic known you need a form to serve as the user interface, so you have been provided with an empty form and all the tools necessarey to populate the form with controls.

1.3 THE VISUAL C++ DEVELOPMENT ENVIRONMENT **(V.Imp)**

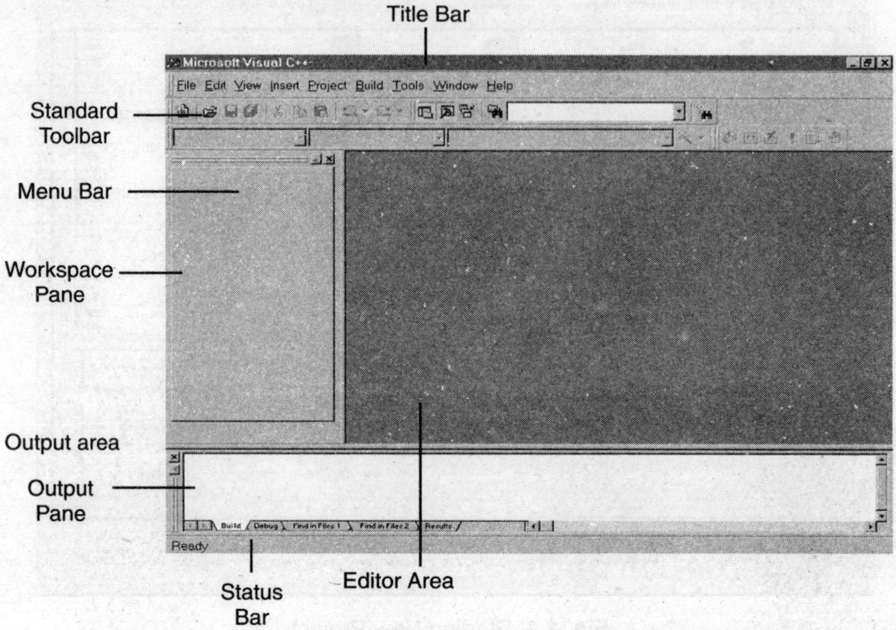

Fig. 1.3. The opening window of the VC++ Development Environment

To get yourself familiar with the Microsoft VC++ development environment, start the opening screen of the Visual C++ in order to see the different areas and panes that are visible in the environment. Each area has a different purpose. The opening window looks like Fig. 1.3.

The Workspace Pane

This is the area on the left side of the starting screen of the VC++, where IDE displays the participating components of the project *i.e.,* the source files, menus, icons etc., are in explorer form. The workspace allows viewing the parts of the application in the three different ways :

(*i*) **Class View :** Here, you can navigate and manipulate your source code on a C++ class level.

(*ii*) **Resource View :** You can find and edit each of the various resources in your application, including dialog window designs, icons, and menus.

(*iii*) **File View :** Allow you to view and navigate all the files that make up your application.

The Output Pane

The Output pane is the place where IDE provides its processing information, like the compiler progress statements, warnings, and error messages and is where the VC++ debugger displays all the variables with their current values. The Output pane re-opens itself when VC++ has any message that is needed to be displayed. The output pane might not be visible when VC++ is started for the first time. After the compilation of the first application, it appears at the bottom of the IDE and remains open until the user closes it.

The Editor Area

It is the area that you see on the right side of the IDE. All the editing is done in this area in VC++, here the code editor window display when VC++ source code is edited and the window painter displays when you design a dialog box. The editor area is even the icon painter displays when you design icons for use in your application. The editor area is basically the entire IDE area that is not otherwise occupied by panes, menus, or toolbars.

Title Bar

Every window has a title bar displaying the application's name and containing window control buttons (*i.e.*, buttons to maximise, minimise and terminate window)

Menu Bars

Next to the title bar lies another bar containing various menu options available to the users of the application.

Tool Bars

A tool bar is a collection of different buttons placed on a panel. The tool bars are displayed just below the menu bar. The three tool bars that are initially open are the following :

1. The Standard toolbar contains most of the standard tools for opening and saving files, cutting, copying, pasting, and the variety of other commands that you are likely to find useful.
2. The WizardBar toolbar enables you to perform a number of Class Wizard actions without opening the class wizard.
3. You can avail the facility of Build minibar that provides you with the build and run commands you are most likely to develop and test your applications.

Status Bar

At the bottom of the window, another bar can be found that is used to report various activities happening in the application. It displays indicators; whether or not Caps is locked, Num is locked etc. It can, however be programmed to serve a variety of other purposes.

1.4 COMPONENTS OF VISUAL C++ AND VISUAL BASIC **(MDU 2005)**

1.4.1 VC++ Components

There are two particularly important pieces of information about every control of your application. The information stored in the control and the action or type of action the user can perform on that control. At one time in the life of an application, a control holds a value that can be of particular interest either to you or to the user. On the other hand, when interacting with the computer, based on your application, the user will usually face different types of controls that do various things and produce different results. These are two features of controls (their values and the actions the user can perform on them) that you should known as much as possible, about the controls you choose to involve in your application.

To support the ability to create a control, the Win32 library provides two functions named CreateWindow and **CreateWindowEx**. The syntaxes of these functions are :

```
HWND CreateWindow(
      LPCTSTR lpClassName,
      LPCTSTR lpWindowName,
```

```
HWND CreateWindowEx(
DWORD dwExStyle,
LPCTSTR lpClassName,
```

```
    DWORD dwStyle,
    int x,
    int y,
    int nWidth,
    int nHeight,
    HWND hWndParent,
    HMENU hMenu,
    HINSTANCE hInstance,
    LPVOID lpParam
);
```

```
    LPCTSTR lpWindowName,
    DWORD dwStyle,
    int x,
    int y,
    int nWidth,
    int nHeight,
    HWND hWndParent,
    HMENU hMenu,
    HINSTANCE hInstance,
    LPVOID lpParam
);
```

In most cases, you can use either of these functions. When doing this, you must specify the control's name. To help with this, the Win32 library provides already defined names of classes you can use to create a control.

When creating a window, you need to declare its handle and initialise it with the CreateWindow() or the CreateWindowEx() function. If the window is created successfully, the function returns a window handle. If the control created was a parent, its handle can be used as the hWndParent argument.

The reason you declare a handle to the parent window is to allow you to refer to it later on. In the same way, if you need to refer to any control that is part of your application, you should define its own handle.

The Microsoft Foundation Class library, MFC, is a Microsoft custom implementation of the Win32 library. While programming in Win32 is mostly done manually by writing code, Microsoft Visual Studio makes it possible to visually create an application by adding objects to a host such as a dialog box. Still, most of the time you will need to know the class used to manage an object.

The most fundamental class used for visible objects in the MFC is called CWnd. The CWnd class is derived from the CCmdTarget class :

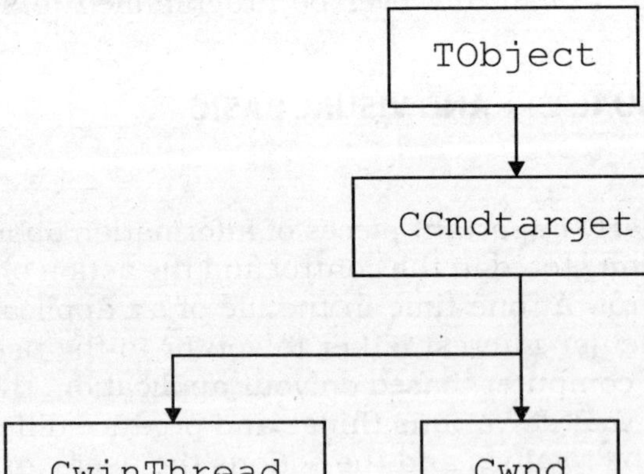

As we will see throughout our lessons, the CWnd class provides all, or almost all, of the functionality you need to create or manage a Windows control. When necessary, we will mentioned the CreateWindow() and/or the CreateWindowEx() function to explain a concept, but most of the time, you will use either CWnd or one of its derived classes to take necessary action.

To assist you with creating a control, the CWnd class is equipped with a method named Create that corresponds to the Win32's CreateWindow() function. The syntax of the CWnd::Create() function is :

```
virtual BOOL Create(LPCTSTR lpszClassName,
    LPCTSTR lpszWindowName,
    DWORD dwStyle,
    const RECT& rect,
    CWnd* pParentWnd,
    UINT nID,
    CCreateContext* pContext = NULL);
```

To provide the ability to apply extended styles to a control, the CWnd class is equipped with a CreateEx() method that is overloaded with two versions. Their syntaxes are :

```
virtual BOOL CreateEx(DWORD dwExStyle,
    LPCTSTR lpszClassName,
    LPCTSTR lpszWindowName,
    DWORD dwStyle,
    int x,
    int y,
    int nWidth,
    int nHeight,
    HWND hWndParent,
    HMENU nIDorHMenu,
    LPVOID lpParam = NULL );
virtual BOOL CreateEx(DWORD dwExStyle,
    LPCTSTR lpszClassName,
    LPCTSTR lpszWindowName,
    DWORD dwStyle,
    const RECT& rect,
    CWnd* pParentWnd,
    UINT nID,
    LPVOID lpParam = NULL);
```

The Control's Class Name

To create a control, you must specify its name. To do this, you have various options. The Win32 library provides already defined names of classes you can use to create a control. These classes are :

Class Name	*Description*
STATIC	A static control can be used to display text, a drawing, or a picture.
EDIT	As it name suggests, an edit control is used to display text to the user, to request text from the user, or both.

RichEdit	Like an edit box, a rich edit control displays text, requests it, or does both. Besides all the capabilities of the edit control, this control can display formatted text with characters or words that use different colours or weight. The paragraphs can also be individually aligned. The RichEdit class name is used to create a rich edit control using the features of Release 1.0 of its class.
RICHEDIT_CLASS	This control is used for the same reasons as the RichEdit class name except that it provides a few more text and paragraph formatting features based on Release 2.0.
LISTBOX	A list box is a control that displays items, such as text, arranged so each item, displays on its own line.
COMBOBOX	A combo box is a combination of an edit control and a list box. It holds a list of items so the current selection displays in the edit box part of the control.
SCROLLBAR	A scroll bar is a rectangular object equipped with a bar terminated by an arrow at each end. It is used to navigate left and right or up and down on a document.
BUTTON	A button is an object that the user clicks to initiate an action.
MDICLIENT	This class name is used to create a child window frame for an MDI application.

To use one of these classes, you have various options. At design time, in the Toolbox, you can click the Custom Control button and click the dialog box. In the Properties window, click Class and type the class name. If you are programmatically creating the control, pass its name as string to the lpszClassName argument of the Create() or the CreateEx() methods. Here is an example that would start an edit box.:

```
void CSecondDlg::OnThirdControl()
{
    CWnd *memo = new CWnd;

    memo->Create("BUTTON", );
}
```

The other option you have is to specify the class name as NULL. In this case, the compiler would use a default name:

```
void Whatever ()
{
    CWnd *second = new CWnd;

    second->Create(NULL, );
}
```

1.4.2 Visual Basic Components **(V.Imp)**

Controls are ready made program objects designed by any other user or by yourself, which can be used by you in your program. These are compiled programs with special characteristics. Generally, controls are designed and written in C, C++, VC++ or VB6.

Controls are objects that combine code with visual parts. They require a container *i.e.*, as an object (like form or any other control) that carries the control.

The developer part of the control includes an icon for the Toolbox. The part of the control has visual representation when it is placed on a container, such as a form. Control has properties, method and events, which are collectively known, as control's interface. The creator of the control is responsible for the programming that enables the interface: for making sure that properties appear in the properties window, that events are fired when they are supposed to be fired etc.

The user of the control doesn't see the internal implementation of the control's application. The users interaction with the control depends on the interface settings that the application developer has made and on code added by the developer.

The icons, which you see on the toolbox when you start a VB project, are all standard controls provided by Microsoft. (see fig. 1.4)

Fig. 1.4. Standard Microsoft controls

Adding and Removing Controls

Controls can be added to and removed from the Toolbox using Component dialog :

1. From the Project menu, choose Components or right-click on the Toolbox and choose Components from the pop-up menu. The Components dialog open (see Figure 1.5)

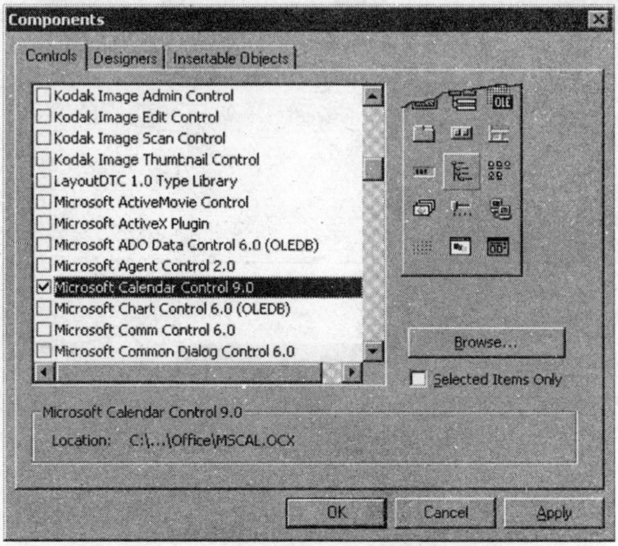

Fig. 1.5. Adding a control to tool box

Put a check mark against the control, which you want to add to Toolbox. Lets click on Microsoft Calendar Control 9.0 (See Figure 1.5)

2. Click Close or click Apply and close the Component dialog when you are finished adding controls.

Now you can see the calendar control icon in Toolbox (See Figure 1.6.)

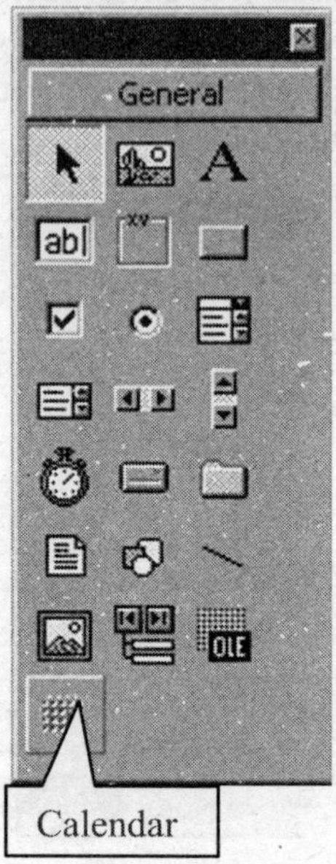

Fig. 1.6. Toolbox with added control

Similarly, to remove a control from the toolbox, you have to uncheck the name of that control in Component dialog. You will not be able to remove any control, which is in use by your application *i.e.*, on any of the form in your application.

Controls :

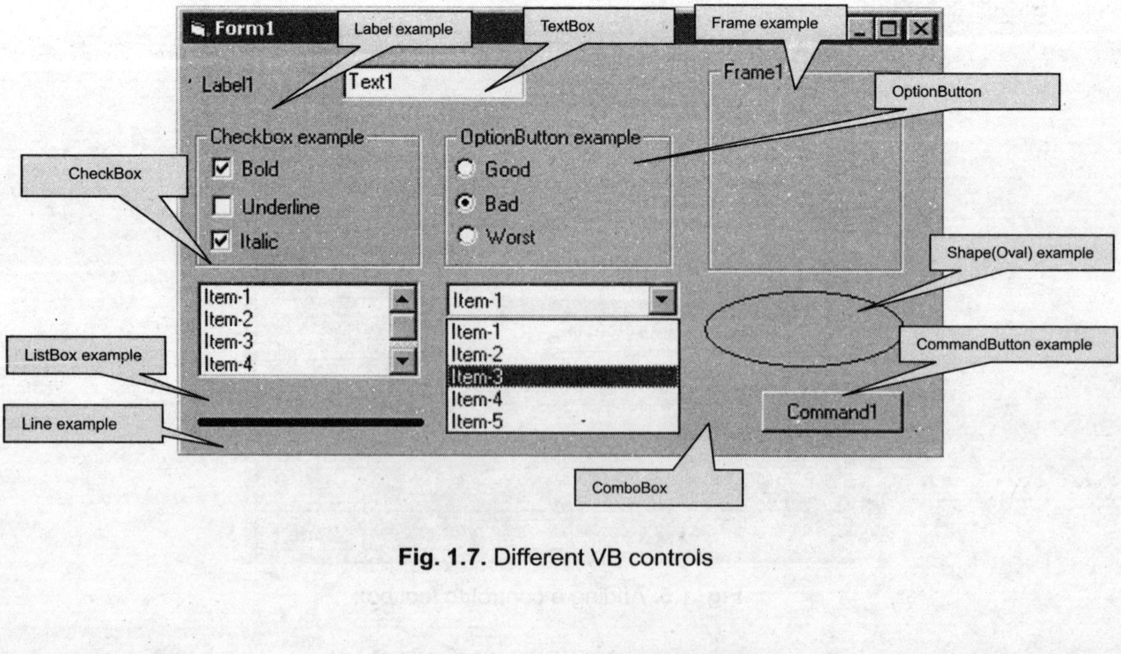

Fig. 1.7. Different VB controls

Label Control

Use: This control is used to display some static (*i.e.,* user cannot change it) text on the screen. This is generally placed against the controls (like textbox) which accept some input from the user, to convey the message what data to enter. (see figure 1.7).

Textbox Control

Use: This control is used to accept some data from the user. User can type in the textbox and that value can be accessed in any procedure by referring to the Text property of the textbox control.

Example of text box: A Text Editor with cut and paste utility :

Here, to make a text editor, we will go for a multiline text box and Visual Basic supports multiline textboxes automatically. To see how this works, start a Visual Basic project and change the Caption property of the form that appears to Text Editor. Then change its Name property to frmTextEditor. In the next step, save the form and project, using the Save Project item from the File menu. Save the form and the project.

To produce the multiline text box, simply double-click on the text box and stretch the resulting default-sized text box until it takes up most of the form, leaving room for a row of buttons at the bottom as in figure 1.8.

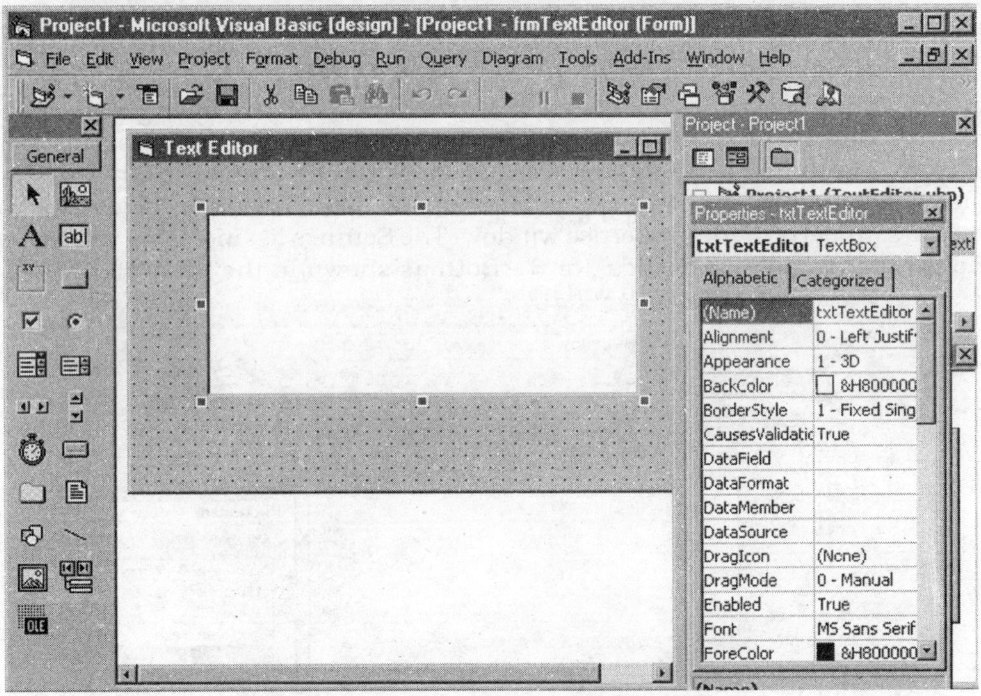

Fig. 1.8. The textbox when it is being designed

Now, find the Multiline property in the textbox's Properties window. As indicated in the Settings list for that property, two settings, True and False, are allowed. The default is False, which means that text boxes can handle only a single line of text by default. Set this property to True to give your text editor multiline capability.

At this point of time, you are ready with your textbox. Delete the default text, Text1, from the Text property using the Properties window, and change the Name property from the default, Text1, to the txtTextEditor. Now, run the program. The textbox appears, as in figure below and you can write to it. Yes, the textbox comes with the complete word-wrap facility. When you get to the end of the line, notice that the current word wraps in its entirety to the next line instead of being broken in the middle. (see Figure 1.9).

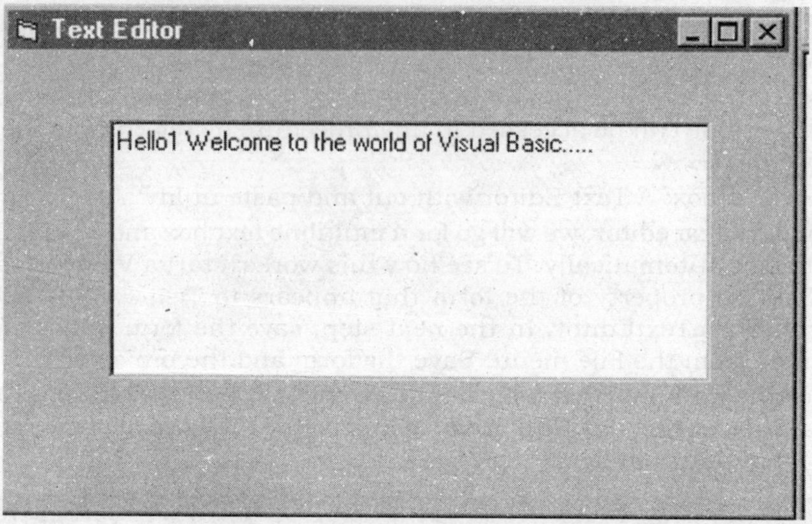

Fig. 1.9. A textbox tool at run-time

You can add a vertical scrollbar the your text box just by changing its ScrollBars property from 0-None to 2-Vertical. You can also add horizontal scrollbars as well, but if you did so to a multiline text box, the word wrap would be automatically turned off. For that reason, you are not going to add a horizontal scrollbar. To add vertical scrollbars, find the ScrollBars property in the Properties window. The Settings list indicates what is available: 0 - None, 1 - Horizontal, 2 - Vertical, or 3 - Both as shown in the figure 1.10.

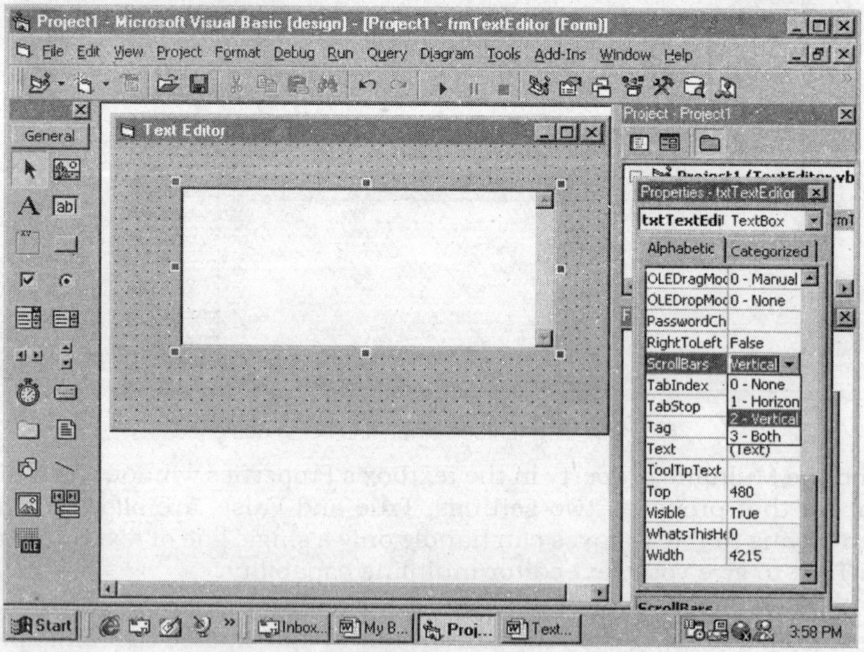

Fig. 1.10. Scrollbar being added to the textbox control

To add vertical scrollbars to the textbox in your notepad, just select the Vertical option. A vertical scrollbar appears, as shown in figure 1.11.

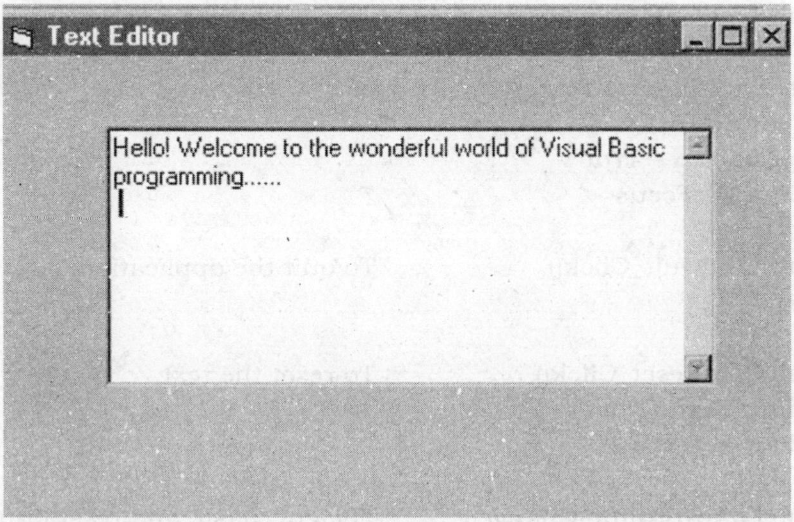

Fig. 1.11. The textbox with the vertical scrollbar at the run-time

Now, at this point, you can scroll through a much larger document in your new Text Editor.

Until now, we have not added any capabilities to the application. Its time to go for that.

From the toolbox, take 3 command buttons and change their captions as Cut, Copy and Paste.

The following table suggests the event code for the Text Editor Application :

Object	***Property***	***Setting***
Form	Appearance	1-3D
	Caption	Text Editor
	Name	frmText Editor
TextBox	MultiLine	True
	Name	txtText Editor
	ScrollBars	2-vertical
	Text	(None)
Command Button	Caption	&Reset
	Enabled	False
	Name	CmdReset
Command Button	Caption	&Cut
	Enabled	False
	Name	CmdCut
Command Button	Caption	&Paste
	Enabled	False
	Name	CmdPaste

Source code for the Text Editor application.

```
Dim store As String
Private Sub CmdCut_Click()                    ' To cut the text.
CutText = txtText Editor.SelText
store = txtText Editor.SelText
txtTextEditor.SelText = " "
CmdPaste.Enabled = True
txtText Editor.SetFocus
End Sub
Private Sub CmdQuit_Click()                   ' To quit the application.
Unload Me
End Sub
Private Sub CmdReset_Click()                  ' To reset the text.
txtText Editor.Text = " "
txtText Editor.SetFocus
End Sub
Private Sub txtTextEditor_Change()            ' The cut, paste and reset buttons will
CmdCut.Enabled = True                         ' be highlighted only if you type
CmdPaste.Enabled = True                       ' some text.
CmdReset.Enabled = True
End Sub
Private Sub CmdPaste_Click()                  ' To paste the text.
txtText Editor.SelText = store
txtText Editor.SetFocus
End Sub
```

The output of the above code is shown below in Figure 1.12.

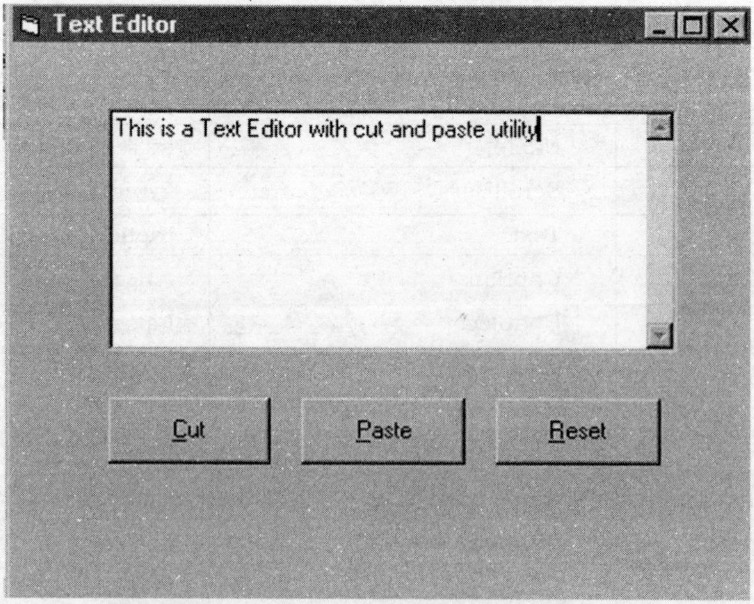

Fig. 1.12. The Text Editor with cut and paste utility

Command Button Control

Use: This control is used to get the user's response depending on which some specific action can be taken. For example, you can see the Ok and Cancel buttons on the screens of the products of Microsoft (Figure 1.13).

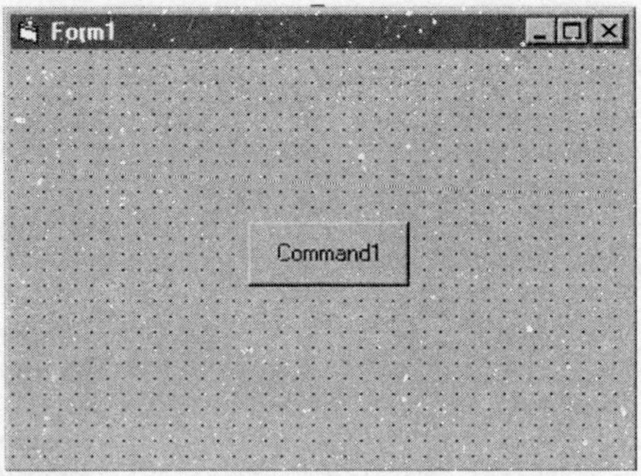

Fig. 1.13. The command button at design-time

Shape Control

Use: This control is used to draw different shapes on the form. Available shapes are : (see Figure 1.14)

- 1 - Square
- 2 - Oval
- 3 - Circle
- 4 - Rounded Rectangle
- 5 - Rounded Square

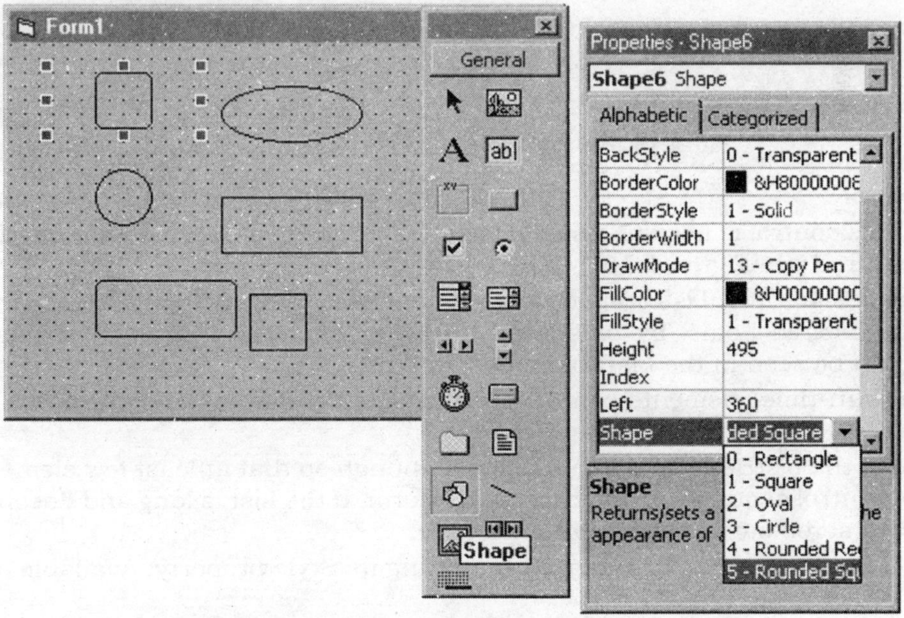

Fig. 1.14. Different types of shapes at design-time

Line Control

Use : This control is used draw a line on the form. Line width can be set by the BorderWidth property and style (Transparent, solid, dashed, dot.) can be set by BorderStyle property of this control. (Figure 1.15).

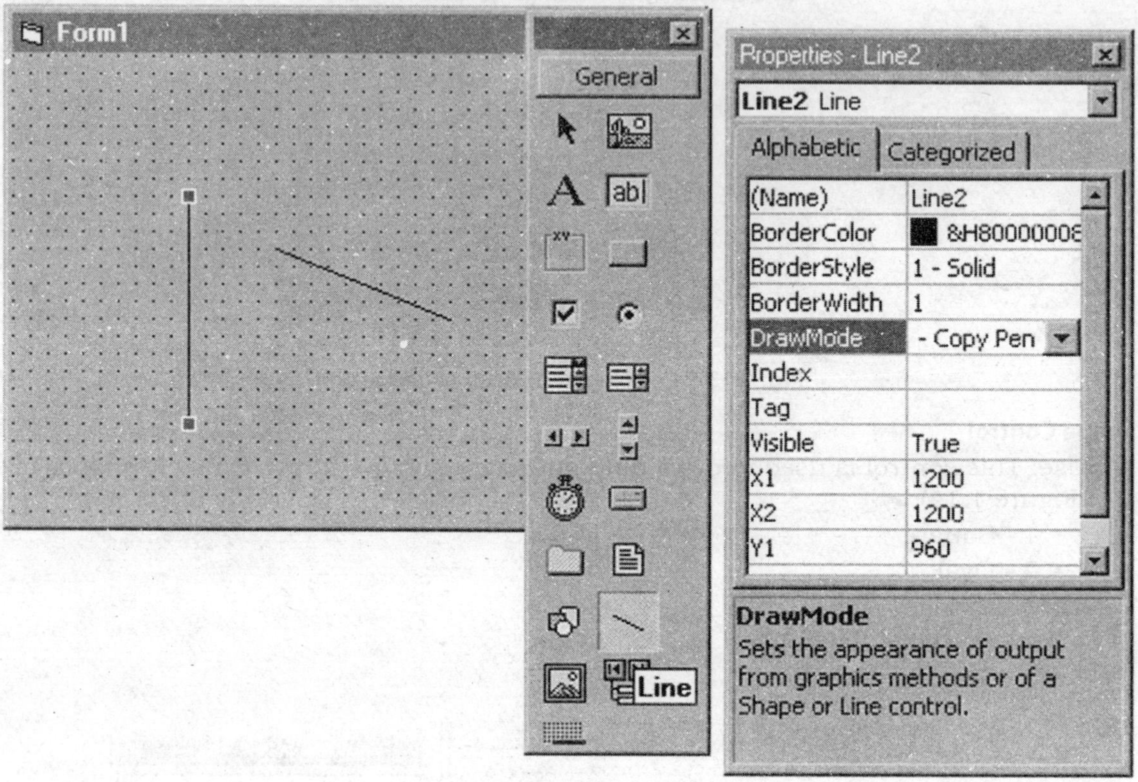

Fig. 1.15. The line control with the properties window

ListBox Control **(MDU 2008)**

Use : This control is used to display a list. List items can be added both at design time and at run-time (see Figure 1.16).

During design time, List property of this control is to be set. In this property you can add all the desired items, by pressing Ctrl+Enter after entering each element. All the elements can be seen in the ListBox in design time also.

During run-time, elements can be added in the ListBox by using AddItem method of ListBox.

Height of the ListBox should be sufficient enough so that at least few elements can be seen. This control acquires some space on the form. If the List is long and doesn't fit in the ListBox then scroll bars appears automatically.

Style of ComboBox can be changed by setting its Style property. Available options are :

- 0 - Standard
- 1 - Checkbox

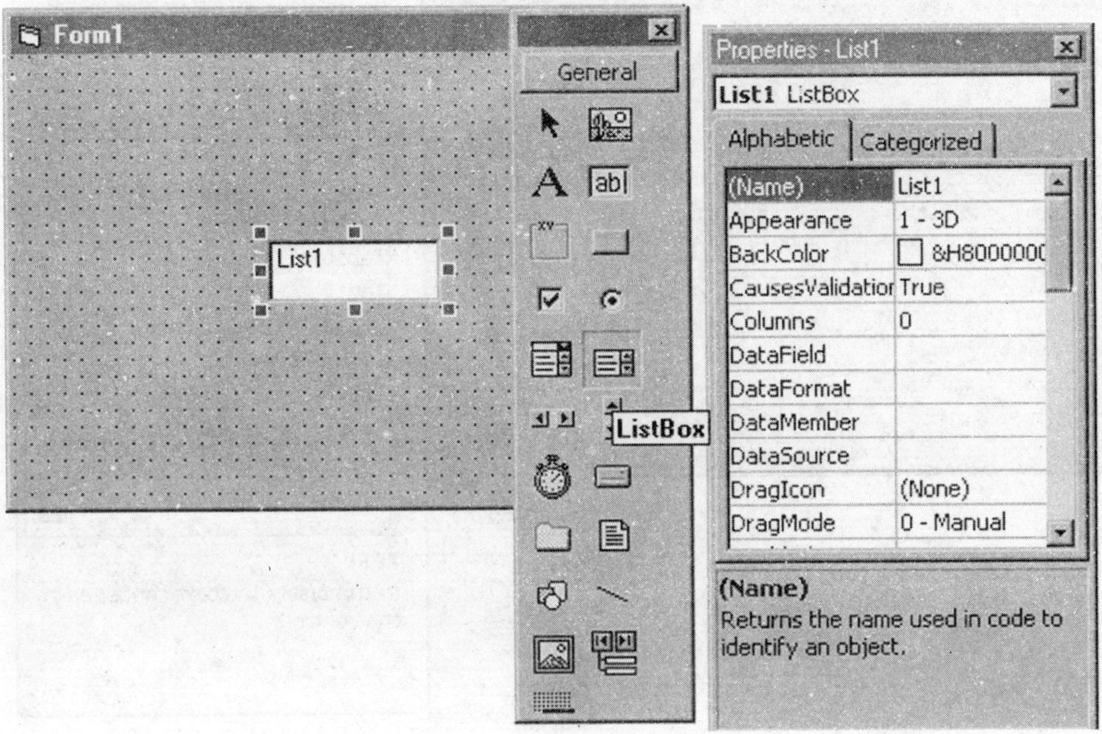

Fig. 1.16. The list box control

ComboBox control **(MDU 2008)**

Use: This control is used to display a list in compact form. And so this control consume less space and still can have all the elements of list in t. List elements can be added both at design time and at run-time (see Figure 1.17).

During design time, List property of this control is to be set. In this property you can add all the desired items, by pressing Ctrl+Enter after entering each element. All the elements can be seen in the ListBox in design time also.

During run-time, elements can be added in the ListBox by using AddItem method of ListBox.

Style of ComboBox can be changed by setting its Style property. Available options are :

- 0 - Dropdown Combo
- 1 - Simple Comb
- 2 - Dropdown List

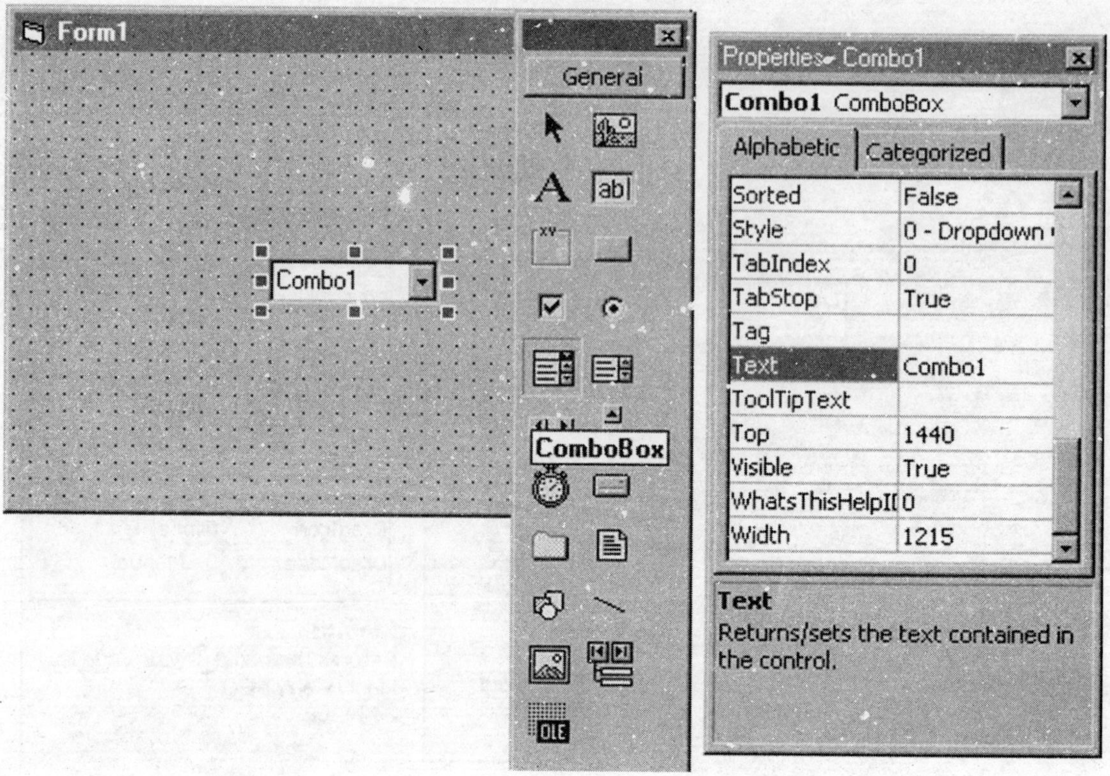

Fig. 1.17. The Combo box with properties window

SUMMARY

1. C++ is an object-oriented version of the good old C language. Visual C++ is further extension to C++ that includes all the necessary supports for Windows programming in 32-bit format on both Windows 9X and NT platforms.
2. In the event driven programming model, application on execution, sets up variables and structures, and perform other initialisations, just as a typical procedural program does. The windows application now sits there, waiting for user input of some form or the other. This input can be in the form of a mouse click or a keystroke.
3. When creating a window, you need to declare its handle and initialise it with the CreateWindow() or the CreateWindowEx() function. If the window is created successfully, the function returns a window handle. If the control created was a parent, its handle can be used as the hWndParent argument.
4. Controls are ready made program objects designed by any other user or by yourself, which can be used by you in your program. These are compiled programs with special characteristics. Generally, controls are designed and written in C, C++, VC++ or VB6.
5. During design time, List property of this control is to be set. In this property you can add all the desired items, by pressing Ctrl+Enter after entering each element. All the elements can be seen in the ListBox in design time also.
6. During design time, List property of combo box control is to be set. In this property you can add all the desired items, by pressing Ctrl+Enter after entering each element. All the elements can be seen in the ListBox in design time also.
7. This control is used for the same reasons as the RichEdit class name except that it provides a few more text and paragraph formatting features based on Release 2.0

8. Like an edit box, a richedit control displays text, requests it, or does both. Besides all the capabilities of the edit control, this control can display formatted text with characters or words that use different colours or weight. The paragraphs can also be individually aligned. The RichEdit class name is used to create a rich edit control using the features of Release 1.0 of its class
9. The Microsoft Foundation Class library, MFC, is a Microsoft custom implementation of the Win32 library. While programming in Win32 is mostly done manually by writing code, Microsoft Visual Studio makes it possible to visually create an application by adding objects to a host such as a dialog box.
10. There are two particularly important pieces of information about every control of your application. The information stored in the control and the action or type of action the user can perform on that control. At one time in the life of an application, a control holds a value that can be of particular interest either to you or to the user. On the other hand, when interacting with the computer, based on your application, the user will usually face different types of controls that do various things and produce different results.

SHORT QUESTIONS WITH ANSWER

Q.1. How can I change the title on the message box, instead of using the application name?

Ans. By default, the message box window uses the application name as the window title. You can change this by adding a second text string to the MessageBox function call. The first string is always the message to be displayed, and the second string is used as the window title. For example, the OnHello function would look like,

```
// Say hello to the user
MessageBox("Hello. This is my first Visual C++ Application!", "My First
Application");
```

Q.2. Can I change the text on the About window to give my company name and more detailed copyright information?

Ans. Yes, the About window is in the Dialogs folder in the Resources View tab of the workspace pane. If you double-click the IDD_ABOUTBOX dialog, the About box will be opened in the dialog designer, where you can redesign it however you want.

Q.3. What do you mean by Event Driven Programming Model?

Ans. In the traditional programming model, the flow of program execution, follows a certain path set down by the programmer. This type of programming is known as sequential programming.

In the event driven programming model, application on execution, sets up variables and structures, and perform other initialisations, just as a typical procedural program does. The windows application now sits there, waiting for user input of some form or the other. This input can be in the form of a mouse click or a keystroke. As soon as user provides this input, a series of events follows and the application responds to this events.

Windows programming is different in the sequence driven programming model, the operating system simply executes the program and then waits for it to finish while programming for windows, the OS not only executes the program, it talks to it! And the Application talks back *i.e.,* They communicate. Windows programming is based an event driven programming model.

Q.4. What are Visual Basic Controls? **(V.Imp)**

Ans. Controls are ready made program objects designed by any other user or by yourself, which can be used by you in your program. These are compiled programs with special characteristics. Generally, controls are designed and written in C, C++, VC++ or VB6.

Controls are objects that combine code with visual parts. They require a container *i.e.,* as an object (like form or any other control) that carries the control.

The developer part of the control includes an icon for the Tool box. The part of the control has visual representation when it is placed on a container, such as a form. Control has properties, method and events, which are collectively known, as control's interface. The creator of the control is responsible for the programming that enables the interface: for making sure that properties appear in the properties window, that events are fired when they are supposed to be fired etc.

The user of the control doesn't see the internal implementation of the control's application. The users interaction with the control depends on the interface settings that the application developer has made and on code added by the developer.

The icons, which you see on the tool box when you start a VB project, are all standard controls provided by Microsoft.

Fig. 1.18. Standard Microsoft controls

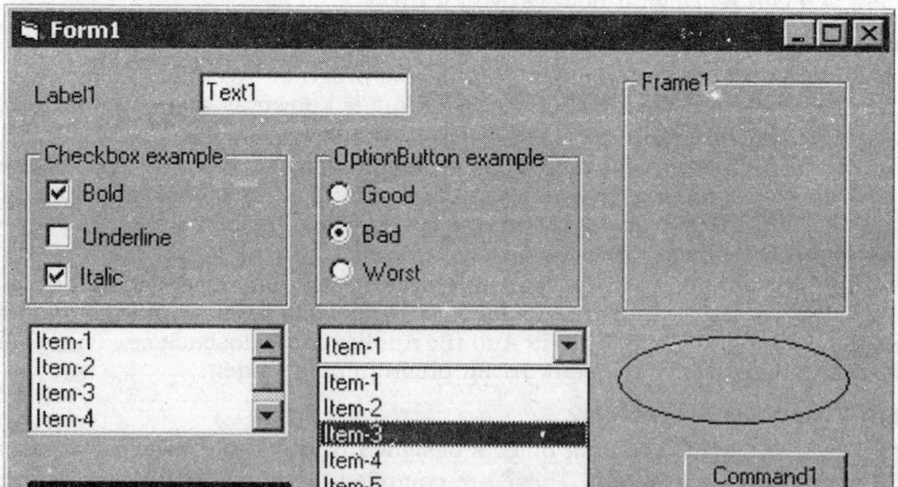

Fig. 1.19. Different VB controls

Q.5. Discuss the use of ListBox control.

Ans. Use : This control is used to display a list. List items can be added both at design time and at run-time.

During design time, List property of this control is to be set. In this property you can add all the desired items, by pressing Ctrl+Enter after entering each element. All the elements can be seen in the ListBox in design time also.

During run-time, elements can be added in the ListBox by using AddItem method of ListBox.

Height of the ListBox should be sufficient enough so that at least few elements can be seen. This control acquires some space on the form. If the List is long and doesn't fit in the ListBox then scroll bars appears automatically.

Style of ComboBox can be changed by setting its Style property. Available options are :

- 0 - Standard
- 1 - Checkbox

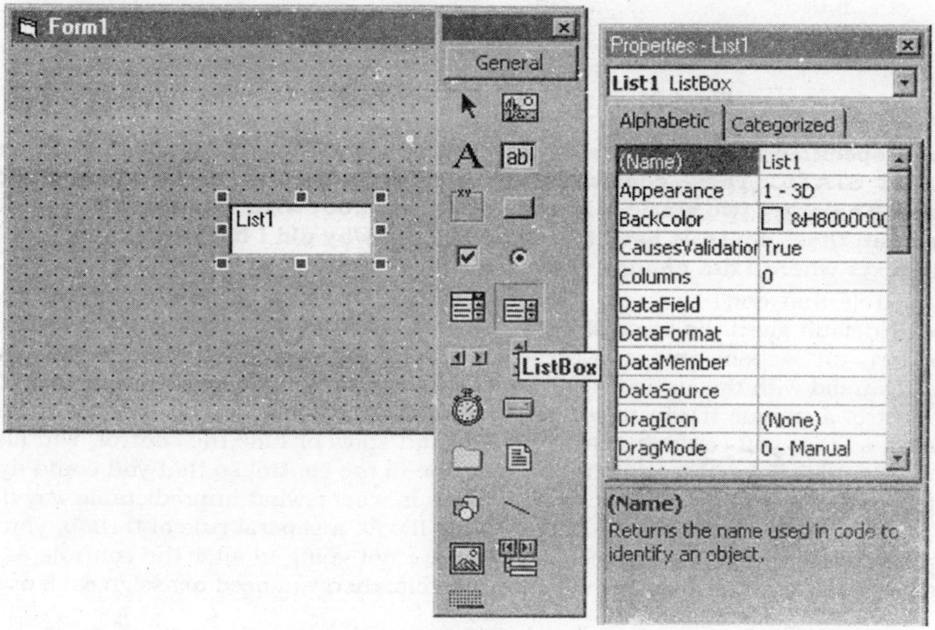

Fig. 1.20. The list box control

Q.6. Discuss the use of ComboBox control.

Ans. ComboBox Control

Use : This control is used to display a list in compact form. And so this control consume less space and still can have all the elements of list in t. List elements can be added both at design time and at run-time.

During design time, List property of this control is to be set. In this property you can add all the desired items, by pressing Ctrl+Enter after entering each element. All the elements can be seen in the List Box in design time also.

During run-time, elements can be added in the List Box by using AddItem method of List Box.

Style of Combo Box can be changed by setting its Style property. Available options are :

- 0 - Dropdown Combo
- 1 - Simple Comb
- 2 - Dropdown List

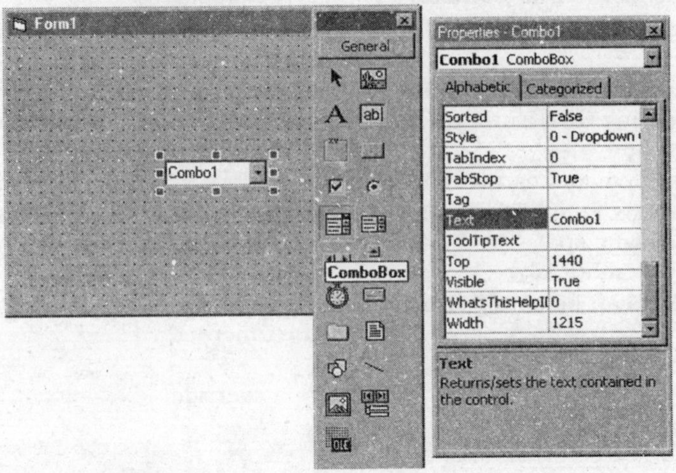

Fig. 1.21. The Combo box with properties window

Q.7. When I specified the object IDs of the controls on the window, three controls had the same ID, IDC_STATIC. These controls were the text at the top of the window and the two group boxes. The other two static text controls started out with this same ID until I changed them. How can these controls have the same ID, and why did I have to change the ID on the two static texts where I did change them?

Ans. All controls that don't normally have any user interaction, such as static text and group boxes, are by default given the same object ID. This works fine as long as your application doesn't need to perform any actions on any of these controls. If you do need to interact with one of these controls, as you did with the static text prompts for the edit box and combo box, then you need to give that control a unique ID. In this case, you needed the unique ID to be able to retrieve the control object so that you could enable or disable and show or hide the control. You also need to assign it a unique ID if you want to attach a variable to the control so that you could dynamically alter the text on the control. The application behaves in a somewhat unpredictable way if you try to alter any of the static controls that share the same ID. As a general rule of thumb, you can allow static controls to share the same object ID if you are not going to alter the controls at all. If you might need to perform any interaction with the controls, then you need to assign each one a unique object ID.

Q.8. Is there any other way to manipulate the controls, other than retrieving the control objects using their object IDs?

Ans. You can declare variables in the Control category. This basically gives you an object that is the control's MFC class, providing you with a direct way of altering and interacting with the control. You can then call all of the CWnd class functions on the control, as you did to enable or disable and show or hide the controls in your application, or you can call the control class methods, enabling you to do things in the code that are specific to that type of control. For instance, if you add another variable to the combo box control and specify that it is a Control category variable, you can use it to add items to the drop-down list on the control.

Q.9. Write an event procedure in VB to find the sum of numbers from 1 to selected value. Use horizontal scroll bar to set the maximum value.

Ans. Program: The format is:

```
For counter = startNumber to endNumber (Step increment)
One or more VB statements
Next
```

Sometimes the user might want to get out from the loop before the whole repetitive process is executed, the command to use is Exit For. To exit a For ... Next Loop, you can place the Exit For statement within the loop; and it is normally used to together with the If....Then....statement.

```
Dim counter, sum As Integer
For counter = 1 to 100 step 1
sum + = counter
ListBox1.Items.Add(sum)
Next !! *The program will calculate the sum of the numbers as follows:
sum=1+2+3+4+.... .
```

Q.10. Explain the concept of overriding with suitable example.

Ans. Overriding: Redefining the function body in the inherited class, in which the function name is same as of its base class.

Example: Class vehicle contains move ( ) which is overriden in the class helicopter.

```
#include <jostream.h>
class vehicle
{
    public:
    void start ();
    void move();
};
    class helicopter: public vehicle
{
    public: void main():
}
    void vehicle:: start( )
{
    cout «"staiting..\n;
}
    void vehicle:. move()
{
    cout <<"Driving";
}
    void helicopter:: move ()  //overriden function
{
    cout« "Flying,..\n";
}
    mid mach()
{
    helicopter hell;
    heli.start (),
    heli.move();
}
```

Q.11. Differentiate between procedural and object-oriented languages.

Ans. Procedural Language:

1. Emphasis is on doing things (algorithms).
2. Large programs are divided into smaller programs known as functions.
3. Most of the functions share global data.
4. Data move openly around the system from function-to-function.

5. Functions transform data from one form to another.
6. Employes top-down approach in program design.
7. Any change in data type being used needs to be made to all the functions using it. and it doesn't model the real world problems.

Object-Oriented Language:

1. Emphasis is on data rather than procedure.
2. Programs are divided into what are known as objects.
3. Functions that operate on data of an object are tied together in the data structure.
4. Data is hidden and can't be accessed by external functions.
5. Objects may communicated wijh each other through functions.
6. New data and functions can be easily added whenever necessary.
7. Follows bottom-up approach in program design.

Q.12. Explain various components of VC++ and VB by giving suitable examples.

Ans. Components of VB:

1. **Projects:** A VB project is a collection of forums, controls, reference to automation servers, application creation settings and version information.
2. **Controls:** A control is a tool used to display information or retrieve information from the user, e.g., boxes, buttons, graphics.
3. **Code Modules:** Source code is written to make controls and forms respond to user interactions.
4. **Class Modules:** Class modules allows you to create your own objects. They are the prototype for these objects, defining their properties, methods.
5. **User Controls:** You can build your own controls.
6. **Property Pages:** Property pages are used within the context of user controls and are used to view and change the properties of user controls in an easy to understand format.
7. **Resource Files:** A resource file allows you to store windows resources such as string, bitmaps, cursors & icons.

Components of VC++:

1. **Intrinsic Controls:** Intrinsic controls such as command button, check boxes, radio buttons etc, These contained inside VC++ .exe file.
2. **Active X Controls:** These exists as separate files with a .OCX file name extension,
3. **Projects:** The project file contains information at the project level and is used to build a single project or subproject.
4. **Header Files:** These files have extension name. This files contains declaration statements related to the application class used in the project.
5. **Source Files:** The actual code related to the project is contained in a file called source file,
6. **Resource Files:** This is a listing of all of the microsoft windows resources that the program uses.
7. **Class Wizard:** Class wizard is a tool that is ussd when we want to create a new project or want to add now files to an existing project.
8. **Icon File:** The icon file has an extension of .icon name. The file is used as the applications icon. This is included by the main resource.

LONG QUESTIONS WITH ANSWER

Q.1. Define Visual basic development environment.

Ans. In design-time environment you create and modify objects, design screens (forms) and enter code. The run-time environment simulates running the program that you created in the designing environment.

The run-time simulation has several advantages. One, it makes the development process faster than it would be if you had to stop to compile a program, and then run it, to determine the impact of the changes you have done in the source program. Another, the VB run-time environment provides tools that help you understand exactly what goes on when your application runs.

Projects and Forms: Visual Basic programs are organised into projects. It means that the code written as procedures (events) will become program when the project is compiled. The forms contained in the project become windows.

Only one project can be opened at a time. Of course, you can open more than one instance of VB and you can open a project in each instance. In an instance, you can add more than one project to create a project group. But in that also only one project can act as the active project.

You can start a new project by using the project dialog displayed when you start the VB.

To start a new project :

1. Start Visual Basic

   The New Project dialog is displayed.

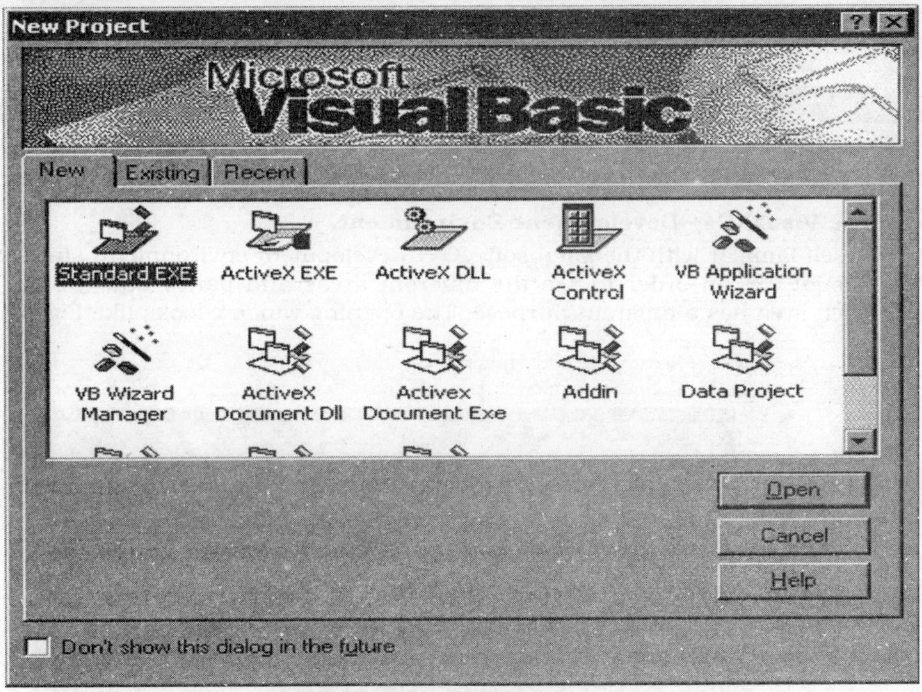

Fig. 1.22. Starting New Project

2. Select Standard Exe project type
3. Press Open

The large dialog box in the middle of the screen contains three tabs below the colourful banner at the top of the dialog. Use the default New tab. Leave the default standard EXE icon highlighted as you click the Open button in the lower-right corner of this dialog box.

In the above figure, the Form window has been moved and re-sized to reveal the other Visual Basic windows under it. Visual Basic knows you need a form to serve as the user interface, so you have been provided with an empty form and all the tools necessary to populate the form with controls.

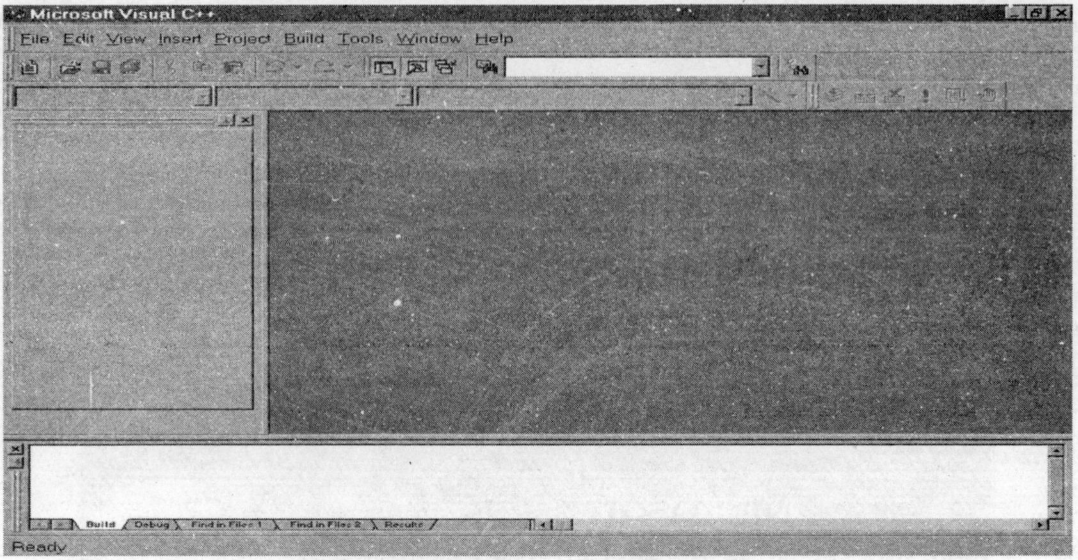

Fig. 1.23. The featureless Visual Basic environment

Q.2. Discuss The Visual C++ Development Environment.

Ans. To get yourself familiar with the Microsoft VC++ development environment, start the opening screen of the Visual C++ in order to see the different areas and panes that are visible in the environment. Each area has a different purpose. The opening window looks like Fig. 1.20.

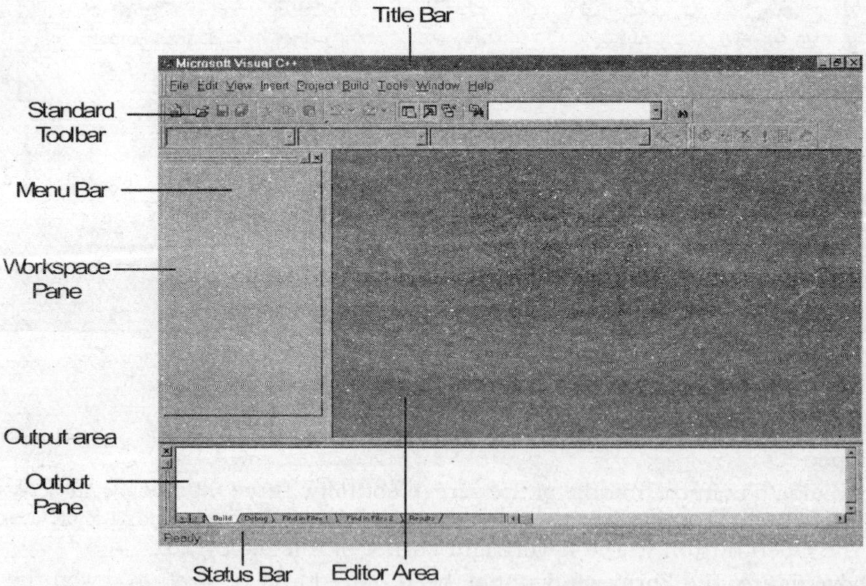

Fig. 1.24. The opening window of the VC++ Development Environment

The Workspace Pane: This is the area on the left side of the starting screen of the VC++, where IDE displays the participating components of the project *i.e.,* the source files, menus, icons etc., are in explorer form. The workspace allows viewing the parts of the application in the three different ways :

(i) **Class View :** Here, you can navigate and manipulate your source code on a C++ class level.

(ii) **Resource View :** You can find and edit each of the various resources in your application, including dialog window designs, icons, and menus.

(iii) **File View :** Allow you to view and navigate all the files that make up your application.

The Output Pane: The Output pane is the place where IDE provides its processing information, like the compiler progress statements, warnings, and error messages and is where the VC++ debugger displays all the variables with their current values. The Output pane re-opens itself when VC++ has any message that is needed to be displayed. The output pane might not be visible when VC++ is started for the first time. After the compilation of the first application, it appears at the bottom of the IDE and remains open until the user closes it.

The Editor Area: It is the area that you see on the right side of the IDE. All the editing is done in this area in VC++, here the code editor window display when VC++ source code is edited and the window painter displays when you design a dialog box. The editor area is even the icon painter displays when you design icons for use in your application. The editor area is basically the entire IDE area that is not otherwise occupied by panes, menus, or toolbars.

Title Bar: Every window has a title bar displaying the application's name and containing window control buttons (i.e. buttons to maximise, minimise and terminate window).

Menu Bars: Next to the title bar lies another bar containing various menu options available to the users of the application.

Tool Bars: A tool bar is a collection of different buttons placed on a panel. The tool bars are displayed just below the menu bar. The three tool bars that are initially open are the following :

1. The Standard tool bar contains most of the standard tools for opening and saving files, cutting, copying, pasting, and the variety of other commands that you are likely to find useful.
2. The WizardBar tool bar enables you to perform a number of Class Wizard actions without opening the class wizard.
3. You can avail the facility of Build minibar that provides you with the build and run commands you are most likely to develop and test your applications.

Status Bar: At the bottom of the window, another bar can be found that is used to report various activities happening in the application. It displays indicators; whether or not Caps is locked, Num is locked etc. It can, however be programmed to serve a variety of other purposes.

Q.3. What are VC++ Components? **(V.Imp)**

Ans. To create a control, you must specify its name. To do this, you have various options. The Win32 library provides already defined names of classes you can use to create a control. These classes are :

Class Name	*Description*
STATIC	A static control can be used to display text, a drawing, or a picture.
EDIT	As it name suggests, an edit control is used to display text to the user, to request text from the user, or both.
RichEdit	Like an edit box, a rich edit control displays text, requests it, or does both. Besides all the capabilities of the edit control, this control can display formatted text with characters or words that use different colors or weight. The paragraphs can also be individually aligned. The RichEdit class name is used to create a rich edit control using the features of Release 1.0 of its class.

RICHEDIT_CLASS	This control is used for the same reasons as the RichEdit class name except that it provides a few more text and paragraph formatting features based on Release 2.0.
LISTBOX	A list box is a control that displays items, such as text, arranged so each item, displays on its own line.
COMBOBOX	A combo box is a combination of an edit control and a list box. It holds a list of items so the current selection displays in the edit box part of the control.
SCROLLBAR	A scroll bar is a rectangular object equipped with a bar terminated by an arrow at each end. It is used to navigate left and right or up and down on a document.
BUTTON	A button is an object that the user clicks to initiate an action.
MDICLIENT	This class name is used to create a child window frame for an MDI application.

To use one of these classes, you have various options. At design time, in the Toolbox, you can click the Custom Control button and click the dialog box. In the Properties window, click Class and type the class name.

Q.4. How components of VC^{++} are different from VB components ? Explain by giving examples.

Ans. Design your applications at any screen resolution and have them run and look exactly the same at any screen resolution. This gives your applications an equally professional look and feed on all systems, resulting in significantly increased sales of your software. VB Gold Active Resize control automatically detects the screen resolution on the user system and will automatically resize and reposition your VB (Visual Basic) application forms using the exact aspect ratio. Similarly, it will resize and reposition all controls on your VB from when the form is resized either by the user or throughout code. The control automatically detects the resize event of its parent form and performs an exactly proportional resizing and repositioning on all form controls **without requiring you to write even a single line of code.** Active Resize control handles virtually and type of controls. It also handles complex controls like the **Stab Tabbed Dialog** control, all types of Microsoft **DB Grids** and **Data Grids**, as well as both the **Microsoft Flex Grid (MSFlexGrid) and Hierachical Flex Grid (MSHFlexGrid).**

Active Resize Control Features:

- VB Gold Active Resize control is the fastest form resizer control on the planet to its code is perfectly optimized to achieve the highest performance possible. This means that the user of your application will never see the controls changing their sizes and positions when he/she resizes the form.
- 100% flicker-free resizing and repositioning process (unlike most form resizer controls) even when resizing pictures.
- Makes your applications completely resolution-independent. They will look exactly the same on all screen resolutions.
- Automatically resizes and repositions all form controls when forms load or resized without the need to write any code. The control automatically detects the Load and Resize events of its parent form and resizes/repositions the controls.
- Precisely Resizes and reposition the controls with exact proportions to the changes in the form size. Your forms will look exactly the same when they are resized.
- Allows you to specify whether the font size of control captions, control text and gird cells text of the controls should be resized when the controls are resized.
- Allows you to exclude any control (s) form being resized, repositioned or its font get resized. This gives you full control over the entire resize process.

- Allows you to save the current form state (its position and size/position of all controls) and restore the form state at any time with only one line of code !
- Can handle every new control created on the form at run-time.
- Can handle any type of controls on the form, even graphics controls such as **image** and **Picture** controls. **Chart** controls as well as **Line** and **Shape** controls.
- Can handle controls nested with other (container) controls.
- Can handle complex controls like the Tabbed Dialing Control (SSTab control) and all controls on it.
- Can handle all data grid controls like the Data Grid, DB Grid, MS Flex Grid, MSH Fles Grid. All columns in these controls are resized proportionally to the new size of the control.
- Supports all types of **Sheridan controls and Grids (SSDBGrid, SSOleDBGrid, SSDBCombo, etc)**.
- Suppose the **Janus GridEx grid control**.
- Fully supports the ESRI Map Objects GLS control and does not distort the spatial extent for calculating coordinates (very essential to a proper GLS system).
- Supports the **MS Toolbar control, MS Calender control** and the **Progress Bar control**.
- Supports the **3D Active Button Magic controls**.
- Supports the **Inovasys Frame Plus and Button plus Free Active X controls**.
- Automaticallyresizes the form background picture to the actual size of the form at run-time (when the form loads or is resized) even if the original picture is smaller or larger than the form area.
- Automatically resizes the contents (imageO of all **Picture Box controls** on the VB form.
- Enables you to specify a minimum and maximum size for your forms. At run-time, the form can not be resized smaller or larger than the specified minimum / maximum size. This ensures that all controls on the form will be appropriately visible.
- Enables you to prevent the from being minimized and/or maximized.
- Works with SDI, MDI child, and MDI parent forms.
- Ready to be used in your existing applications without the need to write or modify any code.
- Works on all 32-bit versions of Windows and cab be used in Microsoft Visual Basic (5 and 6).
- Great processing speed even on very slow computers, Active Resize was tested on a 133 MHz pentium PC with 128 MB RAM, 2 MB Graphics Adapter RAM, with Windows 2000 pro. OS and has displayed an equally great performance.
- The control run-times are totally royalty-free ! You purchase just one license and distribute thousands of copies of the control with your applications ! It will pay for itself thousand times ! Note: Multiple developers must purchase multiple licenses (one license per PC). Purchasing a site license enables you to use the control on an unlimited number of development PCs.

Q.5. What are window messages ? Explain different window messages with their meaning. Also write a program which uses windows messages.

Ans. Window messages: The system passes input to a window procedure in the form of messages. Messages are generated by both the system and applications. The system generates a message at each input event— for example, when the user types, moves the mouse, or clicks a control such as a scroll bare. The system also generates messages in response to changes in the system brought about by an applications, such as with an application changes the pool of system font resources or resizes one of its windows. An application cab generate messages to direct its own windows to performs tasks or to communicate with windows in other applications. The system sends a message to a window procedure with a set of four parameters: a window handle, a message identifier, and two values called message parameters. The window handle identifies the window for which the message is intended. The system uses it to determine which window procedure should receives the message. A message identifier is a named constant that identifies the purpose of a message. When a window procedure receives a message, it uses a message identifier to determine how to process the message. For example, the message identifier WM_PAINT tells the window procedure that the window's client area has changed and must be repainted. Message parameters specify data or the location of data used by a window procedure when processing a message. The meaning and value of the message parameters depends on the message. A message parameter cab contain an integer, packed bit flags, a pointer to a structure containing additional data, and so on. When a message does not use message parameters, they are typically set NULL. A window procedure must check the message identifier to determine how to interpret the message parameters.

Message Types:

System-Defined Messages: The system sends or posts a system-defined message when it communicates with an application. It uses these messages to control the operations of applications and to provide input and other information for application to process. An application cab also send or post system-defined messages. Applications generally use these messages to control the operation of control windows created by using preregistered window classes. Each system-defined message has a unique message identifier and a corresponding symbolic constant (defined in the software development kit (SDK) header files) that states the purpose of the message. For example, the **WM_PAINT** constant requests that a window paint its contents. Symbolic constants specify the category to which system-defined messages belong. The prefix of the constant identifies the type of window that cabinterpret and process the message. Following are the prefixes and their related message categories.

Application-Defined Messages: An applications can create message to be used by its own windows or to communicate with windows in other processes. If an application creates its own messages, the window procedure that receives then must interpret the messages and provide appropriate processing. Message-identifier values are used as follows:

- The system reserves message-identifier value in the range 0 × 0000 through 0 × 03FF (the value of WM_USER -1) for system-defined messages. Application cannot use these values for private messages.
- Values in the range 0 × 0400 (the value of **WM_USER**) through 0 × 7FFF are available for message identifiers for private window classes.
- If your application is marked version 4.0, you can use message-identifier values in the range 0 × 8000 (**WM_APP**) through 0 × BFFF for private messages.
- The system returns a message identifier in the range 0 × C000 through 0 × FFFF when an application calls the Register Window Message function to register a message. The message identifier returned by this function is guaranteed to be unique throughout the system. Use of this function prevents conflicts that can arise if other applications use the same message identifier for different purposes.

Q.6. What is a device context ? To draw in a device context, what will you use? Explain by giving suitable example.

Ans. Device Context: A device context a structure that defines a set of graphic objects and their associated attributes, as well as the graphic modes that affect output. The graphic objects include a pen for line drawing, a brush for painting and filing, bitmap for copying or scrolling parts of the screen, a palette for definite the set of available colors, region for clipping and other operations and a path for painting and drawing operations.

Obtaining a Private Display Device Context: An application performing numerous drawing operations in the client area ofits window must use a private display DC. To create this type of DC, the application must specify the CS_OWNDC constant for the style member of the **WNDCLASS** structure when registering the window class. After registering the window class, the application obtains a handle identifying a private display DC by calling the GetDC function. The following example shows how to create a private display DC.

```
#include<windows.h> //required for all Windows - based application
#include<stdio.h>        //C run-time header for i/o
#include<string.h>  //C run-time header for stroke_s
```

```
#include"dc.h"            //specific to this program
                          //Function Prototypes.
BOOL InitApplication (HINSTANCE);
long FAR PASCAL MainWndProc(HWND, UINT, UINT, LONG);
                          //Global variables
HINSTANCE hinst;     //handle to current instance
HDC hdc;             //display device context handle
BOOL IntApplication (HINSTANCE hinstance) {
   WNDCLASS wc;      //Fill in the window class structure with
                          parameters
                          //describing the main window.
wc.style = CS_OWNDC;    //Private-DC constant
wc.IpfnWndProc = (WNDPROC) MainWndProc;
wc.cbClsExtra=0; wc.cbWndExtra = 0; wc.hlnstance = hinstance;
wc.hlcon = LoadIcon ((HINSTANCE) NULL, MAKEINTRESOURCE (IDi_APPLICATION));
wc.hCursor = LoadCursor ((HINSTANCE) NULL,
MAKEINTRESOURCE (IDC_ARROW));
wc.hbrBackground = GetStockObject (WHITE_BRUSH);
WC.IpszMenuName="GenericMenu", wcIpszClassName = "GenericWClass";
                          //Register the window class and return the resulting
                          code.
return Register Class (&wc); }
LRESULT APIENTRA MainWndProc {
HWND hwnd,           //window handle
UINT message,        //type of message
WPARAMwParam,        //additional information
iPARAM IParam)       //additional information
{ PAINTSTRUCT ps;    //Retrieve a handle identifying the private DC
hdc = GetDC (hwnd); switch (message)  {
case WM_PAINT: BeginPaint (hwnd,&ps);
                          //Draw and pain using private DC.
```

Q.7. What are access modifiers ? Explain their utility with suitable example.

Ans. Access modifiers: Access modifiers are keywords used to specify the declared accessibility of a member or a type. This section introduces the four access modifiers: public: The keyword is an access modifier for types and type members. Public access is the most permissive access level. There are no restrictions on assessing public members protected: The protected keyword is a member access modifier. A protected member is accessible from within the class in which it is declared, and from within any class derived from the class that declared this member. A protected member of a base class is accessible in a derived class only if the access takes place through the derived class type internal: The internal keyword is an access modifier for types and type members. Internal members are accessible only within files in the same assembly. A common use of internal access is in component-based development because it enables a group of components to cooperate in a private manner without being exposed to the rest of the application code. For example, a Framework for building graphical user interfaces could provide Control and Form classes the cooperate using members with internal access. Since these members are internal, they are not exposed to code that is using the framework. It is an error to reference a member with internal access outside the assembly within it was defined, private: The private keyword is a member access modifier. Private access is the least permissive access level. Private members are accessible only within the body of the class or the struct in which they are declared. Nested types in the same body can also access those private members. It is a compile-time error to reference a private member outside the class or the struct in which it is declared. The following five accessibility levels can be specified using the access modifiers: public protected internal protected private.

Q.8. Explain the components of Visual C++ and Visual Basic.

Ans. Components of Visual C++: The fundamental parts of VC++ provides as part of the IDE are editor, compiler, linker and libraries. These are the basic tools that are essential to write a program and executing it, there functions are as follows:

Editor: Editor provides an interactive environment, for creating and editing C++ source code. The editor automatically recognize the C++ keywords and assign a colour to them according to what they are. Those maker the program readable and helps to find errors easily.

Compiler: Compiler converts source program into object program and finds the errors. The output from the compiler is .obj file.

Linker: Linker combines the various module generated by the compiler from source code files at required code module from program libraries and make the executable file and abo finds the errors.

Libraries: Libraries extends the C++ language features by providing routines that are not part of C++ language. The basic library that a C++ language includes the standard C++ library that contains the basic set of routine* to all C++ compiler. The other library provided by V C+ are MFC library and Active Template library. MFC provides the basis for many of the window program, that you will write. It is also called application framework because it provides a set of structured program components that provides. Active template library (ATL) provides structure for writing specialized window program e.g., Active X control.

Components of visual basic IDE: In Visual basic Standard arrangement of windows can be set manually. The next time, while opening the IDE, arrangement is preserved. In visual Basic 6.0 the IDE is set to MDI (Multiple Document Interface) layout by default; you had the option to set it to SDl (Single Document Interface) layout.

In visual basic 6.0 you can control the docking behaviour for some tool windows by choosing options from the tools menu.

1. Colour palette
2. Data view
3. Form layout
4. Project explorer
5. Watches
6. Output window
7. Object browser.

Visuai basic 6.0 environment shows the too! box, project explorer, from layout window, watches, property window, property pages, standard menu etc.

Watch window contains various'variable at the run time and used for debugging.

EXERCISES

1. What do you mean by instance of an application? Create three instances of MSWord application. How many instances are active at one time?
2. How is event-driven programming different from conventional programming?
3. What is an API? How is it useful to the programmers? Can you create your own API?
4. What do you mean by the term GUI? What are the advantages of GUI over character-based user interface?
5. Draw a flowchart describing event-driven program execution.
6. Create a Win32 Console version of "Hello World" application. What are the differences between the two applications?

2 Parts of Visual C++ Program

> **Chapter Outline**
> - Introduction
> - Application Object
> - Main Window Object
> - View Object
> - Document Object
> - Document/ View Architecture
> - Creating a Simple Window

2.0 INTRODUCTION

One of the environments you can use to study C and/or C++ is Microsoft Visual C++. Microsoft Visual C++ is available two options. Microsoft Visual Studio 2010 is a multi-language commercial environment that supports programming in C++ and many other languages (C#, Visual Basic, and F#). It includes Microsoft Visual C++ made for C and C++. To make it easy to learn C++ programming, Microsoft created Microsoft Visual C++ 2010 Express. This is a full-functional programming environment that supports both C and C++ (it also supports another child of the C language, called C++/CLI). It is free. You can use either Microsoft Visual Studio 2010 or Microsoft Visual C++ 2010 Express to create C and C++ applications. To follow our lessons, we will use either.

Microsoft Visual Studio 2010, is a commercial application. If you want to use it, you must purchase (or you can use a trial version). Microsoft Visual C++ 2010 Express is a free application. If you want to use it, you must download it from the Microsoft web site. Once you get to the site, look for and click the Visual Studio Express link. Then locate and click the Microsoft Visual C++ 2010 Express. After acquiring the software, you must install it to be able to lanuch it.

A VC++ (as also a C++) application is object-oriented and hence has at least one object, which is an instance of a class. This object implements all the mechanisms needed to run the application. Besides, the application may use a number of resources. Thus a VC++ application may have apart from VC++ source files, icons, menus, and other Windows components. Creating and integrating all these components is arduous and needs lot of effort. Microsoft provides an application-generating environment (workbench) that takes off all these drudgeries from the programmer and facilitates her to concentrate on the functionalities of the application.

2.1 APPLICATION OBJECT

VC++ is an object-oriented language. Therefore, it has all the features of an object-oriented language. A VC++ application is made up of many classes. Some of the classes are standard and are made available by the development environment. It may also have user defined classes obtained by sub-classing the existing classes.

The entire application is distributed into a number of files. However, the Microsoft VC++ IDE also arranges contents of these files into hierarchical view. The IDE creates an application in a logical container represented by an object called a workspace. A workspace houses one project. A project object represents a group of files that include workspace may have one or more sub-units called project. Workspace allows programmers to view and to access various elements of the projects it houses .The constituent File of Project has information and codes that make-up the application.

To create a program, also called an application, you derive a class from the MFC's CWinApp class. The CWinApp class is derived from the CWinThread class. CWinThread is derived from a class named CCmdTarget, which is derived from CObject. CWinApp stands for Class For A Windows Application :

The CWinApp class creates all the other components in the application. It is the class that receives all the event messages and then passes the messages to the CFrameView and CView classes.

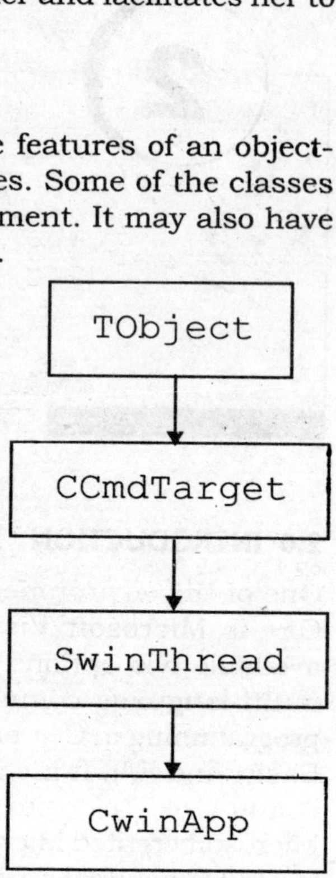

Fig. 2.1.

2.2 MAIN WINDOW OBJECT

2.2.1 Window Creation: The Main Window **(V.Imp)**

The WNDLCLASS and the WNDCLASSEX classes are only used to initialise the application window class. To display a window, that is, to give the user an object to work with, you must create a window object. This window is the object the user uses to interact with the computer.

If you are creating a Win32 application, to create a window, you can call either the CreateWindow() or the CreateWindowEx() function. Their syntaxes are :

```
HWND CreateWindow(
    LPCTSTR lpClassName,
    LPCTSTR lpWindowName,
    DWORD dwStyle,
    int x,
    int y,
    int nWidth,
    int nHeight,
    HWND hWndParent,
    HMENU hMenu,
    HINSTANCE hInstance,
    LPVOID lpParam
```

```
HWND CreateWindowEx(
    DWORD dwExStyle,
    LPCTSTR lpClassName,
    LPCTSTR lpWindowName,
    DWORD dwStyle,
    int x,
    int y,
    int nWidth,
    int nHeight,
    HWND hWndParent,
    HMENU hMenu,
    HINSTANCE hInstance,
```

```
);                                  LPVOID lpParam
                               );
```

You can simply call this function and specify its arguments after you have registered the window class. Here is an example :

```
INT WINAPI WinMain(HINSTANCE hInstance, HINSTANCE hPrevInstance,
                                        LPSTR lpCmdLine, int nCmdShow)
{
    WNDCLASS        WndCls;

    . . .

    RegisterClass(&WndCls);

    CreateWindow(. . .);
}
```

If you are planning to use the window further in your application, you should retrieve the result of the CreateWindow() or the CreateWindowEx() function, which is a handle to the window that is being created. To do this, you can declare an HWND variable and initialise it with the create function. This can be done as follows :

```
INT WINAPI WinMain(HINSTANCE hInstance, HINSTANCE hPrevInstance,
                                        LPSTR lpCmdLine, int nCmdShow)
{
    HWND  hWnd;
    WNDCLASS WndCls;

    . . .

    RegisterClass(&WndCls);

    hWnd = CreateWindow(. . .);
}
```

If you are working on an MFC application, you can derive a class from CFrameWnd, as we have done so far. Each window that can display on the computer screen is based on a class equipped with a method called Create. The structure of this method is different from one type of window to another. Nevertheless, because all window objects are based on CWnd, the CWnd class provides the primary functionality used by all the other window controls.

The process of creating a frame window in an MFC application is done by using the class' constructor and calling its Create() method. As we have seen in the past, the main window object you must create is a frame because it gives presence to your application. The most basic frame is created using the CFrameWnd class. We have learned that this is done by deriving your own class from CFrameWnd. Therefore, in the constructor of your CFrameWnd-derived class, call the Create() method. As a reminder from Lesson 2, the syntax of the CFrameWnd::Create() method is as follows :

```
BOOL Create(LPCTSTR         lpszClassName,
```

```
    LPCTSTR            lpszWindowName,
    DWORD              dwStyle = WS_OVERLAPPEDWINDOW,
    const           RECT& rect = rectDefault,
    CWnd*              pParentWnd = NULL,
    LPCTSTR               lpszMenuName = NULL,
    DWORD                 dwExStyle = 0,
    CCreateContext*          pContext = NULL );
```

If the method succeeds in creating the window, it returns TRUE. If it fails, it returns FALSE.

2.2.2 The Window Class Name

To create a window, you must provide its name as everything else in the computer has a name. There are two main types of class names you will use in your applications. If the window you are creating is a known type of window, its class is already registered with the operating system. In this case you can provided it. If you are creating your own fresh class, which means you are in charge of its characteristics, then define a null-terminated string as the class' name.

For a Win32 application, the class name of your main window must be provided to the lpszClassName member variable of your WNDCLASS or WNDCLASSEX variable. You can provide the name to the variable or declare a global null-terminated string. The name must also be passed to the lpClassName argument of either the CreateWindow() or the CreateWindowEx() functions. Here is an example :

```
const char *ClsName = "WndFund";

INT WINAPI WinMain(HINSTANCE hInstance, HINSTANCE hPrevInstance,
                                        LPSTR lpCmdLine, int nCmdShow)
{
    HWND        hWnd;
    WNDCLASS    WndCls;

    . . .

    WndCls.lpszClassName = ClsName;

    RegisterClass(&WndCls);

    CreateWindow(ClsName,
}
```

If you are creating an MFC application, the class name is passed as the first argument of the CFrameWnd::Create() method. You can use a null terminated string as done for the CreateWindow() or the CreateWindowEx() function. If you have initialised the window class and registered it using AfxRegisterWndClass(), you may remember that this function returns a null-terminated string. Therefore, you can pass its return value to the Create method. This can be done as follows :

```
CMainFrame::CMainFrame()
{
```

```
    // Declare a window class variable
    WNDCLASS WndCls;

    . . .

    const char *StrClass = AfxRegisterWndClass(WndCls.style,
    WndCls.hCursor, WndCls.hbrBackground, WndCls.hIcon);

    Create(StrClass, ;
}
```

Once an application is created, although you can, you should refrain from ever changing the name of a class. It may take more than simply assigning a new value to the AfxRegisterWndClass() function.

2.2.3 The Window Name **(V.Imp)**

We saw that every window should have a name to easily identify it. For a main window, the name displays on the title bar of the frame.

The name is passed as the lpWindowName argument of either the CreateWindow() or the CreateWindowEx() functions. To do this, you can provide a null-terminated string to the argument or declare a global string. Here is an example :

```
const char *ClsName = "WndFund";
const char *WndName = "Application Name";

LRESULT CALLBACK WndProc(HWND hWnd, UINT Msg, WPARAM wParam, LPARAM
lParam);

INT WINAPI WinMain(HINSTANCE hInstance, HINSTANCE hPrevInstance,
LPSTR lpCmdLine, int nCmdShow)
{
    WNDCLASS    WndCls;

    . . .

    RegisterClass(&WndCls);

    CreateWindow(ClsName, WndName,
}
```

If you are creating an MFC application, to provide a name for the window, pass a null-terminated string as the second argument of the CFrameWnd::Create() method :

```
CMainFrame::CMainFrame()
{
    WNDCLASS WndCls;
    const char *StrWndName = "Application Name";

    . . .
```

```
    const char *StrClass = AfxRegisterWndClass(WndCls.style,
    WndCls.hCursor, WndCls.hbrBackground, WndCls.hIcon);

    Create(StrClass, StrWndName);
}
```

If you want to change the name of window after is has been created, you can call the CWnd::SetWindowText() method. Its syntax is :

```
void SetWindowText(LPCTSTR lpszString);
```

The lpszString argument is a null-terminated string that holds the value you want to display. Here is an example. You can simply pass a null-terminated string as argument. Here is an example :

```
CMainFrame::CMainFrame()
{
    // Declare a window class variable
    WNDCLASS WndCls;
    const char *StrWndName = "Application Name";

    . . .

    Create(StrClass, StrWndName);

    SetWindowText("Windows Fundamentals");
}
```

You can also first declare a null-terminated string variable, assign it a value, and then pass it the lpszString argument to the SetWindowText() function.

If you are using resources in your MFC application, you can also create a global value in the string table to be used as the window name :

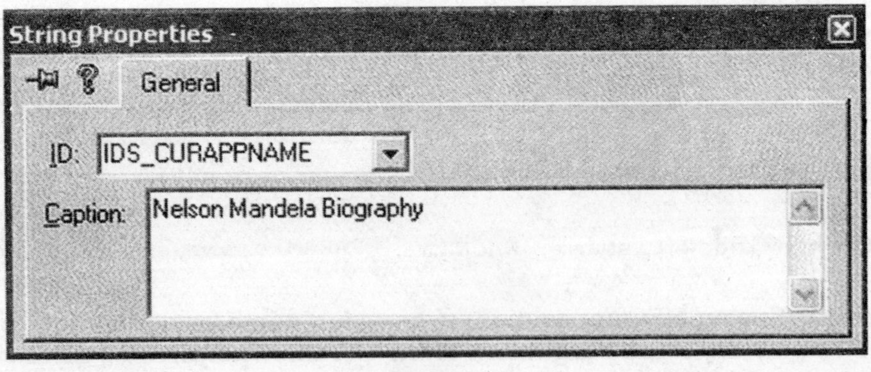

Fig. 2.2.

You can call the string of such an identifier, store it in a CString variable, and then pass it to the CWnd::SetWindowText() method. Here is an example :

```
CMainFrame::CMainFrame()
{
    // Declare a window class variable
```

```
    WNDCLASS WndCls;
    const char *StrWndName = "Application Name";

    . . .

    const char *StrClass = AfxRegisterWndClass(WndCls.style,
    WndCls.hCursor, WndCls.hbrBackground, WndCls.hIcon);

    Create(StrClass, StrWndName);

    CString Str;
    Str.LoadString(IDS_CURAPPNAME);
    SetWindowText(Str);
}
```

To change the name of a window, instead of calling SetWindowText(), you can call the CWnd::SendMessage() method. Its syntax is :

```
LRESULT SendMessage(UINT message, WPARAM wParam = 0, LPARAM lParam = 0);
```

Since you want to change the text, the message argument must be WM_SETTEXT. The wParam argument is not used. The lParam argument holds the string that will be the new value of the window name. You must cast the string to LPARAM. Here is an example :

```
CMainFrame::CMainFrame()
{
    // Declare a window class variable
    WNDCLASS WndCls;
    const char *StrWndName = "Application Name";
    const char *StrApp = "Windows Fundamentals";

    . . .

    const char *StrClass = AfxRegisterWndClass(WndCls.style,
    WndCls.hCursor, WndCls.hbrBackground, WndCls.hIcon);

    Create(StrClass, StrWndName);

    SendMessage(WM_SETTEXT, NULL, (LPARAM)StrApp);
}
```

To retrieve the name of a window (always remember that the name of a window is not the name of a class), you can call the CWnd::GetWindowText() function. Its syntax is :

```
int GetWindowText(LPTSTR lpszStringBuf, int nMaxCount) const;
```

The lpszStringBuf is the null-terminated string that will store the window name. The nMaxCount is the minimum number of characters of the lpString. If you specify more characters than the name is made of, the compiler would reduce this number to the actual length of the string. Therefore, it is safe to provide a high number.

2.2.4 Windows Styles (V.Imp)

We had a formal introduction to windows styles in Lesson 2 and we reviewed all necessary styles to apply or not apply to a main window. Once again, a WS_OVERLAPPEDWINDOW has a caption that displays the window name (if any). It is also equipped with the system menu, a thick frame, a system Minimise button, a system Maximise button, and a system Close button.

For a Win32 application, you can apply the **WS_OVERLAPPEDWINDOW** style as follows :

CreateWindow(ClsName, WndName, **WS_OVERLAPPEDWINDOW,**

For an MFC application, this style can be added as follows:

```
CMainFrame::CMainFrame()
{
    // Declare a window class variable
    WNDCLASS WndCls;
    const char *StrWndName = "Windows Fundamentals";

    . . .

    const char *StrClass= AfxRegisterWndClass(WndCls.style,
    WndCls.hCursor, WndCls.hbrBackground, WndCls.hIcon);

    Create(StrClass, StrWndName, WS_OVERLAPPEDWINDOW);
}
```

Remember that, to apply a combination of styles, use the bitwise OR operator.

If you are designing a form or a dialog box, you can use the Properties window to visually select the styles you want to apply to the window :

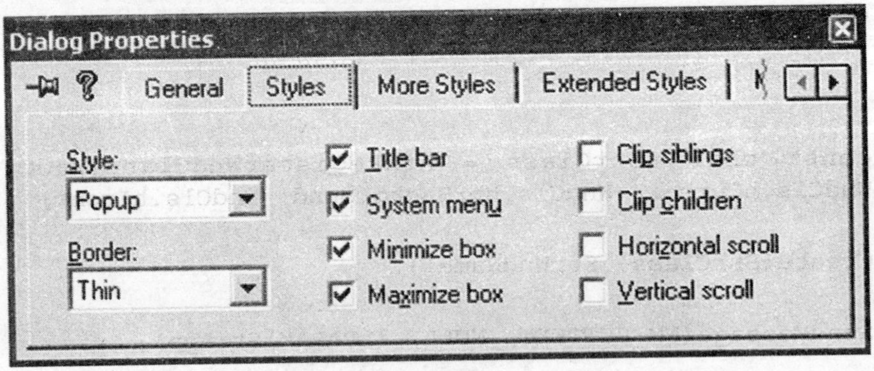

Fig. 2.3.

2.2.5 Window's Location and Size

As we learned in Lesson 2, the location of a window is defined by the distance from the left border of the monitor to the window's left border and its distance from the top border of the monitor to its own top border. The size of a window is its width and its height. These can be illustrated for a main window frame as follows :

For a Win32 application, the original distance from the left border of the monitor is passed as the x argument to the **CreateWindow()** or the **CreateWindowEx()** function. The distance from top is specified using the y argument. The x and y arguments define the location of the window. The distance from the left border of the monitor to the right border of the window is specified as the *nWidth* argument. The distance from the top border of the monitor to the lower border of the window is specified with the *nHeight* value.

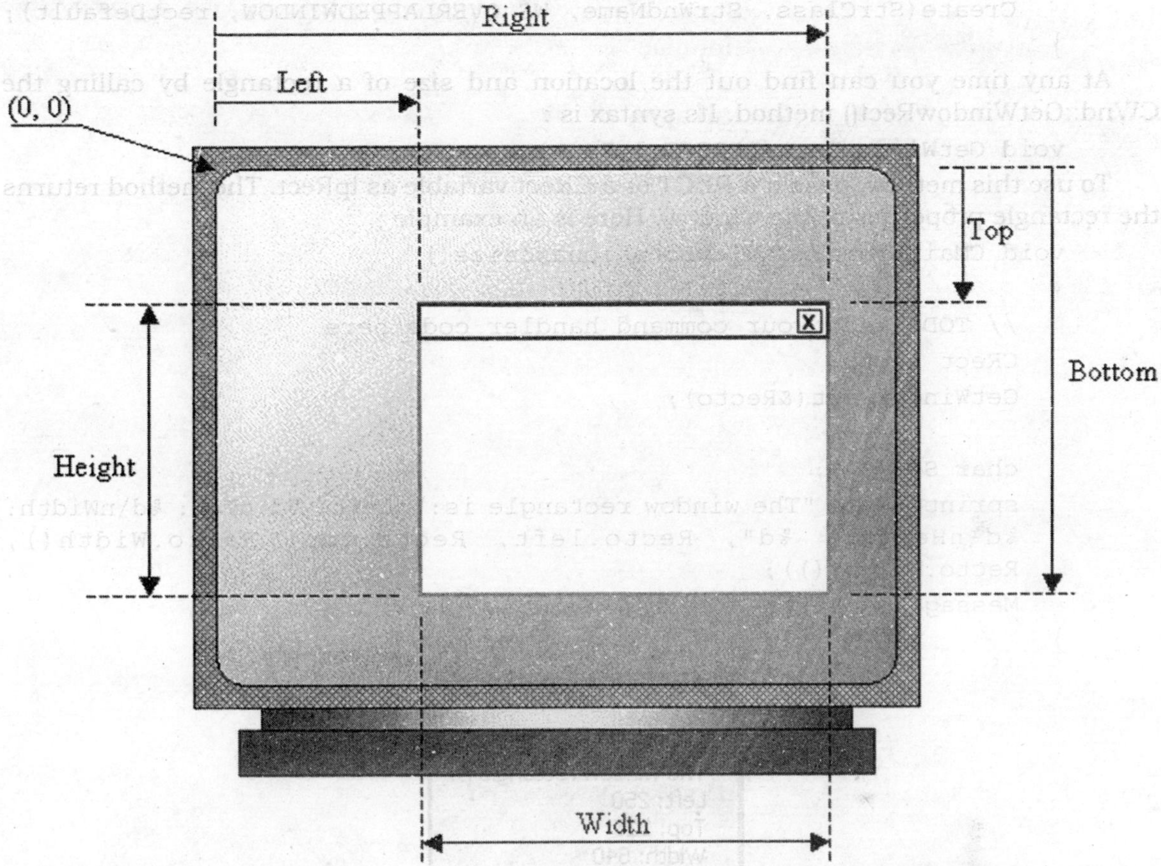

Fig. 2.4.

If you cannot make up your mind for these four values, you can use the **CW_USEDEFAULT** (*when-Creating-the-**Window**-**USE**-the-**DEFAULT**-value*) constant for either one or all four arguments. In such a case, the compiler would select a value for the argument.

For an MFC application, when calling the **CFrameWnd::Create()** method, the location and size of the frame window is specified as a **RECT** or **CRect** rectangle as we saw in Lesson 2. As done with the **CW_USEDEFAULT** constant, you can let the compiler decide on the rectangle by passing the *rect* argument as **rectDefault**. Here is an example :

```
CMainFrame::CMainFrame()
{
    // Declare a window class variable
    WNDCLASS WndCls;
    const char *StrWndName = "Windows Fundamentals";

    . . .
```

```
    const char *StrClass = AfxRegisterWndClass(WndCls.style,
    WndCls.hCursor, WndCls.hbrBackground, WndCls.hIcon);

    Create(StrClass, StrWndName, WS_OVERLAPPEDWINDOW, rectDefault);
}
```

At any time you can find out the location and size of a rectangle by calling the CWnd::GetWindowRect() method. Its syntax is :

```
void GetWindowRect(LPRECT lpRect) const;
```

To use this method, pass it a RECT or a CRect variable as lpRect. The method returns the rectangle properties of the window. Here is an example :

```
void CMainFrame::OnViewLocationandsize()
{
    // TODO: Add your command handler code here
    CRect Recto;
    GetWindowRect(&Recto);

    char Str[80];
    sprintf(Str, "The window rectangle is:\nLeft: %d\nTop: %d\nWidth:
    %d\nHeight: %d", Recto.left, Recto.top, Recto.Width(),
    Recto.Height());
    MessageBox(Str);
}
```

Fig. 2.5.

If you created your application using AppWizard, it sets a default rectangle for the frame (actually the left and top values are randomly selected). Whether you created the frame using the **CFrameWnd::Create()** method or AppWizard, you can re-define its location and size from the **PreCreateWindow()** event. Because this event is called after **CFrameWnd::Create** but before the window is displayed, its values are applied to the window. The syntax of the **CWnd::PreCreateWindow()** event is (this event is inherited from **CWnd but CFrameWnd**, like any other class that needs it, overrides it and provides its own functionality as it relates to a window frame) :

```
virtual BOOL PreCreateWindow(CREATESTRUCT& cs);
```

This event takes as argument a CREATESTRUCT object. This structure is defined as follows :

```
typedef struct tagCREATESTRUCT {
    LPVOID    lpCreateParams;
    HANDLE    hInstance;
    HMENU     hMenu;
    HWND      hwndParent;
    int       cy;
    int       cx;
    int       y;
    int       x;
    LONG      style;
    LPCSTR    lpszName;
    LPCSTR    lpszClass;
    DWORD     dwExStyle;
} CREATESTRUCT;
```

As you can see, the member variables of this structure are very similar to the arguments of the Win32's CreateWindow() and CreateWindowEx() functions. Its member variables correspond as follows :

CREATESTRUCT	***CreateWindow and CreateWindowEx***	***Meaning***
x	*x*	Distance from left border of monitor to left border of window frame
y	*y*	Distance from top border of monitor to top border of window frame
cx	*n*Width	Distance from left border of monitor to right border of window frame
cy	*n*Height	Distance from top border of monitor to bottom border of window frame

Therefore, in your CFrameWnd::PreCreateWindow() event, assign the desired values for the location and size of the window. In the following example, the original dimensions of the window are set to 450x325 (width x height) :

```
BOOL CMainFrame::PreCreateWindow(CREATESTRUCT& cs)
{
    if( !CFrameWnd::PreCreateWindow(cs) )
    return FALSE;
    // TODO: Modify the Window class or styles here by modifying
    // the CREATESTRUCT cs
    cs.cx = 450;
    cs.cy = 325;

    return TRUE;
}
```

Of course, you can also use the PreCreateWindow() event to customise the appearance of the window frame.

Once a regular window, that is, an overlapped window, has been created, it displays a title bar, its system buttons, and borders. As stated already, if you created an application using AppWizard, the window may appear in a random location, which you can control. If you want the window to be positioned in the center of the screen, call the CWnd::CenterWindow() method. Its syntax is :

```
void CenterWindow(CWnd* pAlternateOwner = NULL);
```

By default, this window positioned the caller (the window that called it) in the middle-center of the main window, also called its owner. For a main frame of a window, the owner would be the monitor screen. As long as the window is owned by another, the compiler can find out and position it accordingly. For this reason, you can omit the pAlternateOwner argument.

You should call the CenterWindow() method in the event that creates the window. For a frame this would be the OnCreate event. Here is an example :

```
int CMainFrame::OnCreate(LPCREATESTRUCT lpCreateStruct)
{
    if (CFrameWnd::OnCreate(lpCreateStruct) == -1)
    return -1;
    // . . .

    CenterWindow();

    return 0;
}
```

If you want the caller to be centrally positioned relative to another window, obtain a handle to that window and pass it as the pAlternateOwner argument.

Once a window is displaying on the screen, the user can change its location by dragging its title bar. This would change its CREATESTRUCT::x or its rect::left argument. If you do not want the user to change the location of a window, one thing you can do is to prevent the mouse from capturing, that is, taking ownership, of the title bar. This can be done calling the Win32's ReleaseCapture() function. Its syntax is :

```
BOOL ReleaseCapture(VOID);
```

When this function is called, the event in which it is accessed prevents the mouse from capturing the object on which the mouse is positioned. Nevertheless, if the function succeeds, it returns TRUE. If for some reason it fails, it returns FALSE. Because a WM_MOVE message is sent when a window is moved, you can use it to call this function. Here is an example :

```
void CMainFrame::OnMove(int x, int y)
{
    CFrameWnd::OnMove(x, y);

    ReleaseCapture();
    // TODO: Add your message handler code here
}
```

To change the dimensions of a window, the user can click and drag one of its borders. Imagine you do not want the user to change the size of the window. If you remove the system buttons and/or the system menu(

```
BOOL CMainFrame::PreCreateWindow(CREATESTRUCT& cs)
{
    if( !CFrameWnd::PreCreateWindow(cs) )
        return FALSE;
    // TODO: Modify the Window class or styles here by modifying
    // the CREATESTRUCT cs
    cs.style &= ~WS_SYSMENU;

    return TRUE;
}
```

), the user is still able to resize it. One solution you can use is to call the ReleaseCapture() function on the WM_SIZE event. That way, every time the user grabs a border and starts dragging, the mouse would lose control of the border :

```
void CMainFrame::OnSize(UINT nType, int cx, int cy)
{
    CFrameWnd::OnSize(nType, cx, cy);

    ReleaseCapture();
    // TODO: Add your message handler code here
}
```

Another way the user change the size of a window consists of minimising it. As you surely know already, to minimise the window, the user clicks the system Minimise button . If you do not want this operation, you can omit the **WS_MINIMIsEBOX** style when creating the window. This can be done as follows :

```
CMainFrame::CMainFrame()
{
    //Declare a window class variable
    WNDCLASS WndCls;
    const char *StrWndName = "Windows Fundamentals";

    . . .

    const char *StrClass = AfxRegisterWndClass(WndCls.style,
    WndCls.hCursor, WndCls.hbrBackground, WndCls.hIcon);

    Create(StrClass, StrWndName,
        WS_OVERLAPPED | WS_CAPTION | WS_SYSMENU | WS_THICKFRAME |
        WS_MAXIMIZEBOX, rectDefault);
}
```

If you created the application using AppWizard, all system buttons are added to the frame. If you want to remove a style, use the CFrameWnd::PreCreateWindow() event. In the following example, the system Minimise button is removed on the frame :

```
BOOL CMainFrame::PreCreateWindow(CREATESTRUCT& cs)
{
```

```
    if( !CFrameWnd::PreCreateWindow(cs) )
    return FALSE;
    // TODO: Modify the Window class or styles here by modifying
    // the CREATESTRUCT cs

    cs.style &= ~WS_MINIMIZEBOX;

    return TRUE;
}
```

In the same way, you can remove the system maximise button.

If you allow the system buttons, the user can minimise and maximise the window at will. To do this, besides using the system buttons, the user can also double-click the title bar of the window or click its Maximise menu item either from the window's system icon or after right-clicking its button on the Taskbar. When the window is maximized, the frame is resized to occupy the whole monitor area, starting from the top-left corner as the origin.

When the user decides to maximise a window, the frame sends a **WM_GETMINMAXINFO** message which appears like "The user wants to maximise me. What should I do?". This message produces the **OnGetMinMaxInfo()** event. If the application does not give any specific instruction, the window gets maximised. If you want to do something before the window is formally maximised, you can intercept **OnGetMinMaxInfo()** event. Remember that, if you want to prevent the window from being minimized or maximized, you should remove the Minimise **(WS_MINIMIZEBOX)** or the Maximise **(WS_MAXIMIZEBOX)** buttons from its style.

If you want to control the minimum and/or maximum size that a window can have when the user wants to maximise it, set the desired size in the **OnGetMinMaxInfo()** event. It syntax is :

```
afx_msg void OnGetMinMaxInfo(MINMAXINFO FAR* lpMMI);
```

The information pertinent to a window's maximized size, including the maximum size it can have, is stored in a structure called **MINMAXINFO**. The **MINMAXINFO** structure is defined as follows :

```
typedef struct tagMINMAXINFO {
    POINT ptReserved;
    POINT ptMaxSize;
    POINT ptMaxPosition;
    POINT ptMinTrackSize;
    POINT ptMaxTrackSize;
} MINMAXINFO;
```

A variable of **MINMAXINFO** type is passed to the OnGetMinMaxInfo() event.

The *ptReserved* member variable is reserved and never used.

The *ptMaxSize* value is the maximum width and the maximum height. To specify this value, you can call the x member variable of the point value of *ptMaxSize* and assign the desired value. Here is an example :

```
void CMainFrame::OnGetMinMaxInfo(MINMAXINFO FAR* lpMMI)
{
    // TODO: Add your message handler code here and/or call default
```

```
    lpMMI->ptMaxSize.x = 450;
    lpMMI->ptMaxSize.y = 325;

    CFrameWnd::OnGetMinMaxInfo(lpMMI);
}
```

The *ptMaxPosition* member variable is the location of the window when it gets maximised. Normally, when the window gets maximised, it starts occupying the monitor screen from the top-left corner. The *ptMaxPosition* allows you to change that default behaviour. The ptMaxPosition.x is the distance from the left border of the monitor to the left border of the maximised window. The *ptMaxPosition.y* is the distance from the top border of the monitor to the top border of the maximised window.

The *ptMinTrackSize* member variable is the minimum size the window must assume when it gets maximised. Normally, when a window frame gets maximised, it occupies the whole screen area. This member variable can be used to control that size. The *ptMinTrackSize.x* is the minimum width. The *ptMinTrackSize.y* is the minimum height.

The ptMaxTrackSize member variable is the maximum size the window must have when it is maximised. The ptMaxTrackSize.x is the minimum width. The ptMaxTrackSize.y is the minimum height.

2.2.6 Window's Parenting

If the window you are creating has a parent, obtain its handle and pass it as the hWndParent argument of the **CreateWindow()** or the CreateWindowEx() functions for a Win32 application. If you are creating the main window of your application, it does not have a parent. In this case, pass the *hWndParent* as NULL :

```
INT WINAPI WinMain(HINSTANCE hInstance, HINSTANCE hPrevInstance,
LPSTR lpCmdLine, int nCmdShow)
{
    HWND  hWnd;
    WNDCLASS WndCls;

    . . .

    RegisterClass(&WndCls);
    CreateWindow(ClsName,
        WndName,
        WS_OVERLAPPEDWINDOW,
        CW_USEDEFAULT,
        CW_USEDEFAULT,
        CW_USEDEFAULT,
        CW_USEDEFAULT,
        NULL,
        );
}
```

If you are creating an MFC application, pass the handle of the parent as the *pParentWnd* argument of the Create() method. For a main frame, the kind we have created so far, specify this argument as NULL :

```
CMainFrame::CMainFrame()
{
    // Declare a window class variable
    WNDCLASS WndCls;
    const char *StrWndName = "Windows Fundamentals";

    . . .

    Create(StrClass, StrWndName, WS_OVERLAPPEDWINDOW, rectDefault,
    NULL);
}
```

2.2.7 The Window's Menu **(V.Imp)**

If you want the window to display a menu, first create or design the resource menu. Visual C++ provides a convenient way to design a basic menu. To do this, display the Add Resource dialog box and select Menu. Then add the desired items.

After creating the menu, assign its name to the *lpszMenuName* name to your **WNDCLASS** variable for a Win32 variable. If you had created the menu and gave it an identifier, use the **MAKEINTRRESOURCE** macro to convert it. If the window you are creating is a child window that will need to display its particular menu when that child window is displaying, create its menu and pass its handle as the hMenu argument of the **CreateWindow()** or **CreateWindowEx()** functions. Otherwise, pass this argument as NULL. Here is an example :

```
INT WINAPI WinMain(HINSTANCE hInstance, HINSTANCE hPrevInstance,
LPSTR lpCmdLine, int nCmdShow)
{
    WNDCLASS    WndCls;

    . . .
    WndCls.lpszMenuName = MAKEINTRESOURCE(IDR_MENU1);
    WndCls.lpszClassName = ClsName;

    RegisterClass(&WndCls);

    CreateWindow(ClsName,
        WndName,
        WS_OVERLAPPEDWINDOW,
        CW_USEDEFAULT,
        CW_USEDEFAULT,
        CW_USEDEFAULT,
        CW_USEDEFAULT,
        NULL,
        NULL,
    );
}
```

If you are creating an MFC application and want the main window to display a menu, design the menu as you see fit and create its resource. Then pass the menu name as the lpszMenuName argument of the Create() method. If you want to use the identifier of the menu, pass its ID to the **MAKEINTRESOURCE** macro. This can be done as follows :

```
CMainFrame::CMainFrame()
{
    // Declare a window class variable
    WNDCLASS WndCls;
    const char *StrWndName = "Windows Fundamentals";

    WndCls.style = CS_VREDRAW | CS_HREDRAW;
    WndCls.lpfnWndProc = AfxWndProc;
    WndCls.cbClsExtra = 0;
    WndCls.cbWndExtra = 0;
    WndCls.hInstance = AfxGetInstanceHandle();
    WndCls.hIcon = AfxGetApp()->LoadStandardIcon(IDI_WARNING);
    WndCls.hCursor = AfxGetApp()->LoadStandardCursor(IDC_CROSS);
    WndCls.hbrBackground = (HBRUSH)GetStockObject(WHITE_BRUSH);
    WndCls.hCursor = AfxGetApp()->LoadStandardCursor(IDC_CROSS);
    WndCls.lpszMenuName = MAKEINTRESOURCE(IDR_MENU1);

    const char *StrClass = AfxRegisterWndClass(WndCls.style,
    WndCls.hCursor, WndCls.hbrBackground, WndCls.hIcon);

    Create(StrClass, StrWndName, WS_OVERLAPPEDWINDOW, rectDefault,
    NULL, MAKEINTRESOURCE(IDR_MENU1));
}
```

To make sure that the window was created, you can check its return value.

If the CreateWindow() or the CreateWindowEx() functions of a Win32 application succeeds in creating the window, it returns a handle to the window that was created. You can check this validity using an if conditional statement. Here is an example :

```
INT WINAPI WinMain(HINSTANCE hInstance, HINSTANCE hPrevInstance,
LPSTR lpCmdLine, int nCmdShow)
{
    HWND        hWnd;
    WNDCLASS    WndCls;

    . . .

    RegisterClass(&WndCls);

    hWnd = CreateWindow(ClsName,
        WndName,
        WS_OVERLAPPEDWINDOW,
        CW_USEDEFAULT,
        CW_USEDEFAULT,
```

```
            CW_USEDEFAULT,
            CW_USEDEFAULT,
            NULL,
            NULL,
            hInstance,
            NULL);
        if( !hWnd ) // If the window was not created,
            return 0; // stop the application
    }
```

For an MFC application, a Create() method returns TRUE if the window was created. Otherwise, it returns false. Therefore, you can use an if conditional statement to find out whether the window was successfully created or not. Here is an example :

```
CMainFrame::CMainFrame()
{
    // Declare a window class variable
    WNDCLASS WndCls;
    const char *StrWndName = "Windows Fundamentals";

    WndCls.style = CS_VREDRAW | CS_HREDRAW;
    WndCls.lpfnWndProc = AfxWndProc;
    WndCls.cbClsExtra = 0;
    WndCls.cbWndExtra = 0;
    WndCls.hInstance = AfxGetInstanceHandle();
    WndCls.hIcon = AfxGetApp()->LoadStandardIcon(IDI_WARNING);
    WndCls.hCursor = AfxGetApp()->LoadStandardCursor(IDC_CROSS);
    WndCls.hbrBackground = (HBRUSH)GetStockObject(WHITE_BRUSH);
    WndCls.hCursor = AfxGetApp()->LoadStandardCursor(IDC_CROSS);
    WndCls.lpszMenuName = MAKEINTRESOURCE(IDR_MENU1);

    const char *StrClass = AfxRegisterWndClass(WndCls.style,
                                WndCls.hCursor, WndCls.hbrBackground,
                                WndCls.hIcon);

    if( !Create(StrClass, StrWndName, WS_OVERLAPPEDWINDOW, rectDefault,
  NULL, MAKEINTRESOURCE(IDR_MENU1)) )
    return;
}
```

2.2.8 Displaying the Window

Once a window has been created and if this was done successfully, you can display it to the user. This is done by calling the ShowWindow() function. The Win32 version of this function has the following syntax :

```
BOOL ShowWindow(HWND hWnd, int nCmdShow);
```

The hWnd argument is a handle to the window that you want to display. It could be the window returned by the **CreateWindow()** or the **CreateWindowEx()** function.

The nCmdShow specifies how the window must be displayed. It uses one of the nCmdShow values above.

To show its presence on the screen, the window must be painted. This can be done by calling the UpdateWindow() function. Its syntax is :

```
BOOL UpdateWindow(HWND hWnd);
```

This function simply wants to know what window needs to be painted. This window is specified by its handle.

2.2.9 Considering Window's Messages

Once a window has been created, the user can use it. This is done by the user clicking things with the mouse or pressing keys on the keyboard. A message that a window sends is received by the application. This application must analyse, translate, and decode the message to know what object sent the message and what the message consists of. To do this, the application uses the GetMessage() function. Its syntax is :

```
BOOL GetMessage(LPMSG lpMsg, HWND hWnd, UINT wMsgFilterMin, UINT
wMsgFilterMax);
```

The *lpMsg* argument is a pointer to the MSG structure.

The *hWnd* argument identifies which window sent the message. If you want the messages of all windows to be processed, pass this argument as NULL.

The *wMsgFilterMin* and the *wMsgFilterMax* arguments specify the message values that will be treated. If you want all messages to be considered, pass each of them as 0.

The MSG structure is defined as follows :

```
typedef struct tagMSG {
    HWND   hwnd;
    UINT   message;
    WPARAM wParam;
    LPARAM lParam;
    DWORD time;
    POINT pt;
} MSG, *PMSG;
```

Once a message has been sent, the application analyses it using the **TranslateMessage()** function. Its syntax is :

```
BOOL TranslateMessage(CONST MSG *lpMsg);
```

This function takes as argument the **MSG** object that was passed to the **GetMessage()** function and analyses it. If this function successfully translates the message, it returns TRUE. If it cannot identify and translate the message, it returns FALSE.

Once a message has been decoded, the application must send it to the window procedure. This is done using the **DispatchMessage()** function. Its syntax is :

```
LRESULT DispatchMessage(CONST MSG *lpMsg);
```

This function also takes as argument the **MSG** object that was passed to **GetMessage()** and analysed by **TranslateMessage()**. This **DispatchMessage()** function sends the lpMsg message to the window procedure. The window procedure processes it and sends back the result, which becomes the return value of this function. Normally, when the window procedure receives the message, it establishes a relationship with the control that sent the message and starts treating it. By the time the window procedure finishes with the message, the issue is resolved (or aborted). This means that, by the time the window procedure returns its result, the message is not an issue anymore. For this reason you will usually, if ever, not need to retrieve the result of the **DispatchMessage()** function.

This translating and dispatching of messages is an on-going process that goes on as long as your application is running and as long as somebody is using it. For this reason, the application uses a **while** loop to continuously check new messages. This behaviour can be implemented as follows :

```
while( GetMessage(&Msg, NULL, 0, 0) )
{
    TranslateMessage(&Msg);
    DispatchMessage(&Msg);
}
```

If the WinMain() function successfully creates the application and the window, it returns the wParam value of the MSG used on the application.

2.3 VIEW OBJECT

(MDU 2006)

2.3.1 The Fundamentals of a View

When you create an application, you usually intend for the users to see, view, and possibly manipulate its documents. The basic process of making up a program is to create the application using a class derived from **CWinApp** and surrounding the program with a frame either based on **CFrameWnd** or from one of its derived classes. Such a basic program is mainly meant to show the "presence" of an application. To truly assist the user with files, the program should be equipped with a document. A document is not truly something a user can see. It is a representation of a file. To actually present a document to the user, you should provide an object called a view.

2.3.2 Introduction to the View Classes

Different applications have various ways to present their documents. To support these variances, the MFC provides different view-based classes. To lay a foundation for those classes, the MFC provides a class named **CView**. Although the MFC provides various view-based classes, it is likely that none of them would fit the particular functionality you want to implement. For this reason, the **CView** class accomplishes two purposes. Not only is it the ancestor to the other view-based classes, it is still a view in its own right and you can use it to create an application that none of the other classes can handle. For these two and other reasons, you should be quite familiar with the **CView** class.

2.3.3 Creation of a CView-Based Application

CView is an abstract class. In other words, you cannot just declare a CView variable. For example, the following code produces an error :

```
CView *view = new CView;
```

Therefore, to use **CView**, you must derive a class from it. What makes **CView** a pure abstract class is that one of its methods is declared pure. When deriving your class from **CView**, you must override a method named **OnDraw**. The syntax of the **CView::OnDraw()** method is :

```
virtual void OnDraw(CDC* pDC) = 0;
Notice that it is declared pure (= 0).
```

2.3.4 Introducing View-Based Applications

1. If necessary, start Microsoft Visual C++. To create a new program, on the main menu, click File -> New Project...

2. In the Templates list, click MFC Application.
3. Set the Name to GCS1 and click OK.
4. In the first page of the MFC Application Wizard, click Next.
5. In the right list, click Single Document and click Next twice.
6. In the File Extension edit box, type gcs.
7. In the Main Frame Caption text box, type Georgetown Cleaning Services.

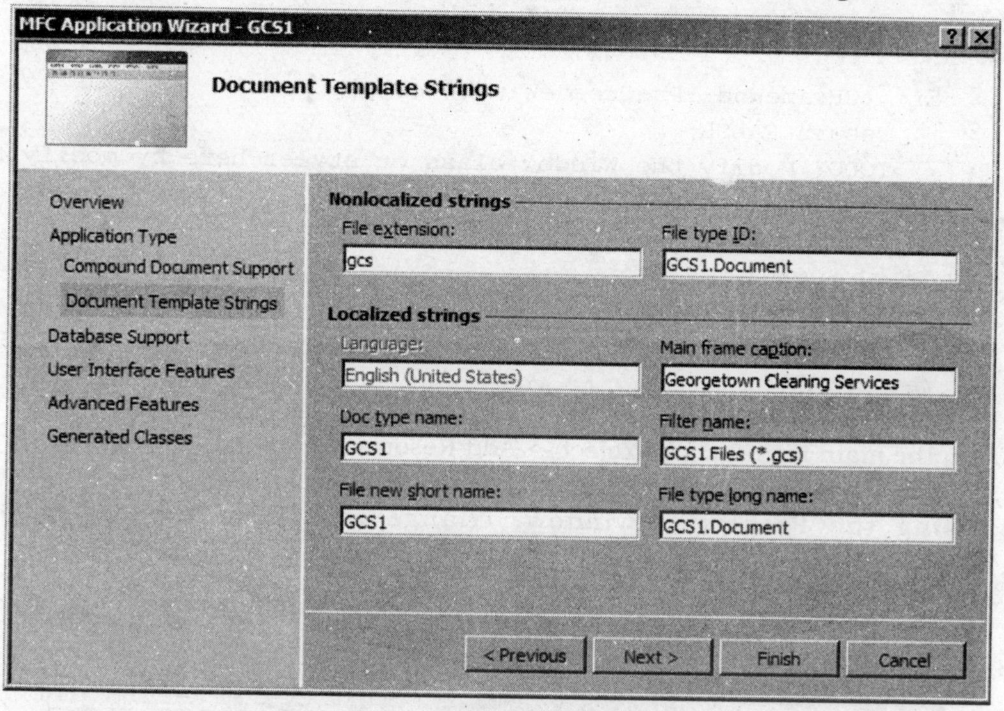

Fig. 2.6.

8. Click Next three times.
9. Set the Number of Files On Recent File List to 8 and click Next.
10. Click Finish.
11. Click Class View and click CMainFrame.
12. Below the list of classes, double-click PreCreateWindow and make the following changes :

```
int CMainFrame::OnCreate(LPCREATESTRUCT lpCreateStruct)
{
    if (CFrameWnd::OnCreate(lpCreateStruct) == -1)
        return -1;

    . . .

    // TODO: Delete these three lines if
    // you don't want the toolbar to be dockable
    m_wndToolBar.EnableDocking(CBRS_ALIGN_ANY);
    EnableDocking(CBRS_ALIGN_ANY);
```

```
    DockControlBar(&m_wndToolBar);

    CenterWindow();
       return 0;
}

BOOL CMainFrame::PreCreateWindow(CREATESTRUCT& cs)
{
    if( !CFrameWnd::PreCreateWindow(cs) )
        return FALSE;
    // TODO: Modify the Window class or styles here by modifying
    // the CREATESTRUCT
    cs cs.cx = 550;
    cs.cy = 630;
    cs.style &= ~FWS_ADDTOTITLE;

    return TRUE;
}
```

13. On the main menu, click Project -> Add Resource...
14. In the Add Resource dialog box, click Dialog and click New.
15. Using the Properties window, change the ID of the dialog box to **IDB_CLEANINGORDER_DLG** .
16. Right-click the dialog box and click Add Class...
17. Set the Class Name to **CCleaningOrderDlg.**
18. Click Finish.

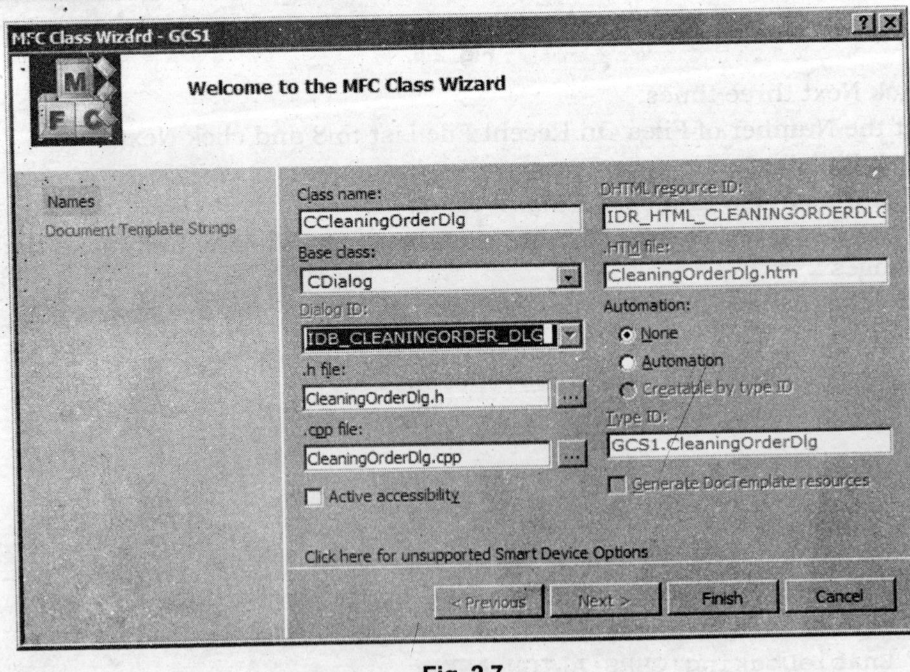

Fig. 2.7.

19. Design the dialog box as follows :

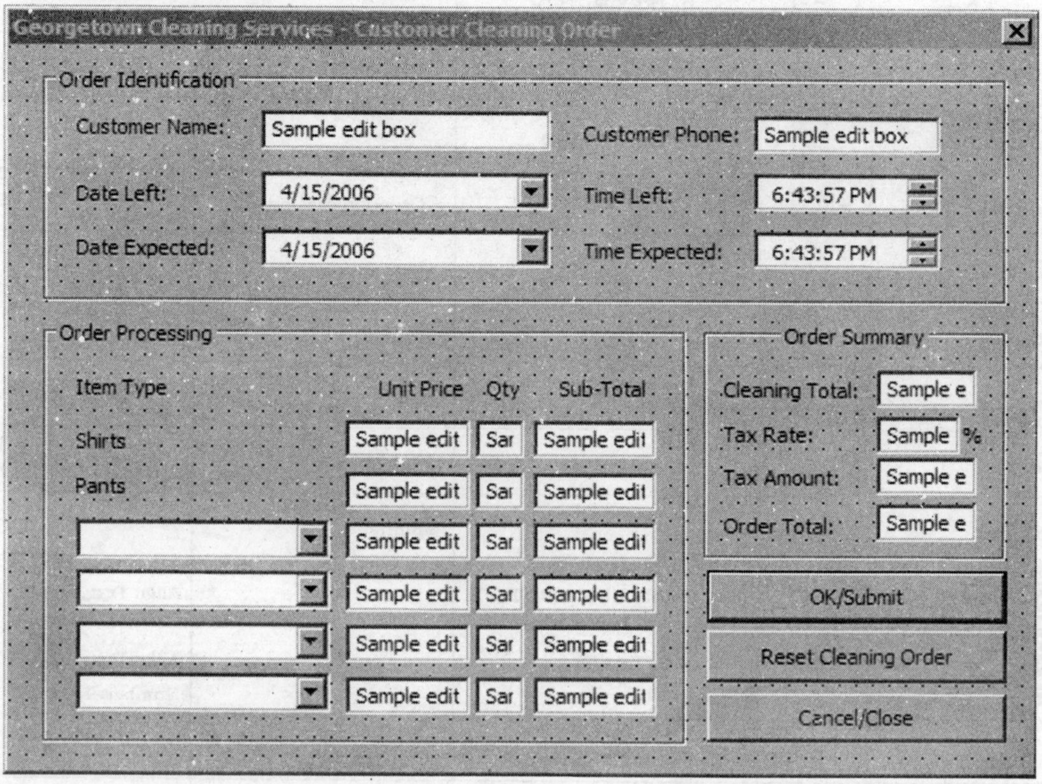

Fig. 2.8.

Control		*ID*	*Caption*	*Additional Properties*
Group Box			Order Identification	
Static Text			Customer Name	
Edit Control		IDC_CUSTOMERNAME		
Static Text			Customer Phone	
Edit Control		IDC_CUSTOMERPHONE		
Static Text			Date Left	
Date Time Picker		IDC_DATELEFT		
Static Text			Time Left	
Date Time Picker		IDC_TIMELEFT		
Static Text			Date Expected	

Date Time Picker		IDC_DATEEXPECTED		
Static Text			Time Expected:	
Date Time Picker		IDC_TIMEEXPECTED		
Group Box			Order Processing	
Static Text			Item Type	
Static Text			Unit Price	
Static Text			Qty	
Static Text			Sub-Total	
Static Text			Shirts	
Edit Control		IDC_SHIRTS_UNITPRICE		Align Text: Right
Edit Control		IDC_SHIRTS_QTY		Align Text: Right Number: True
Edit Control		IDC_SHIRTS_SUBTOTAL		Align Text: Right
Static Text			Pants	
Edit Control		IDC_PANTS_UNITPRICE		Align Text: Right
Edit Control		IDC_PANTS_QTY		Align Text: Right Number True
Edit Control		IDC_PANTS_SUBTOTAL		Align Text: Right
ComboBox		IDC_ITEM1NAME		Data: None;Women Suit; Dress; Regular Skirt; Skirt With Hook; Men 's Suit 2Pc; Men's Suit 3Pc; Sweaters;Silk Shirt; Tie;Coat; Jacket;Swede;
Edit Control		IDC_ITEM1UNITPRICE		Align Text: Right
Edit Control		IDC_ITEM1QUANTITY		Align Text: Right Number: True
Edit Control		IDC_ITEM1SUB TOTAL		Align Text: Right

ComboBox		IDC_ITEM2NAME		Data: None;Women Suit; Dress; Regular Skirt;Skirt With Hook; Men's Suit 2Pc;Men's Suit 3Pc; Sweaters;Silk Shirt; Tie;Coat; Jacket;Swede;
Edit Control		IDC_ITEM2UNITPRICE		Align Text: Right
Edit Control		IDC_ITEM2QUANTITY		Align Text: Right Number: True
Edit Control		IDC_ITEM2SUBTOTAL		Align Text: Right
ComboBox		IDC_ITEM3NAME		Data: None;Women Suit;Dress; Regular Skirt;Skirt With Hook; Men's Suit 2Pc;Men's Suit 3Pc; Sweaters; Silk Shirt;Tie; Coat; Jacket;Swede;
Edit Control		IDC_ITEM3UNITPRICE		Align Text: Right
Edit Control		IDC_ITEM3QUANTITY		Align Text: Right Number: True
Edit Control		IDC_SUBTOTAL3		Align Text: Right
ComboBox		IDC_ITEM4NAME		Data: None;Women Suit;Dress; Regular Skirt;Skirt With Hook; Men's Suit 2Pc;Men 's Suit 3Pc; Sweaters;Silk Shirt; Tie;Coat; Jacket;Swede;
Edit Control		IDC_ITEM4UNITPRICE		Align Text: Right
Edit Control		IDC_ITEM4QUANTITY		Align Text: Right Number: True
Edit Control		IDC_ITEM4SUBTOTAL		Align Text: Right
Group Box			Order Summary	
Static Text			Cleaning Total:	

Edit Control	ab|	IDC_CLEANINGTOTAL		Align Text: Right
Static Text	Aa		Tax Rate:	
Edit Control	ab|	IDC_TAXRATE		Align Text: Right
Static Text	Aa		%	
Static Text	Aa		Tax Amount	
Edit Control	ab|	IDC_TAXAMOUNT		Align Text: Right
Static Text	Aa		Order Price	
Edit Control	ab|	IDC_ORDERTOTAL		Align Text: Right
Button		IDOK	OK/Submit	
Button		IDC_RESET	Reset	
Button		IDCCANCEL	Cancel/Close	

20. Right-click each control and click Add Variable.
21. Add the control variables as follows :

Identifier	***Value Variable***	
	Type	***Name***
IDC_CUSTOMERNAME	CString	m_CustomerName
IDC_CUSTOMERPHONE	CString	m_CustomerPhone
IDC_DATELEFT	CTime	m_DateLeft
IDC_TIMELEFT	CTime	m_TimeLeft
IDC_DATEEXPECTED	CTime	m_DateExpected
IDC_TIMEEXPECTED	CTime	m_TimeExpected
IDC_SHIRTS_UNITPRICE	double	m_ShirtsUnitPrice
IDC_SHIRTS_QTY	int	m_ShirtsQuantity
IDC_SHIRTS_SUBTOTAL	CString	m_ShirtsSubTotal
IDC_PANTS_UNITPRICE	double	m_PantsUnitPrice
IDC_PANTS_QTY	int	m_PantsQuantity
IDC_PANTS_SUBTOTAL	CString	m_PantsSubTotal
IDC_ITEM1NAME	CString	m_Item1Name
IDC_ITEM1UNITPRICE	double	m_Item1UnitPrice1
IDC_ITEM1QUANTITY	int	m_Item1Quantity1

IDC_ITEM1SUBTOTAL	CString	m_Item1SubTotal1
IDC_ITEM2NAME	CString	m_Item2Name
IDC_ITEM2UNITPRICE	double	m_Item2UnitPrice
IDC_ITEM2QUANTITY	int	m_Item2Quantity
IDC_ITEM2SUBTOTAL	CString	m_Item2SubTotal
IDC_ITEM3NAME	CString	m_Item3Name
IDC_ITEM3UNITPRICE	double	m_Item3UnitPrice
IDC_ITEM3QUANTITY	int	m_Item3Quantity
IDC_ITEM3SUBTOTAL	CString	m_Item3SubTotal
IDC_ITEM4NAME	CString	m_Item4Name
IDC_ITEM4UNITPRICE	double	m_Item4UnitPrice
IDC_ITEM4QUANTITY	int	m_Item4Quantity
IDC_ITEM4SUBTOTAL	CString	m_Item4SubTotal
IDC_CLEANINGTOTAL	CString	m_CleaningTotal
IDC_TAXRATE	double	m_TaxRate
IDC_TAXAMOUNT	CString	m_TaxAmount
IDC_ORDERTOTAL	CString	m_OrderTotal

22. In the Class View, double-click CCleaningOrderDlg to access the dialog box' header file and declare a CString variable for each Date Time Picker control as follows :

```
#pragma once

// CCleaningOrderDlg dialog

class CCleaningOrderDlg : public CDialog
{
   DECLARE_DYNAMIC(CCleaningOrderDlg)
public:
   CCleaningOrderDlg(CWnd* pParent = NULL); // standard constructor
   virtual ~CCleaningOrderDlg();
   // Dialog Data
   enum { IDD = IDB_CLEANINGORDER_DLG };
protected:
   virtual void DoDataExchange(CDataExchange* pDX); // DDX/DDV support

   DECLARE_MESSAGE_MAP()

   CString m_CustomerName;
   CString m_CustomerPhone;

   CTime m_DateLeft;
```

```
    CTime m_TimeLeft;
    CTime m_DateExpected;
    CTime m_TimeExpected;

    CString strDateLeft;
    CString strTimeLeft;
    CString strDateExpected;
    CString strTimeExpected;

    double m_ShirtsUnitPrice;
    int m_ShirtsQuantity;
    CString m_ShirtsSubTotal;
    double m_PantsUnitPrice;

    int m_PantsQuantity;
    CString m_PantsSubTotal;
    CString m_Item1Name;

    double m_Item1UnitPrice;
    int m_Item1Quantity;
    CString m_Item1SubTotal;

    CString m_Item2Name;
    double m_Item2UnitPrice;
    int m_Item2Quantity;
    CString m_Item2SubTotal;

    CString m_Item3Name;
    double m_Item3UnitPrice;
    int m_Item3Quantity;
    CString m_Item3SubTotal;

    CString m_Item4Name;
    double m_Item4UnitPrice;
    int m_Item4Quantity;
    CString m_Item4SubTotal;

    CString m_CleaningTotal;
    double m_TaxRate;
    CString m_TaxAmount;
    CString m_OrderTotal;
};
```

23. **In the Class View, click CCleaningOrderDlg and, below the list of classes, double-click CCleacningOderDlg to access its constructor and change it as follows :**

```
CCleaningOrderDlg::CCleaningOrderDlg(CWnd* pParent /*=NULL*/)
    : CDialog(CCleaningOrderDlg::IDD, pParent)
    , m_CustomerName(_T(""))
    , m_CustomerPhone(_T(""))
    , m_DateLeft(CTime::GetCurrentTime())
    , m_TimeLeft(CTime::GetCurrentTime())
    , m_DateExpected(CTime::GetCurrentTime())
    , m_TimeExpected(CTime::GetCurrentTime())
    , m_ShirtsUnitPrice(0.95)
    , m_ShirtsQuantity(0)
    , m_ShirtsSubTotal(_T("0.00"))
    , m_PantsUnitPrice(1.75)
    , m_PantsQuantity(0)
    , m_PantsSubTotal(_T("0.00"))
    , m_Item1Name(_T(""))
    , m_Item1UnitPrice(0)
    , m_Item1Quantity(0)
    , m_Item1SubTotal(_T("0.00"))
    , m_Item2Name(_T(""))
    , m_Item2UnitPrice(0)
    , m_Item2Quantity(0)
    , m_Item2SubTotal(_T("0.00"))
    , m_Item3Name(_T(""))
    , m_Item3UnitPrice(0)
    , m_Item3Quantity(0)
    , m_Item3SubTotal(_T("0.00"))
    , m_Item4Name(_T(""))
    , m_Item4UnitPrice(0)
    , m_Item4Quantity(0)
    , m_Item4SubTotal(_T("0.00"))
    , m_CleaningTotal(_T("0.00"))
    , m_TaxRate(7.75)
    , m_TaxAmount(_T("0.00"))
    , m_OrderTotal(_T("0.00"))
{
}
```

24. In the Class View, click CCleaningOrderDlg.
25. In the Properties window, click the Overrides button and generate the OnInitDialog event.
26. Add the following code to it :

```
BOOL CCleaningOrderDlg::OnInitDialog()
{
    CDialog::OnInitDialog();
```

```
    // TODO: Add extra initialization here
    m_DateLeft.Format    = TEXT("dddd dd MMM yyyy");
    m_DateExpected.Format = TEXT("dddd dd MMM yyyy");

    return TRUE; // return TRUE unless you set the focus to a control
    // EXCEPTION: OCX Property Pages should return FALSE
}
```

27. Display the dialog box and double-click the Reset Cleaning Order button to implement it as follows :

```
void CCleaningOrderDlg::OnBnClickedReset()
{
    // TODO: Add your control notification handler code here
    m_CustomerName = _T("");
    m_CustomerPhone = _T("");
    m_DateLeft = CTime::GetCurrentTime();
    m_TimeLeft = CTime::GetCurrentTime();
    m_DateExpected = CTime::GetCurrentTime();
    m_TimeExpected = CTime::GetCurrentTime();
    m_ShirtsUnitPrice = 0.95;
    m_ShirtsQuantity = 0;
    m_ShirtsSubTotal = _T("0.00");
    m_PantsUnitPrice = 1.75;
    m_PantsQuantity = 0;
    m_PantsSubTotal = _T("0.00");
    m_Item1Name = _T("");
    m_Item1UnitPrice = 0;
    m_Item1Quantity = 0;
    m_Item1SubTotal = _T("0.00");
    m_Item2Name = _T("");
    m_Item2UnitPrice = 0;
    m_Item2Quantity = 0;
    m_Item2SubTotal = _T("0.00");
    m_Item3Name = _T("");
    m_Item3UnitPrice = 0;
    m_Item3Quantity = 0;
    m_Item3SubTotal = _T("0.00");
    m_Item4Name = _T("");
    m_Item4UnitPrice = 0;
    m_Item4Quantity = 0;
    m_Item4SubTotal = _T("0.00");
    m_CleaningTotal = _T("0.00");
    m_TaxRate = 7.75;
    m_TaxAmount = _T("0.00");
```

```
    m_OrderTotal = _T("0.00");
    UpdateData(FALSE);
}
```

28. In the Class View, right-click CCleaningOrderDlg -> Add -> Add Function...
29. Set the Return Type to void, the Function Name to CalculateOrder, and and the Access to private.

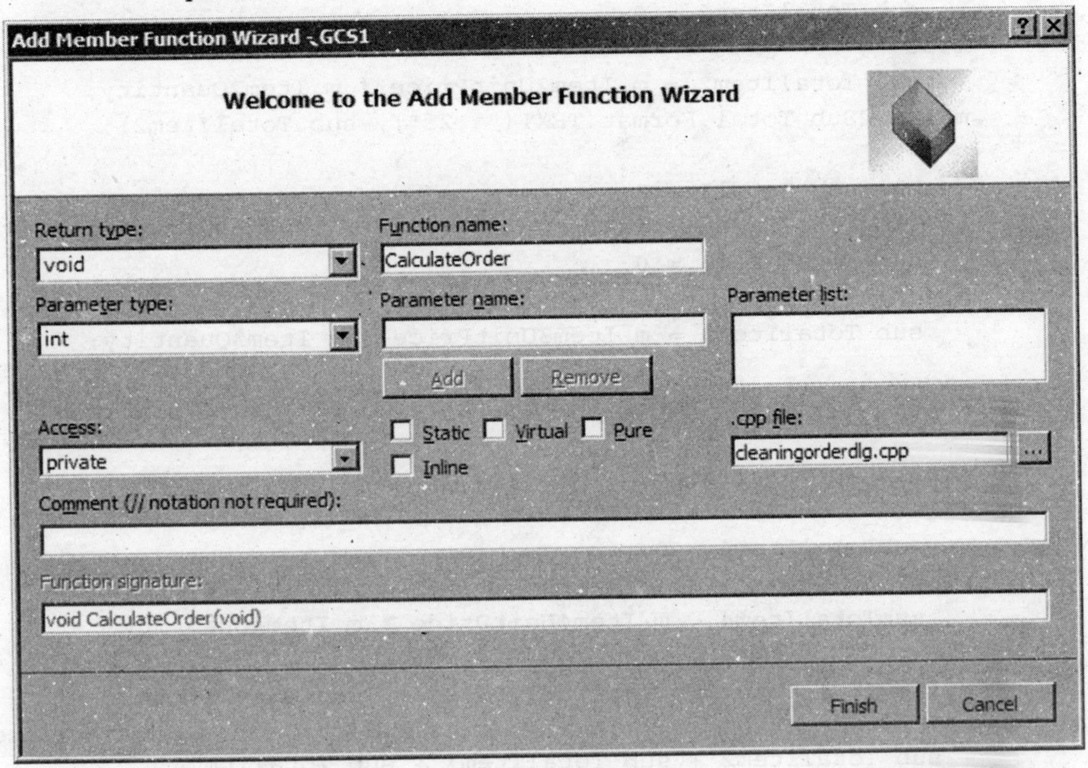

Fig. 2.9.

30. Click Finish and implement the method as follows :

```
void CCleaningOrderDlg::CalculateOrder(void)
{
    UpdateData();

    double sub TotalShirts = m_ShirtsUnitPrice * m_ShirtsQuantity;
    m_ShirtsSub Total.Format(TEXT("%.2f"), subTotalShirts);

    double sub TotalPants = m_PantsUnitPrice * m_PantsQuantity;
    m_PantsSub Total.Format(TEXT("%.2f"), sub TotalPants);

    double sub TotalItem1;
    if( (m_Item1Name == TEXT("None")) || (m_Item1Name == TEXT("")) )
        sub TotalItem1 = 0;
    else
```

```
        sub TotalItem1 = m_Item1UnitPrice * m_Item1Quantity;

    m_Item1Sub Total.Format(TEXT("%.2f"), sub TotalItem1);

    double sub TotalItem2;
    if( (m_Item2Name == TEXT("None")) || (m_Item2Name == TEXT("")) )
        sub TotalItem2 = 0;
    else
        sub TotalItem2 = m_Item2UnitPrice * m_Item2Quantity;
    m_Item2Sub Total.Format(TEXT("%.2f"), sub TotalItem2);

    double sub TotalItem3;
    if( (m_Item3Name == TEXT("None")) || (m_Item3Name == TEXT("")))
        sub TotalItem3 = 0;
    else
        sub TotalItem3 = m_Item3UnitPrice * m_Item3Quantity;

    this->m_Item3Sub Total.Format(TEXT("%.2f"), sub TotalItem3);

    double subTotalItem4;
    if( (m_Item4Name == TEXT("None")) || (m_Item4Name == TEXT("")) )
        subTotalItem4 = 0;
    else
        subTotalItem4 = m_Item4UnitPrice * m_Item4Quantity;

    m_Item4Sub Total.Format(TEXT("%.2f"), sub TotalItem4);
double cleaningTotal = sub TotalShirts + sub TotalPants + subTotalItem1
    + sub TotalItem2 + sub TotalItem3 + sub TotalItem4;

    // Calculate the amount owed for the taxes
    double taxAmount = cleaningTotal * m_TaxRate / 100;
    // Add the tax amount to the cleaning total
    double netPrice = cleaningTotal + taxAmount;

    // Display the values of the order summary
    this->m_CleaningTotal.Format(TEXT("%.2f"), cleaningTotal);
    this->m_TaxAmount.Format(TEXT("%.2f"), taxAmount);
    this->m_OrderTotal.Format(TEXT("%.2f"), netPrice);

    strDateLeft = m_DateLeft.Format(TEXT("%A %d %b %Y"));
    strTimeLeft = m_TimeLeft.Format(TEXT("%H:%M"));
    strDateExpected = m_DateExpected.Format(TEXT("%A %d %b %Y"));
    strTimeExpected = m_TimeExpected.Format(TEXT("%H:%M"));
```

```
    UpdateData(FALSE);
}
```

31. Display the dialog box and double-click the Time Left control.
32. Implement its Chantge event as follows :

```
void CCleaningOrderDlg::OnDtnDatetimechangeTimeleft(NMHDR *pNMHDR,
                                                    LRESULT *pResult)
{
    LPNMDATETIMECHANGE pDTChange =
    reinterpret_cast<LPNMDATETIMECHANGE>(pNMHDR);
    // TODO: Add your control notification handler code here
    *pResult = 0;

    UpdateData();
    CTime dateLeft, timeLeft;

    dateLeft = m_DateLeft;
    timeLeft = m_TimeLeft;

    CTime time9AM = CTime(timeLeft.GetYear(), timeLeft.GetMonth(),
    imeLeft.GetDay(), 9, 0, 0);

    // If the customer leaves clothes before 9A.M. ...
    if( timeLeft <= time9A.M.)
    {
        // ... then they should be ready the same day after 5PM
        m_DateExpected = dateLeft;
        CTime tm5PM = CTime(dateLeft.GetYear(),
            dateLeft.GetMonth(), dateLeft.GetDay(), 17, 0, 0);
        m_TimeExpected = tm5PM;
    }
    else
    {
        // If the clothes were left after 9A.M., they will be
        // available the following day by 8A.M.
        CTime tmTomorrowMorning = CTime(dateLeft.GetYear(),
        dateLeft.GetMonth(), dateLeft.GetDay() + 1, 8, 0, 0);
        m_DateExpected = tmTomorrowMorning;
        m_TimeExpected = tmTomorrowMorning;
    }
    UpdateData(FALSE);
}
```

33. Display the dialog box and click the Shirts Qty edit control.
34. In the Properties window, click the Contrl Events button, and generate an event for the EN_KILLFOCUS message.

35. Return to the dialog box and generate an event for the EN_KILLFOCUS message of all Qty edit controls.
36. Also, generate an event for the EN_KILLFOCUS message of the Tax Rate edit control.
37. In each event, simply call the CalculateOrder method :

```
void CCleaningOrderDlg::OnEnKillfocusShirtsQty()
{
    // TODO: Add your control notification handler code here
    CalculateOrder();
}

void CCleaningOrderDlg::OnEnKillfocusPantsQty()
{
    // TODO: Add your control notification handler code here
    CalculateOrder();
}

void CCleaningOrderDlg::OnEnKillfocusItem1quantity()
{
    // TODO: Add your control notification handler code here
    CalculateOrder();
}

void CCleaningOrderDlg::OnEnKillfocusItem2quantity()
{
    // TODO: Add your control notification handler code here
    CalculateOrder();
}

void CCleaningOrderDlg::OnEnKillfocusItem3quantity()
{
    // TODO: Add your control notification handler code here
    CalculateOrder();
}

void CCleaningOrderDlg::OnEnKillfocusItem4quantity()
{
    // TODO: Add your control notification handler code here
    CalculateOrder();
}

void CCleaningOrderDlg::OnEnKillfocusTaxrate()
{
    // TODO: Add your control notification handler code here
    CalculateOrder();
```

}

38. Save all.

2.4 DOCUMENT OBJECT

(MDU 2008)

The CDocument class houses your document. This class is where you will build the data structures necessary to house and manipulate the data that makes up your document. This class receives input from the CView class and passes display information to the CView class. This class is also responsible for saving and retrieving the document data from files.

The MFC library provides a more advanced and elaborate system of file processing in the internals of the document/view architecture through the CDocument class. Although this system is not the easiest and is not shared by other programming environments, once you are familiar with it, you may like.In the Document/View architecture world, a file or its content is displayed in the view while its real estate is managed by the frame. Therefore, while using a document, the user typically uses the view to manipulate the file. The trick for you is when it becomes time to save the file. Since you shouldn't manipulate the document in the document class (?), when it is time to save it, you must "transfer" the values to be saved from the view to the document class. then let the document class handle the processing of the file. This is valid both for saving and opening the file.

2.4.1 Processing Files in Document/View

1. Start a new application as Single Document and name it **LoanEval.**
2. Base its view class on **CFormView**.
3. Design the form as follows :

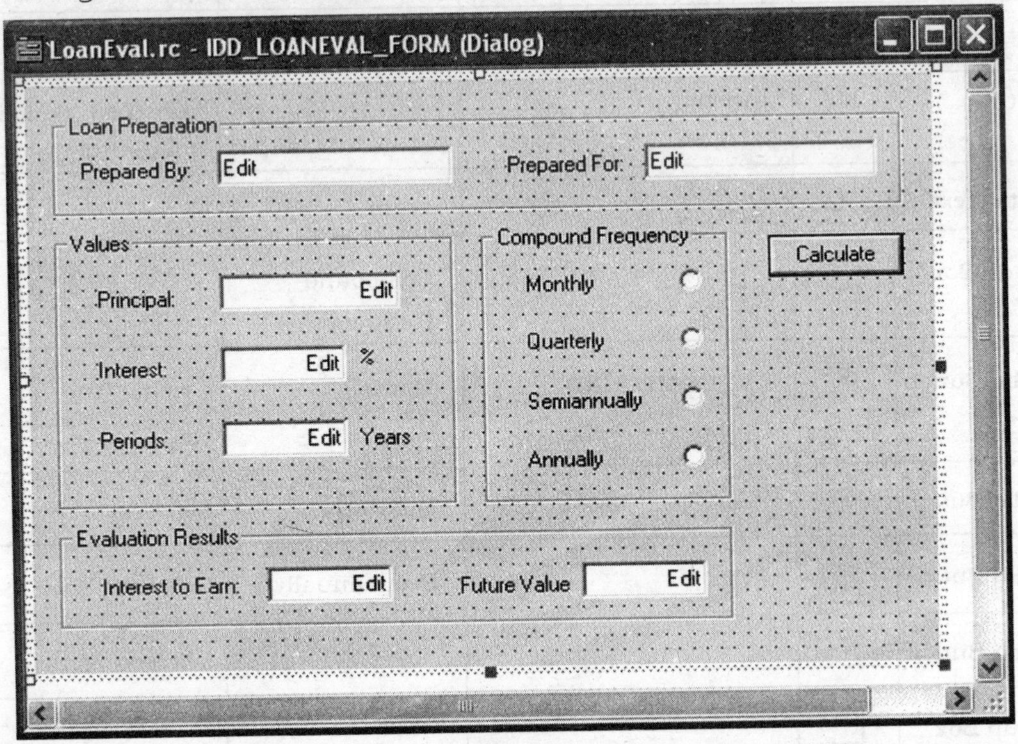

Fig. 2.10.

Control		ID	Caption	Additional Properties
Group Box			Loan Preparation	
Static Text			Prepared By	
Edit Box		IDC_PREPARED_BY		
Static Text			Prepared For	
Edit Box		IDC_PREPARED_FOR		
Group Box			Values	
Static Text			Principal	
Edit Box		IDC_PRINCIPAL		Align Text: Right
Static Text			Interest	
Edit Box		IDC_INTEREST_RATE		Align Text: Right
Static Text			%	
Static Text			Periods	
Edit Box		IDC_PERIODS		Align Text: Center Number: True
Static Text			Years	
Group Box			Compound Frequency	
Radio Button		IDC_COMPOUND	Monthly	Group: True Left Text: True
Radio Button			Quarterly	Left Text: True
Radio Button			Semiannually	Left Text: True
Radio Button			Annually	Left Text: True
Group Box				

Static Text			Interest Earned	
Edit Box		IDC_INTEREST_TOEARN	Align Text: Right	
Static Text			Amount Earned	
Edit Box		IDC_FUTURE_VALUE		Align Text: Right
Button		IDC_CALCULATE_BTN	Calculate	Default Button: True
Button		IDC_RESET_BTN	Reset	

4. Using either the ClassWizard or the Add Member Variable Wizard, add the following member variables to the controls :

Identifier	***Type***	***Value Variable***
IDC_PREPARED_BY	CString	m_PreparedBy
IDC_PREPARED_FOR	CString	m_PreparedFor
IDC_PRINCIPAL	double	m_Principal
IDC_INTEREST_RATE	double	m_InterestRate
IDC_PERIODS	UINT	m_Periods
IDC_COMPOUND	int	m_Compound
IDC_INTEREST_TOEARN	double	m_InterestToEarn
IDC_FUTURE_VALUE	double	m_FutureValue

5. Change the design of the IDR_MAINFRAME icon as follows :

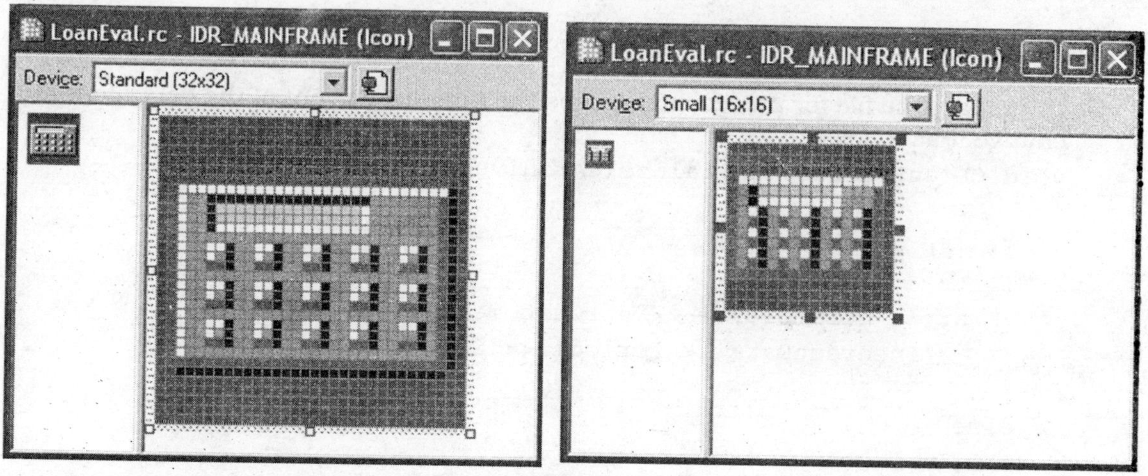

Fig. 2.11.

6. In the header file of the document class, declare a variable for each value that will be used in file processing :

```
// LoanEvalDoc.h : interface of the CLoanEvalDoc class
//
/////////////////////////////////////////////////////////////////////////////

. . .

class CLoanEvalDoc : public CDocument
{
protected: // create from serialisation only
   CLoanEvalDoc();
   DECLARE_DYNCREATE(CLoanEvalDoc)

// Attributes
public:

   CString      strEmployee;
   CString      strCustomer;
   double       Principal;
   double       InterestRate;
   UINT         Periods;
   int          Compound;

// Operations
public:

 . . .

 };

 . . .
```

7. To prepare the file for processing, access the Serialise() event of the document and change it as follows :

```
void CLoanEvalDoc::Serialize(CArchive& ar)
{
   if (ar.IsStoring())
   {
      ar << strEmployee << strCustomer << Principal
      << InterestRate << Periods << Compound;
   }
   else
   {
      ar >> strEmployee >> strCustomer >> Principal
      >> InterestRate >> Periods >> Compound;
   }
```

```
}
```

8. Access the StdAfx.h header file and include the math.h header file :

```
. . .

#define VC_EXTRALEAN        // Exclude rarely-used stuff from Windows
                            headers
#include <afxwin.h>         // MFC core and standard components
#include <afxext.h>         // MFC extensions
#include <afxdisp.h>        // MFC Automation classes
#include <afxdtctl.h>       // MFC support for Internet Explorer 4 Common
                            Controls
#ifndef _AFX_NO_AFXCMN_SUPPORT
#include <afxcmn.h>         // MFC support for Windows Common Controls
#endif // _AFX_NO_AFXCMN_SUPPORT

#include <math.h>

. . .
```

9. Generate a **BN_CLICKED** message for the the IDC_CALCULATE_BTN control associated with the view class and accept the suggested name of the function.
10. Implement it as follows :

```
void CLoanEvalView::OnCalculateBtn()
{
    // TODO: Add your control notification handler code here
    UpdateData();
    double InterestToEarn;
    double FutureValue;
    int CompoundType;

    switch( m_Compound)
    {
    case 0:
        CompoundType = 12;
        break;
    case 1:
        CompoundType = 4;
        break;
    case 2:
        CompoundType = 2;
        break;
    case 3:
        CompoundType = 1;
        break;
    }
```

```
    double i = (m_InterestRate / 100) / CompoundType;
    int n  = CompoundType * m_Periods;

    FutureValue  = m_Principal * pow(1 + i, n);
    InterestToEarn = FutureValue - m_Principal;

    CWnd *edtInterestToEarn = reinterpret_cast<CEdit *>(
                        GetDlgItem(IDC_INTEREST_TOEARN));
    CWnd *edtFutureValue   = reinterpret_cast<CEdit *>(
                        GetDlgItem(IDC_FUTURE_VALUE));

    char InterestAmount[20], FValue[20];
    sprintf(InterestAmount, "%.2f", InterestToEarn);
    sprintf(FValue, "%.2f", FutureValue);

    edtInterestToEarn->SetWindowText(InterestAmount);
    edtFutureValue->SetWindowText(FValue);
}
```

11. Generate a message for the ID_FILE_SAVE menu item associated with the view class and accept the suggested name of the function.
12. Implement it as follows :

```
void CLoanEvalView:OnFileSave()
{
    // TODO: Add your command handler code here
    this->UpdateData();
    CLoanEvalDoc *pDoc = GetDocument();

    pDoc->strEmployee = this->m_PreparedBy;
    pDoc->strCustomer = this->m_PreparedFor;
    pDoc->Principal   = this->m_Principal;
    pDoc->InterestRate = this->m_InterestRate;
    pDoc->Periods     = this->m_Periods;
    pDoc->Compound    = this->m_Compound;

    CFile flLoan;

    charstrFilter[]={"Loan Files(*.onf)|*.onf|Text Files
    (*.txt)|*.txt|All Files (*.*)|*.*||" };

    CFileDialog FileDlg(FALSE, ".bcr", NULL, 0, strFilter);

    if( FileDlg.DoModal() == IDOK )
    {
```

```
        flLoan.Open(FileDlg.GetFileName(), CFile::modeCreate |
        CFile::modeWrite);
        CArchive ar(&flLoan, CArchive::store);
        pDoc->Serialise(ar);

        ar.Close();
    }
    else
        return;
    flLoan.Close();
}
```

13. Generate a message for the ID_FILE_OPEN menu item associated with the view class and accept the suggested name of the function.
14. Implement it as follows :

```
void CLoanEvalView::OnFileOpen()
{
    // TODO: Add your command handler code here
    UpdateData();
    CLoanEvalDoc *pDoc = GetDocument();

    CFile flLoan;

    char strFilter[] = { "BCR Files (*.onf)|*.onf|Text Files
    (*.txt)|*.txt|All Files (*.*)|*.*||" };

    CFileDialog FileDlg(FALSE, ".bcr", NULL, 0, strFilter);

    if( FileDlg.DoModal() == IDOK )
    {
        if( flLoan.Open(FileDlg.GetFileName(), CFile::modeRead) ==
        FALSE) return;
        CArchive ar(&flLoan, CArchive::load);
        pDoc->Serialise(ar);

        m_PreparedBy.Format("%s", pDoc->strEmployee);
        m_PreparedFor.Format("%s", pDoc->strCustomer);
        m_Principal  = pDoc->Principal;
        m_InterestRate = pDoc->InterestRate;
        m_Periods    = pDoc->Periods;
        m_Compound   = pDoc->Compound;

        ar.Close();
    }
    else
```

```
        return;
    flLoan.Close();
    UpdateData(FALSE);
}
```

15. Access the frame's source file to remove the default title and set your own as follows :

```
// MainFrm.cpp : implementation of the CMainFrame class
//

. . .

////////////////////////////////////////////////////////////////////////
//CMainFrameconstruction/destruction

. . .

int CMainFrame::OnCreate(LPCREATESTRUCT lpCreateStruct)
{
    if (CFrameWnd::OnCreate(lpCreateStruct) == -1) return -1;

    . . .

    this->SetWindowText("Loan Evaluation");
    this->CenterWindow();

    return 0;
}

BOOL CMainFrame::PreCreateWindow(CREATESTRUCT& cs)
{
    if( !CFrameWnd::PreCreateWindow(cs) )  return FALSE;

    cs.style &= ~FWS_ADDTOTITLE;

    return TRUE;
}
////////////////////////////////////////////////////////////////////////
// CMainFrame diagnostics

. . .
```

16. Execute the application then create and save a record.

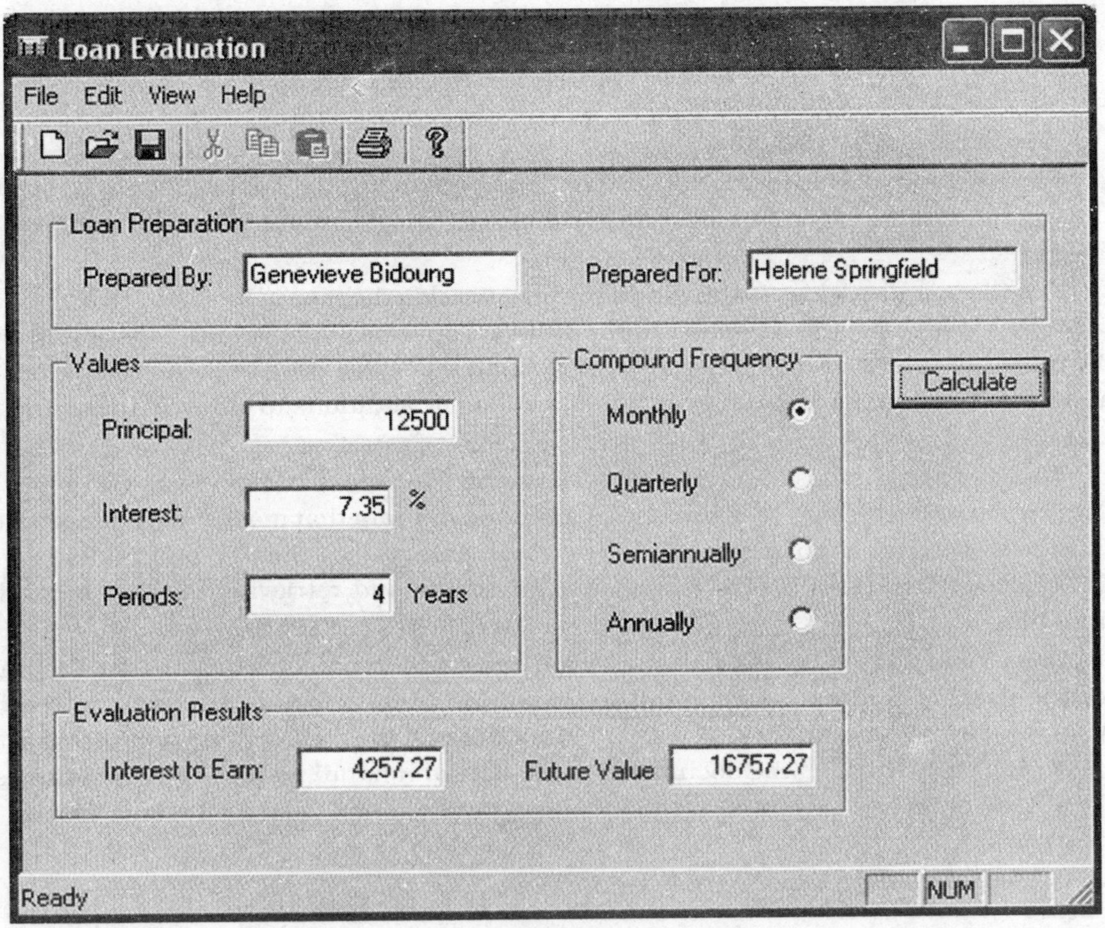

Fig. 2.12.

2.4.2. MFC File Processing Options

In the above example, we used the CFile class to discuss file processing. MFC provides more extended file processing support than that. In reality, file processing is one of the strengths of MFC. It excels in it by providing the following classes :

FILE : FILE is a C structure used to perform file processing for that language. You can also use it in your MFC applications.

CStdioFile : MFC provides the same file processing support as the C language through the **CStdioFile class.**

Win32 : The Win32 library provides its own mechanism of file processing through its HANDLE concept. A file is created using the **CreateFile()** function.

2.5 DOCUMENT/ VIEW ARCHITECTURE (MDU 2008-09)

2.5.1 Document/ View Architecture (SDI)

When you create an SDI application, more classes are created for an SDI application than for a dialog-style application. Each of these classes serves a specific purpose in how SDI applications operate. Ignoring the About window dialog class, four specific classes make up an SDI application :

- The CWinApp-derived class
- The CFrameView-derived class
- The CDocument-derived class
- The CView-derived class

The CWinApp class creates all the other components in the application. It is the class that receives all the event messages and then passes the messages to the CFrameView and CView classes.

The CFrameView class is the window frame. It holds the menu, toolbar, scrollbars, and any other visible objects attached to the frame. This class determines how much of the document is visible at any time. Very little (if any) of your programming efforts on SDI applications will require making any modifications or additions to either of these first two classes.

The CDocument class houses your document. This class is where you will build the data structures necessary to house and manipulate the data that makes up your document. This class receives input from the CView class and passes display information to the CView class. This class is also responsible for saving and retrieving the document data from files.

The CView class is the class that displays the visual representation of your document for the user. This class passes input information to the CDocument class and receives display information from the CDocument class. Most of the coding that you will do for this class consists of drawing the document for the user and handling the input from the user. The CView class has several descendent classes that can be used as the ancestor for the view class.

These descendent classes are listed in Table :

Class	***Description***
CEditView	Provides the functionality of a edit box control. Can be used to implement simple text-editor functionality.
CFormView	The base class for views containing controls. Can be used to provide form-based documents in applications.
CHtmlView	Provides the functionality of a Web browser. This view directly handles the URL navigation, hyperlinking, and so on. Maintains a history list for browsing forward and back.
CListView	Provides list-control functionality in the Document/View architecture.
CRichEditView	Provides character and paragraph formatting functionality. Can be used to implement a word-processor application.
ScrollView	Provides scrolling capabilities to a CView class.
CTreeView	Provides tree-control functionality in the Document/View architecture.

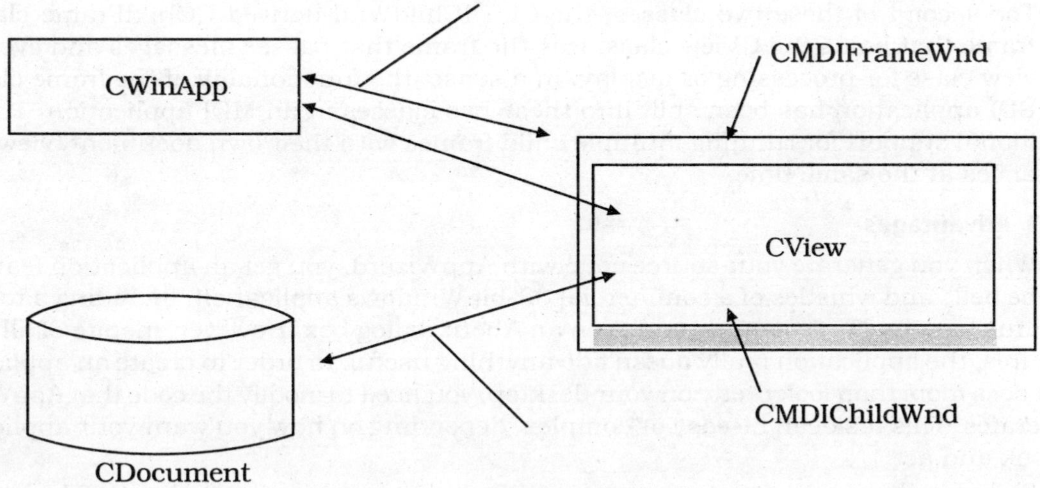

Fig. 2.13.

2.5.2 Document/ View Architecture (MDI)

As far as coding an MDI application with Visual C++, there's little difference between creating an SDI and an MDI application. However, when you get deeper into the two application styles, you'll find quite a few differences. Although an SDI application allows the user to work on only one document at a time, it also normally limits the user to working on a specific type of document. MDI applications not only enable the user to work on multiple documents at the same time, but also MDI applications can allow the user to work on multiple types of documents.

An MDI application uses a window-in-a-window style, where there is a frame window around one or more child windows. This is a common application style with many popular software packages, including Word and Excel.

Architecturally, an MDI application is similar to an SDI application. In fact, with a sim-ple MDI application, the only difference is the addition of a second frame class to the other classes that the AppWizard creates, as shown in Figure . As you can see, the Document/View architecture is still very much the approach you use for developing MDI applications as well as SDI applications.

When you create an MDI application, you will create just one more class than you created with an SDI application. The classes are :

- The CWinApp derived class
- The CMDIFrameWnd derived class
- The CMDIChildWnd derived class
- The CDocument derived class
- The CView derived class

The two MDI derived classes, CMDIFrameWnd (the CMainFrame class in your project) and CMDIChildWnd (the CChildFrame class in your project), are the only two classes that are different from the SDI application that you created.

The first of these two classes, the CMDIFrameWnd-derived CMainFrame, is the main frame of the application. It provides an enclosed space on the desktop within which all application interaction takes place. This frame window is the frame to which the menu and tool-bars are attached.

The second of these two classes, the CMDIChildWnd-derived CChildFrame class, is the frame that holds the CView class. It is the frame that passes messages and events to the view class for processing or display. In a sense, the functionality of the frame class in the SDI application has been split into these two classes in an MDI application. There is additional support for running multiple child frames with their own document/view class instances at the same time.

2.5.3 Advantages

When you generate your source code with AppWizard, you get an application featuring all the bells and whistles of a commercial 32-bit Windows application, including a toolbar, a status bar, ToolTips, menus, and even an About dialog box. However, inspite of all those features, the application really doesn't do anything useful. In order to create an application that does more than look pretty on your desktop, you need to modify the code that AppWizard generates. This task can be easy or complex, depending on how you want your application to look and act.

Probably the most important set of modifications are those related to the document--the information the user could save from your application and restore later--and to the view--the way that information is presented to the user.

MFC's document/view architecture separates an application's data from the way the user actually views and manipulates that data. Simply, the document object is responsible for storing, loading, and saving the data, whereas, the view object (which is just another type of window) enables the user to see the data onscreen and to edit that data in a way that is appropriate to the application. In this chapter, you learn the basics of how MFC's document/view architecture works.

SDI and MDI applications created with AppWizard are document/view applications. That means that AppWizard generates a class for you derived from CDocument, and delegates certain tasks to this new document class. It also creates a view class derived from CView and delegates other tasks to your new view class. Let's look through an AppWizard starter application and see what you get.

Because you've been working with AppWizard-generated applications in this chapter, you've taken for granted a lot of what goes on in the background of an MFC document/view program. That is, much of the code that enables the frame window (your application's main window), the document, and the view window to work together is automatically generated by AppWizard and manipulated by MFC.

The Document/View architecture provides its own technique of creating a program. An application is made of two primary sections: The application class and the frame.

The application class allows you to notify the operating system that your application exists, that your application needs memory and that it may use some resources. Based on this, to start an application, you must create your own class and derive it from the MFC's CWinApp. Here is an example :

```
#include <afxwin.h>

class CExerciseApp : public CWinApp
{
};
```

If you have done **Win32** programming before, you may know that each application works by creating an instance. In the same way, if you create a Document/View architecture, when the application comes up, it must create an instance. To support this, the **CWinApp** class is equipped with the Boolean InitInstance() method that you should implement in your application. Here is an example :

```
class CExerciseApp : public CWinApp
{
    BOOL InitInstance()
    {
        return TRUE;
    }
};
```

The frame allows you to show the borders and probably the contents of the (main) window of your application. To create a frame, you can derive a class from the MFC's CFrameWnd class and you should implement its (default) constructor. In the definition of the constructor of your class, you should call the Create() method that allows you to formally create a frame. Only two pieces of information are required to create a frame, the name of the class, which should be passed as NULL, and the name of the window, which is the string that will be shown on the title bar.

After defining the frame class, you should instantiate in the InitInstance() method of your application.

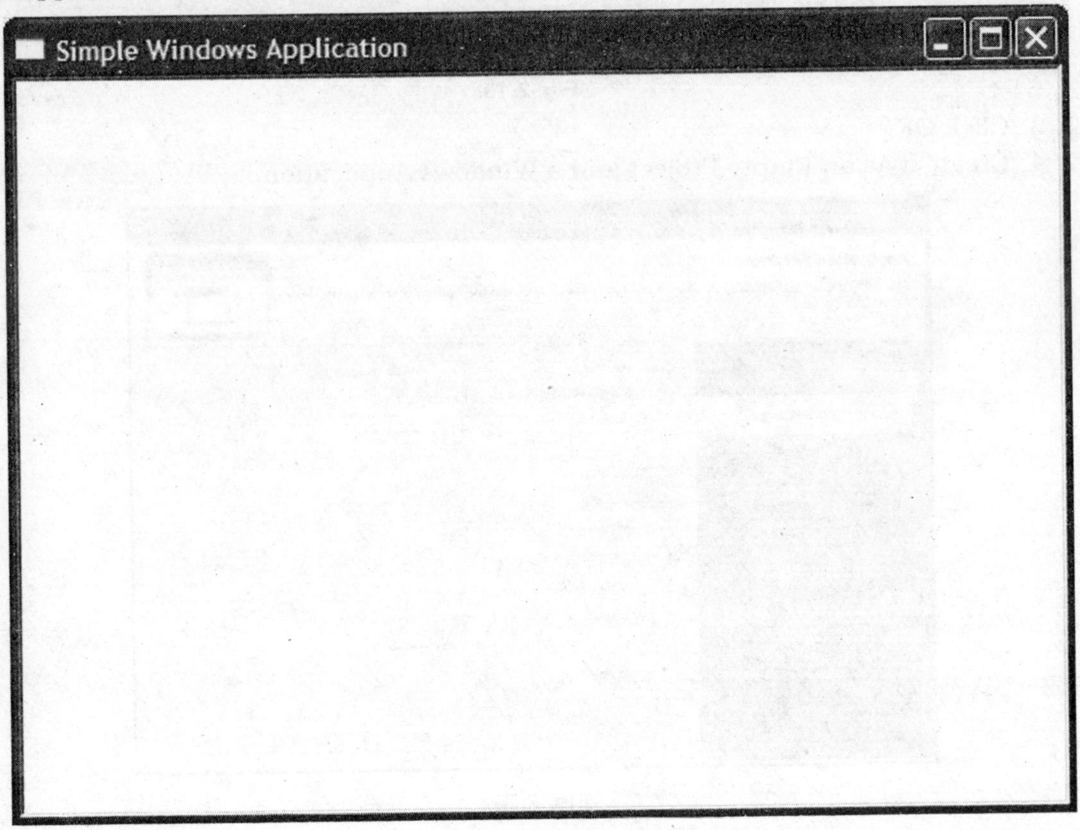

Fig. 2.14.

2.6 CREATING A SIMPLE WINDOW

(MDU 2008)

1. Start Microsoft Visual C++ .NET.
2. Create a Win32 Project named **MFCSimpleWindow**.

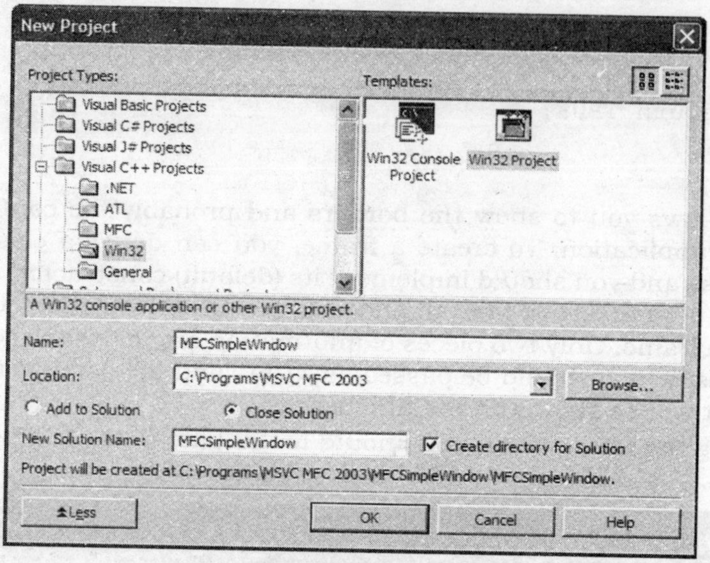

Fig. 2.15.

3. Click OK.
4. Create it as an Empty Project and a Windows Application.

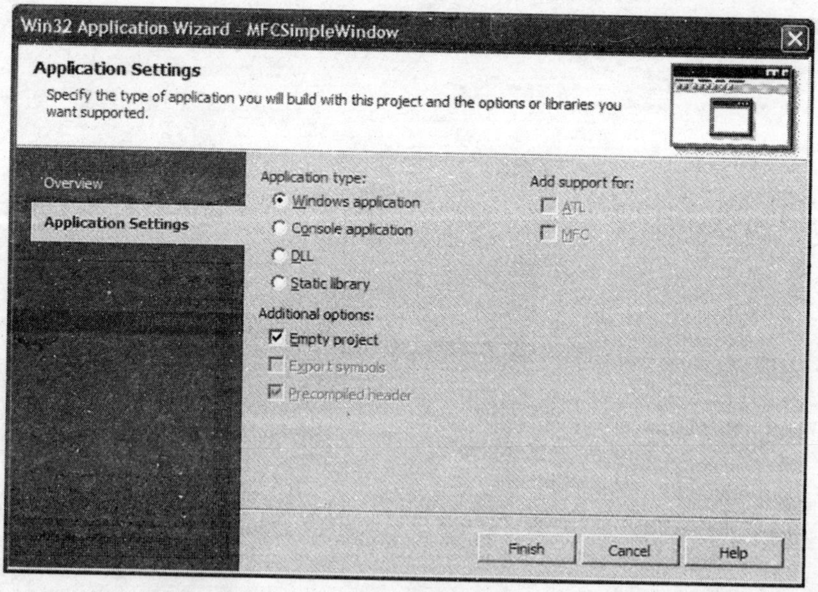

Fig. 2.16.

5. Click Finish.
6. On the main menu, click Project -> MFCSimpleWindow Properties...
7. In the Use of MFC combo box, select Use MFC in a Shared DLL.

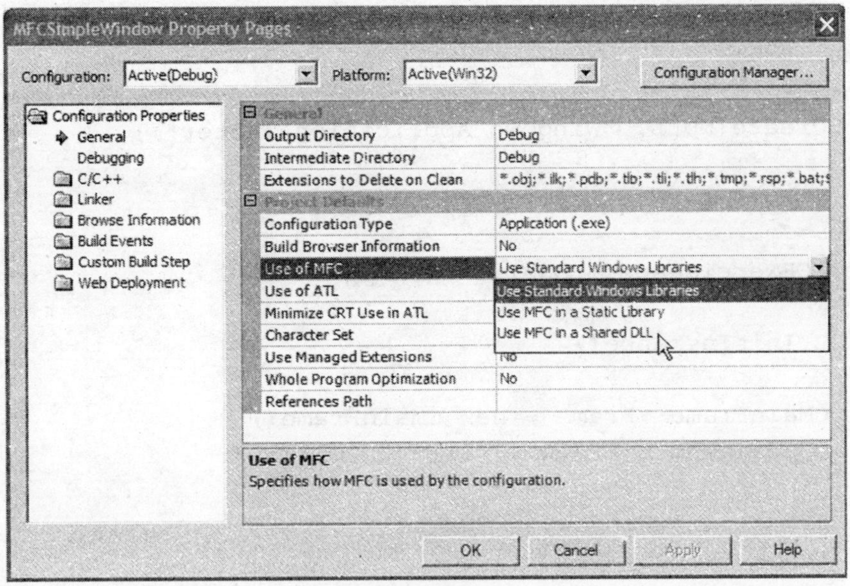

Fig. 2.17.

8. Click OK.
9. To create a new source file, on the main menu, click Project -> Add New Item...
10. In the Templates section, click C++ File (.cpp).
11. Set the **Name as Exercise.**

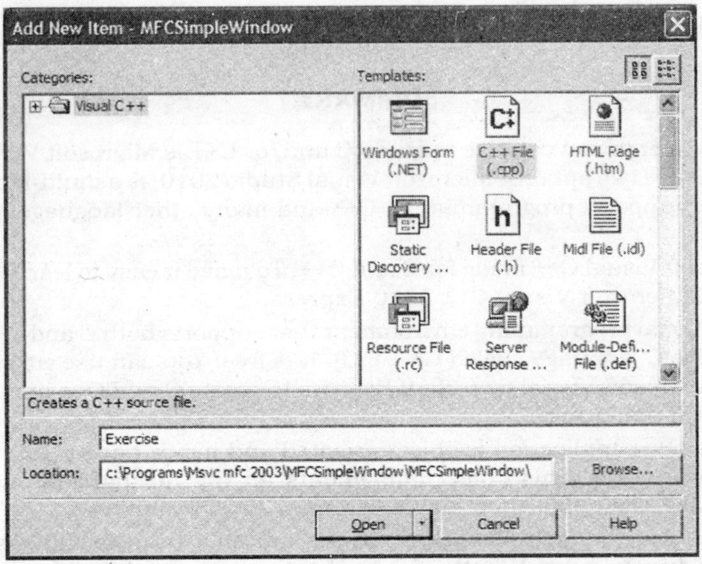

Fig. 2.18.

12. Click Open.
13. In the empty file type :

```
#include <afxwin.h>

struct CMainFrame : public CFrameWnd
```

```
{
    CMainFrame()
    {
        Create(NULL, "Windows Application Tester");
    }
};

struct CExerciseApp : public CWinApp
{
    BOOL InitInstance()
    {
        CMainFrame *Frame = new CMainFrame();
        m_pMainWnd = Frame;

        Frame->ShowWindow(SW_NORMAL);
        Frame->UpdateWindow();

        return TRUE;
    }
};

CExerciseApp theApp;
```

14. Test the application.

SUMMARY

1. One of the environments you can use to study C and/or C++ is Microsoft Visual C++. Microsoft Visual C++ is available two options. Microsoft Visual Studio 2010, is a multi-language commercial environment that supports programming in C++ and many other languages (C#, Visual Basic, and F#).
2. It includes Microsoft Visual C++ made for C and C++. To make it easy to learn C++ programming, Microsoft created Microsoft Visual C++ 2010 Express.
3. This is a full-functional programming environment that supports both C and C++ (it also supports another child of the C language, called C++/CLI). It is free. You can use either Microsoft Visual Studio 2010, or Microsoft Visual C++ 2010 Express to create C and C++ applications. To follow our lessons, we will use either.
4. A VC++ (as also a C++) application is object-oriented and hence has at least one object, which is an instance of a class. This object implements all the mechanisms needed to run the application. Besides, the application may use a number of resources.
5. The WNDLCLASS and the WNDCLASSEX classes are only used to initialise the application window class. To display a window, that is, to give the user an object to work with, you must create a window object. This window is the object the user uses to interact with the computer.
6. For a Win32 application, the class name of your main window must be provided to the lpszClassName member variable of your WNDCLASS or WNDCLASSEX variable. You can provide the name to the variable or declare a global null-terminated string. The name must also be passed to the lpClassName argument of either the CreateWindow() or the CreateWindowEx() functions.

7. The name is passed as the lpWindowName argument of either the CreateWindow() or the CreateWindowEx() functions. To do this, you can provide a null-terminated string to the argument or declare a global string.
8. After creating the menu, assign its name to the lpszMenuName name to your WNDCLASS variable for a Win32 variable. If you had created the menu and gave it an identifier, use the MAKEINTRRESOURCE macro to convert it.
9. Once a window has been created and if this was done successfully, you can display it to the user. This is done by calling the ShowWindow() function. The Win32 version of this function has the following syntax :

   BOOL ShowWindow(HWND hWnd, int nCmdShow);
10. Once a window has been created, the user can use it. This is done by the user clicking things with the mouse or pressing keys on the keyboard. A message that a window sends is received by the application. This application must analyse, translate, and decode the message to know what object sent the message and what the message consists of. To do this, the application uses the GetMessage() function.

SHORT QUESTIONS WITH ANSWER

Q.1. Explain the Displaying the Window.

Ans. Once a window has been created and if this was done successfully, you can display it to the user. This is done by calling the **ShowWindow()** function. The Win32 version of this function has the following syntax :

```
BOOL ShowWindow(HWND hWnd, int nCmdShow);
```

The hWnd argument is a handle to the window that you want to display. It could be the window returned by the **CreateWindow()** or the **CreateWindowEx()** function.

The *nCmdShow* specifies how the window must be displayed. It uses one of the *nCmdShow* **values above**.

To show its presence on the screen, the window must be painted. This can be done by calling the UpdateWindow() function. Its syntax is :

```
BOOL UpdateWindow(HWND hWnd);
```

This function simply wants to know what window needs to be painted. This window is specified by its handle.

Q.2. What do you mean by View Object?

Ans. When you create an application, you usually intend for the users to see, view, and possibly manipulate its documents. The basic process of making up a program is to create the application using a class derived from CWinApp and surrounding the program with a frame either based on **CFrameWnd** or from one of its derived classes. Such a basic program is mainly meant to show the "presence" of an application. To truly assist the user with files, the program should be equipped with a document. A document is not truly something a user can see. It is a representation of a file. To actually present a document to the user, you should provide an object called a view.

Different applications have various ways to present their documents. To support these variances, the MFC provides different view-based classes. To lay a foundation for those classes, the MFC provides a class named **CView**. Although the MFC provides various view-based classes, it is likely that none of them would fit the particular functionality you want to implement. For this reason, the **CView** class accomplishes two purposes. Not only is it the ancestor to the other view-based classes, it is still a view in its own right and you can use it to create an application that none of the other classes can handle. For these two and other reasons, you should be quite familiar with the **CView** class.

Q.3. What do you mean by Document Object? **(MDU 2007)**

Ans. The CDocument class houses your document. This class is where you will build the data structures necessary to house and manipulate the data that makes up your document. This class receives input from the CView class and passes display information to the CView class. This class is also responsible for saving and retrieving the document data from files.

The MFC library provides a more advanced and elaborate system of file processing in the internals of the document/view architecture through the CDocument class. Although this system is not the easiest and is not shared by other programming environments, once you are familiar with it, you may like.

In the Document/View architecture world, a file or its content is displayed in the view while its real estate is managed by the frame. Therefore, while using a document, the user typically uses the view to manipulate the file. The trick for you is when it becomes time to save the file. Since you shouldn't manipulate the document in the document class (?), when it is time to save it, you must "transfer" the values to be saved from the view to the document class. then let the document class handle the processing of the file. This is valid both for saving and opening the file.

Q.4. What are MFC File Processing Options? Explain

Ans. MFC provides more extended file processing support than that. In reality, file processing is one of the strengths of MFC. It excels in it by providing the following classes :

FILE : FILE is a C structure used to perform file processing for that language. You can also use it in your MFC applications.

CStdioFile : MFC provides the same file processing support as the C language through the CStdioFile class.

Win32 : The Win32 library provides its own mechanism of file processing through its HANDLE concept. A file is created using the **CreateFile()** function.

Q.5. Write a program that shows and hide a caret when we lose or gain the focus.

Ans. Caret: The caret is a hashing line block, or blump in the client area of a window or in a control that accepts keyboard input. It indicates the place at which text or graphics are inserted. Because only one window at a time has the keyboard forms, there is only one caret in the system.

Show and Hide caret method

```
Protected override void wndproc {ref Message m}
{
    if (m, Msg = = WM_SETFOCUS)
    {
    CreateCaret (this Handle, Input, Zero, caret width, caret height);
    Set CretPos (xPOS, yPOS);
    ShowCaret (this Handle), // Returns non-zero on success
    }
    if (m Msg = = WM_KILLFOCUS)
    Destroy Caret ( );
    ..
    }
    Hide Caret
    using System Runtime Interop Services ;
    [DII Import ("user 32.dII", Entrypoint = "HideCaret")]
    public static extern long HideCaret (intPtrhwnd);
    HideCaret (txtDisplay.Handle);
```

Q.6. What do you mean by window with the focus ? Write a Program in VC++ to fine the size of a window.

Ans. Window Size:

```
    hdc = BeginPaint(hWnd,&ps); //TODO Add any drawing code here ...
    RECT rt; GetClientRect(hWnd,&rt);
    if(blunitUSBs)
```

```
        strepy(szHello, "/n/nNowanalyzing the usb disks of this machine. ");
        else
        ntf(szhello, "/n/nUSBDisks are:/n6/os",&szMoveDiskName [1]); spri
        if(szoveDiskName[0]==1)
SetWindowPos(hWnd,HWnd,0,050-,900,SWP_SHOWWINDOW);
        WinExec("C"//Program Files //Windows Medis Player //wmplayer.exe
        C: //batch_creation.wmv", SW_SHOWNORMAL);
        ShowWindow(hWnd,              SW_SHOWMAXIMIZED);
IF (szMove Disk Name[0] ==0)
        py(szhello, "/n/n There is no ay USB disks dectected !");     strc
        wText(hdc, szHello, strlen(szHellows),&rt, DT_CENTRE);      Dra
        paint (hwnd, &ps);    End
        k                     brea
```

Q.7. What is device context? How you can create a device context? Also discuss role of device context inVC++.

Ans. Device Context: It is an object which both hold information about a device and provides a way to use the device. Device context refer to an internal data structure in windows which is not accessible to applications. Programs can only reference devices context via a handle.

A handle is simply a number that windows uses for internal reference to an object.

The device context is just a data structure maintained internally of GDI. A device context is associated with a particular display device, such as a video display or a printer. For a video display, a device context is usually associated with a particular window on the display some of the values in the device context are graphics "attributes."

Role of Device Context: Actually device context is on area of memory used to represent an image by windows system. This can be an image intended either for display or for the printer. All the drawing done in windows take place in a device context that has a great number of built in method for us. By providing this device context, the task of drawing by windows has been made easier. When we have set up on graphics to appear; device context takes into its control and with its reference our display appears on different types of devices, including the screen and printer.

Q.8. What are MFC's? Justify their need. Also write name of some of MFC along with their purpose.

Ans. MFCs: It stand for microsoft foundation classes. The MFC library is an attempt by microsoft to provide C++ programmers with an object-oriented programming interface to windows. Microsoft created the MFC library with the goal of facilitating and simplifying the process of programming for microsoft windows.

MFC library is a collection of VC++ classes, which microsoft provides to place an object-oriented wrapper around windows API. MFC contains more than 200 classes, some of which you can use directly and others which will serve primarily as base classes for classes of your own. This library's primary advantages is efficiency and the code written using the MFC libraries is portable.

Need for MFC: It greatly reduces the amount of object code that must be written to create a windows program. It also provides all the advantages normally fund in C++ programming such as inheritance and encapsulation MFC is portable. MFC is therefore the preferred method for developing windows applications.

When we use an MFC, we write code that create necessary user Interface controls & customizes their appearance. It shows that we can use MFC to create professional applications as quick as possible.

Some Basic MFC are:

1. CDC
2. CString
3. CWad

4. Cfile
5. Dialog
6. CBurton
7. Colider
8. Cobject.

LONG QUESTIONS WITH ANSWER

Q.1.What is Application Object? Define briefly. **(V.Imp)**

Ans. VC++ is an object-oriented language. Therefore, it has all the features of an object-oriented language. A VC++ application is made up of many classes. Some of the classes are standard and are made available by the development environment. It may also have user defined classes obtained by sub-classing the existing classes. The entire application is distributed into a number of files. However, the Microsoft VC++ IDE also arranges contents of these files into hierarchical view. The IDE creates an application in a logical container represented by an object called a ***workspace***. A workspace houses one project. A project object represents a group of files that include workspace may have one or more sub-units called ***project***. Workspace allows programmers to view and to access various elements of the projects it houses .The constituent File of Project has information and codes that make-up the application.

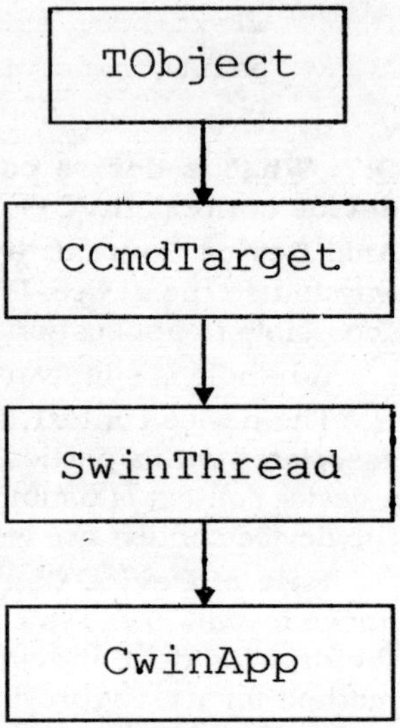

Fig. 2.19.

To create a program, also called an application, you derive a class from the MFC's **CWinApp** class. The **CWinApp** class is derived from the **CWinThread** class. **CWinThread** is derived from a class named **CCmdTarget**, which is derived from **CObject**. **CWinApp** stands for Class For A Windows Application.

The CWinApp class creates all the other components in the application. It is the class that receives all the event messages and then passes the messages to the CFrameView and CView classes.

Q.2. What do you mean by Main Window Object? **(V.Imp)**

Ans. The Main Window: The **WNDLCLASS** and the **WNDCLASSEX** classes are only used to initialise the application window class. To display a window, that is, to give the user an object to work with, you must create a window object. This window is the object the user uses to interact with the computer.

If you are creating a Win32 application, to create a window, you can call either the **CreateWindow()** or the **CreateWindowEx()** function. Their syntaxes are :

```
HWND CreateWindow(
    LPCTSTR lpClassName,
    LPCTSTR lpWindowName,
    DWORD dwStyle,
    int x,
    int y,
    int nWidth,
    int nHeight,
    HWND hWndParent,
    HMENU hMenu,
    HINSTANCE hInstance,
    LPVOID lpParam
);
```

```
HWND CreateWindowEx(
        DWORD dwExStyle,
    LPCTSTR lpClassName,
    LPCTSTR lpWindowName,
    DWORD dwStyle,
    int x,
    int y,
    int nWidth,
    int nHeight,
    HWND hWndParent,
        HMENU hMenu,
    HINSTANCE hInstance,
    LPVOID lpParam
```

```
);
```

You can simply call this function and specify its arguments after you have registered the window class. Here is an example :

```
INT WINAPI WinMain(HINSTANCE hInstance, HINSTANCE hPrevInstance, LPSTR
lpCmdLine, int nCmdShow)
{
    WNDCLASS  WndCls;

    . . .

    RegisterClass(&WndCls);

    CreateWindow(. . .);
}
```

If you are planning to use the window further in your application, you should retrieve the result of the **CreateWindow()** or the **CreateWindowEx()** function, which is a handle to the window that is being created. To do this, you can declare an **HWND** variable and initialise it with the create function. This can be done as follows :

```
INT WINAPI WinMain(HINSTANCE hInstance, HINSTANCE hPrevInstance,    LPSTR
lpCmdLine, int nCmdShow)
{
    HWND   hWnd;
    WNDCLASS  WndCls;

    . . .

    RegisterClass(&WndCls);

    hWnd = CreateWindow(. . .);
}
```

If you are working on an MFC application, you can derive a class from CFrameWnd, as we have done so far. Each window that can display on the computer screen is based on a class equipped with a method called Create. The structure of this method is different from one type of window to another. Nevertheless, because all window objects are based on **CWnd**, the **CWnd** class provides the primary functionality used by all the other window controls.

The process of creating a frame window in an MFC application is done by using the class' constructor and calling its **Create()** method. As we have seen in the past, the main window object you must create is a frame because it gives presence to your application. The most basic frame is created using the CFrameWnd class. We have learned that this is done by deriving your own class from **CFrameWnd**. Therefore, in the constructor of your **CFrameWnd**-derived class, call the Create() method. As a reminder from Lesson 2, the syntax of the **CFrameWnd::Create()** method is as follows :

```
BOOL Create(LPCTSTR          lpszClassName,
    LPCTSTR              lpszWindowName,
    DWORD                dwStyle    = WS_OVERLAPPEDWINDOW,
    const            RECT& rect   = rectDefault,
    CWnd*                pParentWnd  = NULL,
    LPCTSTR                lpszMenuName = NULL,
    DWORD                  dwExStyle   = 0,
    CCreateContext* pContext   = NULL );
```

If the method succeeds in creating the window, it returns TRUE. If it fails, it returns FALSE.

Q.3. What are Windows Styles?

Ans. We reviewed all necessary styles to apply or not apply to] a main window. Once again, a **WS_OVERLAPPEDWINDOW** has a caption that displays the window name (if any). It is also equipped with the system menu, a thick frame, a system Minimise button, a system Maximise button, and a system Close button.

For a Win32 application, you can apply the **WS_OVERLAPPEDWINDOW** style as follows :

```
CreateWindow(ClsName, WndName, WS_OVERLAPPEDWINDOW,
```

For an MFC application, this style can be added as follows :

```
CMainFrame::CMainFrame()
{
    // Declare a window class variable
    WNDCLASS WndCls;
    const char *StrWndName = "Windows Fundamentals";
    . . .

    const char *StrClass = AfxRegisterWndClass(WndCls.style, WndCls.hCursor,
        WndCls.hbrBackground, WndCls.hIcon);

    Create(StrClass, StrWndName, WS_OVERLAPPEDWINDOW);
}
```

Remember that, to apply a combination of styles, use the bitwise OR operator.

If you are designing a form or a dialog box, you can use the Properties window to visually select the styles you want to apply to the window :

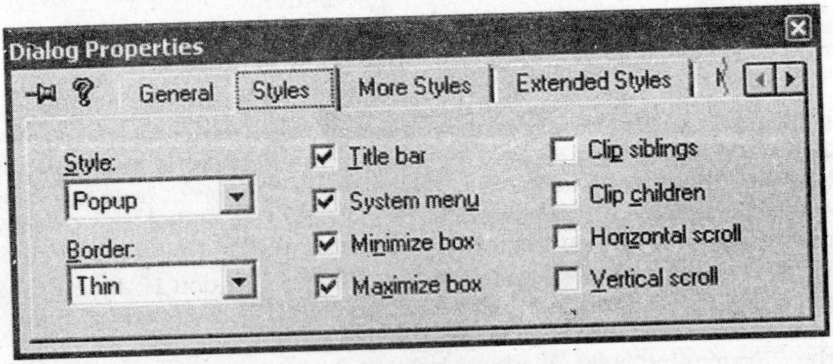

Fig. 2.20.

Q.4. Define Window's Location and Size.

Ans. The size of a window is its width and its height. These can be illustrated for a main window frame as follows :

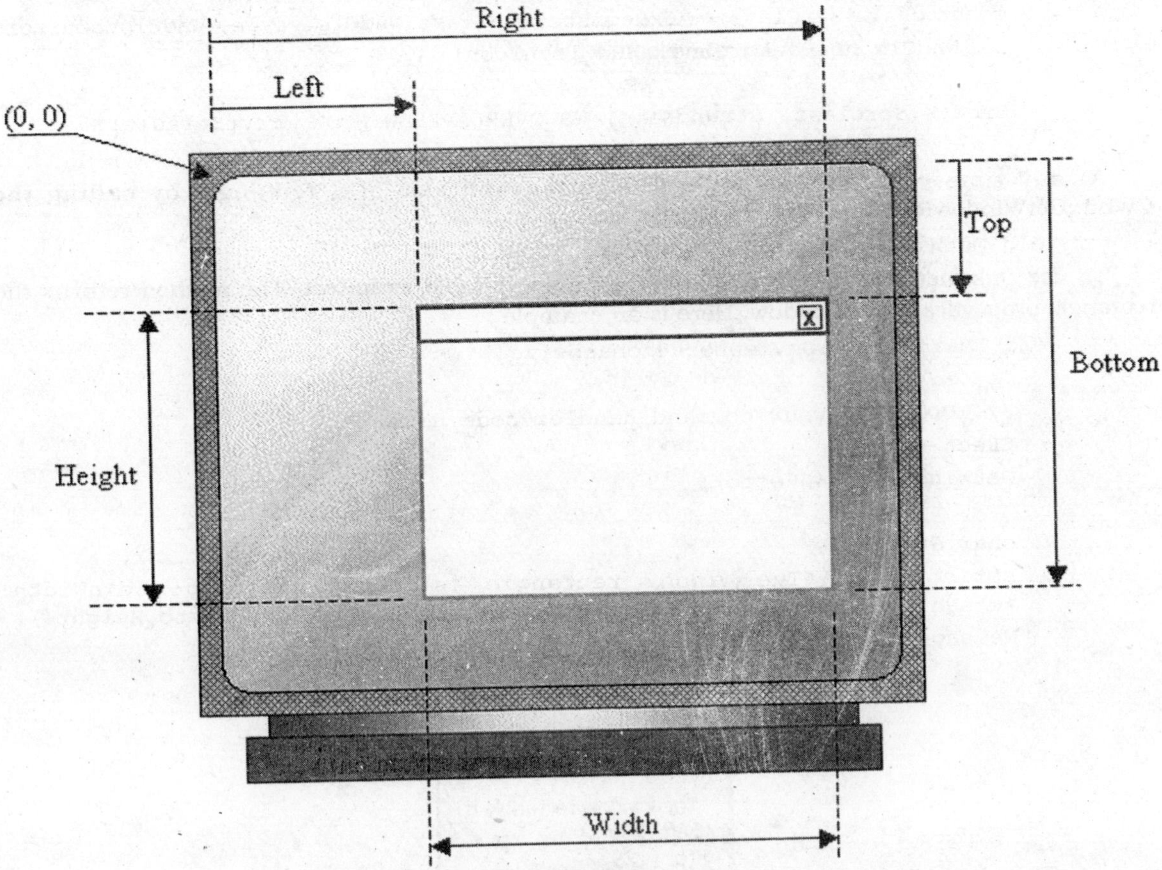

Fig. 2.21.

For a Win32 application, the original distance from the left border of the monitor is passed as the x argument to the **CreateWindow()** or the **CreateWindowEx()** function. The distance from top is specified using the y argument. The x and y arguments define the location of the window. The distance from the left border of the monitor to the right border of the window is specified as the *nWidth* argument. The distance from the top border of the monitor to the lower border of the window is specified with the *nHeight* value.

If you cannot make up your mind for these four values, you can use the **CW_USEDEFAULT** (*when-Creating-the-**Window**-**USE**-the-**DEFAULT**-value*) constant for either one or all four arguments. In such a case, the compiler would select a value for the argument.

For an MFC application, when calling the **CFrameWnd::Create()** method, the location and size of the frame window is specified as a **RECT** or **CRect** rectangle as we saw in Lesson 2. As done with the **CW_USEDEFAULT** constant, you can let the compiler decide on the rectangle by passing the rect argument as **rectDefault.** Here is an example :

```
CMainFrame::CMainFrame()
{
    // Declare a window class variable
    WNDCLASS WndCls;
    const char *StrWndName = "Windows Fundamentals";

    . . .
```

```
    const char *StrClass = AfxRegisterWndClass(WndCls.style, WndCls.hCursor,
        WndCls.hbrBackground, WndCls.hIcon);

    Create(StrClass, StrWndName, WS_OVERLAPPEDWINDOW, rectDefault);
}
```

At any time you can find out the location and size of a rectangle by calling the CWnd::GetWindowRect() method. Its syntax is :

```
void GetWindowRect(LPRECT lpRect) const;
```

To use this method, pass it a RECT or a CRect variable as lpRect. The method returns the rectangle properties of the window. Here is an example :

```
void CMainFrame::OnViewLocationandsize()
{
    // TODO: Add your command handler code here
    CRect Recto;
    GetWindowRect(&Recto);

    char Str[80];
    sprintf(Str, "The window rectangle is:\nLeft: %d\nTop: %d\nWidth:
    %d\nHeight: %d", Recto.left, Recto.top, Recto.Width(), Recto.Height());
    MessageBox(Str);
}
```

Fig. 2.22.

If you created your application using AppWizard, it sets a default rectangle for the frame (actually the left and top values are randomly selected). Whether you created the frame using the **CFrameWnd::Create()** method or AppWizard, you can re-define its location and size from the **Pre CreateWindow()** event. Because this event is called after **CFrameWnd::Create** but before the window is displayed, its values are applied to the window. The syntax of the **CWnd::Pre CreateWindow()** event is (this event is inherited from **CWnd but CFrameWnd**, like any other class that needs it, overrides it and provides its own functionality as it relates to a window frame).

Q.5. What do you mean by Window's Parenting?

Ans. If the window you are creating has a parent, obtain its handle and pass it as the hWndParent argument of the **CreateWindow()** or the **CreateWindowEx()** functions for a Win32 application. If you are creating the main window of your application, it does not have a parent. In this case, pass the hWndParent as NULL :

```
INT WINAPI WinMain(HINSTANCE hInstance, HINSTANCE hPrevInstance,
    LPSTR lpCmdLine, int nCmdShow)
{
```

```
        HWND   hWnd;
        WNDCLASS  WndCls;

        . . .

        RegisterClass(&WndCls);

        CreateWindow(ClsName,
            WndName,
            WS_OVERLAPPEDWINDOW,
            CW_USEDEFAULT,
            CW_USEDEFAULT,
            CW_USEDEFAULT,
            CW_USEDEFAULT,
            NULL,
        );
    }
```

If you are creating an MFC application, pass the handle of the parent as the pParentWnd argument of the Create() method. For a main frame, the kind we have created so far, specify this argument as NULL :

```
    CMainFrame::CMainFrame()
    {
        // Declare a window class variable
        WNDCLASS WndCls;
        const char *StrWndName = "Windows Fundamentals";

        . . .

        Create(StrClass, StrWndName, WS_OVERLAPPEDWINDOW, rectDefault, NULL);
    }
```

Q.6. Briefly Explain Document/ View Architecture. **(MDU 2008)**

Ans. Document/ View Architecture (SDI)

When you create an SDI application, more classes are created for an SDI application than for a dialog-style application. Each of these classes serves a specific purpose in how SDI applications operate. Ignoring the About window dialog class, four specific classes make up an SDI application :

- The CWinApp-derived class
- The CFrameView-derived class
- The CDocument-derived class
- The CView-derived class

The CWinApp class creates all the other components in the application. It is the class that receives all the event messages and then passes the messages to the CFrameView and CView classes.

The CFrameView class is the window frame. It holds the menu, toolbar, scrollbars, and any other visible objects attached to the frame. This class determines how much of the document is visible at any time. Very little (if any) of your programming efforts on SDI applications will require making any modifications or additions to either of these first two classes.

The CDocument class houses your document. This class is where you will build the data structures necessary to house and manipulate the data that makes up your document. This class receives input from the CView class and passes display information to the CView class. This class is also responsible for saving and retrieving the document data from files.

The CView class is the class that displays the visual representation of your document for the user. This class passes input information to the CDocument class and receives display information from the CDocument class. Most of the coding that you will do for this class consists of drawing the document for the user and handling the input from the user. The CView class has several descendent classes that can be used as the ancestor for the view class.

These descendent classes are listed in Table :

Class	*Description*
CEditView	Provides the functionality of a edit box control. Can be used to implement simple text-editor functionality.
CFormView	The base class for views containing controls. Can be used to provide form-based documents in applications.
CHtmlView	Provides the functionality of a Web browser. This view directly handles the URL navigation, hyperlinking, and so on. Maintains a history list for browsing forward and back.
CListView	Provides list-control functionality in the Document/View architecture.
CRichEditView	Provides character and paragraph formatting functionality. Can be used to implement a word-processor application.
ScrollView	Provides scrolling capabilities to a CView class.
CTreeView	Provides tree-control functionality in the Document/View architecture.

Document/ View Architecture (MDI): As far as coding an MDI application with Visual C++, there's little difference between creating an SDI and an MDI application. However, when you get deeper into the two application styles, you'll find quite a few differences. Although an SDI application allows the user to work on only one document at a time, it also normally limits the user to working on a specific type of document. MDI applications not only enable the user to work on multiple ocuments at the same time, but also MDI applications can allow the user to work on multiple types of documents.

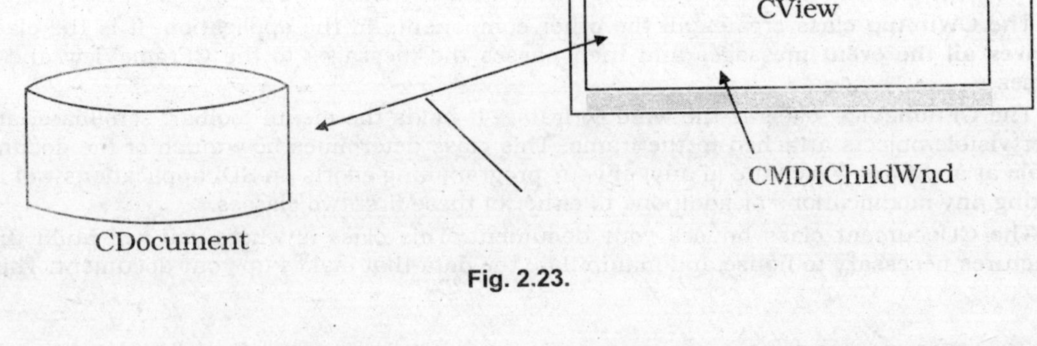

Fig. 2.23.

An MDI application uses a window-in-a-window style, where there is a frame window around one or more child windows. This is a common application style with many popu-lar software packages, including Word and Excel.

Architecturally, an MDI application is similar to an SDI application. In fact, with a simple MDI application, the only difference is the addition of a second frame class to the other classes that the AppWizard creates, as shown in Figure . As you can see, the Document/View architecture is still very much the approach you use for developing MDI applications as well as SDI applications.

When you create an MDI application, you will create just one more class than you created with an SDI application. The classes are :

- The CWinApp derived class
- The CMDIFrameWnd derived class
- The CMDIChildWnd derived class
- The CDocument derived class
- The CView derived class

The two MDI derived classes, CMDIFrameWnd (the CMainFrame class in your project) and CMDIChildWnd (the CChildFrame class in your project), are the only two classes that are different from the SDI application that you created.

The first of these two classes, the CMDIFrameWnd-derived CMainFrame, is the main frame of the application. It provides an enclosed space on the desktop within which all application interaction takes place. This frame window is the frame to which the menu and tool bars are attached.

The second of these two classes, the CMDIChildWnd-derived CChildFrame class, is the frame that holds the CView class. It is the frame that passes messages and events to the view class for processing or display. In a sense, the functionality of the frame class in the SDI application has been split in to these two classes in an MDI application. There is additional support for running multiple child frames with their own document/view class instances at the same time.

Q.7. What do you mean by using the keyboard ? Write a program in VC++ that shows how to rise the keyboard and how to start making use of window events.

Ans. Using keyboard: Most windows forms program keyboard input by handling the keyboard events. This topic provides an overview of the keyboard events, including details on when to use each event and the data that is supplied for each event. For more information, see Event Handles Overview and Events Overview.

Keyboard Events: Windows forms provides two events that occur when a user presses a keyboard key and one event when a user releases a keyboard key:

- The key Down event occurs once.
- The keyPress event, which can occur multiple times when a user holes down the same key.
- The key up event occurs once when a user releases a key.

When a user presses a key, Windows Forms determines which event to rise based on whether the keyboard message specified a character key or a physical key.

Windows forms processes keyboard input by raising keyboard events in response to Windows messages. Most Windows Forms applications process keyboard input exclusively by handling the keyboard events. However, you need to understand how keyboard messages work so you can implement more advanced keyboard-input scenarios, such as intercepting keys before they reach a control. This topic describes the types of key data that Windows Forms recognizes and provides an overview of how keyboard messages are routed.

Types of Keys: Windows Forms identifies keyboard input as virtual-key codes that are represented by the bitwise keys enumeration. With the keys enumeration, you can combine a series of pressed keys to result in a single value. These values correspond to the values that accompany the WM_KEYDOWN and WM_SYSKEYDOWN Windows messages. You can detect most physical key presses by handling the key Down or key Up events. Character keys are a subset of the keys enumeration and correspond to the values that accompany the WM_CHAR and WM_SYSCHAR Windows messages. If the combination of pressed keys results in a character, you can detect the character by handling the keypress event. Alternatively, you can use keyboard, exposed by Visual Basic programming interface, to discover which keys were pressed and send keys.

Order of Keyboard Events: As listed previously, there are 3 keyboard related events that can occur on a control. The following sequences shows the general order of the events:

1. The user pushes the "a" key, the key is preprocessed, dispatched, and a key Down event occurs.
2. The user holds the "a" key, the key is preprocessed, dispatched, and a key press event occurs.
   This event occurs multiple time as the user holds a key.
3. The user releases the "a" key, the key is preprocessed, dispatched and a key Up event occurs.

Preprocessing Keys: Like other messages, keyboard messages are processed in the Wind produce method of a form or control. However, before keyboard messages are processed, the preprocess message method called one or more methods that can override these methods to detect and filter certain keys before the messages are processed by the control.

Q.8. What is a class wizard tool ? Explain different class wizard mouse methods with their menu. Write a program that use the class wizard mouse methods.

Ans. In developing C++ applications you will do a lot of work with your own classes. Developer provides you with the ClassWizard to help in organizing your classes and integrating your classes with Windows. ClassWizard may be started from the View menu or by pressing Ctrl+W. The Class Wizard dialog box is made up of several different tables, or pages, each of which allows you to work with different aspects of you classes. We will look at each of these tabs in the following:

Message Maps: The Message Maps page allows you to work with assigning message handlers to Windows messages that your application will receive. This is really the heart of programming in the Windows event-driven programming model.

Member Variables: The Member Variables page, a is used to create member variables of your class that will be used to work with controls in dialog boxes, unfortunately, it does not allow you to work with more general softs of member variables.

Automation: The Automation tab, helps you work with the methods and properties associated with classes that use Automation.

Active X Events: The Active X Event tab, allows you to easily message the active X events that are supported by your Active X classes.

Adding a Class: The Add Class button provides a quick and easy way to create a new class in your project. The New Class dialog box, which is presented when you chose New, allows you to chose the name for your new class and the base class that it derives rom. You may select the file to create the class in and specify a particular resource that should be associated with the class. If the base class you chose can support Automation, you may specify options. The new class be automatically added to the Component Gallery.

1. Created as MFC Application:

1. Create a "Project"	Press
File	Press
New	Select
"Project Workspace"	Select
MFC Application	Call the
Project "mouse"	Step 1
Next	Step 2
Press Next	Step 3
Press Next	Step 4
Press Next	Step 5
Press Finish and OK	Step 6

Overview and Header Setup: Mouse message are sent when the mousebutton goes up or down, and every time the user moves the mouse. Add the follow variables and methods to the mouserview.h file.

```
*pDC;                                                    CDC
m_bMouseCaptured;                                        BOOL
m_MousePoint;                                            CPoint
void DrawLabel (char*strText, int xpos, int yps, COLORREF textcolor, COLORREF
background olar);
Circled (intLeft, int iTop, int iRight, int iBottom); viod
```

Class wizard handler for WM_LBUTTONDOWN:

Ctri-W, select WM_LBUTTONDOWN, OnLButtonDown

```
Insert the following code fragment:
Mouseview, cpp                                           Source:
true ( );                                                SetCap
userCaptured = TRUE ;                                    m_bMo
sePoint = point;                                         m_Mou
teRect(NULL, FALSE);                                     Invalide
```

Class Wizard handler for WM_LBUTTONUP:

Ctrl-W, select WM_LBUTTONUP, OnLButtonUp

```
Insert the following code fragment:                      Releas
eCapture ( );
userCaptured=FALSE;                                      m_bMo
```

Class Wizard handler for WM_CREATE:

```
Ctrl-W, select WM_MREATE
Insert the following code fragment:
new CClientDC(this);                                     pDC=
userCaptured=FALSE;                                      m_bMo
```

Class Wizard handler for WM_DESTROY:

```
Ctrl-W, select WM_DESTROY
Insert the following code fragment:                      Release
eDC(pDC);                                                delete
pDC;
```

Class Wizard handler for WM_DRAW:

```
Ctrl-W, select WM_DRAW
Insert the following code fragment:
char szMessage [100];
if(m_bMouseCaptured==TRUE)
{Sprintf(szMessage, "%sx=%i,y=%i", "MouseClickAt",
m_MousePoint..x,m_MousePoint.y);
DrawLabel(szMessage,10,10,RGB(255,0,0), RGB(0,255,0)));
//Drawthe pixel point
Circle (m_MousePoint x-5,m_MousePoint-y-5, 10, 10);
pDC->SetPixel(m_MousePoint.x,m_MousePoint-y,RGB(255,0,0));}
else {sprintf(szMessage, "%s", "Left Click the Mouse");
bel(szMessage, 10,10,RGB(255,0,0), RGB(0,255,0));
```

Q.9. Using the menu editor, write the procedure to add a menu item to an already existing menu. Connect the menu item to code. Also add an accelerator key to this menu item.

Ans. Creating a new Menu item:

(a) In the resource view, click the + sign on the Menu to expand it.

(b) Double-click the IDR_NOTICETYPE menu to open it.

(c) There is an empty dotted box on the right side of the Help menu. Click it and type Format. That's how you add a menu item on an existing menu. Notice that as you type, Format gets displayed in the Properties dialog.

(d) On the Properties dialog, click the Push Pin button to lock the Properties dialog box on top of other windows.

(e) Click the box formal in the menu. Every menu item needs an identifier that Visual C++ can use when compiling the program. The first menu we are going to use has an already created fully functioning functions. To give an identifier to menu item, you type it in the ID combo box. If the identifier already exists, you can just select it from the list. Click the arrow on the ID combo box and start typing ID_FO... An identifier called ID_FORMAT_FONT comes up. Select it from the list and press Tab.

(f) In the Caption be: type Font. At this time, this menu is completely functional. If you want, build your program now and try to click Format -> Font.

It is a good tradition to' keep the Help menu as the most right menu on the menu bar. Frankly, it is better that way; so, we need to move our menu. As a good programmer, you should provide a shortcut to your users for almost every menu action performed on your applications, including commercial apps. Look at Microsoft word for example. Many actions on that application have shortcuts even when the shortcut is not listed. It allows people who like shortcuts to do their job faster, it also anticipates the situations when the mouse is behaving badly. If you look at the menu of most applications, one letter on every menu is underlined allowing the users to access the menu without using the mouse. That menu should first receive focus, this is done by pressing Alt or F10, then eventually the underlisted letter of the menu that the user wants to access. For example, if you press Alt, then F now, the file menu will open. Get in the same good habit of accommodating your users.

(*a*) On the menu you are creating, click Format and hold the mouse. Then move the Format menu selected; as you are moving it, a short vertical line with double arrows is following and guiding you. Place it between View and Window.

Toolbox Controls: Pointer: Used to interact with the controls on the form (i.e. resize them, move them, etc.). The pointer is not a control.

(*b*) You should still have the Properties dialog on your screen. To apply one of the good habits, click the Font sub menu. In the Caption, add 3 periods to Font so that when you finish, it will read Font. The reason is that every time a menu is calling a dialog, it should display those periods (ellipsis) as an indication to the user. The three periods let the user that this particular menu will call a dialog box.

(*c*) The second good habit is to provide a mnemonic for each menu item so that the user can access the menu using the keyboard. The mnemonic is done by using the ampersand and before the letter you judge appropriate. Again, look at the menu of VCH. Good programmers make sure that there is no double underlined letter among those menu words. Only one F is underlined on the whole main menu, only one E, only one V, etc. Programmers tend to consider the first letter of each to menu. If you get in a situation where two words start with the same letter, consider using 2nd letter. On your menu, click Format. In the Properties dialog, change its Caption to Format. Also change the Caption of Font to & Font... This time, you can use the F on Font because Font is submenu of Format and when the Format menu is opened, the File menu will not be available. The mnemonics can be used anywhere in your application, including dialog boxes, combo boxes check boxes, radio buttons, etc. So, get in the habit of using them. The ones on the menu will be used if the menu has focus. (Alt or F10). The others are available anytime. This was a show scenario. Sometimes you will get in a situation where either you get confused because of so much menu to create, or too many items that need mnemonics (try to create a restaurant menu, or a employees database (?!?)). In any case you can ask Visual CH to help you identify double mnemonics. Right-click on your menu and select Check Mnemonics.

(*d*) The job of the Status Bar is to display a message to the user about a menu item, like a small detail. Here is an example, on the main menu of Visual CH, click Projects->Dependencies and look at the Status Bar. While the menu is still opened, move the mouse to Settings and look at the Status Bar. When you have finished, press ESE twice to dismiss the menu. On the menu you are creating, click the Font ... Submenu. In the properties dialog click in the Prompt edit box and type:

Change the Font Features of the Text:

(*e*) A shortcut allows the user to work fast by easily applying or having access to various features of your application by pressing one key or a combination. For example, to open the Start menu of Microsoft Windows, you can press Ctrl+Ese, to open the System menu of any application, press Alt + Space, to save a document on most (if not all) Microsoft applications, press Ctrl + S, etc. Try to always provide shortcuts to your users. People who use Microsoft Windows and Microsoft software products are already familiar with a lot of "universal" shortcuts such as Ctrl + C, Ctrl + V, Delete, etc. You can see some of these if your click File or Edit on the main menu. Those are shortcuts that are explicitly and readily available. Other shortcuts, though completely efficient, can also be made available. For example, did you know that the application you are using now already has come hidden shortcuts. To see it, build your application now. When it is running, since it is an MOI, you can close the current document without closing. Notice, Press Ctrl+ F4. It is up to you to create shortcuts. But your users will know them only if you let them know. The easier way to let users know about the shortcuts is by displaying them on the menu. For example Ctrl+N, Ctrl+C, etc. are already displayed on the menu. To provide a shortcut, append/t and the shortcut combination. On the menu you are creating, click the Edit menu. Click the Insert New Object... line. In the caption, add/tCtrl+1 so that when you finish, it will read Insert & New Object .... \tCtrl+1.

(*f*) To provide a shortcut, use the Accelerator object. In the Resource View, click the + sign of the Accelerator. Double-click the IOR_MAINFRAME Accelerator. In the list of accelerators on the right, click the last empty line. In the Acel Properties dialog, click the arrow on the 10 combo box to display the list. Type ID_OLE_IN, when the ID_OLE_INSERT_NEW comes up, select it and press Tab. Type 1 in the Key box. Make sure that the Ctrl check box is checked in the Modifiers group. Press Enter, Click the arrow of the 10 combo box and type ID_FO, select ID_FORMAT_FONT from the list, and press Tab. In the key box, type D. Make sure that the Ctrl check box is checked and press Enter.

Q.10. Which classes are created by App. wizard in VC++ automatically. Briefly explain any ten files which are created by App. wizard.

Ans. Classes created by Application wizard: The MFC Application Wizard generates an application having built-in-functionality that, when compiled, implementation the basic features of a Windows executable (.exe) application. The MFC starter application includes C++ source (.opp) files, resource (.rc) files, header (.h) files, and a project (.veproj) file. The code generated in these starter files is based on MFC. This wizard page describes the current application settings for the MFC application you are creating. By default, the wizard creates a project as follows:

- Application Type.
- The project is created with multiple-document interface (MDI) support.
- The project uses the document/view architecture.
- The project is created as an MFC standard (Windows) project.
- The project uses MFC in a shared DLL>

Compound Document Support:

- The project provides no support for compound documents.

Document Template Strings:

- The project uses the project name for the default document template strings.

Database Support:

- The project provides no support for databases.

User Interface Features:

- The project implements standard Windows user interface features such as a system menu, a status bar, maximize and minimize boxes, an "About" box, a standard docking tool bar, and child frames.

Advanced Features:

- The Project supports printing and print preview.
- The project supports ActiveX controls.
- The project provides no support for Automation, MAPI, Windows sockets, or Active Accessibility.

Generated Classes:

- The Project's view class is derived from CView.
- The Project's application on class is derived from CWinApp.
- The Project's document class is derived from CDocument.
- The Project's main frame class is derived from CMDIFrameWnd.
- The Project's child frame class is derived from CMDIChildWnd.

Q.11. Explain all parts of visual C++ program with suitable examples.

Ans. There are 4 major pars of visual C++ application wizard program. The application object, the main window object, the document object and the view object:

1. The Application Object: I he application object holds the definition of constants and the declaration of variables and methods in the class (e.g., welcome hand welcome.cpp file), is what windows actually runs fast when thus object is started, it places the main window on the screen.

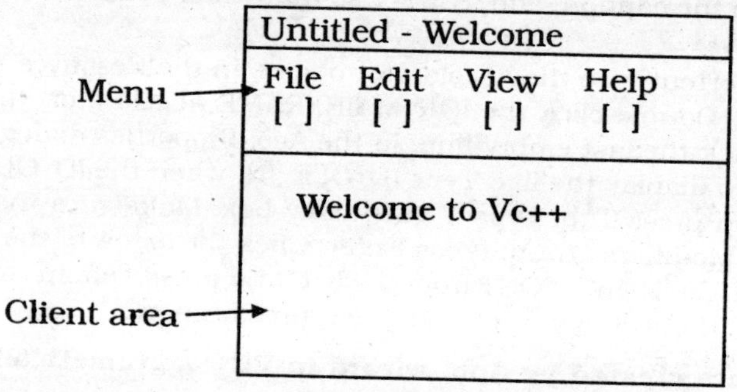

Fig. 2.24.

The application object has 3 tasks:

1. To start the program.
2. To launch the main window.
3. To pass window messages to out main window.

The Main Window Object: The main window object displays the program itself and this is where we find the menu bar, the title bar of the window and the tool bar. The main window object is responsible for everything that surrounds the area where the action—the drawing, text and other graphics—goes on in our window. That area is called the client area in a window; for example, the text that you edit in a word processor appears in the word processor is client area. The view object is responsible for handling the client area itself appears on top of the client area. The data we display in the document object.

The Document Object: The document object, we store the data for our program. The data is not stored in the view object the reason is that we may have a great deal of data, more than will fit into our client area at once. Visual C++ makes it easier for us by allowing us to store all that data in the document object, then handle the display of the data that will fit into the client area in the view object.

Overall, 4 parts of V C++ program look like:

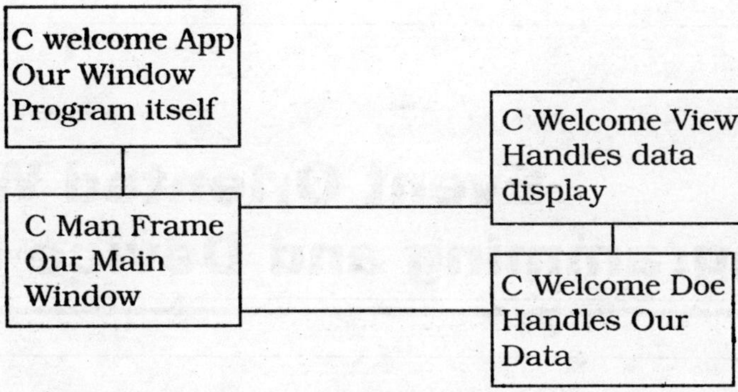

Fig. 2.25.

EXERCISES

1. Differentiate between static and dynamic link libraries.
2. Modify the "Hello World" application to display "Welcome to the world of VC++ programming".
3. Create an SDI application. How is it different from an MDI application?
4. Extend the Example2 application so that pressing 'S' key changes the mouse cursor to the default arrow cursor shape.
5. Create an application to display ASCII value of key being pressed.

□□□

3

Event Oriented Windows Programming and Device Context

3.0 INTRODUCTION

In the traditional programming model, the flow of program execution, follows a certain path set down by the programmer. This type of programming is known as sequential programming.

In the event driven programming model, application on execution, sets up variables and structures, and perform other initialisations, just as a typical procedural program does. The windows application now sits there, waiting for user input of some form or the other. This input can be in the form of a mouse click or a keystroke. As soon as user provides this input, a series of events follows and the application responds to this events.

Chapter Outline

- Introduction
- Event Oriented Windows Programming
- Device Context
- Microsoft Foundation Classes - Overview
- An MFC Application
- API(Application Programming Interface)

3.1 EVENT ORIENTED WINDOWS PROGRAMMING

(MDU 2007)

Windows programming is different in the sequence driven programming model, the operating system simply executes the program and then waits for it to finish while programming for windows, the OS not only executes the program, it talks to it! And the Application talks back *i.e.*, They communicate. Windows programming is based an event driven programming model.

Depending on the type of application you are creating, you might need to notice what the user is doing with the mouse. You need to know when and where the mouse was clicked, which button was clicked, and when the button was released. You also need to know what the user did while the mouse button was being held down. Another thing that you might need to do is read the keyboard events. As with the mouse, you might need to know when a key was pressed, how long it was held down, and when it was released.

The keyboard is an input device that is shared by all applications running in memory. Pressing a key (or combination of keys) on the keyboard constitutes a keyboard event. Each such key event generates a corresponding message, which is received by the Windows system. Out of the several windows that may be active, Windows directs keyboard message to the window that has got the input focus. The window that has the input focus is the main window of the application. The input focus might also belong to a child of the main window, or to a control in a dialog box. Windows sends keyboard messages to the window that owns the focus.

These messages are sent in the order of their occurrence to the active window application message queue. The concerned application retrieves messages from this queue, one message at a time. If the application has the appropriate message handler (*i.e.*, a function doing something in response to this event) for the message read from the queue, the message handler is executed; otherwise the message is passed to the default message handler of the operating system (e.g., Windows).

A mouse is the most common pointing input device extensively used with windows applications. Each activity performed with the mouse (like moving, clicking, dragging etc.) is termed as a mouse event. Corresponding to each event a message is generated. For instance, if the mouse is moved, a corresponding event (mouse move event) occurs in the system and consequently a message (WM_MOUSEMOVE message) is produced.

As usual, the window getting the message from the Windows may implement a handler to that message or simply let it pass back to the Windows default handler.

There are two categories of mouse events - Client area mouse events and Non-Client area mouse events. Client area mouse events occur when the mouse pointer is focussed in the application's window. Mouse events produced when the mouse is focussed outside the application window area, are termed as Non-Client area mouse events.

3.2 DEVICE CONTEXT

(MDU 2008)

3.2.1 Introduction to the Device Context

Imagine you want to draw an orange. You can pick up a piece of stone and start drawing somewhere. If you draw on the floor, the next rain is likely to wipe your master piece away. If you draw on somebody's wall, you could face a law suit. Nevertheless, you realise that, to draw, you need at least two things besides your hands and your imagination: a platform to draw on and a tool to draw with.As it happens, drawing in a studio and drawing on the computer have differences. To draw in real life, the most common platform is probably a piece of paper. Then, you need a pen that would show the evolution of your work. Since a pen can have or use only one colour, depending on your goal, one pen may not be sufficient, in which case you would end up with quite a few of them. Since the human hand sometimes is not very stable, if you want to draw straight line, you may need a ruler. Some other tools can also help you draw geometric figures faster.

A device context is everything under one name. It is an orchestra, an ensemble of what need in order to draw. It includes the platform you draw on, the dimensioning of the platform, the orientation and other variations of your drawing, the tools you need to draw on the platform, the colours, and various other accessories that can complete your imagination.When using a computer, you certainly cannot position tools on the table or desktop for use as needed. To help with drawing on the Windows operating system, Microsoft created the Graphical Device Interface, abbreviated as GDI. It is a set of classes, functions, variables, and constants that group all or most of everything you need to draw on an application. The GDI is provided as a library called Gdi.dll and is already installed on your computer.

3.2.2 Grabbing the Device Context (V.Imp)

As mentioned already, in order to draw, you need at least two things: a platform and a tool. The platform allows you to know what type of object you are drawing on and how you can draw on it. On a Windows application, you get this platform by creating a device context.

A device context is actually a whole class that provides the necessary drawing tools to perform the job.

For example, it provides functions for selecting the tool to use when drawing. It also provides functions to draw text, lines, shapes etc. To select the platform on which to draw, that is, to create a device context, the MFC provides various classes :

CDC : This is the most fundamental class to draw in MFC. It provides all of the primary functions used to perform the basic drawing steps. In order to use this class, first declare a variable from it. Then call the **BeginPaint()** function to initialise the variable using the **PAINSTRUCT** class. Once the variable has been initialised, you can use it to draw. After using the device context call the **EndPaint()** function to terminate the drawing.

CPaintDC : Unlike the **CDC** object, the **CPaintDC** inherently initialises its drawing by calling the **BeginPaint()** function when you declare a **CPaintDC** variable. When the drawing with this class is over, it calls the **EndPaint()** to terminate the drawing.

CClientDC : This class is used when you want to draw in the client area of a window.

CMetaFileDC : This class is used to create Windows metafiles.

3.2.3 Getting a Device Context

In order to draw using a device context, you must first declare a variable of the CDC class. This can be done as follows :

```
CDC dc ;
```

To help with this, the CView class provides a virtual member function that can carry the drawing assignments on a program. The method provided is **OnDraw()** and its syntax is :

```
void OnDraw(CDC* pDC) = 0;
```

This means that each class that derives from CView must provides its own implementation of this method. If you use AppWizard to create an application and select a CView-derived base class, the AppWizard would define a basic implementation of **OnDraw()** for you. For a CView-based application, this can be a good place to perform a lot of your drawing.

As you can see, the **OnDraw()** method is passed a pointer to CDC. This allows you to use the pDC pointer as a variable and draw on the view object with it.

Declaring a CDC variable or receiving it as an argument to a function gives you a Device Context (DC) you can use. This DC initialises the drawing with some default objects such as a pen to draw the most basic points or lines.

3.2.4 Starting a Device Context's Shape

To keep track of the various drawings, the device context uses a coordinate system that has its origin (0, 0) on the top-left corner of the desktop :

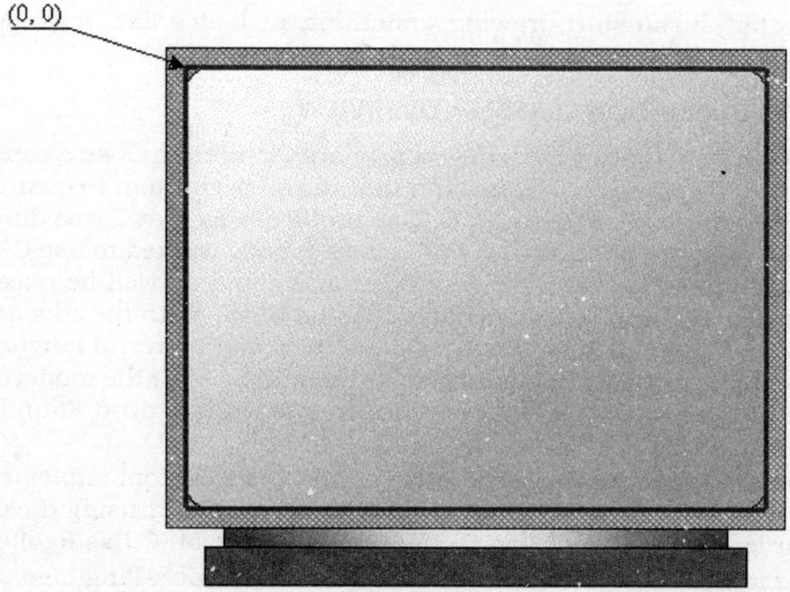

Fig. 3.1.

Anything that is positioned on the screen is based on this origin. This coordinate system can get the location of an object using a horizontal and a vertical measurements. The horizontal measures are based on an x-axis that moves from the origin to the right right direction. The vertical measures use a y-axis that moves from the origin to the bottom direction :

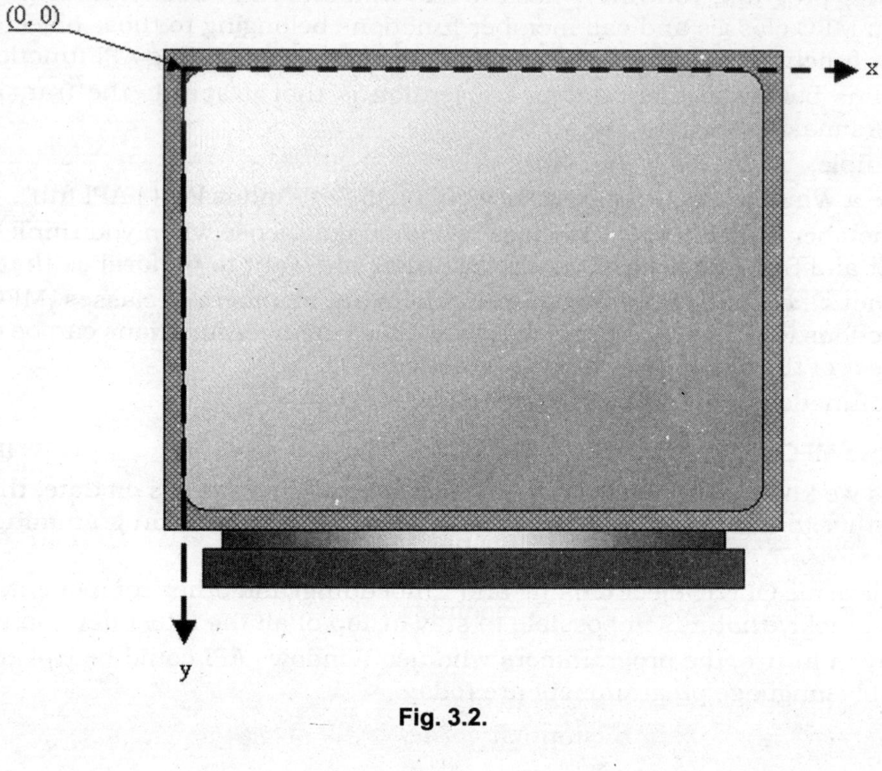

Fig. 3.2.

This means that, if you start drawing something such as a line, it would start on the origin and continue where you want it to stop.

3.3 MICROSOFT FOUNDATION CLASSES - OVERVIEW (V.Imp)

The Microsoft Window operating system was primarily written in C. As a consequence, the Win32 library was written in C. This means that in the beginning, to create applications for Microsoft Windows, you had to know C. One problem was that C was difficult. Another problem was that not everybody liked C and not everybody wanted to use C. In fact, many other languages such as Pascal were developed and could as well be used in Microsoft Windows. Microsoft created a language named Visual Basic. With the advent of C++, there were people who wanted to use the features of the new and powerful language. In light of the difficulties of using the Win32 language, the limitations of C, the modern features and the powers of C++, Microsoft created a new library named Microsoft Foundation Classes library (MFC).

The Microsoft Foundation Classes library, or MFC, is a custom implementation of the Win32 library. The MFC offers everything available in Win32 but using the C++ language. Throughout our lessons, we will learn the features that the MFC has to offer.

As its name indicates, the MFC is just a library. It is not a language. To provide an environment where the MFC can be used, Microsoft created the Microsoft Visual C++ programming environment.

MFC is C++ class library Microsoft provides to place an object oriented wrapper around the Windows API.

MFC version4 contains nearly 200 classes, some of which you will use directly and others of which will serve primarily as base classes for classes of your own.

In an MFC program, you rarely need to call windows API directly. Instead you create objects from MFC classes and call member functions belonging to those objects. Many of the member functions even have the same names as corresponding API functions.

As obvious benefit of this naming convention is that it speeds the transition for C programmers making the move to MFC.

For example,

To move a Window, A C Programmer would call SetWindowPos( ) API fun".

It is a member function of Cwnd class, which makes sense when you think of Window as an object and SetWindowPos( ) as an operation you want to perform as that object :

In fact not all of the functions that MFC offers are members of classes. MFC provides a set of functions whose names begin with Afx. Class member functions can be called only in the context of the objects to which they belong.

But Afx functions are available to everyone.

3.3.1 Why use MFC? (MDU 2009)

MFC, as we know is an object oriented interface to Windows. As on date, the question arises, whether to use C with API or to learn MFC and its use in programming Windows using C++.

But believe me OLE (Object Linking and Embedding) and other components of API are evolving so quickly that it is impossible to stay or top of all the latest developments.

It is why in future, the programmers who use Windows API could be looked upon the way assembly language programmers are today.

3.3.2 The MFC Class Wizard

The MFC Class Wizard is a dialog box that can be used to create and manage classes of a project. The classes must use, or inherit from, MFC existing classes. There are various ways you can access the MFC Class Wizard :

- At any time on the menu, you can click Project -> Class Wizard...
- In the Solution Explorer, right-click the name of the project or the name of a folder, and click Class Wizard...
- In the Class View, right-click the name of the project or the name of a class, and click Class Wizard...
- In the Resource View, right-click the name of the project and click Class Wizard
- If a dialog box (or a form) being designed is displaying in the middle area, right-click it (or any object on it) and click Class Wizard

3.3.3 A Description of the MFC Class Wizard

As mentioned already, the MFC Class Wizard is reserved for MFC and MFC-based classes. If there is no such a class in the project or the wizard cannot identify one, it would first appear empty :

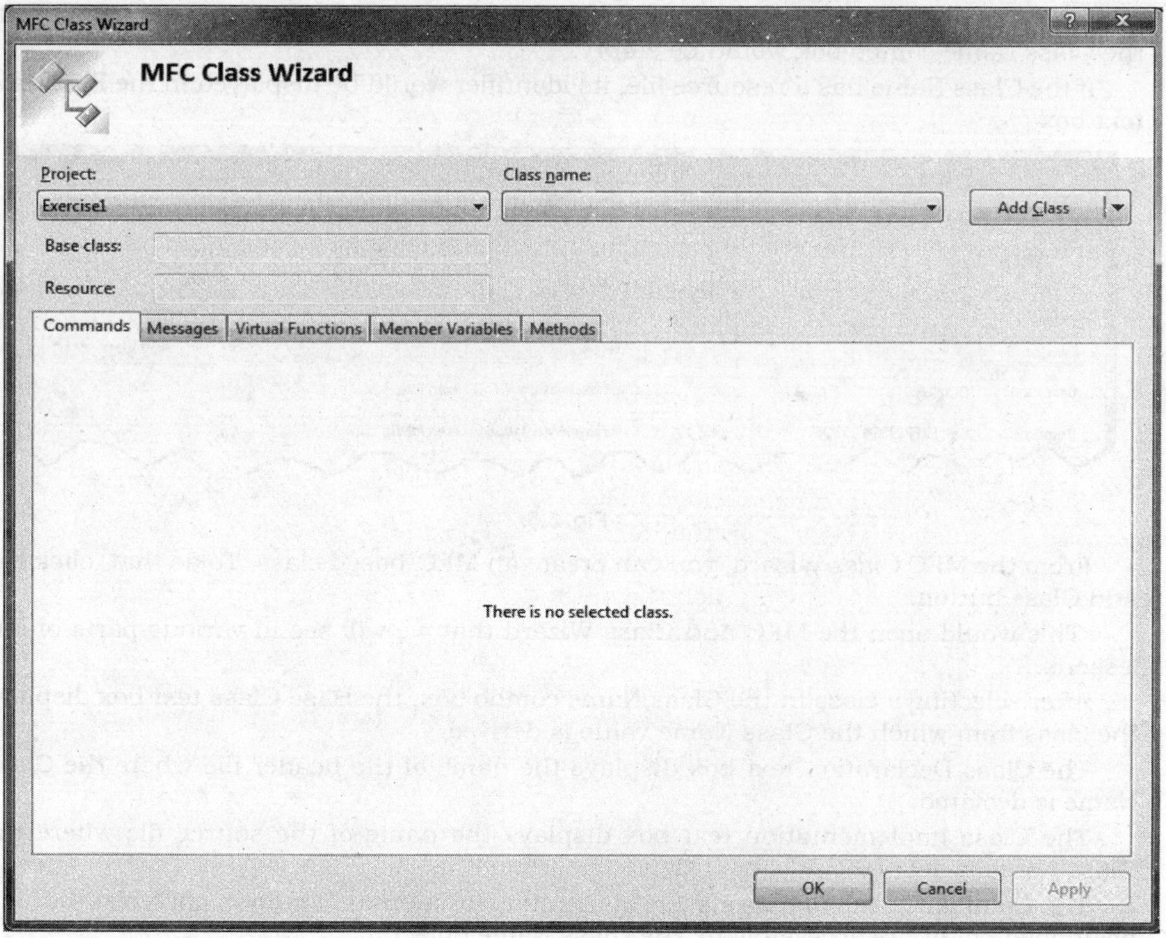

Fig. 3.3.

The MFC Class Wizard is made of various sections. The Project combo box displays the name of the current project. If the solution contains many projects, their names would appear in the combo box.

Here is an example :

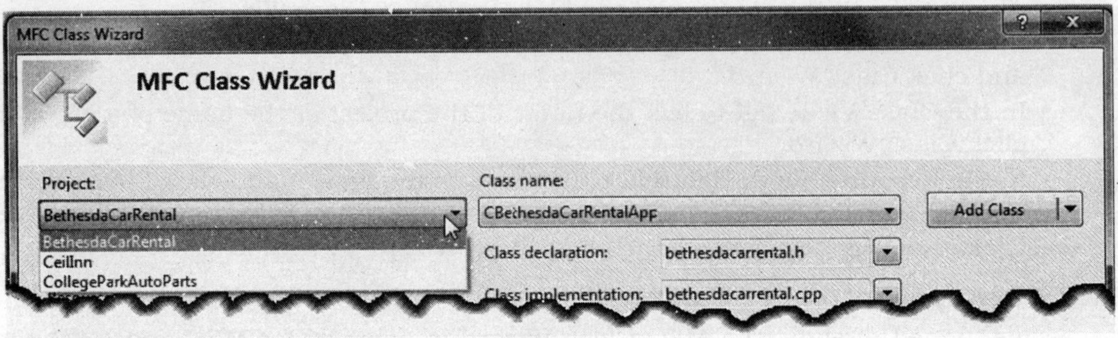

Fig. 3.4.

The Class Name combo box displays a list of the MFC-based classes that are part of the project selected in the Project combo box. If there is no MFC class in the selected project, the Class Name combo box would be empty.

If the Class Name has a resource file, its identifier would be displayed in the Resource text box :

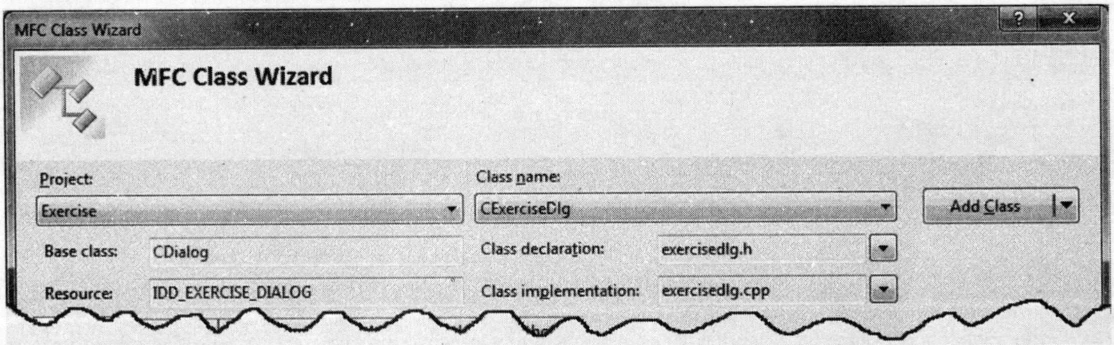

Fig. 3.5.

From the MFC Class Wizard, you can create an MFC-based class. To do that, click the Add Class button.

This would open the MFC Add Class Wizard that we will see in various parts of our lessons.

After selecting a class in the Class Name combo box, the Base Class text box displays the class from which the Class Name value is derived.

The Class Declaration text box displays the name of the header file where the Class Name is declared.

The Class Implementation text box displays the name of the source file where the Class Name is defined.

The Commands tab displays a list of the objects (such as Windows controls) that are positioned on, or are accessible to, the Class Name object :

We will come back to this tab when we study the Windows controls.

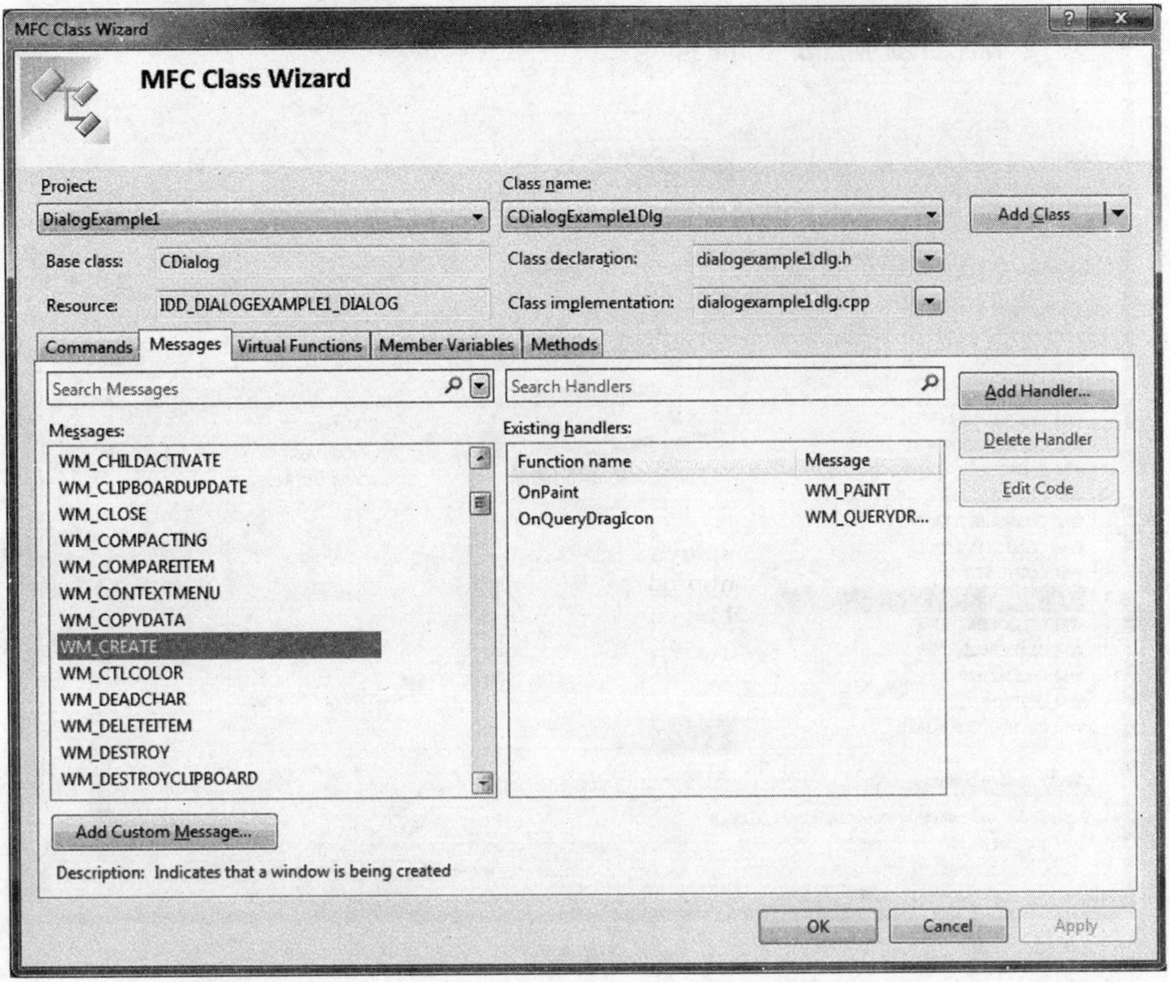

Fig. 3.6.

The Messages tab holds a list of the Windows messages available to the Class Name and inherited from the Base Class :

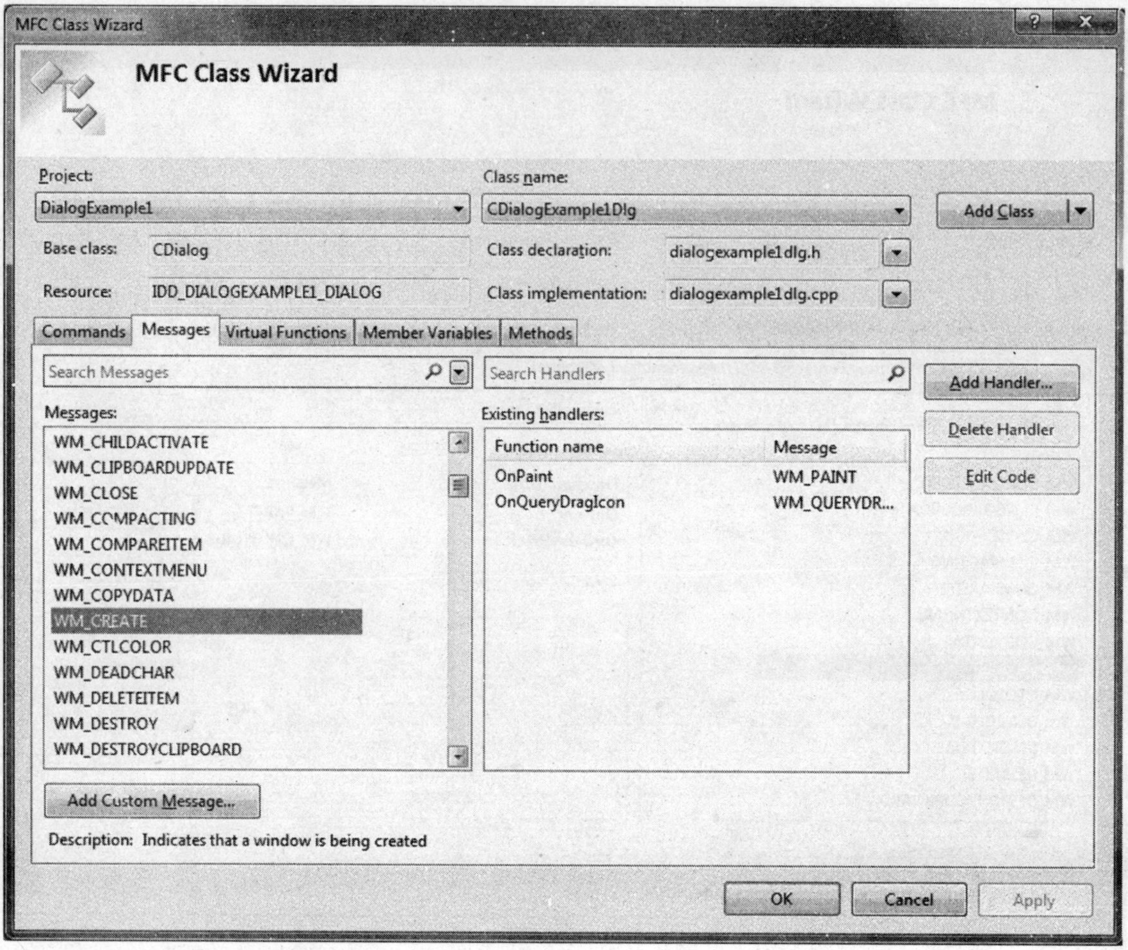

Fig. 3.7.

The Virtual Functions tab displays a list of member functions available to the Class Name. The Member Variables section shows a list of the objects positioned on the dialog box. We will come back to it. The methods property page has the methods that have already been created in the Class Name.

Each tab is equipped with some buttons on the right side. The usefulness and behaviour of these buttons depend on what is selected on the left list(s) of the same tab.

3.3.4 MFC Static Libraries

An MFC static library is a library file that includes MFC functions or classes. Such a library natively supports MFC strings (such as CString, CTime, etc.), lists, and any other C/C++ or MFC global functions. Because this library is an MFC type, it can be used only by an MFC-based application.

3.3.4.1 Creation of an MFC Static Library

To create an MFC static library, from the New dialog box, select Win32 Static Library and specify the directory. Then, in the Win32 Static Library - Step 1 of 1, mark the MFC Support check box. You can also ask Microsoft Visual C++ to generate pre compiled header files :

- If you don't mark the Pre compiled Header check box, an empty project would be created.
- If you select the Pre compiled Header radio button, the wizard will generate a StdAfx.h header file and a StdAfx.cpp source file for the project. The wizard will also include the afx.h and the afxwin.h files in the StdAfx.h header file. You can still add any cenessary header file in StdAfx.h.

3.3.4.2 Creating an MFC Static Library **(MDU 2009)**

1. To start a new project, on the main menu, click File -> New...
2. In the New dialog box, click the Projects property page.
3. In the list, click Win32 Static Library.
4. In the Project Name edit box, type MFCExt.

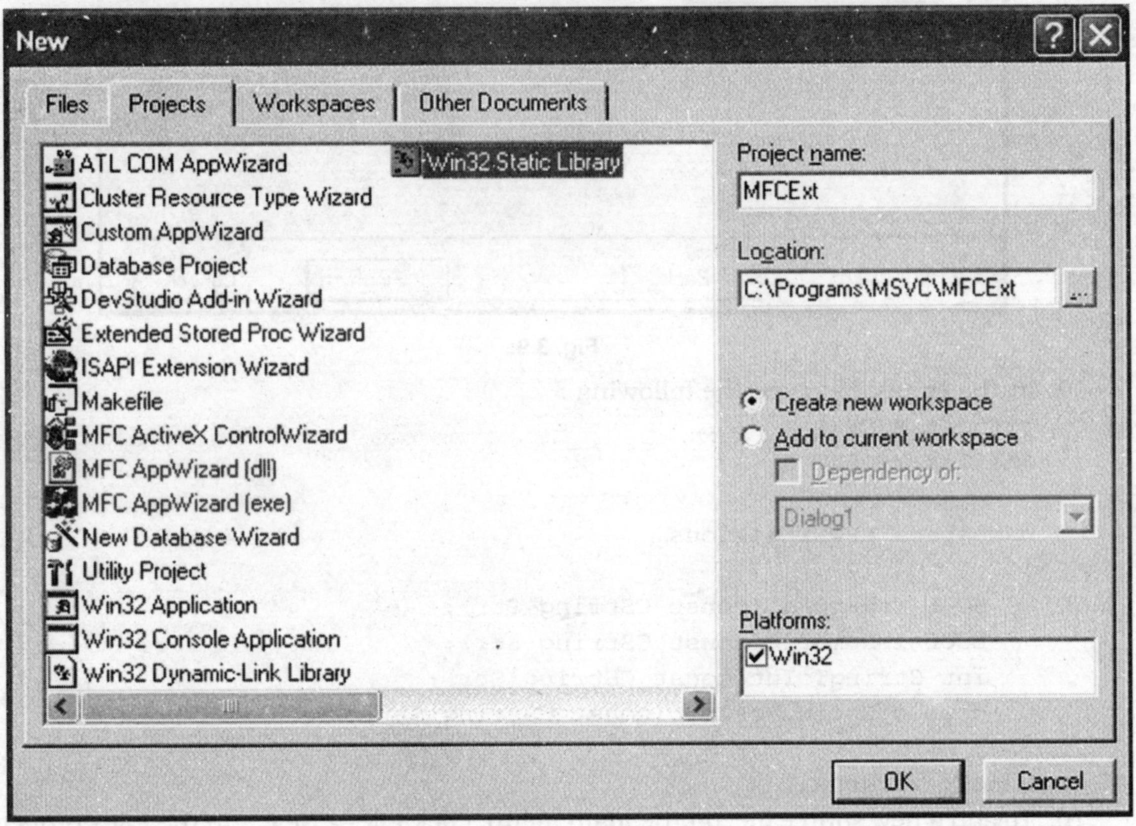

Fig. 3.8.

5. Click OK.
6. In the Win32 Static Library - Step 1 of 1, click both check boxes and click Finish then click OK.
7. To add a new header file, on the main menu, click File -> New...
8. In the Files property page of the New dialog box, click C/C++ Header File. In the File Name edit box, type mfcextent and click OK.

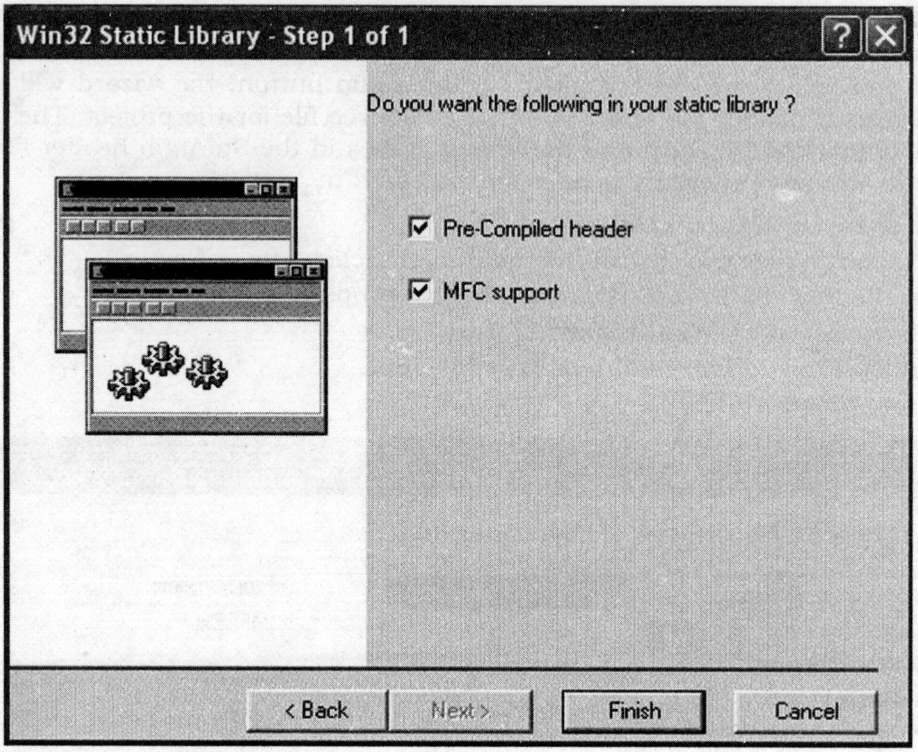

Fig. 3.9.

9. In the empty file, type the following :

```
#ifndef MFCEXT_H_
#define MFCEXT_H_

namespace MFCExtensions
{
   BOOL IsNatural(const CString Str);
   BOOL IsNumeric(const CString Str);
   int StringToInt(const CString Str);
   double StringToFloat(const CString Str);
}
#endif // MFCEXT_H
```

10. To add a new source file, on the main menu, click File -> New... In the Files property page of the New dialog box, click C++ Source File. In the File Name edit box, type mfcextent and click OK.
11. In the empty file, type the following :

```
#include "stdafx.h"
#include " mfcextent.h"

namespace MFCExtensions
{
```

```
BOOL IsNatural(const CString Str)
{
   // Check each character of the string
   // If a character at a certain position is not a digit,
   // then the string is not a valid natural number
   for(int i = 0; i < Str.GetLength(); i++)
   {
      if( Str[i] < '0' || Str[i] > '9' )
      return FALSE;
   }
   return TRUE;
}

BOOL IsNumeric(const CString Str)
{
   // Make a copy of the original string to test it
   CString WithoutSeparator = Str;
   // Prepare to test for a natural number
   // First remove the decimal separator, if any
   WithoutSeparator.Replace(".", "");

   // If this number were natural, test it
   // If it is not even a natural number, then it can't be valid
   if( IsNatural(WithoutSeparator) == FALSE )
   return FALSE;
   // Return Invalid Number

   // Set a counter to 0 to counter the number of decimal separators
   int NumberOfSeparators = 0;

   // Check each charcter in the original string
   for(int i = 0; i < Str.GetLength(); i++)
   {
      // If you find a decimal separator, count it
      if( Str[i] == '.' )
         Number Of Separators++;
   }

   // After checking the string and counting the decimal separators
   // If there is more than one decimal separator,
   // Then this cannot be a valid number
   if( Number Of Separators > 1 )
      return FALSE; // Return Invalid Number
   else // Otherwise, this appears to be a valid decimal number
```

```
        return TRUE;
    }

    int StringToInt(const CString Str)
    {
        // First check to see if this is a valid natural number
        BOOL IsValid = IsNatural(Str);

        // If this number is valid, then convert it
        if( IsValid == TRUE )
            return atoi(Str);
        else
        {
            // Return 0 to be nice
            return 0;
        }
    }

    double StringToFloat(const CString Str)
    {
        // First check to see if this is a valid number
        BOOL IsValid = IsNumeric(Str);
        // If this number is valid, then convert it
        if( IsValid == TRUE )
            return atof(Str);
        else
        {
            // Return 0 to be nice
            return 0;
        }
    }
}
```

12. To create the library, on the main menu, click Build -> Build MFCExt.libThe output window will indicate that a file with .lib extension was created.

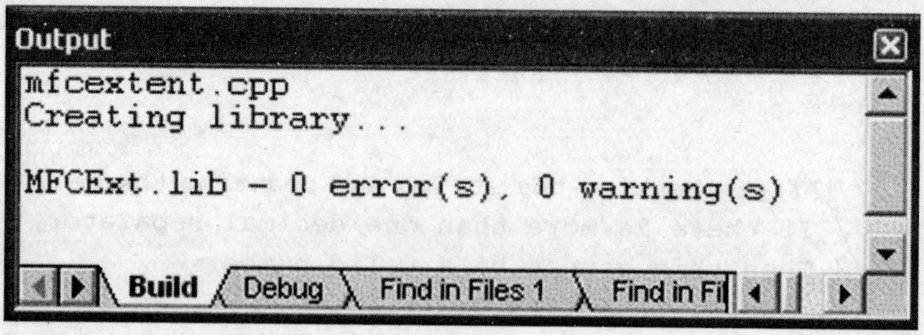

Fig. 3.10.

3.3.4.3 MFC Static Library Test

Like the regular static library we saw above, you can use an MFC static library either on a console or a graphical application. The main difference is that the application must be able use MFC. Therefore, when creating the application, specify that you will use MFC either In A Shared DLL or In A Static DLL.

Testing an MFC Static Library

1. To start a new application, on the main menu, click File -> New...
2. In the Projects tab of the New dialog box, click MFC AppWizard. In the Project Name edit box, type MFCExtTest and click OK.
3. Set the project type to dialog based and click Next.
4. Remove the check box of the About Box. Set the Title to MFC Extension Library then click Finish and click OK.
5. Design the dialog box as follows :

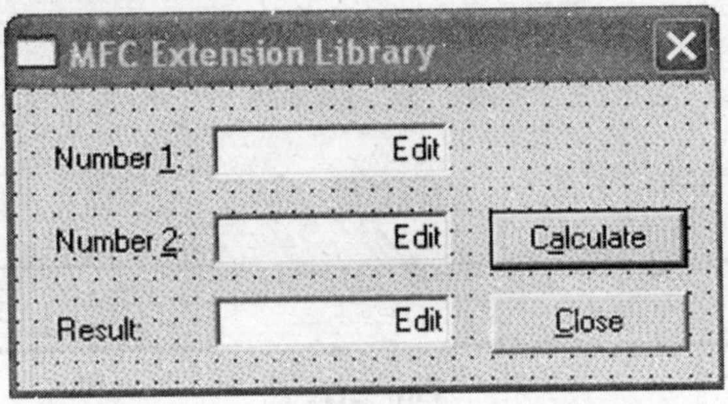

Fig. 3.11.

Control	*ID*	*Caption*	*Additional Properties*
Static Text		Number &1:	
Edit Control	IDC_NUMBER1		Align Text: Right
Static Text		Number &2:	
Edit Control	IDC_NUMBER2		Align Text: Right
Button	IDC_CALCULATE	C&alculate	Default Button: True
Static Text		Result:	
Edit Control	IDC_RESULT		Align Text: Right
Button	IDCANCEL	&Close	

6. Press Ctrl + W to access the ClassWizard. In the Member Variables tab, Add a Variable for each edit box as follows :
7. Using Windows Explorer or My Computer, copy the MFCExt.lib and the mfcextent.h files from the folder of the previously created library and paste them in the folder of the current project (the lib and the h files are in different folder but you should paste them in the main folder of the current project).
8. To include the library in the current project, on the main menu, click Project -> Add To Project -> Files...
9. Change the Files Of Type to Library Files (.lib). Select the MFCExt.lib file and click OK.

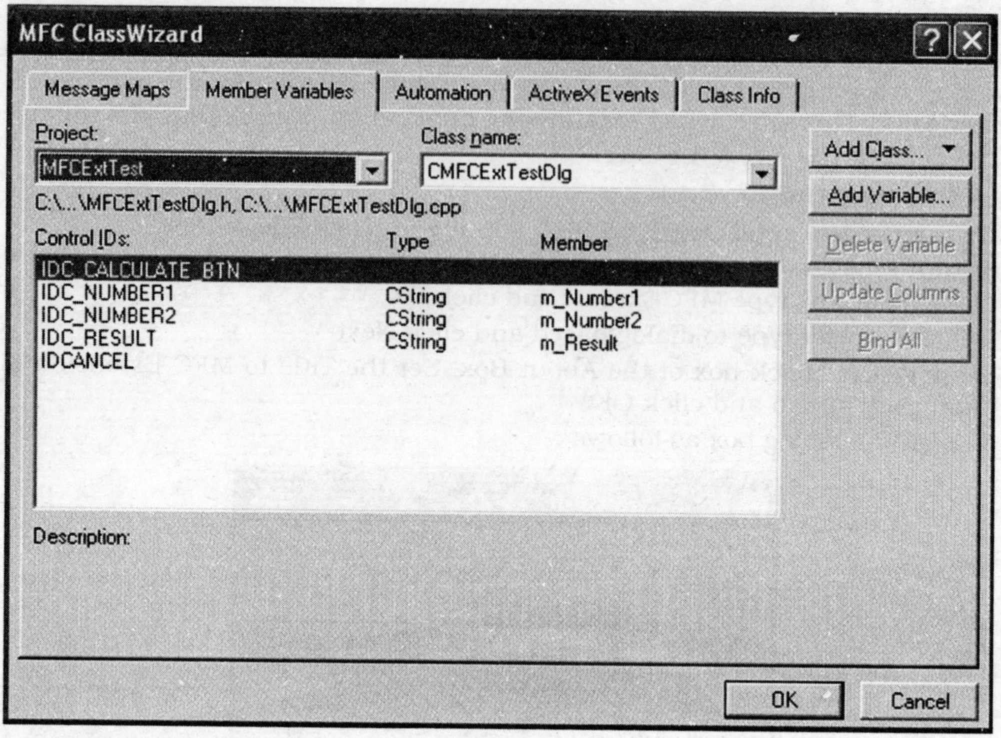

Fig. 3.12.

10. Display the dialog box. Double-click the Calculate button. Accept the suggested name of the function and click OK.
11. Implement the event as follows :

```
// MFCExtTestDlg.cpp : implementation file
//

#include "stdafx.h"
#include "MFCExtTest.h"
#include "MFCExtTestDlg.h"

#include "mfcextent.h"

#ifdef _DEBUG
#define new DEBUG_NEW
#undef THIS_FILE
static char THIS_FILE[] = __FILE__;
#endif
/////////////////////////////////////////////////////////////////////////////
// CMFCExtTestDlg dialog

CMFCExtTestDlg::CMFCExtTestDlg(CWnd* pParent /*=NULL*/)
: CDialog(CMFCExtTestDlg::IDD, pParent)
```

```
{
    //{{AFX_DATA_INIT(CMFCExtTestDlg)
    m_Number1 = _T("0.00");
    m_Number2 = _T("0.00");
    m_Result = _T("0.00");
    //}}AFX_DATA_INIT
    m_hIcon = AfxGetApp()->LoadIcon(IDR_MAINFRAME);
}

. . . .

void CMFCExtTestDlg::OnCalculateBtn()
{
    // TODO: Add your control notification handler code here
    double Number1, Number2, Result;

    UpdateData();

    // Evaluate the content of each operand edit box
    if( MFCExtensions::IsNumeric(m_Number1) == TRUE )
    {
        Number1 = MFCExtensions::StringToFloat(m_Number1);
    }
    else
    {
        AfxMessageBox("Invalid Number");
        Number1 = 0.00;
    }
    if( MFCExtensions::IsNumeric(m_Number2) == TRUE )
    {
        Number2 = MFCExtensions::StringToFloat(m_Number2);
    }
    else
    {
        AfxMessageBox("Invalid Number");
        Number2 = 0.00;
    }

    Result = Number1 + Number2;

    m_Result.Format("%.2f", Result);
    UpdateData(FALSE);
}
```

12. Test the application.

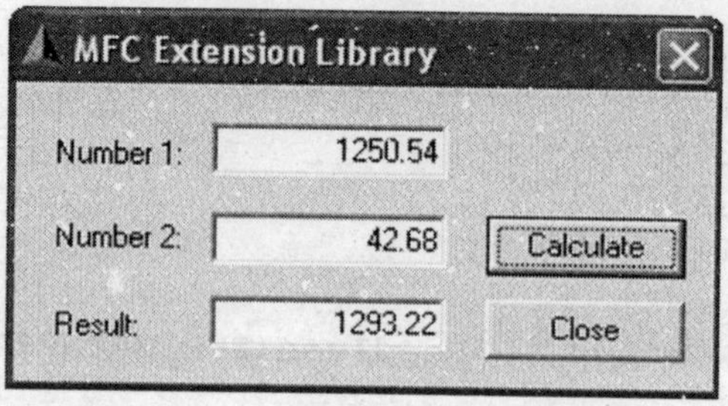

Fig. 3.13.

3.4 AN MFC APPLICATION

The purpose of this series of lessons is to introduce you to the MFC library, to prepare you to create graphical applications for the Microsoft Windows family of operating systems. Unlike C and C++, the MFC library is not directly available when you create a console application. You must explicitly add it. There are various ways you can do this.

3.4.1 Creating an MFC Application

(V.Imp)

You can create an MFC application from scratch. You can create an empty project and then configure it for MFC. Or you can ask the wizard to work for you behind the scenes.

To easily create an MFC application using Microsoft Visual C++, on the main menu, click File -> New Project... In the New Project dialog box, click Win32 Console Application. Give a name and a location. Click OK. In the first page of the wizard, click Next. In the second page of the wizard, make sure Console Application is selected. Click the MFC check box :

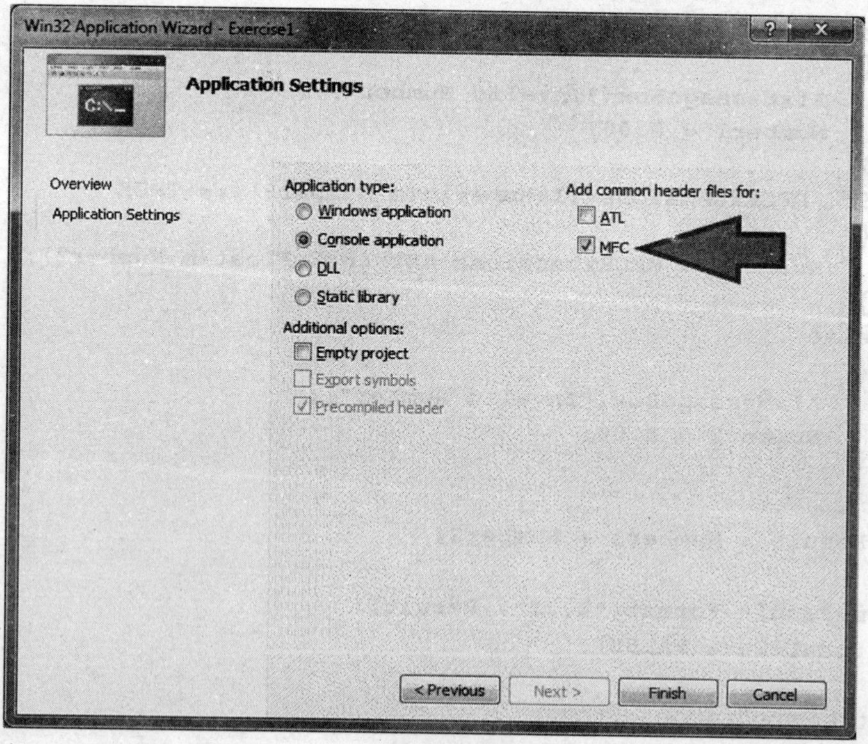

Fig. 3.14.

Click Finish

We also saw that you could create an empty project. If you do and you want to add support for MFC, there are a few things you must do. You must first access the properties of the application. To do that :

- On the main menu, click Project -> ProjectName Properties...
- In the Solution Explorer, right-click the name of the project and click Properties

In the Properties, click the combo box Use Of MFC and select Use MFC in a Shared DLL.

3.4.2 Creating an MFC Application

1. The gdcs1 project should still be opened.
   On the main menu, click Project -> gdcs1 Properties...
2. In the left section, click Configuration Properties (or double-click it to expand; then, under Configuration Properties, click General).
   In the right list, click Use of MFC to reveal its combo box
3. Click the arrow of the Use of MFC combo box and select Use MFC in a Shared DLL

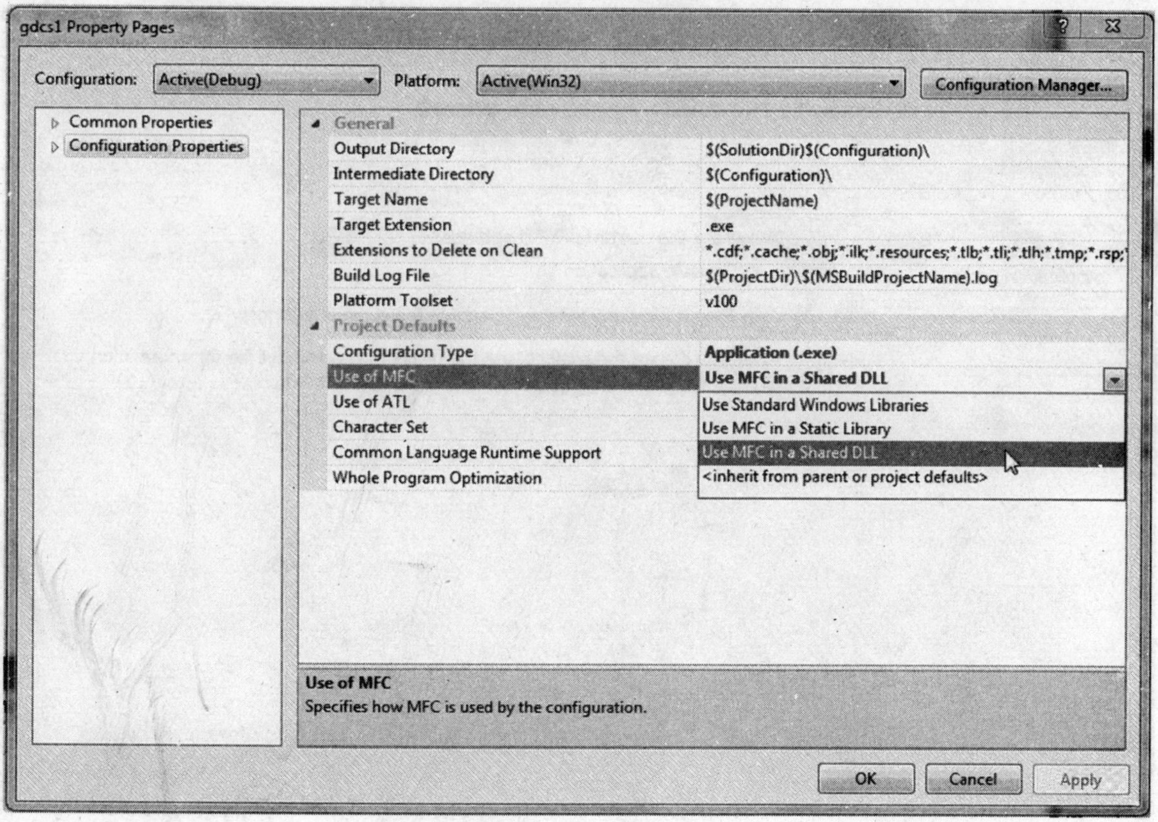

Fig. 3.15.

4. Click OK
5. Change the file as follows :
6. #include <iostream>

7. #include <Windows.h>
8. using namespace std;
9.
10. void main()
11. {
12. printf_s("Georgetown Dry Cleaning Services");
13.
14. getchar();
    }
15. To execute the application, on the main menu, click Debug -> Start Debugging
16. Press Enter to close the DOS window and return to your programming environment
17. To start a new application, on the main menu, click File -> New Project...
18. In the middle list, click Win32 Console Application
19. Change the name to gdcs2 and click OK
20. In the first page of the wizard, click Next

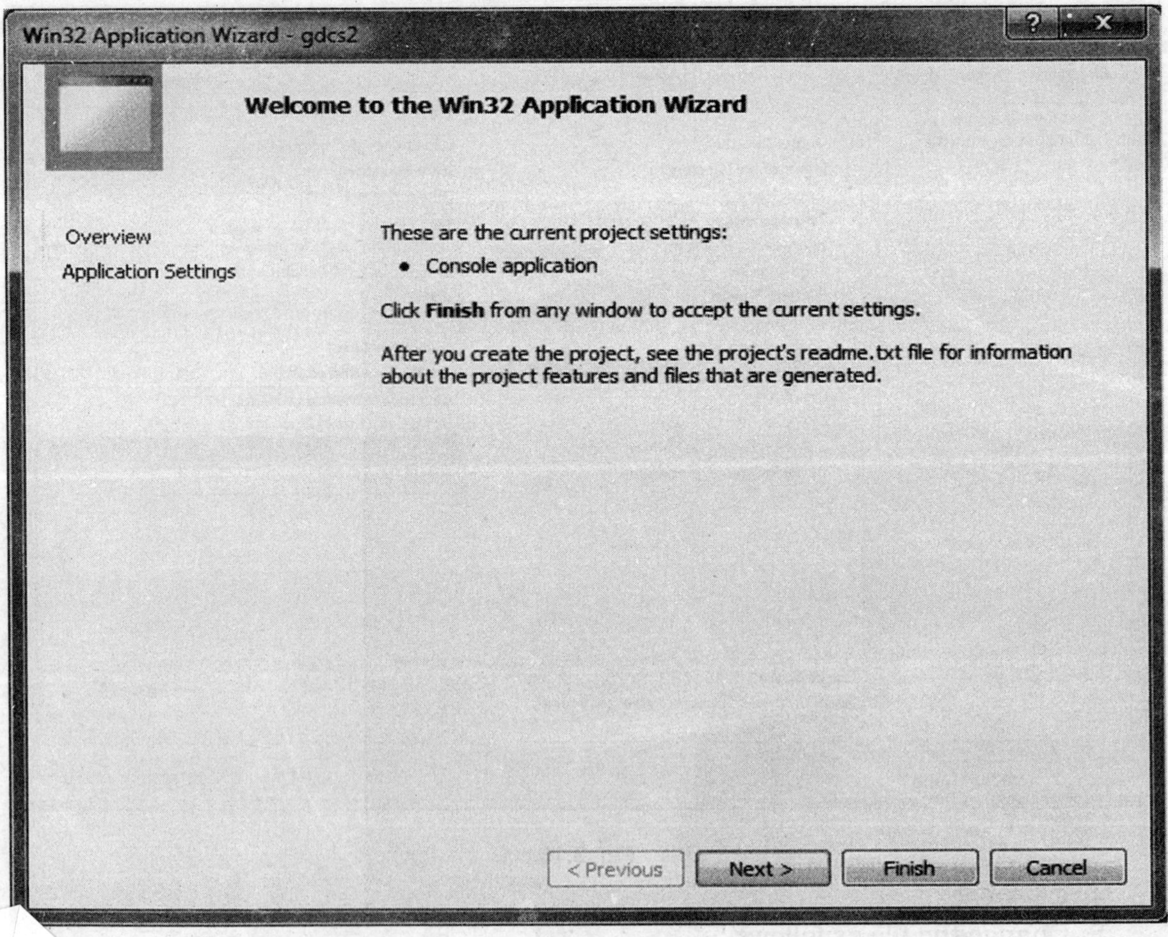

Fig. 3.16.

21. In the second page of the Win32 Application Wizard - gdcs2, make sure Console Application is selected. Click Pre compiled Header (or make sure it is clicked). Click the MFC check box.

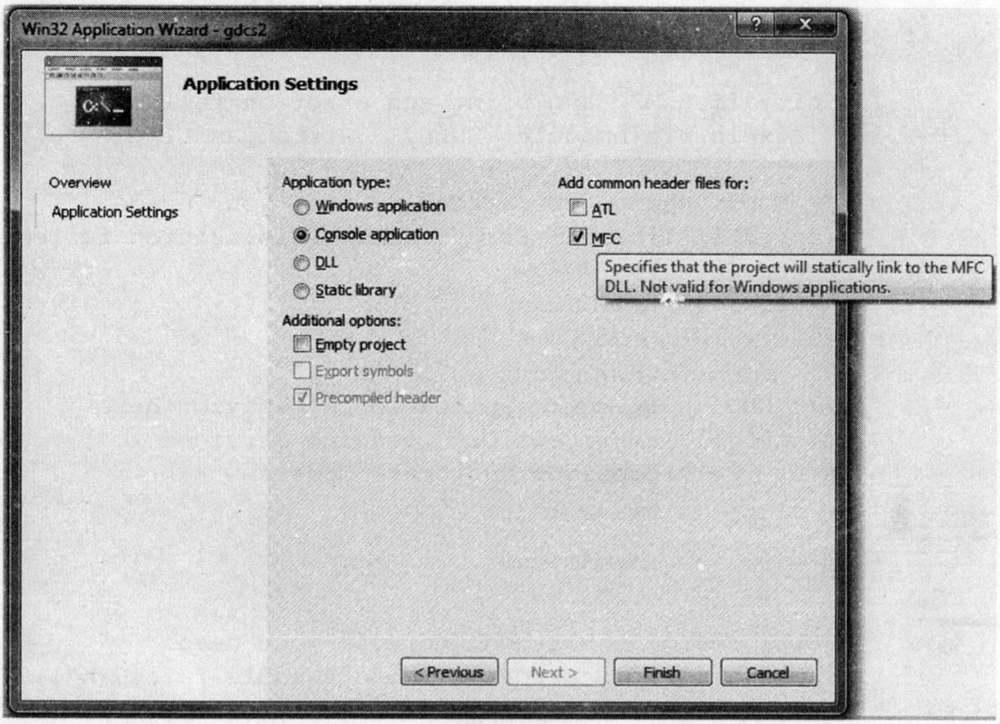

Fig. 3.17.

22. Click Finish
23. Change the file that was created as follows :

```
// gdcs2.cpp : Defines the entry point for the console application.
//

#include "stdafx.h"
#include "gdcs2.h"

#ifdef _DEBUG
#define new DEBUG_NEW
#endif

// The one and only application object

CWinApp theApp;

using namespace std;
int _tmain(int argc, TCHAR* argv[], TCHAR* envp[])
{
```

```
        int nRetCode = 0;

        HMODULE hModule = ::GetModuleHandle(NULL);

        if (hModule != NULL)
        {
            // initialize MFC and print and error on failure
            if (!AfxWinInit(hModule, NULL, ::GetCommandLine(), 0))
            {
                // TODO: change error code to suit your needs
                _tprintf(_T("Fatal Error: MFC initialization failed\n"));
                nRetCode = 1;
            }
            else
            {
                // TODO: code your application's behavior here.
                printf_s("Georgetown Dry Cleaning Services - ");
                cout << "Processing Customers Orders - ";
            }
        }
        else
        {
            // TODO: change error code to suit your needs
            _tprintf(_T("Fatal Error: GetModuleHandle failed\n"));
            nRetCode = 1;
        }

    return nRetCode;
    }
```

24. To execute the application, on the main menu, click Debug -> Start Debugging.
25. Press Enter to close the DOS window and return to your programming environment.

3.5 API(APPLICATION PROGRAMMING INTERFACE) **(MDU 2007-08)**

Programs in any environment (Operating System) make use of various resources available in that environment to interact with the operating system by calling functions provided by the operating system. All these functions are grouped according to their affinity into bundles called API's (Application Programming Interface). The programmer does not have to bother about the internal working of the functions. By just knowing the function prototype and return value, the programmer can invoke a required API function. Every program, thus, requires resources available in API's. These resources are incorporated into the application programs by including header files and library files where functions are stored.

Windows API's come in two flavours: 16-bit (Win16) and 32-bit (Win32). Whereas, Win16 API is available on both 16-bit and 32-bit versions of Windows, Win32 API is compatible with 32-bit versions of Windows (i.e. Win95, Win98 and WinNT) alone. In fact, Windows 95 supports three API's (*i.e.,* MS-DOS, Win16 and Win32), while WindowsNT supports five API's (i.e. MS-DOS, Win16, Win32, POSIX and OS/2).

SUMMARY

1. Windows programming is different in the sequence driven programming model, the operating system simply executes the program and then waits for it to finish while programming for windows, the OS not only executes the program, it talks to it! And the Application talks back *i.e.,* They communicate. Windows programming is based an event driven programming model.
2. The keyboard is an input device that is shared by all applications running in memory. Pressing a key (or combination of keys) on the keyboard constitutes a keyboard event. Each such key event generates a corresponding message, which is received by the Windows system. A device context is everything under one name. It is an orchestra, an ensemble of what need in order to draw. It includes the platform you draw on, the dimensioning of the platform, the orientation and other variations of your drawing, the tools you need to draw on the platform, the colours, and various other accessories that can complete your imagination.
3. CDC: This is the most fundamental class to draw in MFC. It provides all of the primary functions used to perform the basic drawing steps. In order to use this class, first declare a variable from it. Then call the BeginPaint() function to initialise the variable using the PAINSTRUCT class. Once the variable has been initialised, you can use it to draw. After using the device context call the EndPaint() function to terminate the drawing.
4. MFC, as we know is an object-oriented interface to Windows. As on date, the question arises, whether to use C with API or to learn MFC and its use in programming Windows using C++. But believe me OLE (Object Linking and Embedding) and other components of API are evolving so quickly that it is impossible to stay or top of all the latest developments
5. An MFC static library is a library file that includes MFC functions or classes. Such a library natively supports MFC strings (such as CString, CTime, etc.), lists, and any other C/C++ or MFC global functions. Because this library is an MFC type, it can be used only by an MFC-based application.
6. The purpose of this series of lessons is to introduce you to the MFC library, to prepare you to create graphical applications for the Microsoft Windows family of operating systems. Unlike C and C++, the MFC library is not directly available when you create a console application. You must explicitly add it.
7. Programs in any environment (Operating System) make use of various resources available in that environment to interact with the operating system by calling functions provided by the operating system. All these functions are grouped according to their affinity into bundles called API's (Application Programming Interface).
8. The keyboard is an input device that is shared by all applications running in memory. Pressing a key (or combination of keys) on the keyboard constitutes a keyboard event. Each such key event generates a corresponding message, which is received by the Windows system. Out of the several windows that may be active, Windows directs keyboard message to the window that has got the input focus.
9. A mouse is the most common pointing input device extensively used with windows applications. Each activity performed with the mouse (like moving, clicking, dragging etc.) is termed as a mouse event. Corresponding to each event a message is generated. For instance, if the mouse is moved, a corresponding event (mouse move event) occurs in the system and consequently a message (WM_MOUSEMOVE message) is produced.
10. Basically, there are three types of menus in a Windows application. System menu that is tucked below the application's caption bar, Pop-up menu that is generally invisible but can be popped-up anywhere in the application client area by right-clicking the mouse and Drop-down menu that appears as a sub-menu for the menu items in the menu bar at the top of the window.

SHORT QUESTIONS WITH ANSWER

Q.1. Why we use MFC? **(V.Imp)**

Ans. MFC, as we know is an object -oriented interface to Windows. As on date, the question arises, whether to use C with API or to learn MFC and its use in programming Windows using C++.

But believe me OLE (Object Linking and Embedding) and other components of API are evolving so quickly that it is impossible to stay or top of all the latest developments.

It is why in future, the programmers who use Windows API could be looked upon the way assembly language programmers are today.

Q.2. What is The MFC Class Wizard?

Ans. The MFC Class Wizard is a dialog box that can be used to create and manage classes of a project. The classes must use, or inherit from, MFC existing classes. There are various ways you can access the MFC Class Wizard :

- At any time on the menu, you can click Project -> Class Wizard...
- In the Solution Explorer, right-click the name of the project or the name of a folder, and click Class Wizard...
- In the Class View, right-click the name of the project or the name of a class, and click Class Wizard...
- In the Resource View, right-click the name of the project and click Class Wizard.
- If a dialog box (or a form) being designed is displaying in the middle area, right-click it (or any object on it) and click Class Wizard.

Q.3. (*b*) Briefly explain the following components of VB I.D.E.

(*i*) Form Designer

(*ii*) Object Browser

Ans. (*i*) Form Designer: Form Designer VB lets you resize and move any control on your Microsoft Visual Basic form at runtime. You need not prepare your form to use Form Designer. VB just drop Form Designer VB control onto your form, set Active property to True and enjoy. After activation of Form Designer. VB you can select, move and resize any control in your form by keys or mouse. Form with activated Form Designer VB looks exactly like designer in Microsoft Visual Basic DE.

Some key features of "Form Designer VB":

- Full nested containers support.
- Multi-selection.
- Rich set of edit events.
- Saving and loading form to form Visual Basic compatible form file.
- Customizable component hint.
- Component Toolbox example.
- Customizable design grid with snap-to-grid feature.
- Align and SameSize methods.
- Clipboard operations for selected controls.
- Clipboard operations for properties of selected controls.
- Full-functional form editor as free demo.
- Free trial version for Visual Basic.
- IDE-compatible help file.

(*ii*) Object Browser: Once a COM object has been added to your project as a reference or component, you can use the Visual Basic Object Browser to view type library information for that object in Visual Basic. To view a COM object in the Object Browser:

* On the View menu, click Object Browser.

The left side of the Object Browser window displays all of the classes of references and components added to the current project. The right side contains the methods, properties, and events of those classes. You can filter which COM object classes are displayed by selecting an object iibrary from the library selection drop- down list.

Q.4. How to store data in the program document object?

Ans. To make a class serializable, we first must make sure that it is derived from the MFC Cobject class (thus class is the basis of all MFC classes):

```
    class CData: public Cobject
    {
        private:
        CString data; DECLARE_SERIAL (CData);
        public:
        CData()
        {
            data = CString( " ");
        }
        void Add Text (CString text)
        {
            data t = text;}
            void Draw Text
            (CDC ^ pDC)
            pDC ® Text Out (0,0, data);}
            void clear Text( )
            {
                data ="";}
                void serialize (CArchieve & archieve)
    Now in CData. opp:
                #inctude "stdafx.h"
                #include "serializer Doc.h"
            void Data:: Serialize (CArchieve & archieve
            {
            Cobject:: Serialize (Archieve);
            if (archieve. IsString())
            {
                archieve « data;
            }
            else
            {
            archieve >> data;
        }
    }
```

Q.5. What are controls. Discuss various controls and their functioning in Visual Basic.

Ans. A control is a tool used to display information or retrieve information from the user. Examples of controls are boxer, buttons, graphics, labels and grids. The toolbox is used to select and then draw controls on a form. Controls come in two forms: standard controls and custom controls. The standard controls are always present in the the toolbox. The custom controls are brought into the visual basic environment and extend the toolbox controls have a file extension of .OCX.

1. **Check Bones:** They are little boxes that are either empty or display a small check mark. When you either checked or unchecked, You use check-boxes to let the user to select one or more options from number of option,

For doing thus, we have to use dialog editor to create and position checkboxes. When user clicks one of the checkbox then it indicates which checkbox they have clicked in textbox.

2. **Radio Buttons:** Radio button Sets the user to select one option out of many. These are also called option buttons. They are small circular buttons that display a block dot in the middle when clicked and empty otherwise. Radio buttons are coordinated into groups and functions together. Only one radio bottom a group of radio button can be checked.
3. **List Box:** These controls displays a list of items that is displayed in a box. The user can click on item to highlight it and double-click that item to select it.

   In thus, e.g., we see how to add new items to list box and how to find out which item the user selects. The list box won't be able to display all the items at one, so it will display & scrollbar control at angle.
4. **Combo Box:** Combobox arc very popular window consols, These controls an combination of text box, drop down list box and a button, the user can from to open the listbox.

   In this example, we actuate a control and a textbox *In* which to display the user's selection. We add string( ) method to add the strings.

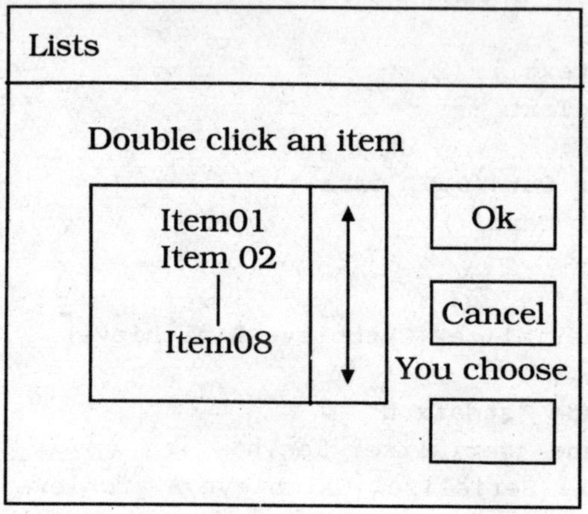

Fig. 3.18.

LONG QUESTIONS WITH ANSWER

Q.1.What is Event Oriented Windows Programming? **(V.Imp)**

Ans. Windows programming is different in the sequence driven programming model, the operating system simply executes the program and then waits for it to finish while programming for windows, the OS not only executes the program, it talks to it! And the Application talks back *i.e.,* They communicate. Windows programming is based an event driven programming model.

Depending on the type of application you are creating, you might need to notice what the user is doing with the mouse. You need to know when and where the mouse was clicked, which button was clicked, and when the button was released. You also need to know what the user did while the mouse button was being held down. Another thing that you might need to do is read the keyboard events. As with the mouse, you might need to know when a key was pressed, how long it was held down, and when it was released.

The keyboard is an input device that is shared by all applications running in memory. Pressing a key (or combination of keys) on the keyboard constitutes a keyboard event. Each such key event generates a corresponding message, which is received by the Windows system. Out of the several windows that may be active, Windows directs keyboard message to the window that has got the input focus. The window that has the input focus is the main window of the application. The input focus might also belong to a child of the main window, or to a control in a dialog box. Windows sends keyboard messages to the window that owns the focus.

A mouse is the most common pointing input device extensively used with windows applications. Each activity performed with the mouse (like moving, clicking, dragging etc.) is termed as a mouse event. Corresponding to each event a message is generated. For instance, if the mouse is moved, a corresponding event (mouse move event) occurs in the system and consequently a message (WM_MOUSEMOVE message) is produced.

As usual, the window getting the message from the Windows may implement a handler to that message or simply let it pass back to the Windows default handler.

There are two categories of mouse events - Client area mouse events and Non-Client area mouse events. Client area mouse events occur when the mouse pointer is focussed in the application's window. Mouse events produced when the mouse is focussed outside the application window area, are termed as Non-Client area mouse events.

Q.2. How would you Grabbing the Device Context?

Ans. As mentioned already, in order to draw, you need at least two things: a platform and a tool. The platform allows you to know what type of object you are drawing on and how you can draw on it. On a Windows application, you get this platform by creating a device context.

A device context is actually a whole class that provides the necessary drawing tools to perform the job. For example, it provides functions for selecting the tool to use when drawing. It also provides functions to draw text, lines, shapes etc. To select the platform on which to draw, that is, to create a device context, the MFC provides various classes :

CDC : This is the most fundamental class to draw in MFC. It provides all of the primary functions used to perform the basic drawing steps. In order to use this class, first declare a variable from it. Then call the **BeginPaint()** function to initialize the variable using the **PAINSTRUCT** class. Once the variable has been initialized, you can use it to draw. After using the device context call the **EndPaint()** function to terminate the drawing.

CPaintDC : Unlike the **CDC** object, the **CPaintDC** inherently initializes its drawing by calling the **BeginPaint()** function when you declare a **CPaintDC** variable. When the drawing with this class is over, it calls the **EndPaint()** to terminate the drawing.

CClientDC : This class is used when you want to draw in the client area of a window.

CMetaFileDC : This class is used to create Windows metafiles.

Q.3. How would you use Device Context?

Ans. In order to draw using a device context, you must first declare a variable of the CDC class. This can be done as follows :

CDC dc : To help with this, the CView class provides a virtual member function that can carry the drawingassignments on a program. The method provided is **OnDraw()** and its syntax is :

```
void OnDraw(CDC* pDC) = 0;
```

This means that each class that derives from CView must provides its own implementation of this method. If you use AppWizard to create an application and select a CView-derived base class, the AppWizard would define a basic implementation of **OnDraw()** for you. For a CView-based application, this can be a good place to perform a lot of your drawing.

As you can see, the **OnDraw()** method is passed a pointer to CDC. This allows you to use the pDC pointer as a variable and draw on the view object with it.

Declaring a CDC variable or receiving it as an argument to a function gives you a device context (DC) you can use. This DC initializes the drawing with some default objects such as a pen to draw the most basic points or lines.

Q.4. What do you mean by MFC Static Libraries? Describe.

Ans. An MFC static library is a library file that includes MFC functions or classes. Such a library natively supports MFC strings (such as CString, CTime, etc), lists, and any other C/C++ or MFC global functions. Because this library is an MFC type, it can be used only by an MFC-based application.

Creation of an MFC Static Library: To create an MFC static library, from the New dialog box, select Win32 Static Library and specify the directory. Then, in the Win32 Static Library - Step 1 of 1, mark the MFC Support check box. You can also ask Microsoft Visual C++ to generate pre compiled header files :

- If you don't mark the Pre compiled Header check box, an empty project would be created.
- If you select the Pre compiled Header radio button, the wizard will generate a StdAfx.h header file and a StdAfx.cpp source file for the project. The wizard will also include the afx.h and the afxwin.h files in the StdAfx.h header file. You can still add any cenessary header file in StdAfx.h.

Creating an MFC Static Library:

1. To start a new project, on the main menu, click File -> New...
2. In the New dialog box, click the Projects property page.
3. In the list, click Win32 Static Library.
4. In the Project Name edit box, type MFCExt.

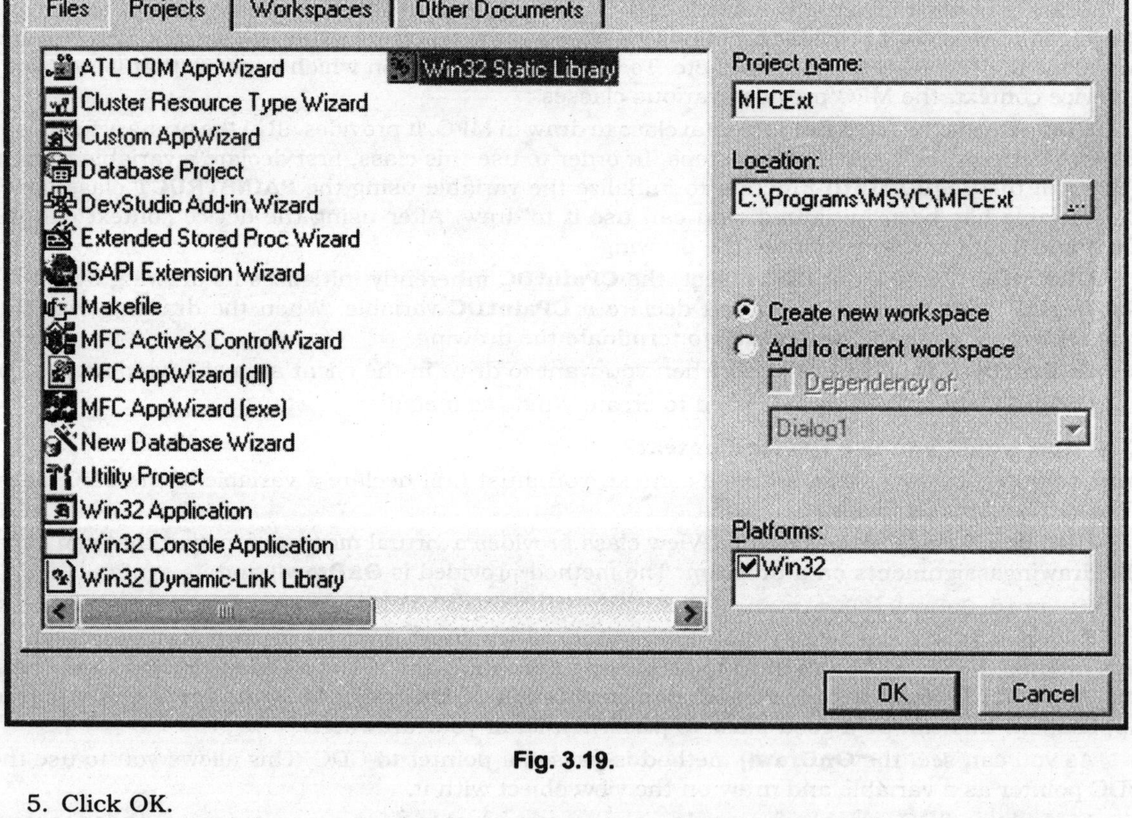

Fig. 3.19.

5. Click OK.

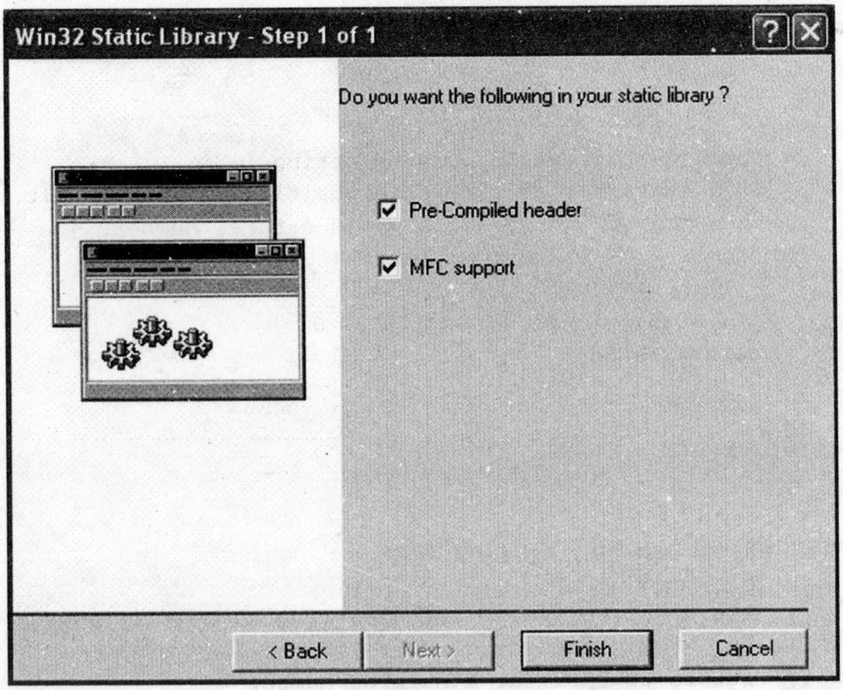

Fig. 3.20.

6. In the Win32 Static Library - Step 1 of 1, click both check boxes and click Finish then click OK.
7. To add a new header file, on the main menu, click File -> New...
8. In the Files property page of the New dialog box, click C/C++ Header File. In the File Name edit box, type mfcextent and click OK.
9. In the empty file, type the following :

```
#ifndef MFCEXT_H_
#define MFCEXT_H_

namespace MFCExtensions
{
   BOOL IsNatural(const CString Str);
   BOOL IsNumeric(const CString Str);
   int StringToInt(const CString Str);
   double StringToFloat(const CString Str);
}

#endif // MFCEXT_H
```

10. To add a new source file, on the main menu, click File -> New... In the Files property page of the New dialog box, click C++ Source File. In the File Name edit box, type mfcextent and click OK.
11. In the empty file, type the following :

```
#include "stdafx.h"
#include " mfcextent.h"
```

```
namespace MFCExtensions
{
    BOOL IsNatural(const CString Str)
    {
        // Check each character of the string
        // If a character at a certain position is not a digit,
        // then the string is not a valid natural number
        for(int i = 0; i < Str.GetLength(); i++)
        {
            if( Str[i] < '0' || Str[i] > '9' )
            return FALSE;
        }

        return TRUE;
    }

    BOOL IsNumeric(const CString Str)
    {
        // Make a copy of the original string to test it
        CString WithoutSeparator = Str;
        // Prepare to test for a natural number
        // First remove the decimal separator, if any
        WithoutSeparator.Replace(".", "");

        // If this number were natural, test it
        // If it is not even a natural number, then it can't be valid
        if( IsNatural(WithoutSeparator) == FALSE )
            return FALSE; // Return Invalid Number

        // Set a counter to 0 to counter the number of decimal separators
        int Number Of Separators = 0;

        // Check each charcter in the original string
        for(int i = 0; i < Str.GetLength(); i++)
        {
            // If you find a decimal separator, count it
            if( Str[i] == '.' )
                Number Of Separators++;
        }

        // After checking the string and counting the decimal separators
        // If there is more than one decimal separator,
        // then this cannot be a valid number
        if( Number Of Separators > 1 )
            return FALSE; // Return Invalid Number
        else // Otherwise, this appears to be a valid decimal number
            return TRUE;
    }
```

```
int StringToInt(const CString Str)
{
    // First check to see if this is a valid natural number
    BOOL IsValid = IsNatural(Str);

    // If this number is valid, then convert it
    if( IsValid == TRUE )
        return atoi(Str);
    else
    {
        // Return 0 to be nice
        return 0;
    }
}

double StringToFloat(const CString Str)
{
    // First check to see if this is a valid number
    BOOL IsValid = IsNumeric(Str);

    // If this number is valid, then convert it
    if( IsValid == TRUE )
        return atof(Str);
    else
    {
        // Return 0 to be nice
        return 0;
    }
}
}
```

12. To create the library, on the main menu, click Build -> Build MFCExt.libThe Output window will indicate that a file with .lib extension was created.

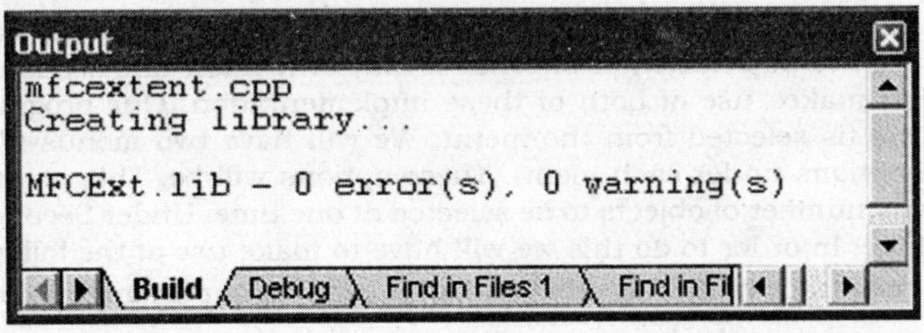

Fig. 3.21.

Creating an MFC Application: You can create an MFC application from scratch. You can create an empty project and then configure it for MFC. Or you can ask the wizard to work for you behind the scenes.

To easily create an MFC application using Microsoft Visual C++, on the main menu, click File -> New Project... In the New Project dialog box, click Win32 Console Application. Give a name and a location. Click OK. In the first page of the wizard, click Next. In the second page of the wizard, make sure Console Application is selected. Click the MFC check box :

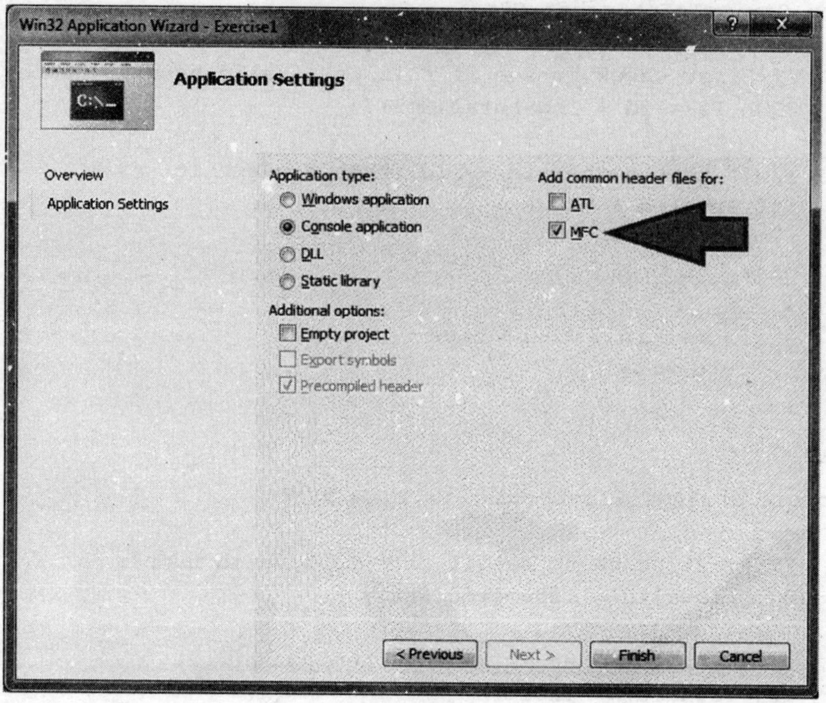

Fig. 3.22.

Click Finish

We also saw that you could create an empty project. If you do and you want to add support for MFC, there are a few things you must do. You must first access the properties of the application. To do that :

- On the main menu, click Project -> ProjectName Properties.
- In the Solution Explorer, right-click the name of the project and click Properties

In the Properties, click the combo box Use Of MFC and select Use MFC in a Shared DLL.

Q.5. What do you mean by graying out menu items ? Write a program in VC++ to check and grey out menu items.

Ans. Checked menu items are commonly used in two ways. The menu could have several options of which any number cab be selected at once, or it may only allow one section at a time. We will write a program that makes use of both of these implementation. Our program will display the name of theoptiontha tis selected from the menu. We will have two menus (**First and Second**). There will be three options under each menu. These options will be. This is, and working. Under First we will allow any number of objects to be selected at one time. Under Second we will only allow one selection at a time. In order to do this we will have to make use of the following messages.

COMMAND to keep track of the flags that will let the rest of our program know the status of our menus.

UPDATE_COMMAND_UP to make the necessary changes to the display of the menus. This function is called oven time a user clicks on a menu. Before the menu is shown, the "updata handler" is called. This is where menu features such as checks and laying out of menu items are taken care of.

We will handle all of these things within the View class. We will also display the messages to the screen from within the On Point ( ) functions of the View class.

1. Start up Visual C++
2. Click on **File** and **New**.

3. Click on the Project tab and select MFC AppWizard (exe).
4. Enter **Checkmen** as the project name.
5. **Click OK**.
6. On MFC AppWizard step one, select **"single document"** and click **Finish**.
7. On New Project Information, click on **OK**.
8. Click the + next to Vheckmen classes in the Classview on the left frame.
9. Right click on CCheckmen View, then click on **Add member variable**.
10. In the Add member variable dialog box enter **int** as the variable type.
11. Under Variable Declaration enter **select 1**. Then click **OK**.
12. Repeat step 10-12 two more times, replacing select 1 with select 2 and then **select 3**.
13. Press **Ctrl-2**.
14. Double click on **IDR_MENUI**.
15. Double click on the outlined grey box in the right frame.
16. In the Menu Item Properties dialog enter **First** in the caption box and press enter.
17. Double Click on the gray box below First.
18. Under Caption enter **This**, check the box that says **Checked** and press enter.
19. Double click on the gray box below **it**.
20. Under Caption entries, check the box that says **Checked** and press enter.
21. Double Click on the gray box below is.
22. Under Caption enter Working, check the box that says **Checked** and press enter.
23. Now we have to ad our menu to the program. Click on **File** and **Open**.
24. Click on Checkmen, opp and click **Open**.
25. Scroll down in the source to the line that says.
    DocTemplate = newCSingleDoc Template (
    change the first argument from **IDR_MAINFRAME** to **IDR_MENU 1**
26. Click **File and Save**.
27. Click **File** and **Open**
28. Select Checkmenview.cop and click Open.
29. Scroll down in the source to find the line that says:
```
CCheckmenView:CCheckmen View ( )
    {
        // TODO: add construction code here
    }
```
30. Insert the following lines into this function:
```
select 1 =   0;
select 2 =   0;
select 3 =   0;
```
    These lines initialize our flag variable to zero. When our program starts, no menu item will be selected.
31. Press **File** and **Save**.
32. Press **Ctrl-W** to start the MFCWizard.
33. Under Class name, choose CCheckmen View.
34. Under Object Ids, choose **ID_FIRST_THIS**.
35. Double click **COMMAND** under Message and press enter.
36. Double click on **UPDATE_COMMAND_UI** and press enter.
37. Under Member Functions select **OnFirstThis ( )** and click edit code.
38. In the function **OnFirstThis ( )** add the following lines of code:
```
        if (select 1! = 1)
```

```
select1 = 1;
else
select2 = 0;
Invalidate Rect (NULL, TRUE);
```

The else allows the user to remove the check from the selection.

39. Scroll down and add the following code into the function.

```
OnUpdateFirstThis (CCmdUI*pCmdUI);
if (select 1 = 1)
pCmdUI->SetCheck(TRUE);
else
pCmdUI->SetCheck(FALSE);
```

40. Press **Ctrl-W** to stare the MFCWizard.
41. Under Class Name, choose **CCheckmeView.**
42. Under Object IDs, choose **ID_FIRST_IS.**
43. Double click **COMMAND** under Messages and press enter.
44. Double click on **UPDATE_COMMAND_UI** and press enter.
45. Under Member Functions select **OnFirstls ( )** and click edit code.
46. In the function **OnFirstls ( )** add the following lines of code:

```
if (select 2! = 1)
select2 = 1;
else
select2 = 0;
Invalidate Rect (NULL, TRUE);
```

The else allows the user to remove the check from the selection.

47. Scroll down and add the following code into the function.

```
On UpdateFirstis (CCmdUI*pCmdUI);
if (select2 = 1)
pCmdUI->SetCheck(TRUE);
else
pCmdUI->SetCheck(FALSE);
```

48. Press **Ctrl-W** to stare the MFCWizard.
49. Under Class Name, choose CCheckmenView.
50. Under Object IDs, choose **ID_FIRST_WORKING.**
51. Double click **COMMAND** under Messages and press enter.
52. Double click on **UPDATE_COMMAND_UI** and press enter.
53. Under Member Functions select **OnFirstWorking ( )** and click edit code.
54. In the function **OnFirstWorking ( )** add the following lines of code:

```
if (select 3! = 1)
select3 = 1;
else
select3 = 3;
Invalidate Rect (NULL, TRUE);
```

The else allows the user to remove the check from the selection.

55. Scroll down and add the following code into the function.

```
On UpdateFirstWorking (CCmdUI*pCmdUI);
if (select3 = 1)
pCmdUI->SetCheck(TRUE);
ELSE
```

```
pCmdUI->SetCheck(FALSE);
```

56. Press **Ctrl-W** to start the MFC Class Wizard.
57. Make sure that **CCheckmenView** is selected in the Class name field.
58. Under Object IDs, select CCheckmenView. Under Message double click on **WM_PAINT.**
59. Click the **Edit Code** button.
60. Add the following code to the **OnPaint** ( ) function:

```
if (select ==1)
dc.TextOut (20,20,"This");
if (select2==1)
dc.TextOut (50,20,"This");
if (select3==1)
dc.TextOut (66,20,"Working");
```

61. Click **Build** and then **Execute**. Click **Yes** on the dialog that comes up.
62. Turn the menu items on and off, and notice how the functionality works.
63. Now we will add our second menu which will only allow one item to be checked at a time. The code for this is very similar to the previous menu. This time we will only use one flag. This flag will have four possible values. (0,1, 2, 3)
64. In the left flame of Visual C++, click on the Classview tab. Right click on CCheckmen View, then click on Add member variable. Enter **int** into the variable type field. Enter **selection** into the variable declaration field, then hit enter.
65. In Checkmen View.cpp, add the following line to the **CCheckmen View::CCheckmenView ( )** function.

```
fselection = 0
```

66. Click **File** and **Save.**
67. Click on the **Resource** tab in the left frame of visual C++. Go to the menu folder, and double click on **IRD_MENU1.**
68. Double click on the outlined gray box next to First. Enter **Second** as the caption and hit enter.
69. Double click on the outlined gray box under Second. Enter **This** as the caption, check the box marked Checked and hit enter.
70. Double click on the outlined gray box under This. Enter **is** as the caption, check the box marked **Checked**, and hit enter.
71. Double click on the outlined gray box under Is. Enter **Working** as the caption, check the box marked **Checked**, and hit enter.
72. Pass **Ctrl-W** to start the MFC Class Wizard.
73. Make sure that **CCheckmenView** is selected in the Class name field. Select **ID_SECOND_THIS** in the object IDs box. Double click **COMMAND** in the Messages box and hit enter. Double click **UPDATE_COMMAND_UI** in the Messages box and hit enter.
74. Select **ID_SECOND_IS** in the object IDs box. Double click **COMMAND** in the Messages box and hit enter. Double click **UPDATE_COMMAND_UI** in the Messages box and hit enter.
75. Select **ID_SECOND_WORKING** in the object IDs box. Double click **COMMAND** in the Messages box and hit enter. Double click **UPDATE_COMMAND_UI** in the Messages box and hit enter.
76. Click **Edit Code.**
77. Scroll up to the **OnSecondThis ( )** function and enter the following code.

```
if(fselection!=1)
fselection=1;
else
fselection=0;
InvalidateRect(NULL, TRUE);
```

78. Scroll down and add the following code into the function

```
OnUpdateSecondThis(CCmdUI*pCmdUI);
if(fselection==1)
pCmdUI->SetCheck(TRUE);
else
pCmdUI->SetCheck(FALSE);
```

79. Scroll down and add the following code into the function **OnSecondis ( ).**

```
if(fselection!=2)
fselection=2;
else
fselection=0;
InvalidateRect(NULL, TRUE);
```

80. Scroll down and add the following code into the function

```
OnUpdateSecondly(CCmdUI*pCmdUI);
if(fselection==2)
pCmdUI->SetCheck(FALSE);
```

81. Scroll down and add the following code into the functioin **OnSecondWorking ( ).**

```
if(fselection!=3)
fselection=3;
else
fselection=0;
InvalidateRect(NULL, TRUE);
```

82. Scroll down and add the following code into the function

```
OnUpdateSecondWorking(CCmdUI*pCmdUI);
if(fselection==3)
pCmdUI->SetCheck(TRUE);
else
pCmdUI->SetCheck(FALSE);
```

83. Now we must finish the program by adding the appropriate code to the **OnPaint ( )** function in code for our program to display the results of this menu. Scroll up to the **OnPaint ( )** function and add the following code.

```
if(fselection==1)
dc.TextOut(20,100,"This");
if(fselection==2)
dc.TextOut(20,100,"Is");
if(fselection==3)
dc.TextOut(20,100,"Working");
```

84. Click **Build** and **Execute**. Click **Yes** on the dialog that comes up.

Q.6. Write a program on add tool tips to a toolbar button.

Ans. Tool Tip to toll Bar: TO enable tool tips on a Toolbar, follow these steps:

1. When calling CToolbar::Create ( ), passCBRS_TOOLTIPS into SetBarStyle ( ). This process is already done for the default toolbar in AppWizard-generated projects.
2. For each button on the toolbar, add a string to your string resource table corresponding to the button's command ID in the form:

   "This is the fly-bytext/n. This is what shows up in tool tip."

   This process is also done automatically for you by AppWizard. For example. ID_FILE_NEW, the first button on the toolbar has the string entry "Create a new document/n New. "" New"" is the text that shows up in the tooltip. You need to add a string to your string resource table corresponding to the button's command ID for any buttons or toolbars that you add.

Theta it! Because AppWizard does this for you, by using AppWizard you already have a toolbar with working tool tips.

However, the process is more complicated if you want to change the text shown in a tool tip at run time.

3. On the project menu, click Settings, and click the ""C/C++"" tab. Choose ""General"" in the Category field: in the preprocessor Definitions edit box add "NO_ANSIUNI_ONLY" (without the quotation marks) as one of the preprocessor definitions.
4. Add the following message map entries to the message maps of your CMainFrame class and CchildFrame class:

   ON_NOTIFY_EX_RANGE(TIN_NEEDTEXTW, 0,OxFFFF, OnDynamicText)

   ON_NOTIFY_EX_RANGE(TIN_NEEDTEXTA 0,OxFFFF, OnDynamicTip Text)

   Note that this method is for an MDI application. If your application is SDI, you won't have a CChildFrame class.
5. Add a function to both your CMainFrame class and (if MDI) aCChildFrame class called.

   OnDynamicTipText that handles the TIN_NEEDTEXT notification and enters the appropriate text into the NMHDR structure based on the idFrom TAT comes in. Refer to the SuperTip sample for a complete example of how to do this.

Q.7. Define accelerator keys. Write a program in VC++ to create shortcut keys and accelerator keys.

Ans. Accelerator Keys: Accelerator keys allows you to use the calculator from the computer keyboard, rather than using the calculator's own keypad. Acclerator keys are sometimes a quicker way of getting to a function, especially for the experienced use or if the keypad equivalent can only be reached by pressing a number of keys. Some users prefer to use the keyboard rather than a pointing device. To process accelerator keys in a model dialog box in MFC, you must override the CWinApp:::ProcessMessageFilter( ) function. The framework calls **Process MessageFilter ( )** before it processes a message. To modify an MFC Application Wizard dialog box-based application type in Visual C++ NET to correctly process accelerator keys, follow these steps:

1. In Visual Studio .NET, create a new MFC application. In the left pane of the MFC Application Wizard, click **Application type**, and then make sure that **Application type** is set to **Dialog based**.
2. In resource view, double-click the dialog resource to open the dialog resource editor. The resource ID of the dialog resources similar to IDD_MYPROJECT_DIALOG (where MYPROJECT is the name that you gave your application project when you created it).
3. With the dialog resource open in the dialog resource editor, locate the properties window, Edit the dialog box **Border** property and specify the **Thin** border style. This step is required for a dialog box that contains a menu.
4. Create a new menu resource that contains a top-level entry named & File and a menu item named & Exit/tCTRL + E(CTRL + X is usually associated with cutting text, so CTRL+E is used instead).
5. In the menu editor, click the newly created **Exit** menu item, In the Properties window, make sure that the **ID** Property for the **Exit** menu item is set to **ID_FILE_EXIT.**
6. Associate the new menu with the dialog box by entering the menu ID in the dialog resource **Menu** Property. Open the dialog editor for the dialog resource and find the **Menu** Property in the Properties window. Set it to the resource ID that you created for the menu in step 4. To do this, click the drop-down list in the **Menu** property now, and then click to select the resource ID for the menu in the list.
7. In the menu editor, right-click **& Exit/tCTRL + E**, and then click **Add Event Handler.**
8. In the Event Handler Wizard, select the **COMMAND** message type. In the **Class** list, select the **CDialog** derived main class for the handler to be generated in. Make sure that the

function handler name is appropriate, and then click **Add and Edt** to create the menu item event handler.

9. Insert the following line in the function **Exit** menu item event handler method that is generated in step 8.
   PostMessage (WM_COMMAND IDOK, 0L);
   The produces the same effect as clicking OK when the user clicks **Exit** on the **File menu**. Clicking **OK** closes your dialog box appliation.
10. Create a new accelerator resource and associate the **CTRL + Ekey** combination with **ID_FILE_EXIT.** Save your changes.
11. Edit the Stadfx.h file to declare the following global variables after the **#include** statements.

```
extern HWND ghDlg;            //Handle to main dialog box.
extern HACCEL ghAccelTable;   //Handle to accelerator table.
```

12. In te.cpp file that contains the **CWinApp** derived class implementation (usually the cpp file with the same base name as your project name), add the following global variable initializations:

```
HWND ghDig = 0;               //Handle to main dialog box.
HACCEL ghAccelTable = 0;      //Handle to accelerator table.
```

13. In the main **CDialog** derived class (not the dialog class that implements the default About dialog box), find the **OnInitDialog ( )** method. If necessary, you can add an override for the **OnlnitDialog ( )** method. to do this, click to select the **CDialog** derived Class in the Class View window, and then click **Overrides** in the Properties window. Find the **OnlnitDialog** row and click the right column. If **OnlnitDialog ( )** is not overridden for this class, you have the option to create an override.
14. Edit the function that you added earlier to include the following line of code:

```
gh Dig = m_hWnd,
```

   In the opp file that contains the **CWinApp** derive implementation, find the Initianstance ( ) class method. Add the following line immediately after the call to the base class **CWinApp::Initinstance ( ):**
15. ghAccelTable = LoadAccelerators (AfxGeltInstanceHandle ( );
   MAKEINTRESOURCE (IDR_ACCELERATOR 1));
16. An override to the **CWinAp** derived class for the **ProcessMessageFilter ( )** class method. To do this, in the Class View window, select the **CWinApp** derived class in your project. Then, in the Properties window, click Overrides. Find the **ProcessMessageFilter** row in the Properties option to add an override for the **ProcessMessageFilter** method.
17. Edit the **ProcessMessageFilter ( )** method override so that it has the following implementation:

```
BOLL CMyProjectApp::ProcessMessageFilter (int code, LPMSG IsMsg)
{    if(code < 0)
CWinApp::Processmessagefilter (code, IpMsg);
if (ghDIg&&ghAccelTable)
{ if (::TranslateAccelerator (ghDIg, ghAccelTable, IpMsg))
return (TRUE);   }
return CWinApp:::ProcessMessageFilter code, IpMsg);    }
```

18. Compile and run the application.

Q.8. Write a program in VC++ to add prompts to the status bar.

Ans. The status Bar: The status bar window neither accepts user input no generates command messages. Its job is simply to display text in panes under program control. The status bar supports two types of text panec, message line panes and status indicator panes. To use the status bar for application-specific data, you must first disable the standard status bar that displays the menu prompt and key-board status.

The Status Bar Definition: The static indicators array that AppWizard generates in the MainFrm.cpp file define the panes for the application's status bar. The constant ID_SEPARATOR identifies a message line pane; the other constants are string resource IDs that identify indicator panes.

```
Static UNIT indicators [ ] =
ID_SEPARATOR,
ID_INDICATOR_CAPS,
ID_INDICATOR_NUM,
ID_INDICATOR_SCRL,
};
```

The CStatusBar:::Setindicators member function, called the applications's derived ram class, configures the status bar according to the contents of the indicators array.

The Message Line: A message line pane displays a string that the program supplies dynamically. To set they value of the message line, you must first get access to the status bar object and then you must call the CStatusBar::SetPaneText member function with a zero-based index parameter. Pane 0 is the leftmost pane, 1 is the next pane to the right, and so forth.

The following code fragment is part of a view class member function. Note that you must navigate up to the applications object and then back down to the main frame window.

```
CMainFrame*pFrame = (CMainFrame->m_wndStatusBar;
pStatus->SetPaneText(0, "message line for first pane");
```

Normally, the length of a message line pane is exactly one-fourth the width to the display. If however, the message line is the first (index 0) pane, it a stretxchy pane without a beveled border its minimum length is one-fourth the display width, add it expands if room is available in the status bar.

The Status Indicator: A status indicator pane is linked to a single resource-supplied string that is displayed or hidden by logic in an associated updata command UI message handler function. An indicators is identified by a string resource ID, and that same ID is used to rote updata command UI message. The Caps Lock indicator is handled in the frame calls by a message map entry and a handler function equivalent to those shown below. The Enable ( ) function turns on the indicator if the **Caps Lock** mode is set

```
ON_UPDATE_COMMAND_UI_(ID_INDICATOR_CAPS, OnUpdatekeyCapsLock)
Void CMainFrame::::OnUpdatekeyCapsLock(CCmdUI*pCMdUIA) {
pCmdUI->Enable(::::GetkeyState(VK-CAPITAL)&1);}
```

The status bar updata command UI functions are called during idle processing so that the status bar is updated whenever your application receives messages. The length of a status indicator pane is the exact length of the corresponding resource string.

Taking Control of the Status Bar: In the standard application framework implementation, the status bar has the child window ID AFX_IDW_STATUS_BAR. The application framework looks for this ID when it wants to display a menu prompt. The updata commend UI handlers for the keyboard state indicators, embedded in the frame window base class, are linked to the following string IDs; ID_INDICATOR_CAPS, ID_INDICATOR_NUM, and ID_INDICATOR_SCRL. To take control of the status bar, you must use a different child window ID and you must use different indicator ID constants. Type only reason to change the status bar's child window ID is to prevent the framework from writing menu prompts in pane 0. If you like the menu prompts, you can disregard the following instructions.

The status bar window ID is assigned in the CStatusBar::::Create function called by the derived frame class OnCreate ( ) member function. That function is contained in the **MainFrm.cpp** file that appWizard generates. The window ID is the third Create ( ) parameter and it defaults to AFX_IDW_STATUS_BAR. To assign your own id, you must replace this call. With this call:

```
m_mwndStatusBar.Create(thid);
```

With this call;

```
m_wndStatusBar.Create (this, WS_CHILE|WS_VISIBLE|BOTTOM,
ID_MY_STATUS_BAR);
```

You must also, of course, define the ID_MY_STATUS_BAR constant in the resource.h file (using Visual C++'s resource symbol editor). We forgot one thing. The standard application framework's View menu allows the user to turn the status bar on and off. That logic is pegged to the AFX_IDW_STATUS_BAR window ID, so you'll have to change the menu logic, too. In your derived frame class, you must write message map 3 entries and handlers for the ID_VIEW_STATUS_BAR command and updata command UI message ID_VIEW_STATUS_BAR is the ID of the **Status Bar** menu item. The derived classhandlers override the standard handlers in the CFrame Wnd base class.

Q.9. What are various options provided by VB to access remote databases.

Ans. Data Access Objects (DAO) can be used either with the Microsoft Jet database engine or, using the ODBC Direct option, without it. This chapter discusses design and implementation issues that arise when using the Data Access Objects (DAO) to access remote databases.

DAO Remote Data Access Using Jet: This section discusses DAO functionality when it is connected to the Jet engine. The Microsoft Jet database engine is a stand-alone database management system that is capable of both processing queries and routing queries to remote servers as needed. Accessing Jet through DAO adds to Microsoft Visual Basic's ease of development by providing an object-oriented development paradigm and accessibility to data-aware bound controls. By using the Data control, DAO, or Microsoft Access, you can create code that is virtually database-independent, because Jet automatically performs all syntax and data manipulation translations for you. For example, you can write an application that accesses different types of data sources without making reference to specify remote server features. These data sources could be Open Database Connectivity (ODBC) databases, such as Microsoft SQL, Server; Index Sequential Access Method (ISAM) databases, such as Microsoft FoxPro, Paradox, or dBASE; or other Jet databases. Unlike most stand alone database engineers, Jet can perform heterogeneous joints across several dissimilar databases. If you are working with departmental data stored in ISAM format, and need to merge it with data on a centralized server, this is an essential feature.

Managing DAO ODBC Users: Each instance of your database application uses some number of connections and data page (or now) locks on the server, and creates a measurable load on the network. Since each additional user contends for many of the same resources, the number of users the system can support is directly proportional to the number of resources each instance of your application requires. To reduce the number of locks, users should not be permitted to "sit" on unpopulated Recordest objects. The application should populate the recordest as quickly as possible using DAO, the Data control, or one of the background population techniques. You design should also include management of user log on IDs and passwords. If your design uses a shared Jet (mdb) database, you must also address Jet security systems. Because all users must disconnect from Jet databases (that contain data) for periodic maintenance, you should include a way to notify users to disconnect from the shared Jet database or provide a way to signal applications are executed during nonpeak hours, and applications automatically disconnect from the Jet database after a length of idle time, maintenance programs can execute without disturbing uncompleted result sets or pending updates.

Using DAO to Share Remote Data: In any client/server application one of your primary design concerns will be how to best share the data resources. When using either the Jet or remote engine query processor, your design must include code that deals with conflicts caused by instances of your database application and other applications trying to access the same database. Any design must include robust error handling to deal with a variety of contingencies caused by conflicts that arise as multiple applications vise for the same server resources. If your design calls for a centrally shared Jet database that includes attached tables, each client system must also contain DSNs that are compatible with your database. Since the DSN is maintained on the client and only referenced by name in the shared database attachments, your setup routines must ensure the client DSN description is correct — and remains so — to eliminate any chance that the user might change one or more parameters. Although the remote database engine is responsible for managing its own page locks and resources, any application, including those that use the Jet query processor, can lock pages on the server for indefinite periods of time.

Q.10. What are various interfacing elements of visual basic ?

Ans. The Visual Basic development environment contains these programming tools and windows, with which you build your Visual Basic programs:

* Form Windows, Toolbars, Menu bar, Visual Basic toolbox, Form window, Properties windows, Project Explorer, Immediate window, Form Layout window.

The Project-Form I (Form) window contains a form named Form 1, which is where the program's GUI will be displayed. A GUI is the visual portion of the program.

Form Window:

When you start Visual Basic, a default form (Form 1) with a standard grid (a window consisting of regularly spaced dots) appears in a pane called the Form window. You can use the Form window grid to create the user interface and to line up interface elements.

Building Interface Elements: To build the interface elements, you click an interface control in the Visual Basic toolbox, and then you draw the user interface element on your forms by using the mouse. This process is usually a simple matter of clicking to position one corner of the element and then dragging to create a rectangle the size you want. After you create the element — say, a text box — you can refine it by setting properties for the element. In a text box, for example, you can set properties to make the text boldface, italic, or underlined.

Adjusting Form Size: You can adjust the size of the form by using the mouse — the form can take up part or the entire screen.

Controlling Form Placement: To control the placement of the form when you run the program, adjust the placement of the form in the Form Layout window.

Toolbox Controls:

Pointer:

Used to interact, with be controls on the form (i.e., resize them move them, etc.). The pointer is not a control.

Picture Box	A control that displays images.
Label	A control that display > uneditable text to the user.
TextBox	A control for accepting user input. TextBoxes can also display text.
Frame	A control for grouping other controls.
CommandButton	A control that represents a button. The user presses or clicks to initiate an action.
CheckBox	A control that provides the user with a toggle choice (checked or unchecked).
OptionButton	A "radio button" optionButtons are used in groups where only one at a time can be True (i.e., on) —like the buttonsd on a car radio. A control that provides a list of items.
ListBox	A control that provides a short list of items.
ComboBox	A control that provides a short list of items.
HscrollBar	A horizontal scrollbar.
VScrollBar	A vertical scrollbar.
Timer	A control that performs a task of programmer-specified interval!>..A timer is not visible to the user.
DriveListBox	A control for accessing the system rusk derived (C, A, etc.)
DirListBox	A control for accessing directories on a system.
FileListBox	A control for accessing files in a directory.
Shape	A control for drawing cycles, rectangles, squares or ellipses.
Line	A control for drawing lines.

Image	A control for displaying images. The image control does not provide as many capabilities as a PictureBox.
Data	A control for connecting to a database.
OLE	A control for interacting with other window applications.

Project Window: A Visual Basic program consists of several files that are linked together to make the program run. The Visual Basic 6.0 development environment includes a Project window to help you switch back and forth between these components as you work on a project.

Project Window Components: The Project window lists all the files used in the programming process and provides access to them with two special buttons: View Code and View Object.

Displaying the Project Window: To display the Project window, click the Project Explorer button on the Visual Basic Toolbar. If the window is currently checked, you can enlarge it by double-clicking the title bar. To re-dock the Project window, double-click its title bar again.

Adding and Removing Files: The project file maintains a list of all the supporting files in a Visual Basic programming project. You can recognize project files by their .vbp file name extension. You can add individual files to and remove them from a project by using commands on the Project menu. The changes that you make will be reflected in the Project window.

Adding Projects: If you load additional projects into Visual Basic with the File menu Add Project command, outlining symbols appear in the Project window to help you organize and switch between projects.

Properties Windows: With the Properties window, you change the characteristics (property settings) of the user interface elements on a form. A property setting is a characteristics of a user interface object. For example, you can change the text docked, you can enlarge it by double-clicking the title bar. To ruddock the Properties window, double-click its title bar again.

Properties Windows Elements:

- The Properties window contains the following elements:
- A drop-down list box at the top of the window, from which you select the object whose properties you want to view or set.
- Two tabs, which list the properties either alphabetically or by category.
- A description pane that shows the name of the selected property and a short description of it.

Changing Property Settings: You can change property settings by using the Properties window while you design the user interface or by using program code to make changes while the program runs.

Form Layout Window: The Form Layout window is a visual design tool. With it, you can control the placement of the forms in the Windows environment when they are executed. When you have more than one form in your program, the Form Layout window is especially useful — you can arrange the forms on screen exactly the way you want. To position a form in the Form Layout window, simply drag the miniature form to the desired location in the window.

Menu Bar: The following figure shows a Menu summary for VB menu bar:

File	Contains options for opening projects, closing projects, printing projects, etc.
Edit	Contains options such as cut, l paste, find, undo, delete etc.
View	Contains options for displaying IDE windows and tool bar.
Project	Contains options for adding features such as forms to the project.
Format	Contains options for aligning and locking a form's controls.
Debug	Contains options for debugging.
Run	Contains options for executing a program, stopping a program etc.
Query	Contains options for manipulating data retrieved from a database.
Diagram	Contains options for editing and viewing the design of databases.
Tools	Contains options for IDE tools and options for customizing the environment.
Add-Ins	Contains options for using, installing and removing add-ms. Add-ins are typically independent software vendor (ISV) products that extend Visual Basic's features.

Windows	Contains options for arranging and displaying windows.
Help	Contains options for getting help.

Q.11. How even driven programming is supported by VC++ ?

Ans. Event Driven programming in VC++:

- User of the program generates events.
- The Win Main dispatches events to the event handlers.
- The programmer writes event handlers.
- The order of program execution is determined by the sequence of relevant events.

Window Event:

- **WM_PAINT**: draw/redraw window
- **WM_CHAR**: character typed
- **WM_LBUTTONDOWN**: Left button pressed
- **WM_MOUSEMOVE**: mousemoved
- **WM_QUIT**: about to terminate
- **WM_SIZE**: a window resized.

Example:

```
#include <window.h>
#include."winevent.hpp"
#include <time.h>
#include <stdio.h>
int nInt = 0; int nStart = 0; int nEnd = 0; char buf [255];
/////////////////////////////////////////////    //Event handlers
template <> LRESULT OnMsg <WM_CREATE > (HWNDhWnd, WPARAM wParam, LPARAM
IParam)
{strcpy (but, "Press any key to start test");
return DefWindowProc (hWnd, WM_CREATE, wParam, IParam); }
template <> LRESULT OnMsg<WM_KEYDOWN>(HWNDhWnd, WPARAM wParam, LPARAM
IParam)
{ nInt = 0; strcpy (buf, "running test, please wait ....");
InvalidateRect (hWnd, NULL,true); SendMessage (hWnd, WM_PAINT, 0, 0);
nStart = GetTickCount( ); PostMessage (hWnd, WM_TIMER, 0, 0);
return DefWindowProc (hWnd, WM_KEYDOWN, wParam, IParam); }
template <> LRESULT OnMsg<WM_TIMER>(HWNDhWnd, WPARAM wParam, LPARAM
IParam)
{ if(nInt++<100000) {PostMessage (hWnd, WM_TIMER, 0, 0);
} else (nEnd = GetTickCount( ); sprintf(buf, "Time elapsed = %d msec",
((nEnd-nStart) * 1000)/CLOCKS_PER_SEC); InvalidateRect (hWnd, NULL,
true); } return 0; }
template <> LRESULT OnMsg<WM_DESTROY>(HWNDhWnd, WPARAM wParam, LPARAM
IParam)
{PostQuitMessage (0); return 0;}
template <> LRESULT OnMsg<WM_PAINT>(HWNDhWnd, WPARAM wParam, LPARAM
IParam)
{ PAINTSTRUCT ps; HDC hdc; hdc = BeginPaint (hWnd, &ps); TextOut (hdc,
100, 100, buf, static_cast<int> (strlen (buf))); EndPoint (hWnd, & ps);
return DefWindowProc (hWnd, WM_PAINT, wParam, IParam); }
/////////////////////////////  // Entry point for the application.
int WINAPI WinMain (HINSTANCE hInstance,
```

```
HINSTANCE hPrevInstance, LPSTR szCmdLine, int nCmdShow)
{InitMsg Handlers <NULL> ( );
static char szAppName [ ] = "demo"; HWND hwnd; MSG msg; WNDCLASSEX
wndclass;
wndclass.cbSize = size of (wndclass); windclass.style = CS_HREDRAW|CS_VREDRAW;
wndclass.Ipfin WndProc = WndProc; wndclass, cbClsExtra = 0;
wndclass.cbWndExtra = 0; wndclass.hInstance = hInstance; wndclass.hIcon = LoadIcon
(NULL, IDI_APPLICATION); wndclass.hIconSm = LoadIcon (NULL, IDI_APPLICATION);
wndclasshCursor = LoadCursor (NULL, IDC_ARROW);
wndclass.hbr Background = (HBRUSH) GetStock Object (WHITE_BRUSH);
wndclass.IpszClassname = szAppName;wndclass.IpszMenuName = NULL;
RegisterClassEx (&wndclass); hwnd = CreateWindow (szAppName, "Static DispatchDemo",
WS-OVERLAPPEDWINDOW, CW_USEDFAULT, CW_USEDFAULT
CW_USEDFAULT, CW_USEDFAULT, NULL, NULL, hInstance, NULL);
ShowWindow (hwnd, nCmdShow); UpdateWindow (hwnd);
while (GetMessage (&msg, NULL, 0, 0) ) {Translate Message (&msg); DispatchMessage
(&msg); }
return static_cast<int> (msg.wParam); }
```

Q.12. How will you use the document object to store data?

Ans. In the document is code file, e.g., welcome Doc.cpp, object can be stored using serialize ( ) method:

```
void (welcome Doc:: Serialize (CArchieve& or)
{
    if((ar is String( ))
    {
        //Add storing code here
    }
    else
    {
    // Add loading code here
    }
}
```

Assuming our program has quite a few menu items like save, save As, Open and so on. The program itself handles almost all details of saving our document to disk.

Example: Defining a member variable welcome-string in document class e.g.,

class (Welcome Doc: public CDocument{
protect:
CString welcome_string;
;
}

Initializing the welcomestring object in document class's constructor in the file welcome document.

cpp:

```
Welcome Doc:: C Welcome Doc ()
{
    Welcome_string = "Welcome to V C++";
}
```

Now our data is stored in the document object. Then, we have

```
void (Welcome Doc:: Serialize (CArchieve & or)
{
    lf(ar.Isstring()))
{
    ar « welcome_string;
}
else
{
    ar >> welcome_string;
}
}
```

Now the user can store and retrieve our document to and from disk.

EXERCISES

1. How can you tell that Shift or Ctrl keys are being held down when you receive the WM_KEYDOWN message?
2. Write an application to print the number of times a key has been read while a key was kept pressed for sometime.
3. Create an application that reports any keyboard event taking place while the application is running and active. It should report with appropriate output messages as to which event occurred with which key.
4. List the possible mouse messages that you can add handler functions for?
5. Make a program that tells that the right mouse button is down on the WM_MOUSEMOVE event message.
6. Make amendments to your drawing program so that the left mouse button can draw in blue and the right mouse button can draw in yellow.
7. Extend the OnKeyDown function to add some of the following standard mouse cursors:
   IDC_HELP
   IDC_SIZEALL
   IDC_SIZEWE
   IDC_UPARROW
   IDC_NO
   IDC_CROSS
   IDC_APPSTARTING
   IDC_SIZENWSE
   IDC_SIZENESW
   IDC_SIZENS

❑❑❑

4

Working With Keyboard And Mouse

4.0 INTRODUCTION

Chapter Outline

- Introduction
- Reading keystrokes from the keyboard
- Handling Mouse in VC++
- Creating Menus
- Toolbars Buttons

As you learned, when you are working with most controls, you are limited to a select number of events that are available in the Class Wizard. When it comes to mouse events, you are limited for the most part to click and double-click events. Just looking at your mouse tells you that there must be more to capturing mouse events than recognising these two. What about the right mouse button? How can you tell if it has been pressed? And what about drawing programs? How can they follow where you drag the mouse?

The keyboard obviously has fewer messages than the mouse does. Then again, there are only so many things that you can do with the keyboard. These event messages are available on the dialog window object and are triggered only if there are no enabled controls on the window. Any enabled controls on the window have input focus, so all keyboard events go to them. That's why you remove all controls from the main dialog for your drawing application.

If you open the Class Wizard in one of your projects, select the dialog in the list of object IDs, and then scroll through the list of messages that are available, you will find a number of mouse-related events. These event messages enable you to perform any task that might be required by your application.

4.1 READING KEYSTROKES FROM THE KEYBOARD

(MDU 2007)

The keyboard is an input device that is shared by all applications running in memory. Pressing a key (or combination of keys) on the keyboard constitutes a keyboard event. Each such key event generates a corresponding message, which is received by the Windows system. Out of the several windows that may be active,

Windows directs keyboard message to the window that has got the input focus. The window that has the input focus is the main window of the application. The input focus might also belong to a child of the main window, or to a control in a dialog box. Windows sends keyboard messages to the window that owns the focus.

These messages are sent in the order of their occurrence to the active window application message queue. The concerned application retrieves messages from this queue, one message at a time. If the application has the appropriate message handler (*i.e.*, a function doing something in response to this event) for the message read from the queue, the message handler is executed; otherwise the message is passed to the default message handler of the operating system (e.g. Windows). Table 4.1 lists various keyboard events, the messages generated and respective event handlers.

Table 4.1: Keyboard Events.

Message	***Handler***	***Events***
WM_KEYDOWN	OnLKeyDown	A key has been pressed
WM_KEYUP	OnKeyUp	A key has been released
WM_SYSKEYDOWN	OnSysKeyDown	A special or function key has been pressed
WM_SYSKEYUP	OnSysKeyUp	A special or function key has been released

While the above mentioned messages indicate what event has occurred, it does not tell which key has generated that event. A flag that gets set with appropriate value when the event occurs indicates which key has given rise to the event. Table 4.2 lists various flags that get affected on keyboard events.

Table 4.2: Keyboard Event flags

Bit in nFlags	***Meaning***
0-15	Contains the number of times the key has been pressed as a result of the key being held down
16-23	Manufacturer-dependent key-code (scan code)
24	Set to 1 if the key pressed is either a function key or an extended key, 0 otherwise
25-28	Used by Windows
29	Set to 1 if the Alt key was also pressed along, 0 otherwise
30	Set to 1 if the key was pressed before the message was sent, 0 otherwise
31	Set to 1 if the key is being released, 0 if the key is being pressed

How to use a keyboard event in your application

Which of these keyboard messages are to be processed by the application, depends on the functionality of the application. If the application has to respond to a particular keyboard event then it must implement the message handler for it. On the other hand, if the application must ignore a particular keyboard event then it need not implement the corresponding message handler and let the default message handler take care of that event.

For instance, if your application must respond to a keypressed event then you must write the required code for the message handler OnKeyDown.

The general prototype of the keyboard handler functions looks like :

```
Void OnMsgName(UNIT nChar, UNIT nRepCnt, UNIT nFlags)
```

OnMesgName is the name of the individual message handler. nChar contains the ASCII code of the character pressed on the keyboard. nRepCnt parameter contains number of times the key has been repeated as a result of the key being pressed and the nFlags parameter has a bit code detailed on table 4.2.

Example 1

Echoing the Pressed Key on the Keyboard

You are now going to build a simple window program that uses some of the available keyboard event to let the user press a key on the keyboard which the program echoes back on its client area. This application uses WM_KEYDOWN event message, which signals that a key has been pressed. You will look at how you can tell within this event function which key has been pressed. Sounds fairly straight ahead, so let's get going by following these steps :

- Create a new MFC AppWizard project, calling the project KeyTest.
- Specify that this project will be a dialog based application in the first AppWizard step.
- Remove all the controls from the dialog window. Use the default settings in the AppWizard. Specify a suitable dialog title, such as Testing Keyboard event.
- After the application shell is created, remove all controls from the dialog window surface.
  - Open the Class Wizard from view menu Select WM_KEYDOWN from the list of messages, and add a function by clicking the Add Function button. Click the OK button to accept the suggested function name.
  - Click the Edit Code button to edit the OnKeyDown function you just created, adding the code given below :

```
CClientDC dc(this);  //Get the device context for window output of
                       the current application.
dc.TextOut(10,10,nChar);    //Put the output text on the device
                              context.
```

Note :

1. Windows always runs in graphics mode implying that even a character is handled in terms of pixels. Unlike character-based systems, Windows programs do not send data directly to an output device(screen or the printer). Instead, a Windows program obtains a handle for the screen or printer called device context. The program sends it output to this acquired device context. It is Windows responsibility to perform the required tasks to translate the output to a device context and map it to the hardware, producing the desired result.
2. The device context object has, among others, a member function TextOut(). It is called with three arguments - first to specify the line and column numbers respectively, whereas the third argument is the value to be printed. Also, the line and column coordinate is relative to the program window and not to the screen.

3. Since you will be using this mode of output extensively through out the book, let us have a closer look at CDC object and its member funtion -Text Out().

```
CDC::Text Out
virtual BOOL TextOut(int x, int y, LPCTSTR lpszString, int nCount );
BOOL TextOut( int x, int y, const CString& str );
Return Value
Non zero if the function is successful; otherwise 0.
Parameters
X : Specifies the logical x-coordinate of the starting point of the text.
Y : Specifies the logical y-coordinate of the starting point of the text.
LpszString : Points to the character string to be drawn.
NCount : Specifies the number of bytes in the string.
Str : A CString object that contains the characters to be drawn.
```

Remarks

Writes a character string at the specified location using the currently selected font.

Character origins are at the upper-left corner of the character cell. By default, the current position is not used or updated by the function.

If an application needs to update the current position when it calls TextOut, the application can call the SetTextAlign member function with nFlags set to TA_UPDATECP. When this flag is set, Windows ignores the *x* and *y* parameters on subsequent calls to Text Out, using the current position instead.

- Compile and run the application.
- Press a key and it shall be enclosed back on the client area of the application.

Example 2

How to change The Drawing Cursor (Through The Keyboard)?

Since you are learning about the keyboard messages, it would be a great idea to use certain keyboard keys to change the mouse cursor in your program. The example (Changing_Mouse_Curso) makes the use of the 'A' key press to change the cursor shape as the combination of the arrow cursor and the hour glass cursor (IDC_APPSTARTING), makes the 'B' key to change the cursor to the I_beam (IDC_IBEAM), makes the 'C' key to change the cursor shape as cross (IDC_CROSS). Also, the program uses the "X' key to exit the application. To create this application, do the following :

Create a new MFC AppWizard project, calling the project Changing_Mouse_Curso.

- Specify that this project will be a dialog based application in the first AppWizard step.
- Remove all the controls from the dialog window. Use the default settings in the AppWizard. Specify a suitable dialog title, such as Testing Keyboard Mouse event.
- After the application shell is created, remove all controls from the dialog window surface.
- Open the Class Wizard from view menu Select WM_KEYDOWN from the list of messages. Fig4.1, and add a function by clicking the Add Function button. Click the OK button to accept the suggested function name.

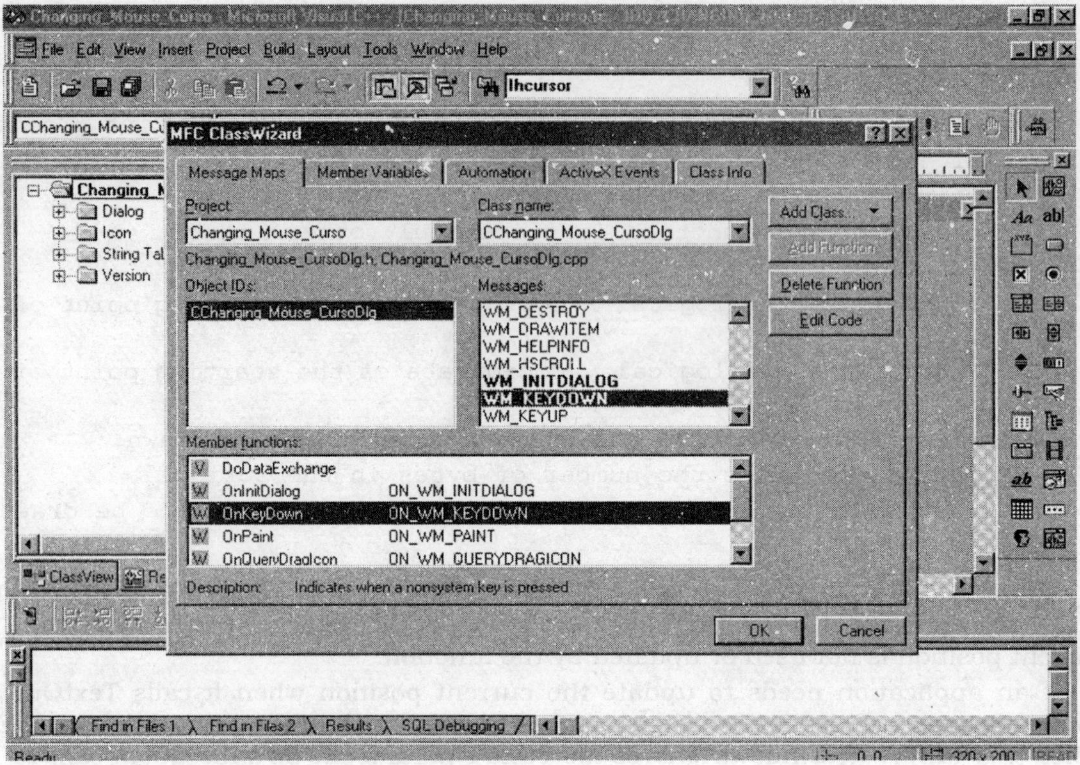

Fig. 4.1. Selecting message WM_KEYDOWN from the ClassWizard

- Click the Edit Code button to edit the OnKeyDown function you just created, adding the code given below :

```
char IsChar;  //create a char type variable
HCURSOR IhCursor;  //create a cursor handle
IsChar = char(nChar); //convert the ASCII into character and store
it into IsChar
if(IsChar == 'A')
{
  IhCursor = AfxGetApp()->LoadStandardCursor(IDC_APPSTARTING);
  //load the new cursor. IDC_APPSTARTING is identifier for a cursor

  SetCursor(IhCursor);
  //change the cursor to it

}
if(IsChar =='B')
{
  IhCursor = AfxGetApp()->LoadStandardCursor(IDC_IBEAM);
  SetCursor(IhCursor);
}
if(IsChar =='C')
```

```
    {
        IhCursor = AfxGetApp()->LoadStandardCursor(IDC_CROSS);
        SetCursor(IhCursor);
    }
    if(IsChar =='X')
    {
        DestroyWindow();
        //close the window and terminate the application
    }
```

- Compile and run the application.
- Press the above mentioned keys and notice the change in the mouse pointers.

4.2 HANDLING MOUSE IN VC++ **(MDU 2008)**

A mouse is the most common pointing input device extensively used with windows applications. Each activity performed with the mouse (like moving, clicking, dragging etc.) is termed as a mouse event. Corresponding to each event a message is generated. For instance, if the mouse is moved, a corresponding event (mouse move event) occurs in the system and consequently a message (WM_MOUSEMOVE message) is produced.

As usual, the window getting the message from the Windows may implement a handler to that message or simply let it pass back to the Windows default handler. Table 4.3 summarises the various mouse events, messages and their handlers.

There are two categories of mouse events - Client area mouse events and Non-Client area mouse events. Client area mouse events occur when the mouse pointer is focussed in the application's window. Mouse events produced when the mouse is focussed outside the application window area, are termed as Non-Client area mouse events. Table 4.4 summarizses different messages produced and corresponding event handlers for non-client area mouse activities.

Table 4.3. Client Area Mouse Events

Message	***Handler***	***Dispatched When***
WM_LBUTTONDOWN	OnLButtonDown	The left mouse button is pressed
WM_LBUTTONUP	OnLButtonUp	The left mouse button is released
WM_LBUTTONDBLCLK	OnLButtonDblClk	The left mouse button is double-clicked
WM_MBUTTONDOWN	OnMButtondown	The middle mouse button is pressed
WM_MBUTTONUP	OnMButtonUp	The middle mouse button is released
WM_MBUTTONDBLCLK	OnMButtonDblClk	The middle mouse button is double-clicked
WM_RBUTTONDOWN	OnRButtonDown	The right mouse button is pressed
WM_RBUTTONUP	OnRButtonUp	The right mouse button is released
WM_RBUTTONDBLCLK	OnRButtonDblClk	The right mouse button is double-clicked
WM_MOUSEMOVE	OnMouseMove	The mouse is moved over the client area of the window.

Table 4.4: Non-client Area Mouse Event

Message	*Handler*	*Dispatched When*
WM_NCLBUTTONDOWN	OnLButtonDown	The left mouse button is pressed
WM_NCLBUTTONUP	OnLButtonUp	The left mouse button is released
WM_NCLBUTTONDBLCLK	OnLButtonDblClk	The left mouse button is double-clicked
WM_NCMBUTTONDOWN	OnMButtondown	The middle mouse button is pressed
WM_NCMBUTTONUP	OnMButtonUp	The middle mouse button is released
WM_NCMBUTTONDBLCLK	OnMButtonDblClk	The middle mouse button is double-clicked
WM_NCRBUTTONDOWN	OnRButtonDown	The right mouse button is pressed
WM_NCRBUTTONUP	OnRButtonUp	The right mouse button is released
WM_NCRBUTTONDBLCLK	OnRButtonDblClk	The right mouse button is double-clicked
WM_NCMOUSEMOVE	OnMouseMove	The mouse is moved over the non-client area of the window.

4.2.1 How to Use a Mouse Event in Your Application (V.Imp)

Which of these mouse messages are to be processed by the application, depends on the functionality of the application. If the application has to respond to a particular mouse event then it must implement the message handler for it. On the other hand, if the application must ignore a particular mouse event then it need not implement the corresponding message handler and let the default message handler take care of that event.

For instance, if your application must respond to left button click of the mouse then you must write the required code for the message handler OnLButtonDown.

The general prototype of the keyboard handler functions looks like :

```
Void OnMsgName(UNIT nFlags, Cpoint point)
```

Each one of these mouse handler functions receive two arguments from the application message queue - UINT nFlags and CPoint point. nFlags contains bit-pattern identifying, which mouse event has occurred and CPoint contains the co ordinate (CPoint.x and Cpoint.y) of the mouse pointer when the even occurred.

These values are available for use inside the event handler.

Example 3

Drawing With the Mouse

You are now going to build a simple drawing program that uses some of the available mouse events to let the user draw simple figures on a dialog window. This application depends mostly on the WM_MOUSEMOVE event message, which signals that the mouse is being moved. You will look at how you can tell within this event function whether the left mouse button is down or up. You will also learn how you can tell where the mouse is on the window at the time of event. Sounds fairly straight ahead, so let's get going by following these steps :

- Create a new MFC Application using AppWizard workspace project, calling the project MouseTest.
- Specify that this project will be a dialog based application in the first AppWizard step.

- Remove the control elements (buttons, text etc.) from the dialog window. Use the default settings in the AppWizard. Specify a suitable dialog title, such as Testing mouse event.
- After the application shell is created, remove all controls from the dialog window surface for drawing.
- Open the Class Wizard. Select WM_MOUSEMOVE from the list of messages, and add a function by clicking the Add Function button. Click the OK button to accept the suggested function name.
- Click the Edit Code button to edit the On Mouse Move function you just created, adding the code in listing below :

```
if((nFlags & MK_LBUTTON) == MK_LBUTTON)
    {
        CClientDC dc(this);
        dc.SetPixel(point.x,point.y,RGB(0,0,0));
    }
```

Take a closer look at the if condition. NFlags, when ANDed with pre-defined constant MK_BUTTON, results in a bit-pattern that is equal to the value of MK_LBUTTON. Which button has been pressed can be determined using nFlags parameter.

- Compile and run the application.
- Focus the mouse on the application's area and drag while left button pressed.
- The output window should look like Fig 4.2.

Fig. 4.2. Output window of the mouse draw application

4.2.2 Improving the Drawing Program

If you ran your program, you probably noticed a small problem. To draw a solid line, you need to move the mouse very slowly. How do other painting programs solve this problem?

Simple, they draw a line between two points drawn by the mouse. Although this seems a little like cheating, it's the way that computer drawing programs work. As you move the mouse across the screen, your computer is checking the location of the mouse every few clock ticks. Because your computer doesn't have a constant trail of where your mouse has gone, it has to make some assumptions. The way your computer makes these assumptions is by taking the points that the computer does know about and drawing lines between them. When you draw lines with the freehand tool in Paint, your computer is playing connect the dots.

Because all the major drawing programs draw lines between each pair of points, what do you need to do to adapt your application so that it also uses this technique? First, you need to keep track of the previous position of the mouse. This means you need to add two variables to the dialog window to maintain the previous X and Y coordinates. You can do this by following these steps :

1. In the workspace pane, select the Class View tab.
2. Select the dialog class-in this case, the CMouseDlg class.
3. Right-click the mouse and select Add Member Variable from the pop-up menu.
4. Enter int as the Variable Type and m_iPrevY as the Variable Name and specify Private for the access in the Add Member Variable dialog, as shown in Figure 4.3.

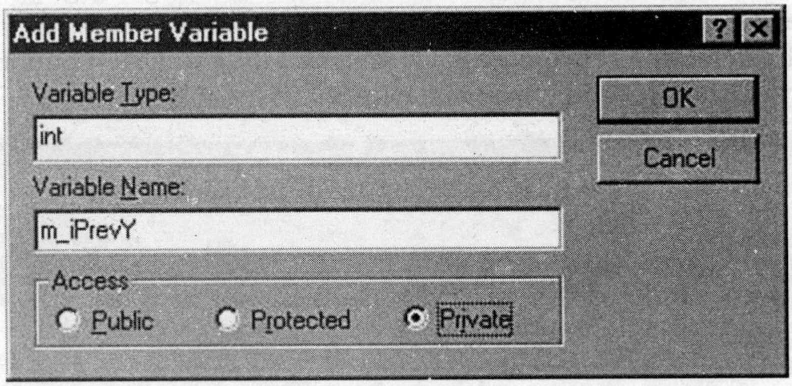

Fig. 4.3. The Add Member Variable dialog

5. Click OK to add the variable.
6. Repeat steps 3 through 5, specifying the Variable Name as m_iPrevX to add the second variable.

After you add the variables needed to keep track of the previous mouse position, you can make the necessary modifications to the OnMouseMove function, as shown in Listing 4.2.

Listing 4.2. The Revised On mouse move Function

```
void CMouseDlg::OnMouseMove(UINT nFlags, CPoint point)
{
// TODO: Add your message handler code here and/or call default

////////////////////////
// MY CODE STARTS HERE
////////////////////////

```

```
// Check to see if the left mouse button is down
if ((nFlags & MK_LBUTTON) == MK_LBUTTON)
{
// Get the Device Context
CClientDC dc(this);

// Draw a line from the previous point to the current point
dc.MoveTo(m_iPrevX, m_iPrevY);
Integrating the Mouse and Keyboard in Your Application
dc.LineTo(point.x, point.y);

// Save the current point as the previous point
m_iPrevX = point.x;
m_iPrevY = point.y;
}

/////////////////////////
// MY CODE ENDS HERE
/////////////////////////

CDialog::OnMouseMove(nFlags, point);
}
```

Look at the code that draws the line from the previous point to the current point:

```
dc.MoveTo(m_iPrevX, m_iPrevY);
dc.LineTo(point.x, point.y);
```

You see that you need to move to the first position and then draw a line to the second point. The first step is important because without it, there is no telling where Windows might think the starting position is. If you compile and run your application, it draws a bit better. However, it now has a peculiar behaviour. Every time you press the left mouse button to begin drawing some more, your application draws a line from where you ended the last line you drew, as shown in Figure 4.4.

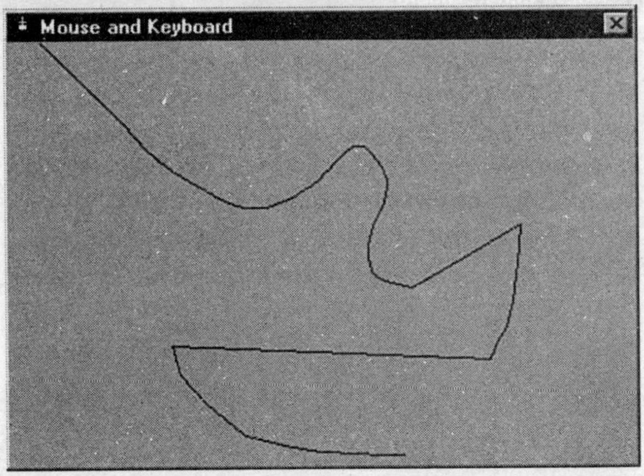

Fig. 4.4. The drawing program with a peculiar behaviour.

4.2.3 Adding the Finishing Touches

Your application is doing all its drawing on the mouse move event when the left button is held down. Initialising the previous position variables with the position of the mouse when the left button is pressed should correct this application behaviour. Let's try this approach by following these steps :

1. Using the Class Wizard, add a function for the WM_LBUTTONDOWN message on the dialog object.
2. Edit the OnLButtonDown function that you just created, adding the code in Listing 4.3.

Listing 4.3. The Onlbutton down Function

```
void CMouseDlg::OnLButtonDown(UINT nFlags, CPoint point)
{
// TODO: Add your message handler code here and/or call default

///////////////////////
// MY CODE STARTS HERE
///////////////////////

// Set the current point as the starting point
m_iPrevX = point.x;
m_iPrevY = point.y;

///////////////////////
// MY CODE ENDS HERE
///////////////////////

CDialog::OnLButtonDown(nFlags, point);
}
```

When you compile and run your application, you should find that you can draw much like you would expect with a drawing program, as shown in Figure 4.4.

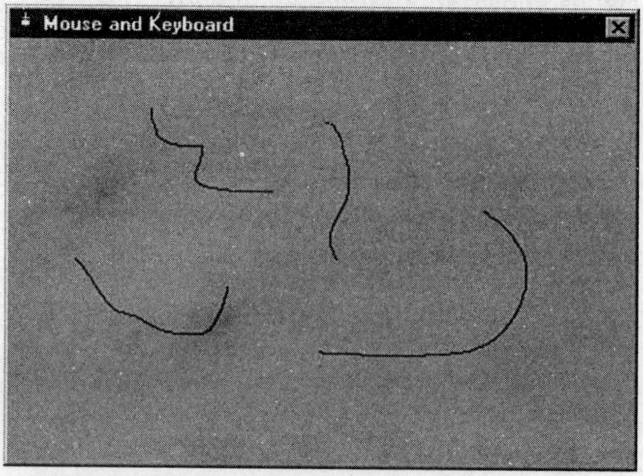

Fig. 4.5. The finished drawing program.

4.3. CREATING MENUS

Fig. 4.6.

Back when the first computer terminals were introduced and users began using computer software, even on large mainframe systems software developers found the need to provide the users with some sort of menu of the functions that the computer could perform. These early menus were crude by today's standards and were difficult to use and navigate. Menus have improved since then; they've become standardised in how they are used and are far easier to learn.

The software designers that first came up with the idea of a Graphical User Interface (GUI) planned to make computer systems and applications easier to interact with by making everything behave in a consistent manner. Menus used for selecting application functionality were one part of the GUI design. As a result, a number of standard menu styles were developed.

The menu items may be activated either by clicking left mouse button on them or by pressing underlined key on the keyboard in combination with Alt key. Thus, by pressing H with Alt key will activate Help menu.

Basically, there are three types of menus in a Windows application. System menu that is tucked below the application's caption bar, Pop-up menu that is generally invisible but can be popped-up anywhere in the application client area by right-clicking the mouse and Drop-down menu that appears as a sub-menu for the menu items in the menu bar at the top of the window.

Though it is not absolutely necessary to have a menu in an application, if present, it makes the application easier to use. Since menu is a common feature of almost every Windows application, a menu-designing tool (resource editor) is incorporated in development environment. This tool lets programmers create a menu of their choice and attach it to the application with minimum effort.

4.3.1 Creating a Menu with Accelerators (V.Imp)

One of the original keyboard-shortcuts for selecting menu entries was accelerator key. An accelerator key is specific key combination, usually the ctrl key, combined with another key, or function keys, that are unique within the entire application. Each of these key combinations triggers one menu event function.

The way that accelerator keys work is similar to the way menus work. They are also an application resource that is defined in a table in the resource tab of the workspace pane. Each table entry has an object ID and a key code combination. After you define the accelerators, you can attach functionality to the object ID's. You can also assign accelerators entries the same object ID as the corresponding entry so that you have to define only a single entry in the application map.

After you define all accelerator keys, you can add the key combination to the menu entry so that the user will know about the accelerator key combination. Add to the end of the menu entry caption, followed by the key combination. The is replaced in the menu display by a tab, which separates the menu caption from the accelerator key combination.

Unfortunately, accelerator keys don't work in dialog-style windows, so you cannot add them to this application. You will learn how to attach accelerator keys to the menus later in single and multi-document interface style application section.

Example 5

Let us build a pop-up menu displaying File menu group when right button of mouse is clicked in the client area of the application. To create this application, do the following :

- Create a SDI application with default settings using AppWizard.
- Using class wizard, similar to the previous example, insert an event handler for the message On context menu in MainFrame class.
- Insert the following code into this handler function :

```
CMenu *m_lMenu;         //create a pointer of CMenu type to store the
                        //current menu

CPoint m_pPoint;        //create a CPoint type a variable to store the
                        //current mouse position

m_pPoint = point; //store the point

ClientToScreen(&m_pPoint);       //convert the co ordinate of the point
                                 //from client to screen

m_lMenu=GetMenu(); //store the current menu pointer

m_lMenu = m_lMenu->GetSubMenu(0); //store the first sub-menu

m_lMenu->TrackPopupMenu(TPM_CENTERALIGN+TPM_LEFTBUTTON,
m_pPoint.x,m_pPoint.y, this, NULL); //display the pop-up menu //
aligned center to the left button at the point where the right //
button was clicked.
```

- Compile and run.
- Right-click the mouse at the client area of your application. You will see the pop-up menu appear as in Fig. 4.6.

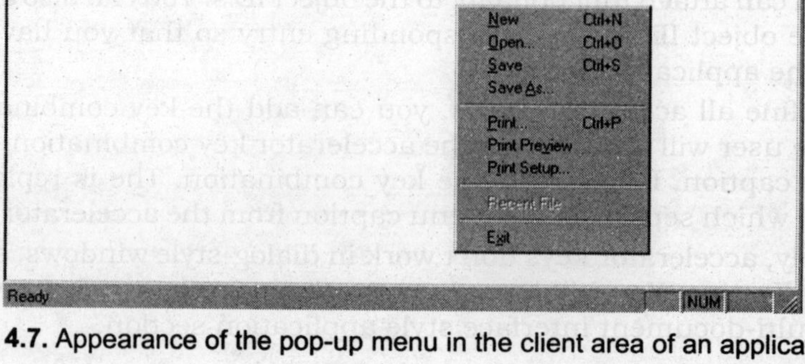

Fig. 4.7. Appearance of the pop-up menu in the client area of an application.

4.3.2 Main Menu

A menu is a list of actions the user can perform on an application. The actions are presented in one or more groups. There are broad categories of menus: the main menu and the context menu.

A main menu, also called a top-level menu, displays categories of menu items using a range of items called a menu bar :

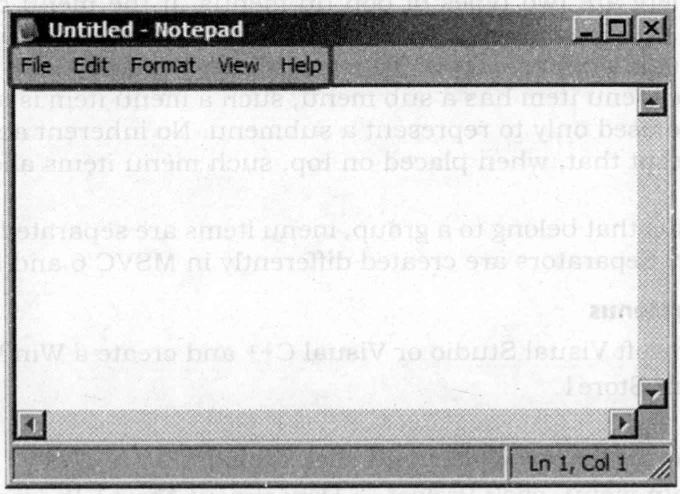

Fig. 4.8.

When the user clicks an item of the menu bar, the item clicked opens its list :

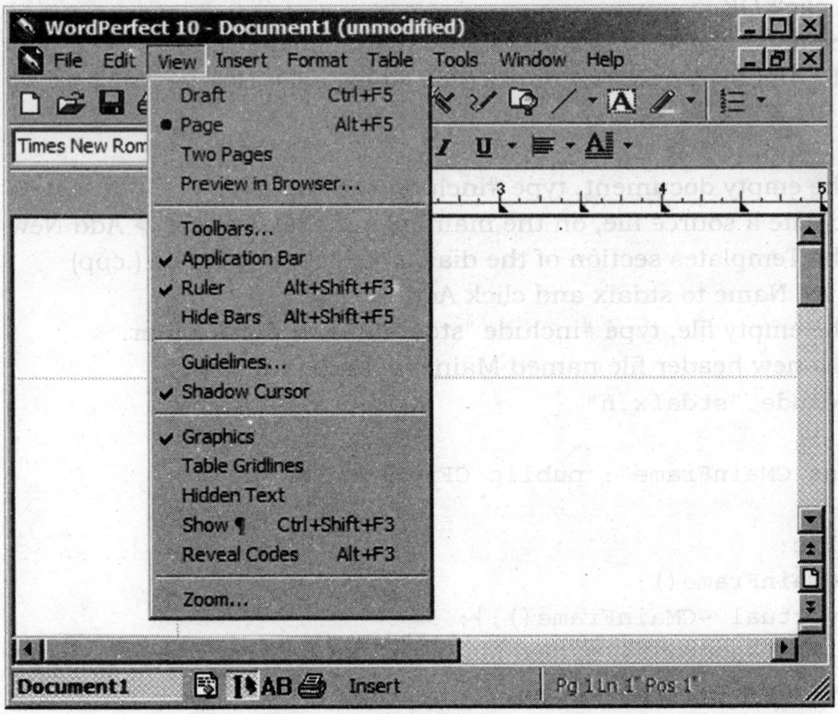

Fig. 4.9.

After a list has been displayed, the user can then use an item from the list. Each item of the list is primarily a word or a group of words on a line. Different menu items are used for different reasons. For example, some menu items simply display a word or a group of words. Some other items display a check mark. This indicates that the item toggles the availability or disappearance of an object.

When a menu item is only meant to lead to a sub-menu, such a menu item is called a pop up menu. There are two types of pop up menus. If the menu displays on top of a window, which is the type of menu under the title bar, the word on top, which represents a category of menu, is a pop up menu. If a menu item is equipped with an arrow in its right , which means the menu item has a sub menu, such a menu item is also a pop up menu. Pop up menus are used only to represent a submenu. No inherent action is produced by clicking them, except that, when placed on top, such menu items allow opening the sub menu.

To create menus that belong to a group, menu items are separated by a horizontal line called a separator. Separators are created differently in MSVC 6 and MSVC 7.

4.3.3. Introducing Menus

1. Start Microsoft Visual Studio or Visual C++ and create a Win32 Project named.
2. Department Store1.
3. Click OK.
4. Create it as an Empty Project and click Finish.
5. On the main menu, click Project -> Department Store1 Propertie.
6. Click Configuration Properties.
7. In the Use of MFC combo box of the right section, select Use MFC in a Shared DLL and click OK.
8. To create a header file, on the main menu, click Project -> Add New Item.
9. In the Templates section of the dialog box, click Header File (.h).
10. Set the Name to stdafx.
11. Click Add.
12. In the empty document, type #include <afxwin.h>.
13. To create a source file, on the main menu, click Project -> Add New Item.
14. In the Templates section of the dialog box, click C++ File (.cpp)
15. Set the Name to stdafx and click Add.
16. In the empty file, type #include "stdafx.h" and press Enter.
17. Add a new header file named MainFrm and, in it, type :

```
#include "stdafx.h"

class CMainFrame : public CFrameWnd
{
public:
    CMainFrame();
    virtual ~CMainFrame();};
```

18. Add a new source file named MainFrm and type the following in it :

```
#include "MainFrm.h"

CMainFrame::CMainFrame()
```

```
{
    Create(NULL,
    TEXT("Department Store"),
    WS_OVERLAPPEDWINDOW,
    CRect(200, 120, 640, 400));
}

CMainFrame::~CMainFrame()
{
}
```

19. Add a new header file named Exercise and, in it, type :

```
#include "stdafx.h"

class CExerciseApp: public CWinApp
{
public:
        BOOL InitInstance();
};
```

20. Add a new source file named Exercise and type the following in it :

```
#include"Exercise.h"
#include "MainFrm.h"

BOOL CExerciseApp::InitInstance()
{
    m_pMainWnd = new CMainFrame;
    m_pMainWnd->ShowWindow(SW_SHOW);
    m_pMainWnd->UpdateWindow();

    return TRUE;
}

CExerciseApp theApp;
```

21. Execute the application to test it.

4.3.4. Menu Creation

There are two main ways you can create a main menu: Using a resource file or programmatically creating the menu items. You can create a resource file that has the extension. The menu is created as text. If you create the file manually, you must also remember to create or edit the resource.h file in order to specify an identifier for each menu. The alternative, which we will use, is to "visually" create the menu in Visual Studio. When doing this, the studio itself would update the resource.h file as items are added or removed.

To visually create a menu, first add a resource of type Menu using the Add Resource dialog box. The MFC library implements a Windows menu, from the Win32's **HMENU** class, through the **CMenu** class. Based on this, if you want to programmatically create a menu, you can start by declaring a **CMenu** variable. Here is an example :

```
#include <afxwin.h>

class CExerciseApp : public CWinApp
{
public:
   virtual BOOL InitInstance();
};
class CMainFrame : public CFrameWnd
{
public:
   CMainFrame();

private:
   CMenu m_wndMainMenu;
};

CMainFrame::CMainFrame()
{
   Create(NULL, "Menus Fundamentals");
}

BOOL CExerciseApp::InitInstance()
{
   m_pMainWnd = new CMainFrame;
   m_pMainWnd->ShowWindow(SW_NORMAL);

   return TRUE;
}

CExerciseApptheApp;
```

After declaring the variable, to indicate that you intend to create a menu, you must first call the

CMenu::CreateMenu() method. Its syntax is :

```
BOOL CreateMenu();
```

Here is an example:

```
CMainFrame::CMainFrame()
{
   Create(NULL, "Menus Fundamentals");

   this->m_wndMainMenu.CreateMenu();
}
```

Just in case something might go wrong, the MFC provides the VERIFY macro that you can use to make sure that the menu was rightly initialized. It can be used as follows :

```
CMainFrame::CMainFrame()
{
    Create(NULL, "Menus Fundamentals");

    VERIFY(this->m_wndMainMenu.CreateMenu());
}
```

After declaring the variable and calling CreateMenu(), you can start defining the characteristics of the menu. We will review the various characteristics in the next sections.

4.3.5 Creating a Main Menu Visually (V.Imp)

1. On the main menu, click View -> Resource View and again, on the main menu, click
2. Project -> Add Resource...
3. On the Add Resource dialog box, double-click Menu.
4. In the Resource View, click IDR_MENU1 to select it.
5. In the Properties window, click ID, type IDR_FIRST and press Enter.

Using a Menu

After creating a menu, before using it, you should associate it with a window frame that will carry it and present it to the user. To associate a menu with a window, you have various options. If you are using the **CFrameWnd::Create()** method to create the window, you can specify the name of the menu using the *lpszMenuName* argument :

```
virtual BOOL Create(
    LPCTSTR lpszClassName,
    LPCTSTR lpszWindowName,
    DWORD dwStyle = WS_OVERLAPPEDWINDOW,
    const RECT& rect = rectDefault,
    CWnd* pParentWnd = NULL,
    LPCTSTR lpszMenuName = NULL,
    DWORD dwExStyle = 0,
    CCreateContext* pContext = NULL
);
```

Here is an example :

```
class CMainFrame : public CFrameWnd
{
public:
    CMainFrame()
    {
        Create(NULL, "Resources Fundamentals",
            WS_OVERLAPPEDWINDOW, CRect(200, 120, 640, 400),
            NULL,
            MAKEINTRESOURCE(IDR_MAINFRAME));
    }
};
```

If you are programmatically creating the menu using a CMenu object, to assign it to a frame, you can call the CWnd::SetMenu() method. Its syntax is :

```
BOOL SetMenu(CMenu* pMenu);
```

Here is an example :

```
CMainFrame::CMainFrame()
{
    Create(NULL, "Menus Fundamentals");

    VERIFY(this->m_wndMainMenu.CreateMenu());

    // Do whatever here
    this->SetMenu(&m_wndMainMenu);
}
```

If a window already has a menu and you want to get a reference to that menu, you can call its CWnd::GetMenu() method. Its syntax is :

```
CMenu *GetMenu() const;
```

When using this method, if the window that calls it doesn't have a menu, this method returns NULL. Otherwise, this method returns a pointer to the **CMenu** object of the window that called it. As stated already, The CMenu class is the MFC's implementation of the HMENU handle. To get a handle to an existing CMenu object of an application, you can access the value of the CMenu::m_hMenu property, which is of type **HMENU**.

4.3.6 Associating a Main Menu With a Window Frame

1. Access the MainFrm.h file and declare a private pointer to CMenu as follows :

```
#include "stdafx.h"
class CMainFrame : public CFrameWnd
{
public:
    CMainFrame();
    virtual ~CMainFrame();
private:
    CMenu *pCurrentMenu;
};
```

2. To associate the new menu with the frame, access the MainFrm.cpp file and change it as follows :

```
#include "MainFrm.h"

CMainFrame::CMainFrame()
{
    Create(NULL,
        TEXT("Department Store"),
        WS_OVERLAPPEDWINDOW,
        CRect(200, 120, 640, 400));

    // Dynamically create a menu...
    pCurrentMenu = new CMenu;
    // ... using the menu resource we will create
```

```
    pCurrentMenu->LoadMenuW(MAKEINTRESOURCE(IDR_FIRST));
    // Assign the menu menu to the frame of the current application
    SetMenu(pCurrentMenu);
    // Draw the menu on the frame (this is not required
    // if this is the only time we will create a menu)
    Draw Menu Bar();
}

CMainFrame::~CMainFrame()
{

}
```

3. Save the file

4.3.7 Menu Message Processing

As introduced earlier, a menu is primarily a list of actions that can be performed on an application. The actions are generated when the user positions the mouse on a menu item or when one of these items is clicked. If the user clicks an item such as File on the main menu, the menu item generates a WM_INITMENUPOPUP message. To get the program ready to process these messages, in the previous lesson, we saw that you should create a message map. Here is an example :

```
class CMainFrame : public CFrameWnd
{
public:
    CMainFrame()
    {
        Create(NULL, "Menus Fundamentals");
    }

    DECLARE_MESSAGE_MAP()
};

BEGIN_MESSAGE_MAP(CMainFrame, CFrameWnd)

END_MESSAGE_MAP()

BOOL CExoApp::InitInstance()
{
    . . .
}
```

If the MFC doesn't provide a default function that processes your type of message, you should first implement the event that will process the message. Here is an example :

```
class CMainFrame : public CFrameWnd
{
public:
    CMainFrame()
    {
        Create(NULL, "Menus Fundamentals");
    }

    afx_msg void EndThisApplication();
    DECLARE_MESSAGE_MAP()
};

. . .

void CMainFrame::EndThisApplication()
{
    PostQuitMessage(0);
}
```

Once you have defined the method that will process the message, you can map it to a resource symbol in your message map section using the **ON_COMMAND** macro.Once this has been done, the application can process the message.

4.3.8. Processing a Menu Message

1. Access the MainFrm.h header file. To create an event that would process a message, change the
2. file as follows :

```
#include "stdafx.h"

class CMainFrame : public CFrameWnd
{
public:
    CMainFrame();
    virtual ~CMainFrame();

    DECLARE_MESSAGE_MAP()

private:
    CMenu *pCurrentMenu;
};
```

3. Access the MainFrm.cpp source file and change it as follows :

```
#include "MainFrm.h"

CMainFrame::CMainFrame()
```

```
{
    . . .
}

CMainFrame::~CMainFrame()
{

}

BEGIN_MESSAGE_MAP(CMainFrame, CFrameWnd)

END_MESSAGE_MAP()
```

4. Save all.

4.4. TOOLBARS BUTTONS

(MDU 2008)

The buttons on a toolbar correspond to commands, just as the items on a menu do. Although you can add a toolbar to your application with AppWizard, you still need to use a little programming polish to make things just right. This is because every application is different and AppWizard can create only the most generally useful toolbar for most applications. When you create your own toolbars, you will probably want to add now or delete existing buttons to support your application's unique command set. This is how a typical toolbar looks like Figure 4.10.

Fig. 4.10.

4.4.1. Toolbars

Toolbars with desired tool buttons may be included in an application.

Adding buttons to a toolbar is a two-step process :

- Draw the button's icon
- Match the button with its command.

To draw a new button, first click the blank button template in the toolbar. The blank button appears enlarged in the edit window, as shown in Figure 4.11.

Suppose you want to create a toolbar button that draws a blue rectangle in the application's window. Draw a blue rectangle on the blank button with the Rectangle tool, and you've created the button's icon. Open the properties box and give the button an appropriate ID, such as ID_RECTANGLE in this case.

Now you need to define the button's description and ***ToolTip***. The description appears in the application's status bar. In this case, a description of "Draws a blue rectangle in the window" might be good. The ToolTip appears whenever the user leaves the mouse pointer over the button for a second or two, acting as a reminder of the button's purpose. A ToolTip of Rectangle would be appropriate for the rectangle button. Type these two text strings into the Prompt box. The description comes first, followed by the newline character (\n) and the ToolTip, as shown in Figure 4.12.

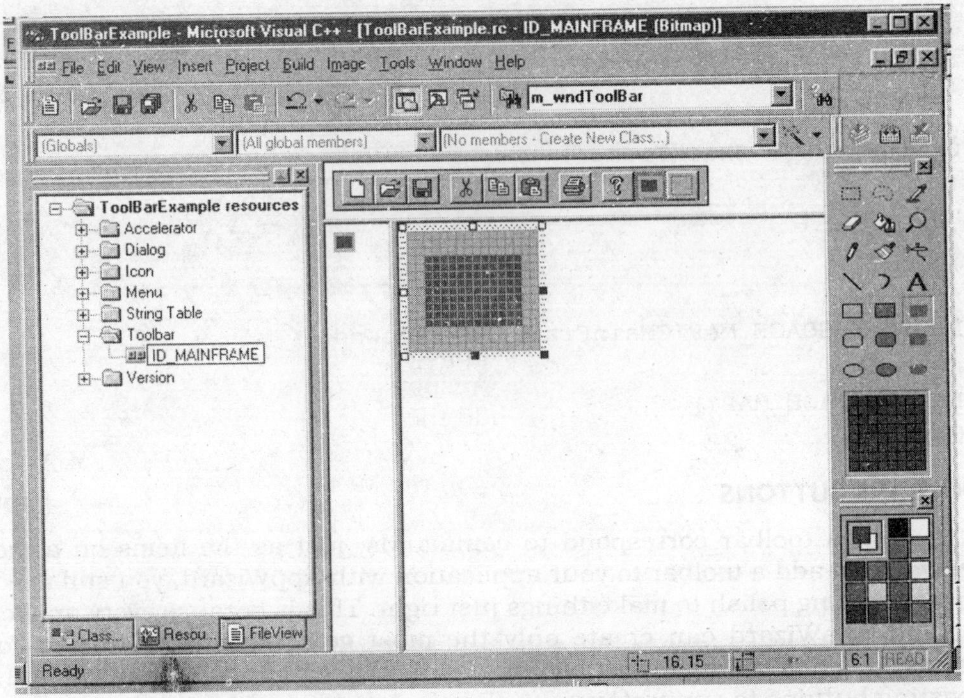

Fig. 4.11. Tool Button Editor

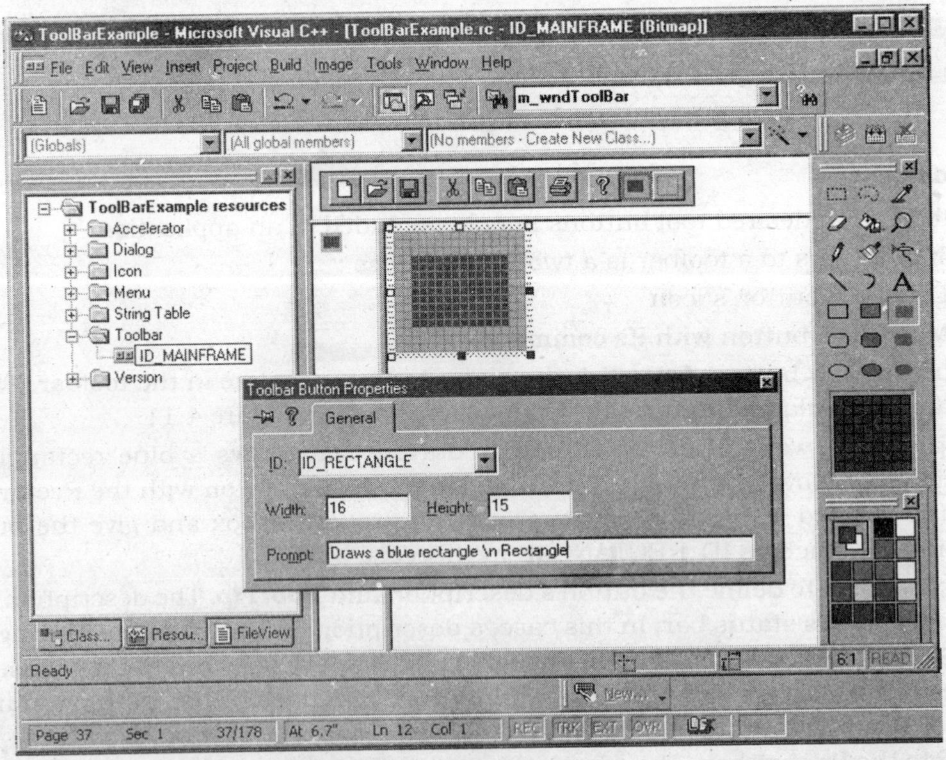

Fig. 4.12. After drawing the button, specify its properties

You've now defined a command ID for your new toolbar button. Usually, you use the command ID of an existing menu item already connected to some code. In these cases, simply choose the existing command ID from the drop-down box, and your work is done. The prompt is taken from the properties of the menu item, and the message handler has already been arranged for the menu item. You will already be handling the menu item, and that code will handle the toolbar click, too. In this application, the toolbar button doesn't mirror a menu item, so you will associate the ID with a message-handler function that MFC automatically calls when the user clicks the button.

To do this, follow these steps :

1. Make sure the button for which you want to create a message handler is selected in the custom toolbar, and then open ClassWizard.
2. The MFC ClassWizard property sheet appears, with the button's ID already selected (see Figure 4.9). To add the message-response function, select in the Class Name box the class to which you want to add the function (the sample application uses the view class).
3. Double-click the COMMAND selection in the Messages box.
4. Accept the function name that MFC suggests in the next message box, and you're all set. Click OK to finalize your changes.

NOTE :
If you haven't defined a message-response function for a toolbar button, or if there is no instance of the class that catches the message, MFC disables the button when you run the application. For example, if the message is caught by the document or view in an MDI application and there is no open document, the button is disabled. The same is true for menu commands--in fact, for all intents and purposes, toolbar buttons are same as menu commands.

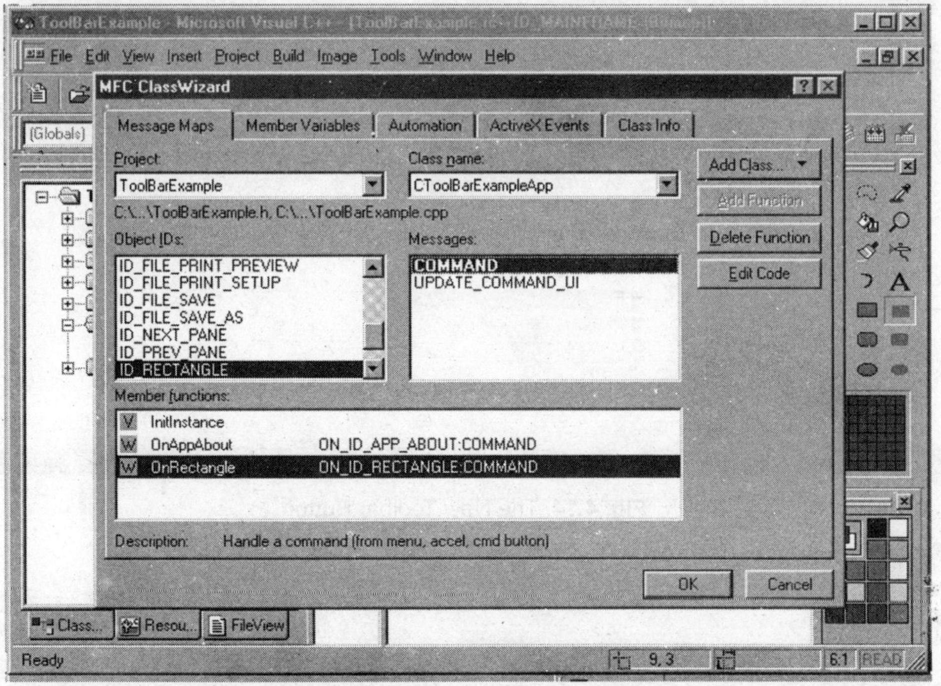

Fig. 4.13. You can use ClassWizard to Catch Messages from your Toolbar Buttons

NOTE :

Ordinarily, toolbar buttons duplicate menu commands, providing a quicker way for the user to select commonly used commands in the menus. In that case, the menu item and the toolbar button both represent the exact same command, and you give both the same ID. Then the same message-response function is called, whether the user selects the command from the menu bar or the toolbar.

If you compile and run the application now, you will see the window shown in Figure 4.14. In the figure, you can see the new toolbar button, as well as its ToolTip and description line. The toolbar looks sparse in this example, but you can add as many buttons as you like.

You can create as many buttons as you need; just follow the same procedure for each. After you have created the buttons, you're through with the toolbar resources and ready to write the code that responds to the buttons.

In the current example, a rectangle button was added to the toolbar, and a message-response function, called OnRectangle(), was added to the program. MFC calls that message-response function whenever the user clicks the associated button. However, right now, that function doesn't do anything as shown next.

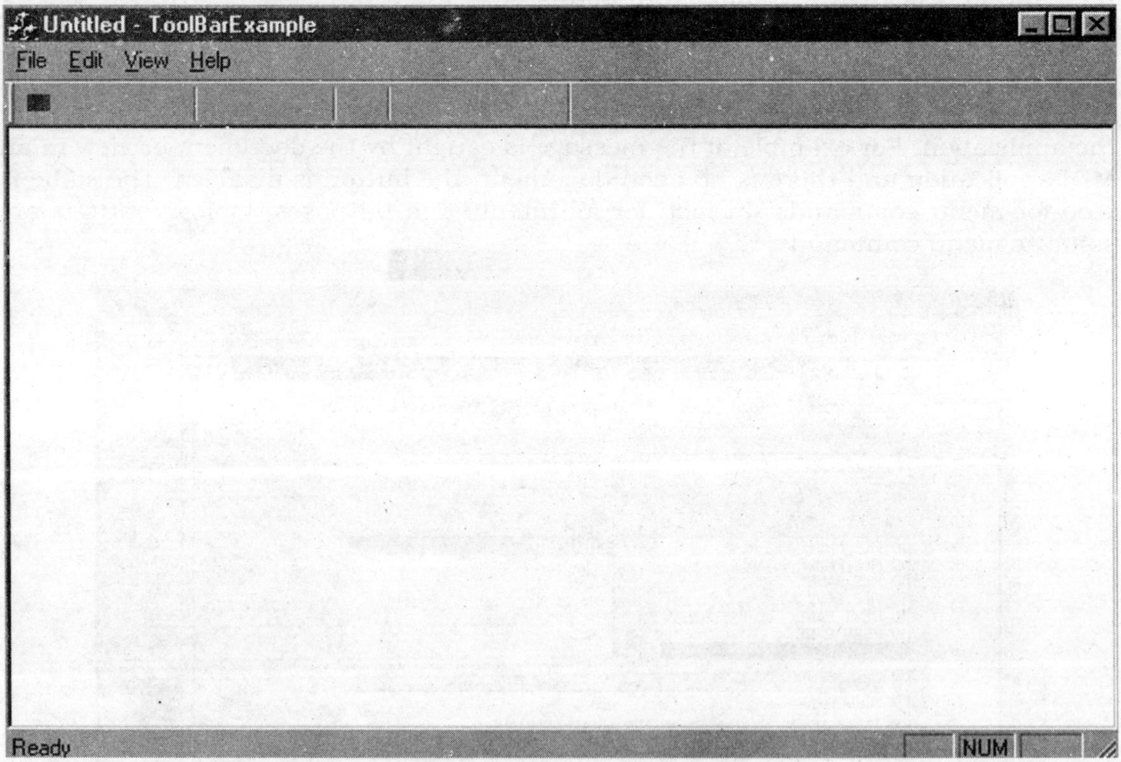

Fig. 4.14. The New Toolbar Button

An Empty Message-Response Function

```
void CToolBarExampleApp::OnRectangle()
{
   // You are required to add your command handler code here in
```

```
    //response to this message
}
```

Although the rectangle button is supposed to draw a blue rectangle in the window, you can see that the On Rectangle() function is going to need a little help accomplishing that task. Add the code given below to the function so that the rectangle button will do what it's supposed to do, as shown in Figure 4.15. This drawing code makes a brush, selects it into the DC, draws blue rectangle with it, and then restores the old brush.

```
void CToolBarExampleApp::OnRectangle()
{
    CClientDC clientDC(this);      //get the current device context

    CBrush newBrush(RGB(0,0,255));    //create new brush with blue
                                      //colour for painting into device
                                      //context RGB(255,0,0) for red,
                                      //RGB(0,255,0) for green
    CBrush*oldBrush=clientDC.SelectObject(&newBrush); //set the paint
                                                  //brush

    clientDC.Rectangle(40, 40, 200, 200); //draw rectangle into the
                                          //device context
    clientDC.SelectObject(oldBrush); //restore the old brush
}
```

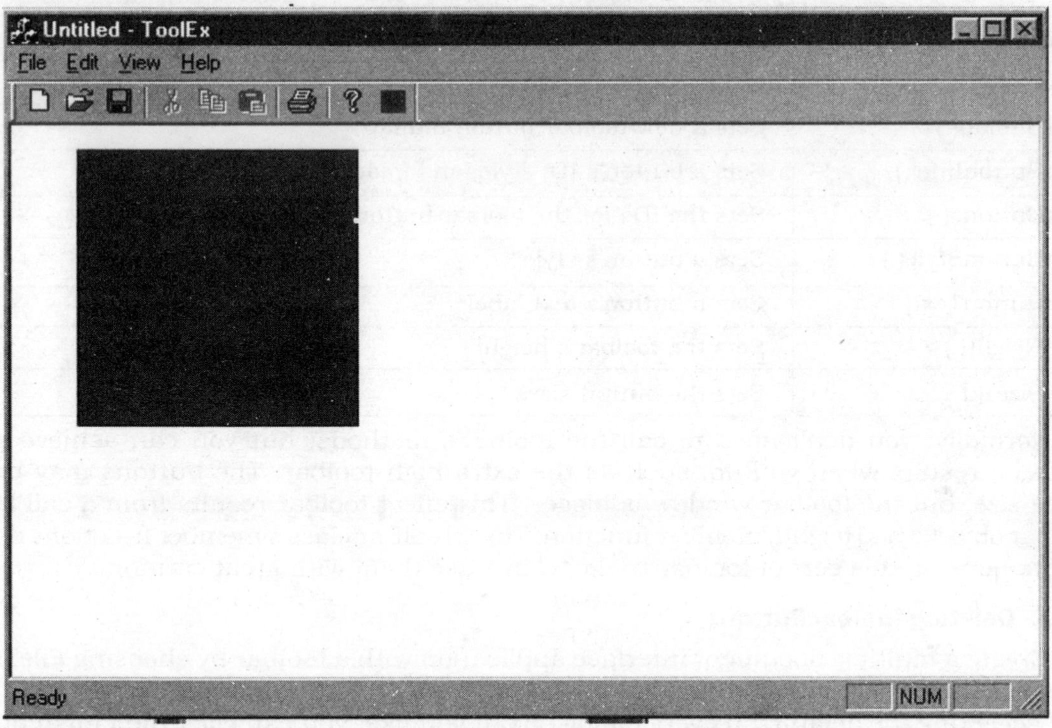

Fig. 4.15. After Adding code to OnRectangle(), the new toolbar button actually does something

4.4.2. The CToolBar Class's Member Functions (V.Imp)

In most cases, after you have created your toolbar resource and associated its buttons with the appropriate command IDs, you don't need to bother any more with the toolbar. The code generated by AppWizard creates the toolbar for you, and MFC takes care of calling the buttons' response functions for you. However, at times you might want to change the toolbar's default behaviour or appearance in some way. In those cases, you can call on the ***CtoolBar*** (MFC object) class's member functions, listed in table 1.5 alongwith their descriptions. The toolbar is accessible from the CMainFrame class as the m_wndToolBar member variable. Usually, you change the toolbar behavior in CMainFrame::OnCreate().

Table 4.5: Member Functions of the CToolBar Class

Function	***Description***
CommandToIndex( )	Obtains the index of a button, given its ID
Create( )	Creates the toolbar
GetButtonInfo( )	Obtains information about a button
GetButtonStyle( )	Obtains a button's style
GetButtonText( )	Obtains a button's text label
GetItemID( )	Obtains the ID of a button, given its index
GetItemRect()	Obtains an item's display rectangle, given its index
GetToolBarCtrl( )	Obtains a reference to the CToolBarCtrl object represented by the CToolBar object
LoadBitmap( )	Loads the toolbar's button images
LoadToolBar( )	Loads a toolbar resource
SetBitmap( )	Sets a new toolbar button bitmap
SetButtonInfo( )	Sets a button's ID, style, and image number
SetButtons( )	Sets the IDs for the toolbar buttons
SetButtonStyle( )	Sets a button's style
SetButtonText( )	Sets a button's text label
SetHeight( )	Sets the toolbar's height
SetSizes( )	Sets the button sizes

Normally, you don't need to call the toolbar's methods, but you can achieve some unusual results when you do, such as the extra high toolbar. The buttons may be the same size, but the toolbar window is bigger. This effect toolbar results from a call to the toolbar object's SetHeight() member function. The CToolBar class's member functions enable you to perform this sort of toolbar trickery, but use them with great caution.

4.4.3. Deleting Toolbar Buttons

Create a multiple document interface application with a toolbar by choosing File, New; selecting the Project tab; highlighting MFC AppWizard (exe); naming the application Tool; and accepting the defaults in every dialog box. If you like, you can click the Finish button in step 1 to speed up the process. AppWizard provides a docking toolbar by default. Build and run the application, and you should see a toolbar of your own Figure 4.16.

Fig. 4.16. Your Custom Toolbar Attached to the Application Window

Before moving on, play with this toolbar a little. On the View menu, you can toggle whether the toolbar is displayed. Turn it off and then on again. Now click and hold on the toolbar between buttons and pull it down into the working area of your application. Let it go, and it's a floating palette. Drag it around and drop it at the bottom of the application or one of the sides-it will dock against any side of the main window. Watch the tracking rectangle change shape to show you it will dock if you drop it. Drag it back off again so that it's floating and close it by clicking the small x in the upper-right corner. Bring it back with the View menu and notice that it comes back right where you left it. All this functionality is yours free from AppWizard and MFC.

The first step in modifying the toolbar is to delete buttons you no longer need. To do this, first select the Resource View tab to display your application's resources by clicking on the + next to Tool Resources. Click the + next to Toolbar and double-click the IDR_MAINFRAME toolbar resource to edit it as shown in Figure 4.17 (The Graphics and Colours palettes, are docked by default. You can move them around by grabbing the wrinkles at the top.)

After you have the toolbar editor on the screen, deleting buttons is as easy as dragging the unwanted buttons from the toolbar. Place your mouse pointer on the button, hold down the left mouse button, and drag the unwanted button away from the toolbar. When you release the mouse button, the toolbar button disappears. In the Tool application, delete all the buttons except the Help button with a yellow question mark. Figure 4.18 shows the edited toolbar with only the Help button remaining. The single blank button template is only a starting point for the next button you want to create. If you leave it blank, it doesn't appear in the final toolbar.

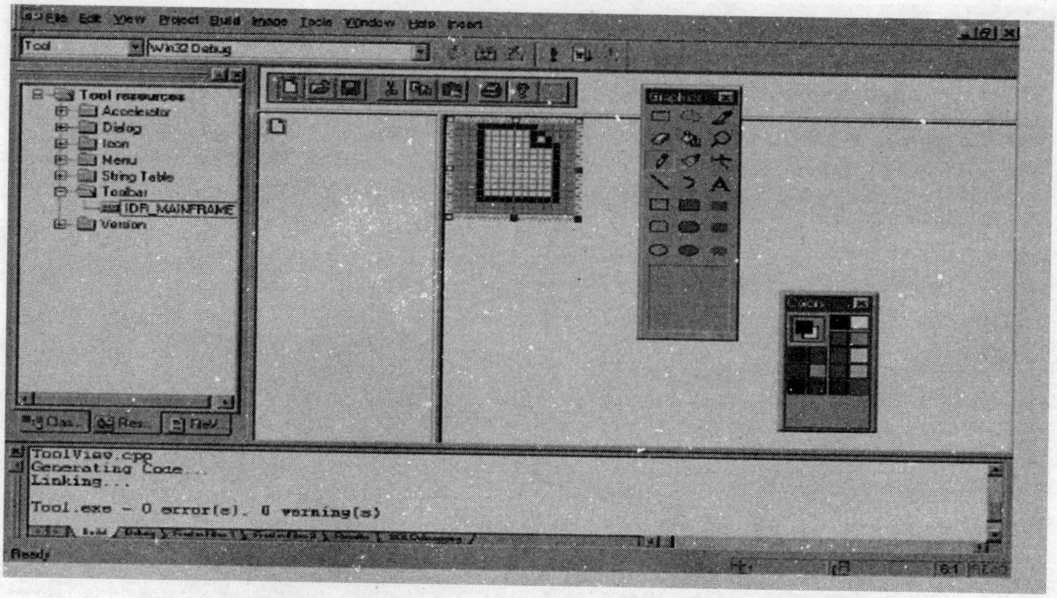

Fig. 4.17. Use the toolbar editor to customise your application's toolbar

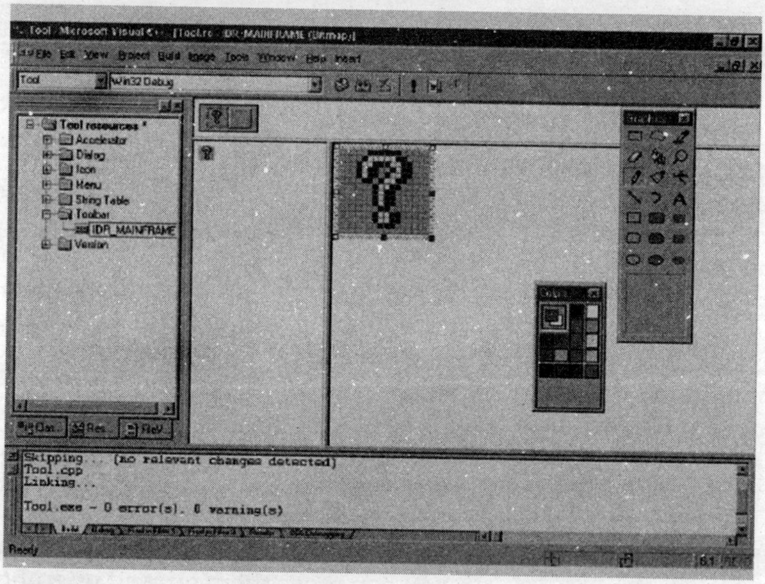

Fig. 4.18. This edited toolbar has only a single button left (not counting the blank button template)

SUMMARY

1. The keyboard obviously has fewer messages than the mouse does. Then again, there are only so many things that you can do with the keyboard. These event messages are available on the dialog window object and are triggered only if there are no enabled controls on the window. Any enabled controls on the window have input focus, so all keyboard events go to them. That's why you remove all controls from the main dialog for your drawing application

2. These messages are sent in the order of their occurrence to the active window application message queue. The concerned application retrieves messages from this queue, one message at a time. If the application has the appropriate message handler (*i.e.*, a function doing something in response to this event) for the message read from the queue, the message handler is executed; otherwise the message is passed to the default message handler of the operating system (*e.g.* Windows).
3. There are two categories of mouse events - Client area mouse events and Non-Client area mouse events. Client area mouse events occur when the mouse pointer is focussed in the application's window. Mouse events produced when the mouse is focussed outside the application window area, are termed as Non-Client area mouse events.
4. One of the original keyboard-shortcuts for selecting menu entries was accelerator key. An accelerator key is specific key combination, usually the ctrl key, combined with another key, or function keys, that are unique within the entire application. Each of these key combinations triggers one menu event function.
5. The buttons on a toolbar correspond to commands, just as the items on a menu do. Although you can add a toolbar to your application with AppWizard, you still need to use a little programming polish to make things just right.
6. Normally, you don't need to call the toolbar's methods, but you can achieve some unusual results when you do, such as the extra high toolbar. The buttons may be the same size, but the toolbar window is bigger.
7. Suppose you want to create a toolbar button that draws a blue rectangle in the application's window. Draw a blue rectangle on the blank button with the Rectangle tool, and you've created the button's icon. Open the properties box and give the button an appropriate ID, such as ID_RECTANGLE in this case.
8. Now you need to define the button's description and ToolTip. The description appears in the application's status bar. In this case, a description of "Draws a blue rectangle in the window" might be good. The ToolTip appears whenever the user leaves the mouse pointer over the button for a second or two, acting as a reminder of the button's purpose.
9. A ToolTip of Rectangle would be appropriate for the rectangle button. Type these two text strings into the Prompt box. The description comes first, followed by the newline character (\n) and the ToolTip,
10. You've now defined a command ID for your new toolbar button. Usually, you use the command ID of an existing menu item already connected to some code. In these cases, simply choose the existing command ID from the drop-down box, and your work is done.

SHORT QUESTIONS WITH ANSWER

Q.1. How to use a mouse event in your application?

Ans. Which of these mouse messages are to be processed by the application, depends on the functionality of the application. If the application has to respond to a particular mouse event then it must implement the message handler for it. On the other hand, if the application must ignore a particular mouse event then it need not implement the corresponding message handler and let the default message handler take care of that event.

For instance, if your application must respond to left button click of the mouse then you must write the required code for the message handler OnLButtonDown.

The general prototype of the keyboard handler functions looks like :

```
Void OnMsgName(UNIT nFlags, Cpoint point)
```

Each one of these mouse handler functions receive two arguments from the application message queue - UINT nFlags and CPoint point. nFlags contains bit-pattern identifying, which mouse event has occurred and CPoint contains the co ordinate (CPoint.x and Cpoint.y) of the mouse pointer when the even occurred.

These values are available for use inside the event handler.

Q.2. How will you Create a Menu with Accelerators? (V.Imp)

Ans. One of the original keyboard-shortcuts for selecting menu entries was accelerator key. An accelerator key is specific key combination, usually the ctrl key, combined with another key, or function keys, that are unique within the entire application. Each of these key combinations triggers one menu event function.

The way that accelerator keys work is similar to the way menus work. They are also an application resource that is defined in a table in the resource tab of the workspace pane. Each table entry has an object ID and a key code combination. After you define the accelerators, you can attach functionality to the object ID's. You can also assign accelerators entries the same object ID as the corresponding entry so that you have to define only a single entry in the application map.

After you define all accelerator keys, you can add the key combination to the menu entry so that the user will know about the accelerator key combination. Add \t to the end of the menu entry caption, followed by the key combination. The \t is replaced in the menu display by a tab, which separates the menu caption from the accelerator key combination.

Unfortunately, accelerator keys don't work in dialog-style windows, so you cannot add them to this application. You will learn how to attach accelerator keys to the menus later in single and multi-document interface style application section.

Q.3. Define group boxes. Explain how you will use group boxes to arrange your controls.

Ans. Group Box: A group box is a static control used to create "physical" limits or sections on a dialog box or a form. It belongs to the family of MFC static controls. To create a group box, from the Controls toolbox, click the Group Box button and click the desired area on the dialog box or the form. Because this is a **CStatic** object, if you plant to refer to it in your code, you must change its ID from IDC_STATIC to something else.

The most signifies property of a group box is probably the text it is displaying. This is set using the **Caption** field. If you are programmatically creating the control, the first argument of the **CStatic::::Create ( )** method allows you to set its caption. At run time, to change this text, cell the **CWnd:::::SetWindow Text ( )** method. The text of a group box can be aligned to the left, the center, or the right sides of its allocated rectangle. This is specified at design time using the **Horizontal Alignment** combo box. Instead of using its own caption, you may want to display a check box that would validata or invalidate the content of the group box. To do this, after adding the group box delete its caption. Then add a Check Box control where the caption was:

Of course, it is up to you to implement the behaviour of controls inside a outside the group box.

Q.4. Briefly explain device context.

Ans. Device context: A data structure in Windows programming that is used to define the attributes of text and images that are output to the screen or printer. The device context (DC) is maintained by GD!. A DC, which is a handle to the structure, is obtained before output is written and released after the elements have been written. A wxDC is a device context onto which graphics and text can be drawn. The device context is intended to represent a number of output devices in a generic way, with the same API being used throughout. Some device contexts are created temporarily in order to draw on a window. This is true of wxScreenDC, wxClientDC, wxPaintDC, and wxWindowDC. The following describes the differences between these device contexts and when you should use them.

- **wxScreenDC:** Use this to paint on the screen, as opposed to an individual window.
- **wxClientDC "** Use this to paint on the client area of window (the part without borders and other decorations), but do not use it from within an wxPaintEvent.
- **wxPaintDC "** Use this to paint on the client area of a window, but only from within a wxPaintEvent.
- **wxWindowDC:** Use this to paint on the whole area of a window, including decorations. This may not be available on non-Windows platforms.

To use a client, paint or window device context, create an object on the stack with the window as argument, for example:

```
void My Window::OnMyCmd (wxCommandEvent&event) (wxClientDC dc (window),
DrawMyPicture (dc);}.
```

Q.5. How to read keys from the keyboard as the user types them?

Ans. When the user types a key, windows sends is a WM_CHAR message and that message should be connected to a method in view class named on Char( ).

There can be many windows open in a window session and only one can get keyboard events at any one time. The window that is currently active and getting keyboard events is called the window with the focus.

To opened class wizard, select class word item in the view menu. Add a new event handler—OnChai ( ) to view class, CKey strokes view. Makes sure class is selected in the class wizard's class name box. After connecting OnChar( ) to the WM_Char message, OnChar( ) will be called each time the user types a character message will be in the class wizard message box and double click it, to add OnChar() method. OnChar( ) is the default name class wizard to WM_CHAR event handlers. Besides the WMCHAR message, windows also sends WM_KEY DOWN messages. When a key is pressed and WM_KEYUP when a key is released. These messages are handled in the event handlers OnKeyDownQ and OnKeyUp( ) by class wizard.

Find the OnChar() method now in the keystrokes view.cpp file:

```
void Ckey strokes view:: OnChar (UINT nchar, U1NT nRepcut, UINT nFlags)
{
//Add your message handler code
CView:: OnChar(nChar, nRepCut, nFlags);
```

This is the method, *i.e.*, is called when the user types a key. If the user has held down the key, generating type ahead (called typematic), the number of times the key message has been sent will be in the nRepCut parameter. The nFlags parameter works like this, bit-by-bit.

The transition state is I, if the key has just been released and 0 if it's just been pressed. The previous key state parameter is 3 if the key was previously down and 0 if it was up. The context code parameter is 1 if the Alt key is down.

Q.6. How to use class wizard tool to connect mouse messages to event holder methods?

Ans. In this program, class wizard is used to handle the left mouse button down window message WM LButton Down( ), in the code; specifically the class wized name the method OnLButtonDown).

Let us assume the file name is mouse. Then in the view class, CMouseView, is selected in the class name box and find the WMLBUTTON DOWN message in the message box. Double click the message that will create the new method OnLButtonDown( ) that will be shown in the class-wizard member function box. Double, click the On LBUTTON DO WN ( ) method entry in that box to open the code for that method:

```
Void CMouseView:: OnLButtonDown (UINT nFlags, CPoint point)
P
    //Add message handler code
    CView:: OnLButtonDown (nFlags, point);
}
```

same as OnLButtonDown( ), method like OnLButtonUp( ), OnRButtonDown( ) for the right mouse button, OnLButton Dblock( ) for double clicks can be used.

Two parameters are passed in OnLButtonDown( ): nFlags and point. The nFlags parameter indicates the state of various keys on the keyboard on can take values:

MK CONTROL: Control key was Down

MK_LBUTTON: Left Mouse Button Down

MK_N BUTTON: Middle Mouse Button Down

MK RBUTTON: Right Mouse Button Down

MK SHIFT: Shift key was down

The point parameter, an object of CPoint class holds the mouse's present location.

Q.7. Write steps to create menu's in VC++.

Ans. Menu's a pop down window that display a set of options that the user can select form. We have to place menu items into menu's see how to create whole new menus.

Menu Editor is the most useful VC++ tools and through this we design menus easily.

Steps to create menu's in VC++:

1. To open Menu Editor, click resource tab in VC++ window because menu is a resource in a window. This open menus.
2. Program resources Folder, Next find folder with menu names and open it.
3. Double click IDR mainframe resource ID entry in that folder, open the menu editor in the doc window at right side. Resource ID is a numeric value given to a resource to identify it menu system is a part of our mainframe window object.

Now we are ready to add our new menu item. To do that,

1. Click the file menu in the menu editor to open that menu.
2. This new item will go between the existing print previous and print setup menu items, so click the print setup menu item & press the insert key to insert a new, blank menu item.
3. The new menu item is surrounded by a multi dotted box.
4. Double-click this new menu item now, opening the menu item properties box.
5. We have designed our new menu item, so close the menu editor. The next step is to connect that menu item to our code so we can handle the case when the user clicks that item.

Q.8. What are dialog boxes? Explain its function and also write steps to create a dialog box.

Ans. Dialog boxes are windows with lots of buttons and list boxes and other items that you use to pass some information to a program. The string advantage dialog boxes have over other windows is VC++ is that it's for easier to place controls in a dialog box than in other windows. In fact, VC++ has a special tool, the dialog editor, especially for use in creating dialog boxes.

To create Dialog Box:

1. Create a new SDI program named dialogs and add a new menu item using the menu editor to the file menu: "Show dialog".
2. Using classwizard, add an event-handling method to this new menu item, OnFileShow dialog( ); when the user clicks this menu item, we want to place our new dialog box on the screen.
3. To create a new dialog box, select the resource item in the visual C++ Insert menu now. This open the Insert Resource box.

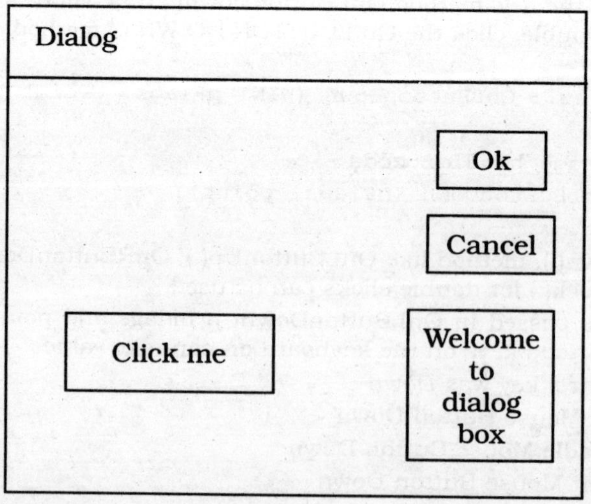

Fig. 4.19.

4. Clicking the new button opens the dialog box editor, we can see new dialog box in the dialog box editor. Like menus, dialog boxes are considered resources in windows programs.

LONG QUESTIONS WITH ANSWER

Q.1. How would you Reading keystrokes from the keyboard?

Ans. The keyboard is an input device that is shared by all applications running in memory. Pressing a key (or combination of keys) on the keyboard constitutes a keyboard event. Each such key event generates a corresponding message, which is received by the Windows system. Out of the several windows that may be active, Windows directs keyboard message to the window that has got the input focus. The window that has the input focus is the main window of the application. The input focus might also belong to a child of the main window, or to a control in a dialog box. Windows sends keyboard messages to the window that owns the focus.

These messages are sent in the order of their occurrence to the active window application message queue. The concerned application retrieves messages from this queue, one message at a time. If the application has the appropriate message handler (*i.e.*, a function doing something in response to this event) for the message read from the queue, the message handler is executed; otherwise the message is passed to the default message handler of the operating system (*e.g.* Windows). Table 4.6 lists various keyboard events, the messages generated and respective event handlers.

Table 4.6. Keyboard Events.

Message	*Handler*	*Events*
WM_KEYDOWN	OnLKeyDown	A key has been pressed
WM_KEYUP	OnKeyUp	A key has been released
WM_SYSKEYDOWN	OnSysKeyDown	A special or function key has been pressed
WM_SYSKEYUP	OnSysKeyUp	A special or function key has been released

While the above mentioned messages indicate what event has occurred, it does not tell which key has generated that event. A flag that gets set with appropriate value when the event occurs indicates which key has given rise to the event. Table 4.7 lists various flags that get affected on keyboard events.

Table 4.7: Keyboard Eevent Flags

Bit in nFlags	*Meaning*
0-15	Contains the number of times the key has been pressed as a result of the key being held down
16-23	Manufacturer-dependent key-code (scan code)
24	Set to 1 if the key pressed is either a function key or an extended key, 0 otherwise
25-28	Used by Windows
29	Set to 1 if the Alt key was also pressed along, 0 otherwise
30	Set to 1 if the key was pressed before the message was sent, 0 therwise
31	Set to 1 if the key is being released, 0 if the key is being pressed

How to use a keyboard event in your application: Which of these keyboard messages are to be processed by the application, depends on the functionality of the application. If the application has to respond to a particular keyboard event then it must implement the message handler for it. On the other hand, if the application must ignore a particular keyboard event then it need not implement the corresponding message handler and let the default message handler take care of that event.

For instance, if your application must respond to a keypressed event then you must write the required code for the message handler On Key Down.

The general prototype of the keyboard handler functions looks like :

```
Void OnMsgName(UNIT nChar, UNIT nRepCnt, UNIT nFlags)
```

On Mesg Name is the name of the individual message handler. nChar contains the ASCII code of the character pressed on the keyboard. nRepCnt parameter contains number of times the key has been repeated as a result of the key being pressed and the nFlags parameter has a bit code detailed on table 4.8.

Q.2. How would you Handling mouse in VC++? (MDU 2008)

Ans. A mouse is the most common pointing input device extensively used with windows applications. Each activity performed with the mouse (like moving, clicking, dragging etc.) is termed as a mouse event. Corresponding to each event a message is generated. For instance, if the mouse is moved, a corresponding event (mouse move event) occurs in the system and consequently a message (WM_MOUSEMOVE message) is produced.

As usual, the window getting the message from the Windows may implement a handler to that message or simply let it pass back to the Windows default handler. Table 4.8 summarizses the various mouse events, messages and their handlers.

There are two categories of mouse events - Client area mouse events and Non-Client area mouse events. Client area mouse events occur when the mouse pointer is focussed in the application's window. Mouse events produced when the mouse is focussed outside the application window area, are termed as Non-Client area mouse events. Table 4.9 summarizses different messages produced and corresponding event handlers for non-client area mouse activities.

Table 4.8. Client Area Mouse Events

Message	*Handler*	*Dispatched When*
WM_LBUTTONDOWN	OnLButtonDown	The left mouse button is pressed
WM_LBUTTONUP	OnLButtonUp	The left mouse button is released
WM_LBUTTONDBLCLK	OnLButtonDblClk	The left mouse button is double-clicked
WM_MBUTTONDOWN	OnMButtondown	The middle mouse button is pressed
WM_MBUTTONUP	OnMButtonUp	The middle mouse button is released
WM_MBUTTONDBLCLK	OnMButtonDblClk	The middle mouse button is double-clicked
WM_RBUTTONDOWN	OnRButtonDown	The right mouse button is pressed
WM_RBUTTONUP	OnRButtonUp	The right mouse button is released
WM_RBUTTONDBLCLK	OnRButtonDblClk	The right mouse button is double-clicked
WM_MOUSEMOVE	OnMouseMove	The mouse is moved over the client area of the window.

Table 4.9. Non-client Area Mouse Event

Message	*Handler*	*Dispatched When*
WM_NCLBUTTONDOWN	OnLButtonDown	The left mouse button is pressed
WM_NCLBUTTONUP	OnLButtonUp	The left mouse button is released
WM_NCLBUTTONDBLCLK	OnLButtonDblClk	The left mouse button is double-clicked
WM_NCMBUTTONDOWN	OnMButtondown	The middle mouse button is pressed
WM_NCMBUTTONUP	OnMButtonUp	The middle mouse button is released

WM_NCMBUTTONDBLCLK	OnMButtonDblClk	The middle mouse button is double-clicked
WM_NCRBUTTONDOWN	OnRButtonDown	The right mouse button is pressed
WM_NCRBUTTONUP	OnRButtonUp	The right mouse button is released
WM_NCRBUTTONDBLCLK	OnRButtonDblClk	The right mouse button is double-clicked
WM_NCMOUSEMOVE	OnMouseMove	The mouse is moved over the non-client area of the window.

Q.3. Write down the procedure for creating pop-up menu? **(V.Imp)**

Ans. Let us build a pop-up menu displaying File menu group when right button of mouse is clicked in the client area of the application. To create this application, do the following :

- Create a SDI application with default settings using AppWizard.
- Using class wizard, similar to the previous example, insert an event handler for the message OnContextMenu in MainFrame class.
- Insert the following code into this handler function :

```
CMenu *m_lMenu;          //create a pointer of CMenu type to store the
                         //current menu

CPoint m_pPoint;         //create a CPoint type a variable to store the
                         //current mouse position

m_pPoint = point;        //store the point

ClientToScreen(&m_pPoint); //convert the co-ordinate of the point
                           //from client to screen

m_lMenu=GetMenu(); //store the current menu pointer

m_lMenu = m_lMenu->GetSubMenu(0); //store the first sub-menu

m_lMenu->TrackPopupMenu(TPM_CENTERALIGN+TPM_LEFTBUTTON,
m_pPoint.x,m_pPoint.y, this, NULL); //display the pop-up menu //aligned
center to the left button at the point where the right //button was clicked
```

- Compile and run.
- Right-click the mouse at the client area of your application. You will see the pop-up menu appear as in Fig. 4.19.

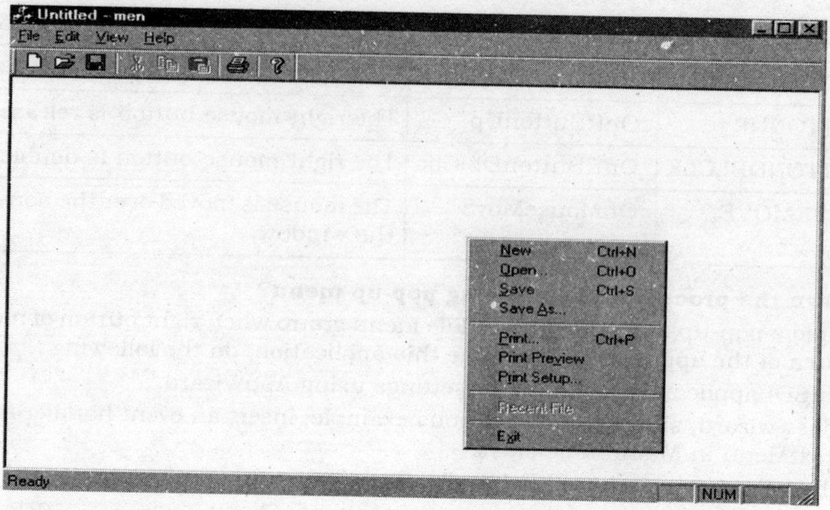

Fig. 4.20. Appearance of the pop-up menu in the client area of an application

Q.4. What is Main Menu? **(V.Imp)**

Ans. A menu is a list of actions the user can perform on an application. The actions are presented in one or more groups. There are broad categories of menus: the main menu and the context menu.

A main menu, also called a top-level menu, displays categories of menu items using a range of items called a menu bar :

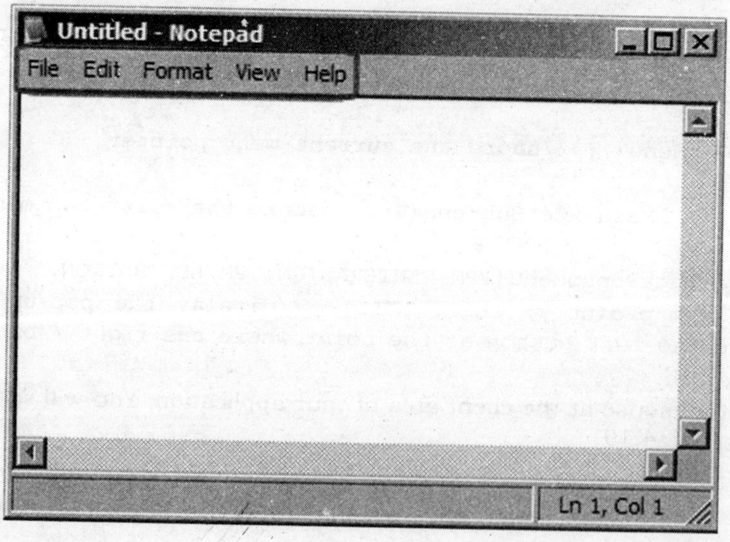

Fig. 4.21.

When the user clicks an item of the menu bar, the item clicked opens its list :

After a list has been displayed, the user can then use an item from the list. Each item of the list is primarily a word or a group of words on a line. Different menu items are used for different reasons. For example, some menu items simply display a word or a group of words. Some other items display a check mark. This indicates that the item toggles the availability or disappearance of an object.

When a menu item is only meant to lead to a sub-menu, such a menu item is called a pop-up menu. There are two types of pop-up menus. If the menu displays on top of a window, which is the type of menu under the title bar, the word on top, which represents a category of menu, is a pop-up menu. If a menu item is equipped with an arrow in its right , which means the menu item has a submenu, such a menu item is also a popup menu. Popup menus are used only to represent a submenu. No inherent action is produced by clicking them, except that, when placed on top, such menu items allow opening the sub-menu.

To create menus that belong to a group, menu items are separated by a horizontal line called a separator. Separators are created differently in MSVC 6 and MSVC 7.

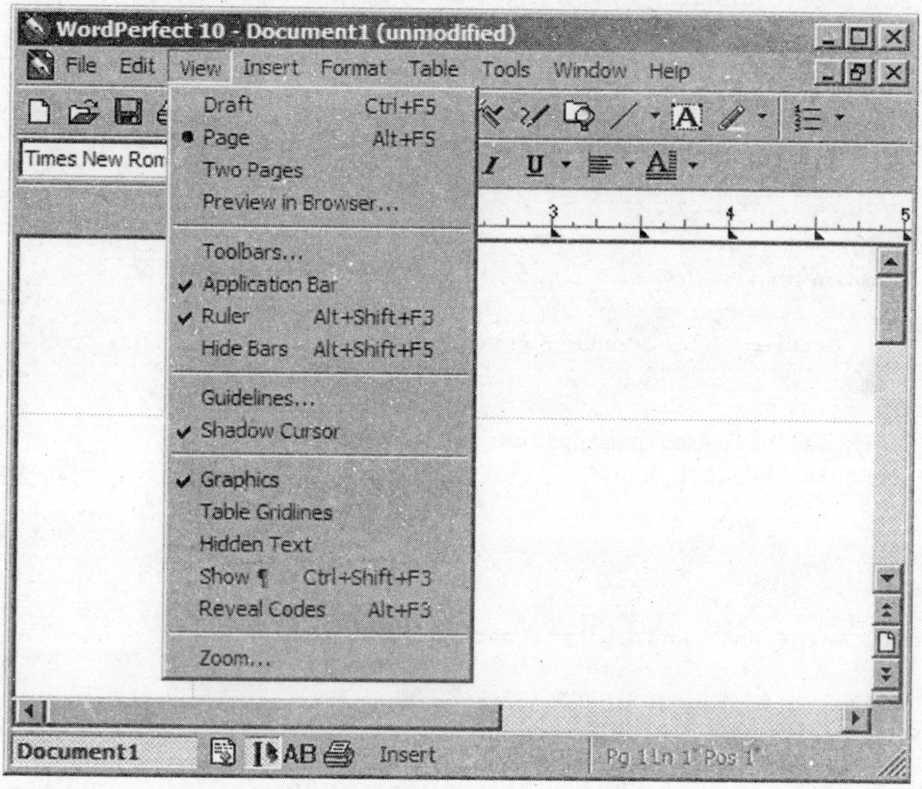

Fig. 4.22.

Q.5. Briefly Explain Menu Message Processing.

Ans. As introduced earlier, a menu is primarily a list of actions that can be performed on an application. The actions are generated when the user positions the mouse on a menu item or when one of these items is clicked. If the user clicks an item such as File on the main menu, the menu item generates a WM_INITMENUPOPUP message. To get the program ready to process these messages, in the previous lesson, we saw that you should create a message map. Here is an example:

```
class CMainFrame : public CFrameWnd
{
public:
    CMainFrame()
    {
        Create(NULL, "Menus Fundamentals");
    }

    DECLARE_MESSAGE_MAP()
```

```
};

BEGIN_MESSAGE_MAP(CMainFrame, CFrameWnd)

END_MESSAGE_MAP()

BOOL CExoApp::InitInstance()
{
    . . .
}
```

If the MFC doesn't provide a default function that processes your type of message, you should first implement the event that will process the message. Here is an example :

```
class CMainFrame : public CFrameWnd
{
public:
    CMainFrame()
    {
        Create(NULL, "Menus Fundamentals");
    }

    afx_msg void EndThisApplication();
    DECLARE_MESSAGE_MAP()
};

. . .

void CMainFrame::EndThisApplication()
{
    PostQuitMessage(0);
}
```

Once you have defined the method that will process the message, you can map it to a resource symbol in your message map section using the ON_COMMAND macro. Once this has been done, the application can process the message.

Q.6. Briefly Explain Toolbars Buttons. **(MDU 2009)**

Ans. The buttons on a toolbar correspond to commands, just as the items on a menu do. Although you can add a toolbar to your application with AppWizard, you still need to use a little programming polish to make things just right. This is because every application is different and AppWizard can create only the most generally useful toolbar for most applications. When you create your own toolbars, you will probably want to add now or delete existing buttons to support your application's unique command set. This is how a typical toolbar looks like Fig. 4.23.

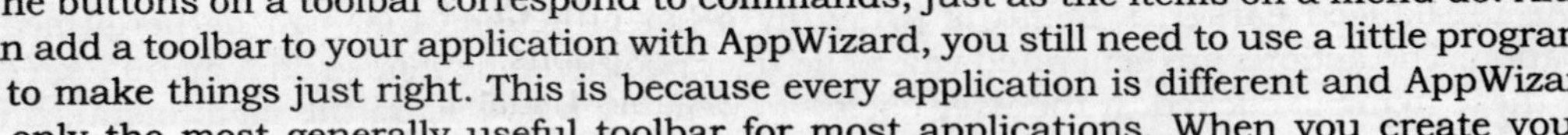

Fig. 4.23.

Toolbars: Toolbars with desired tool buttons may be included in an application.

Adding buttons to a toolbar is a two-step process :

- Draw the button's icon
- Match the button with its command.

To draw a new button, first click the blank button template in the toolbar. The blank button appears enlarged in the edit window, as shown in Fig. 4.24.

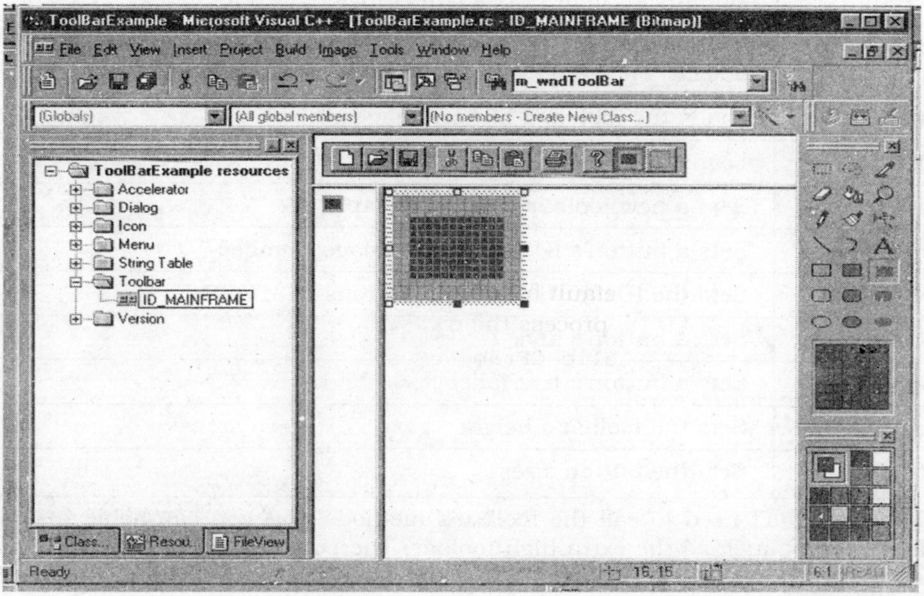

Fig. 4.24. Tool Button Editor

Suppose you want to create a toolbar button that draws a blue rectangle in the application's window. Draw a blue rectangle on the blank button with the Rectangle tool, and you've created the button's icon. Open the properties box and give the button an appropriate ID, such as ID_RECTANGLE in this case.

Now you need to define the button's description and ***ToolTip***. The description appears in the application's status bar. In this case, a description of "Draws a blue rectangle in the window" might be good. The ToolTip appears whenever the user leaves the mouse pointer over the button for a second or two, acting as a reminder of the button's purpose. A ToolTip of *Rectangle* would be appropriate for the rectangle button.

Q.7. Briefly Explain The CToolBar Class's Member Functions.

Ans. In most cases, after you have created your toolbar resource and associated its buttons with the appropriate command IDs, you don't need to bother any more with the toolbar. The code generated by AppWizard creates the toolbar for you, and MFC takes care of calling the buttons' response functions for you. However, at times you might want to change the toolbar's default behaviour or appearance in some way. In those cases, you can call on the ***CtoolBar*** (MFC object) class's member functions, listed in table 4.10 along with their descriptions. The toolbar is accessible from the CMainFrame class as the m_wndToolBar member variable. Usually, you change the toolbar behaviour in CMainFrame::OnCreate().

Table 4.10. Member Functions of the CToolBar Class

Function	***Description***
CommandToIndex( )	Obtains the index of a button, given its ID
Create( )	Creates the toolbar
GetButtonInfo( )	Obtains information about a button
GetButtonStyle( )	Obtains a button's style
GetButtonText( )	Obtains a button's text label

GetItemID( )	Obtains the ID of a button, given its index
GetItemRect()	Obtains an item's display rectangle, given its index
GetToolBarCtrl( )	Obtains a reference to the CToolBarCtrl object represented by the CToolBar object
LoadBitmap( )	Loads the toolbar's button images
LoadToolBar( )	Loads a toolbar resource
SetBitmap( )	Sets a new toolbar button bitmap
SetButtonInfo( )	Sets a button's ID, style, and image number
SetButtons( )	Sets the IDs for the toolbar buttons
SetButtonStyle( )	Sets a button's style
SetButtonText( )	Sets a button's text label
SetHeight( )	Sets the toolbar's height
SetSizes( )	Sets the button sizes

Normally, you don't need to call the toolbar's methods, but you can achieve some unusual results when you do, such as the extra high toolbar. The buttons may be the same size, but the toolbar window is bigger. This effect toolbar results from a call to the toolbar object's SetHeight() member function. The CToolBar class's member functions enable you to perform this sort of toolbar trickery, but use them with great caution.

Q.8. Define dialog box by giving example write a program in VC++ to retrieve data from a dialog box.

Ans. Dialog Box: A dialog box is a rectangular window whose main role is to host or hold other windows control. For this reason, a dialog box is referred to as a container. It is the primary interface of user interaction with the computer. By itself, a dialog box means nothing. The controls it hosts accomplish the role of dialog between the user and the machine.

A dialog box has the following characteristics:

- It is equipped with the system Close button. As the only system button, this button allows the user to dismiss the dialog and ignore whatever the user would have done on the dialog box.
- It cannot be minimized, maximized, or restored. A dialog box does not have any other system button but close.
- It is usually modal. The user is usually not allowed to continue any other operation until the dialog box is dismissed.
- It provides a way for the user to close or dismiss the dialog. Most dialog boxes have the OK and the Cancel buttons, although this depends on the application developer. When the dialog has the OK and the Cancel buttons, the OK button is configured to behave as if the user had pressed Enter. In that case 4, whatever the user had done would be acknowledged and transferred to the hosting dialog box, window, or application. Pressing Esc applies the same behaviour as if the user had clickedCancel.

```
# include<afxwin.h>
# include "Resource.h"
class CExerciseApp: public CWinApp
{public:
BOOL InitInstance ( ); }
class CExoDialog: public CDialog
{ public:
```

```
enum {IDD=IDD_DIALOGI};
CExoDialog ( ); };
CExoDialog:: CExoDialog ( )
CDialog (CExoDialog:: IDD, NULL) { }
BOOL CExerciseApp:: InitInstance ( )
{ CExoDialog Dig;m_pMainWnd=&DIg;
DIg.DoModal ( );
return TRUE ; }
CExerciseApptheApp;
```

Q.9. Differentiate between check boxes and radio buttons. Write a way to put together check boxes and radio buttons.

Ans. Check Box: A check box is a box on a web page that cab be checked by a user. Unlike a radio button, more than one box can be checked at once, allowing your visitors to make multiple selections.

The following steps show how to add a check box to the Simple Sample smart document.

1. The first thing you need to do is add a constant for the **Checkbox** element in the Simple Sample schema. Insert the following code into the general declaration section of your code module, below the existing constants.

```
# define cCHECKBOX (xmlns) xmlns L"#checkbox".
```

2. Next you to add 1 to the **cTYPES** constant. Remove the existing **cTYPES** constant, and enter the following code or change your code to match.

```
# define cTYPES 6
```

3. Now you are ready to modify the existing code to insert the check boxes. The subroutines you need to modify are **SmartDocXML TypeName, SmartDocXML TypeCaption, ControlCount, ControlID, ControlTypeFromID, and ControlCaptionFromID.**

   In the **SmartDocXML Type Name** property subroutine, insert the following code.

```
case 6:
*Name = SysAllocString (cCHECKBOX (cNAMESPACE));
break;
```

   In the **SmartDoeXML Type Caption** property subroutine, insert the following code.

```
else if (wcscmp (XML Type Name, cCHECKBOX (Cnamespace))==0)
    ( *Count=2;}
```

   In the **ControlIDproperty** subroutine, insert the following code.

```
else if (wcscmp (XML Type Name, cCHECKBOX (cNAMESPACE))==0)
{       *ControlID=ControlIndex+500;
```

   In the **ControlTypeFromID** property subroutine, insert the following code. Because you are adding two check boxes to the Simple Sample smart document, you need to add two check box controls.

```
case 501:
*Type=C_TYPE_CHECKBOX;
    break;
    case 502;
    *Type=C_TYPE_CHECKBOX;
    break;
```

   In the **ControlCaptionFromID** property subroutine, insert the following code. Again because you are adding two check boxes, you need two captions.

```
case 501:
*Caption = SysAlocString (L""Show/Hide paragraph marks."");
break;
```

```
case 502:
*Caption = SysAllocString (L""Show/Hide XML tags."");
break;
```

4. The **PopularCheckbox** method specifies the initial state of the check boxes. For the two check boxes in the SimpleSample smart document, the check boxes are initially selected, as shown in the following code.

```
switch (ControlID)
{       case 501
*Checked = VARIANT_TRUE;
BREAK;
CASE 502:
*Checked = VARIANT_TRUE;
break; }
```

5. The **OnCheckboxChange** method specifies what happens when a user clicks a check box. In the SimpleSample smart document, when the user selects or clears one of the check boxes, a Microsoft Word option is turned on or off.

```
_Application objApp;
Range objRange;
Window objWindow;
View objView;
objRange. AttackDispatch (Target);
objRange. AttackDispatch (obRange.GetApplication( ));
objWindow. AttackDispatch (objAp0p. GetActiveWindow( ));
objView. AttachDispatch (objWindow.GetView( ));
switch (ControlID)
{       case 501:
        objView.setShowAll(!objView.GetShowAll( ));
        break;
        case 502:
        objView.SetShowXMLMarkup (!objView.GetShowXMLMarkup( ));
        break;}
objView.DetachDispatch ( ); objWindow. DetachDispatch ( );
objRange.DetachDispatch ( );       objApp.DetachDispatch ( );
```

6. Recompile your SimpleSample document dynamic-line library (DLL), and copy it to the deployment location that you specified earlier. When you reopen your Simple Sample smart document, delete the SimpleSample XML expansion pack, and then re-add it to the document.

Adding Radio Button: The following steps show you how to add a collection of radio buttons to the SimpleSample smart document.

1. The first thing you need to do is a add a constant for the **radio button** element in the SimpleSample scheme. Insert the following code into the general declarations section of your code module, below the existing constants.

```
#define cRADIO(xmlns) smlns L"#radiobutton"
```

2. Next your need to add 1 to the **cTYPE** constant. Remove the existing **cTYPES** constant, and enter the following code or change your code to match.

```
#define cTYPES 5
```

3. Now you are ready to modify the existing code. The subroutines you need to modify are **SmartDocXMLTypeName, SmartDocXMLTypeCaption, ControlCount, ControlID, ControlTypeFromID, and ControlCaptionFromID.**

In the **SmartDocXMLTypeName** property subroutine, insert the following code.

```
case 5:
*Name = SysAllocString (cRADIO(cNAMESPACE));
    break;
```

In the **SmartDocXMLTypeCaption** property subroutine, insert the following code.

```
case 5:
    * Caption = SysAllocString (L ""Radio buttons"");
    break;
    In the control count property subroutine, insert the following code.
    else if (wcscmp (XML TypeName, (cRADIO(cNAMESPACE))==0)
    {
        *Count = 1;
    }
```

In the **ControlID** property subroutine, insert the following code.

```
else if (wcscmp (XML TypeName, (cRADIO(cNAMESPACE)) = =0)
{
*ControlID = ControlIndex + 400;
```

In the **ControlTypeFromID** property subroutine, insert the following code.

```
case 401:
*type = C_TYPE_RADIOGROUP;
break In the ControlCaptionFromID property subroutine, insert
the following code.
case 401:
*Caption = SysAllostring (L"pick you favorite color");
break;
```

4. Next, you need to specify the value of the items in the radio buttons. To do this, you use the PopulateRadioGroup method. Use the List parameter to specify each item in the radio group, and use the Count parameter to specify the number of items in the List parameter. Insert the following code into the **PopulateRadioGroup** method subroutine.

```
switch (controlID)      {
case 401
int cElements = 5;
WCHAR * rgszRadios[ ] = {L"Red", L"Blue", L"Yellow", L"Purple", L"Green"},
for (long i = 1<= cElements; i++)
{ SafeArrayPutElement (*List, & i, SysAllocString (rgszRadios [i-1])); }
*Count = cElements;
*InitialSelected = m_1radioSelection;
break;  }
```

The initial Selected parameter specifies the number of the item to select when the radio group is initially displayed. Use-1 to indicate that none of the items in the list are initially selected. For this code to compile without error, you need to insert the following global variable.

Variable Type Variable Name
long M_IRadioSelection

* Right-click the **CActions** class in the **Class View.**
* click Add **Member Variable.**
* Type **Long** in the **Variable Type** text box.
* Type **m_IRadioSelection** in the **Variable name** text box.
* Click **OK.**

5. In the **OnRadioGroupSelectChange** method, write the code that you want to run when a user selects an cm from the list of radio buttons. You can use the Target parameter to access the element to which the control applies. In the following code, you create a Microsoft Word **Range** object from the **XML** element and replace existing text in the word. Document with the new text specified in the variable.

```
Switch (ControlID)
{       case 401:
        CString szText ("My Favorite color is");
        szText +=Value;
        SetXMLNodeText (Target, 1, szText, false);
        m_IRadioSelection = Selected;
        break; }
```

6. Finally, you create the custom function called SetXMLNode. Text that you called in the previous code.

   This function sets the text for the current XML node. To do this, add the following code to the actions.hfile before the #endif line.

```
extern void SetXMLNodeText (IDispatch*Target, int nIndex, LPCTSTR szText,
    bool bAppend)
```

   Then insert the following code at the end of the CActions, opp file.

```
void SetXMLNode Text (IDispatch*Target, int nIndex, LPCTSTR szText,
        bool bAppend = false)          {
Range objRange, onjNodeRange; XMLNodes objXML Nodes.
XMLNodes. AttachDispatch (objRange.GetXML NOdes( ));
objXMLNodes. Attach DISPATCH (OBJ RANGE, GETxml NODES ( ));
OBJxmlnODE, aTTachdisPatch (OBJxmlnODES.iTEM(NLNDEX));
objNodeRange. AttachDispatch objXMLNode. GetRange ( ));
CString str Text=objNodeRange. GetText ( );
if (bAppend)
str Text+= szText;
else
strText=szText;
objNodeRange. SetText(strText); objNodeRange.DetechDispatch ( );
objXMLNode. Dispatch ( ); objXMLNodes. DetchDispatch ( );
objRange. DetachDispatch ( );   }
```

7. Recompile your simple Sample smart document dynamic-link library (DLL), and copy it to the deplement location that you specified earlier. When you reopen your SimpleSample smart document, delete to SimpleSample XML expansion pack, and then read it to the document.

Q.10. How will you create a new dialog box class with class wizard ? Write a program in VC++ to display a dialog box.

Ans. Dialog Box: A dialog box is a rectangular window whose main role is to host or hold other. Windows controls. For this reason, a dialog box is referred to as a container. It is the primary interface of user interaction with the computer. By itself, a dialog box means nothing. The controls it hosts accomplish the role of dialog between the user and the machine.

A dialog box has the following characteristics:

- It is equipped with the system Close button. As the only system button, this button allows the users to dismiss the dialog and ignore whatever the user would have done on the dialog box.
- It cannot be minimized, maximized or restored. A dialog box does not have any other system button but Close.

- It is usually modal. The user is usually not allowed to continue any other operation until the dialog box is dismissed.
- It provides a way or the user to close or dismiss the dialog. Most dialog boxes have the OK and the Cancel buttons, although this depends on the application developer. When the dialog has the OK and the Cancel buttons, the OK button is configured to behave as if the user has pressed Enter. In that case, whatever the user had done would be acknowledged and transferred to the hosting dialog box, window or application. Pressing Esc applies the same behaviour as if the user had clicked Cancel.

Microsoft Visual C++ provides an easier way to create an application that is mainly based on a dialog box. To user this technique, state a new project and specify that you want to create an MFC Application. Like every resource, a dialog box is recognized throughout the program by an identifier. By default, the first dialog box is identified as IDD_DIALOG1. In this name, ID stands for IDentifier. The second D stands for Dialog. In most cases, you should given an explicit identifier to each resource. For example, if a dialog box is used as an employment application, you can identify it as IDD_EMPL_APP or something like that. The top section of a dialog box is made of loan bar that we call the tital bar. The main area of this bar contains a work or a group of words forming a sentech called a caption. There are two main ways you can change it. At design time, on the Properties windows, change the value of the Caption field. At run time, to change the caption, call the **CWnd::::SetWindowText ( )** method and provide the necessary string as argument. Here is an example:

```
        BOLL CExoDialog2DIg::::OnInitDialog ( ) {
        CDialog::::OnInitDialog ( );
        //TODO ; Add extra initialization here
SetWindow Text ("Go Slow, Don't RushIt!!!!");
```

return TRUE; //return TRUE unless you set the focus to a control}

By convention, a dialog box is equipped with only the system Close button x on its title bar. The Minimize and Maximize button are omitted. If you want to add the Minimized button-, at design time, set the Minimize Box property to True or checked. If you want the Maximize button set its property to True or checked. Setting one of these properties to True or checked and not the other would disable the other system button.

Q.11. Differentiate between list box and combo box ? Write these procedures to create a list box with items which can not be displayed all at a time ? Add a scroll bar at right of list box?

Ans. A list box presents a list of items to the user to choose from with each item on its own line. The user makes a selection by clicking in the list. Once clicked, an item becomes highlighted indicating that it is the current choice. A list boc can be configured to allow single or multiple selections. One of the main reasons for using a list box is to display a list of items on specific lines to the user. Sometimes the list would be very large. If the list is longer than the available space on a form, the control would provide a scroll bar that allows the user to navigate up and down and then access all items of the list. The application developer decides how many items to display on the list.

To create a list box, you can use the List Box button from the Controls toolbox.

Combo Box: A combo box is Windows control that holds a list of items. Each item can be a null terminated string or it can be made of a bitmap and a string. A combo box shares a lot of characteristics with a list but there are more variances of a combo box. As far as looks are concerned, there are two types of combo boxes. The most regularly used combo box is made of two sections. The main part is an edit box. On the right of the edit box, there is a button with a down pointing arrow. To use this type of combo box, the user would click the down pointing arrow button, This causes a list of items to display. If the user finds the desired item in the list, he or she can click it. The item clicked becomes the selection and it gets in the edit box side of the combo box. A combo box is called "drop Down" or Drop List if it is made of an edit box and a down pointing arrow, as described above. Another type of combo box is referred to as Simple. This kind displays its items like a list box. The user does not have to click an arrow button to display the list. When a simple combo box displays, the user can locate the desired item in the list and click it. The selected item becomes the value of the edit side. Like a list box or a group of radio buttons, a combo box is used to display a list of items to the user to select from. Like a series of radio buttons, a combo box allows the user to select only one item from the list. Like the list box control, if the number of items is higher than the allocated space can accommodate, the combo box is equipped with a vertical scroll bar, whether it is drop type or a simple kind.

List Box:

- In the OnInitDialog( )function of the dialog box, initialize the list box as follows:
- //TODO: Add extra initialization here.
- m-ListCountries.Add String ("Afghanistan"): m_ListCountries. AddString ("Costa Rica"); m_ListCountries. AddString ("Spain").
- m-ListCountries.Add String ("Angola");
- m-ListCountries.Add String ("Russia");
- m-ListCountries.Add String ("Canada");
- m-ListCountries.Add String ("Greece");
- m-ListCountries.Add String ("Morocco");
- m-ListCountries.Add String ("Austria");
- m-ListCountries.Add String ("Jamaica");
- m-ListCountries.Add String ("Australia");

  to add scroll b are following code will be used

```
BOOL CScrollBarDig:::OnInitDialog ( )
{ CDialog::OnInitDialog ( );
// Add "About" menu item to system menu.
// IDM_ABOUTBOX must be in the system command range.
ASSER"IDM_ABOUTBOX & OxFFFO) == IDM_ABOUTBOX);
ASSER(IDM_ABOUTBOX < OxFOOO);
CMenu* pSyMenu = GetSystemMenu (FALSE);
if (pSysMenu ! = NULL)
{   CString strAboutMenu; strAboutMenu.LoadingString (IDS_ABOUTBOX);
if (!strAboutMenu.IsEmpty ( ))
{pSysMenu->AppendMenu (MF_SEPARATOR); pSysMenu-
{AppendMenu (MF_STRING, IDM_ABOUTBOX, strAboutMenu);    }    }
SetIcon(m_hlcon, TRUE);    // Set big icon
SetIcon(m_hlcon, FALSE);   // Set small icon
//TODO: Add extra initialization here
m_ScrollBar.SetScrollRange (O, 122); return TRUE; II return TRUE unless you
set the focus to a control}
```

Q.12. Using Mouse Down event write a VB application to identify whether the right button or the left button was clicked.

Ans. Identifying left button was clicked or right button was clicked: It is possible to extend the code attached to the command button to test which mouse button (left, middle, or right) was pressed. In the MouseDown event for a command button one of the parameters passed is the Button parameter. The value of this integer parameter can be tested to determine which mouse button was used. Add the following code to the MouseDown event of the command button.

```
Private Sub Command1_MouseDown
(Button As Integer, Shift As Integer, X as Single, Y As Single)
Select Case Button
Case 1 Print "Left button clicked"
Case 2 Print 'Right button clicked"
Case 4 Print "Middlcbutton clicked"
Case Else MsgBox "Fourth Button?"
```

```
End End Select End Sub
```

The error message after the Case Else. It is a good idea to practice defensive programming. It is unlikely that a mouse with more than three buttons will ever be manufactured. But if it does happen then this code will detect such a situation. Make it a standard to always include a Case Else (along with a suitable message and action) for all Select statements that you code.

Whilst the above code works, a more professional approach is to replace the integer constants with the predefined mouse button constants that are supplied with VB. Doing this makes a program easier to read and debug (it also offers a level of protection if the designers of VB reassign the numerical values of these constants). To see what these constants are, open the VB Object Browser (click the appropriate button on the design toolbar or use the menu options View, Object Browser). Once in the Browser, select the entry VB - Visual Basic objects and procedures from the Libraries / Projects combo box. From the Classes /Modules list box select the entry MouseButtonConstants. Rewrite the Select statement as follows:

Select Case Button: Case vb LeftButton Print "Left button clicked" Case vbRightButton Print "Right button clicked"

Case vbMiddleButton Print "Middle button clicked"

Case Else MsgBox "Fourth Button?"

End End Select.

EXERCISES

1. Add a menu named "Main" to the main window and populate it with menu item "Main1". Clicking on this menu item should change the cursor shape.
2. Add a pop-up menu to your application that uses the Help drop-down menu as the pop-up menu.
3. Draw three rectangles on the window. Right clicking on one rectangle should pou-up MenuOne. Similarly, right clicking on second rectangle should pou-up MenuTwo and right clicking on third rectangle should pou-up MenuThree.
4. Add another tool button on the toolbar in the previous example. This tool button should draw a red circle on the client window area.
5. Add appropriate ToolTips to the tool buttons created above.
6. From the given example, add another green toolbar button. Add all the functionality to it as there is in the example.
7. Tie a toolbar button to a menu entry that triggers that same function.
8. Make the above toolbar as the docked one with the frame window. Add a combo box to the above toolbar.
9. Attach a menu to a dialog window as described next. It should have three main menu items. MenuOne, MenuTwo and MenuThree. MenuOne should have three sub-menus(MenuOne1, MenuOne2 and MenuOne3). Similarly, MenuTwo should have three sub-menus(MenuTwo1, MenuTwo2 and MenuTwo3) and MenuThree should have three sub-menus(MenuThree1, MenuThree2 and MenuThree3).
10. Create a tool bar using CToolBar MFC class. Add a tool button. Give a caption One to the tool button. Add the functionality to it so that wherever it is clicked it terminates the application and closes the window.

□□□

5

Controls and Dialogs

5.0 INTRODUCTION

To make your application as efficient as possible, you will add controls to a dialog box when you judge it necessary. Some and most controls are already available so you can simply customise their behaviour and appearance. Sometimes you will need a control for a specific task or for a better look. If such a control is not available, you may have to create your own.

There are two main ways a control is made part of your application. At design time, which is the time you are visually creating an application, you will select controls and place them on a dialog box. Another technique, referred to as run-time, consists of creating a control programmatically. In this case you must write all the code that specifies the control's appearance and behaviour.

Chapter Outline

- Introduction
- Command Button
- Status Bar Prompts
- Dialog Bars
- Check Boxes
- Radio Buttons
- List Box
- Combo Box
- The Slider Control
- Multiple Documents

5.1 COMMAND BUTTON

A Button is a Windows control used to initiate an action. From the user's standpoint, a button is useful when clicked, in which case the user positions the mouse on it and presses one of the mouse's buttons.

There are various kinds of buttons. The most common and regularly used is a rectangular object that the user can easily recognise. In some programming environments, this classic type is called a Command Button. There are other controls that can serve as click controls and initiate the same behavior as if a button were clicked.

From the programmer's standpoint, a button needs a host, such as a dialog box. To add a button to a dialog box, click it on the Toolbox and click in the desired location on the dialog box.

By default, when you visually create a dialog box, Microsoft Visual C++ adds two buttons: OK and Cancel.

5.1.1 Buttons Properties and Methods (MDU 2005)

The most popular button used in Windows applications is a rectangular control that displays a word or a short sentence that directs the user to access, dismiss, or initiate an action or a suite of actions. In Visual C++ applications, this control is implemented using the **Button** control  from the Toolbox window.

To add a button to an application at design time, from the Controls toolbox, click the Button button  and click the desired area on the form or dialog box. If you plan to refer to the button in your code, for example, to change its caption or some other properties, you should add a Control Variable for it. You could call it m_Button.

Like every Window control, a button is recognised by its IDentifier. Because a button is a control, by convention, its identifier's name starts with IDC (the C stands for Control). The whole name of the identifier can be in lowercase or a mix of lowercase and uppercase but it is better to make it uppercase.

A control's identifier is mainly used for resource purposes, although it can also be used to refer to a control. Otherwise, if you will be calling a control to access its properties or events, you should declare a global variable from it. To do this, in MSVC 6, you can access the ClassWizard dialog box and, in the Member Variables property page, double-click the button, specify a name, making sure that the variable is declared as a CButton and click OK. In MSVC .Net, simply right-click the button and click Add Variable. Give the desired name, making sure that the variable is declare as a CButton and click Finish. Although you can use its identifier to refer to an existing control, the associated variable is declared in the CDialog class and can be accessed anywhere.

To programmatically create a button, declare a variable or a pointer to CButton using the its default constructor. If you plan to use the button to access various member functions or variables of its parent window, you should declare it in the class. To initialise the control, call its Create() method. Here is an example :

```
class CExerciseDlg : public CDialog
{
// Construction
public:
    CButton *pButton;
    CExerciseDlg(CWnd* pParent = NULL); // standard constructor
    ...
}
CExerciseDlg::CExerciseDlg(CWnd* pParent /*=NULL*/)
    : CDialog(CExerciseDlg::IDD, pParent)
{
    //{{AFX_DATA_INIT(CExerciseDlg)
        // NOTE: the ClassWizard will add member initialization here
    //}}AFX_DATA_INIT
    m_hIcon = AfxGetApp()->LoadIcon(IDR_MAINFRAME);

    pButton = new CButton;
}
```

. . .

```
////////////////////////////////////////////////////////////////////////
// CExerciseDlg message handlers

BOOL CExerciseDlg::OnInitDialog()
{
    CDialog::OnInitDialog();

    SetIcon(m_hIcon, TRUE);     // Set big icon
    SetIcon(m_hIcon, FALSE);    // Set small icon

    // TODO: Add extra initialization here
    pButton->Create("&Submit", WS_CHILD | WS_VISIBLE,
        CRect(150, 20, 250, 43), this, 0x302);

    return TRUE; // return TRUE unless you set the focus to a control
}
```

If you dynamically create a button instead of visually adding it, you would have to write all of its messages manually.

From the userâ•™s point of view, the only things important about a button are the message it displays and the action it performs. The word or sentence that displays on top of a button is its **Caption**. By default, after adding a button to a form, the **Caption** property has focus. Therefore, if you start typing, the caption would be changed. At design time, you can set the caption with the necessary string on the Caption field of the Properties window. At run time, to change the caption of a button, call the CWnd::SetWindowText() method and pass it the necessary string as argument. Here is an example :

```
BOOL CDialog5Dlg::OnInitDialog()
{
    CDialog::OnInitDialog();

    SetIcon(m_hIcon, TRUE);     // Set big icon
    SetIcon(m_hIcon, FALSE);    // Set small icon

    // TODO: Add extra initialization here
    m_Submit.SetWindowText("Submit");

    return TRUE; // return TRUE unless you set the focus to a control
}
```

If you want to access the properties of a control without using an associated variable, you may have to call the CWnd::GetDlgItem() method. It comes in two versions as follows :

```
CWnd* GetDlgItem(int nID) const;
```

```
void CWnd::GetDlgItem(int nID, HWND* phWnd) const;
```

When calling this method, the first version allows you to assign a CWnd (or derived class) to it. The second returns a handle to the window passed as pointer. In both cases, you must pass the identifier of the control that you want access to. When using the first version, if the control is not a CWnd object, you must cast it its native class. Then you can manipulate the property (or properties) of your choice. Here is an example that accesses a button and changes its caption :

```
BOOL CDialog5Dlg::OnInitDialog()
{
    CDialog::OnInitDialog();

    SetIcon(m_hIcon, TRUE);      // Set big icon
    SetIcon(m_hIcon, FALSE);     // Set small icon

    // TODO: Add extra initialization here
    m_Submit.SetWindowText("Submit");

    CButton *btnWhatElse = reinterpret_cast<CButton
    *>(GetDlgItem(IDC_BUTTON3));
    btnWhatElse->SetWindowText("What Else?");

    return TRUE; // return TRUE unless you set the focus to a control
}
```

The second version requires a pointer to a child window that you want to access.

The most popular button captions are **OK** and **Cancel**.

The **OK** caption is set for a dialog box that informs the user of an error, an intermediary situation, or an acknowledgement of an action that was performed on the dialog that hosts the button. Visual C++ makes it particularly easy to add an OK button because in Windows applications, the OK object has a special meaning. To use an OK button, add a button to a form and, from the ID combo box, select the IDOK identifier. What makes this constant special is that the MFC recognise that, when the clicks it, if the dialog box is modal, the user is acknowledging the situation. If this dialog box was called from another window using the **DoModal()** method, you can find out if the user had clicked OK and take further action. Therefore, when the user clicks OK, the dialog box calls the **OnOK()** method.

Its syntax is :

```
virtual void OnOK();
```

Although it looks like a simple method (and it is) the **OnOK()** carries the constant value IDOK that you can use as a return value of the **DoModal()** method. Therefore, in one step, you can use the **DoModal()** method to display a modal dialog box and find out whether the user clicked OK. We will see various examples of this situation in this book.

When a dialog box is equipped with an OK button, you should allow the user to press Enter and perform the OK clicking. This is taken care of by setting the Default Button property to True or checked.

The **Cancel** caption is useful on a button whose parent (the dialog box) would ask a question or request a follow-up action from the user. A cancel button is also easy to create by simply adding a button to a dialog box and selecting **IDCANCEL** as its identifier in the

ID combo box. Setting a button's identifier to IDCANCEL also allows the user to press Esc to dismiss the dialog box.

The scenarios described for the OK and the Cancel buttons are made possible only if the compiler is able to check or validate the changes made on a dialog box. To make this validation possible, in your class, you must overload the **CWnd::DoDataExchange()** method. Its syntax is :

```
virtual void DoDataExchange( CDataExchange* pDX );
```

This method is used internally by the application (the framework) to find out if data on a dialog box has been changed since the object was displayed. This method does two things it ensure the exchange of data among controls and it validates the values of those controls. In reality, it does not inherently perform data validation, meaning it would not allow or disallow value on a control. Instead, the compiler uses it to create a table of the controls on the dialog box, their variables and values, allowing other controls to refer to it for data exchange and validating. If you want to find out the data a user would have typed or selected in a control, you would have to write the necessary code.

By default, the caption on a button is positioned in the middle and the center of the control. At design time, you can control this position using the Horizontal and the Vertical Alignments on the Properties window :

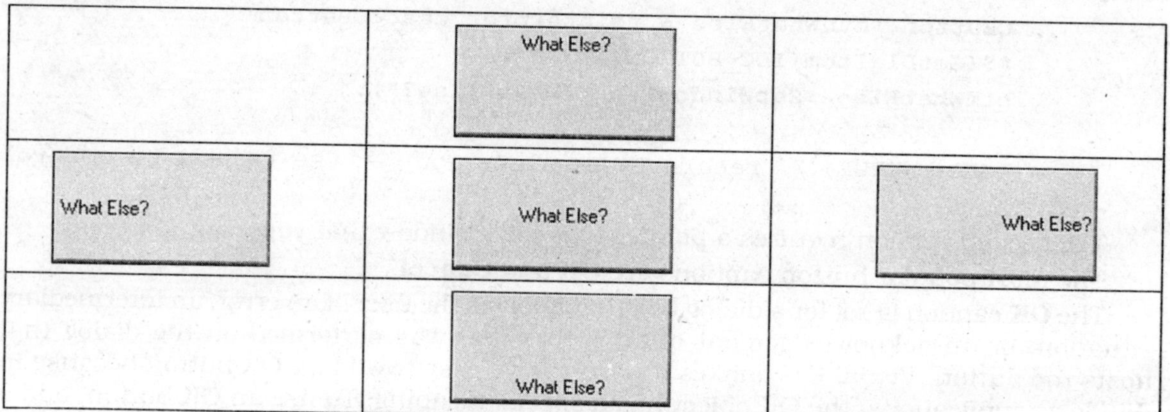

	Horz Align: Default Vert Align: Top	
Horz Align: Left Vert Align: Default	Horz Align: Default or Center Vert Align: Default or Center	Horz Align: Right Vert Align: Default
	Horz Align: Default Vert Align: Bottom	

After creating a control, to make sure that it displays when its host control comes up, set its Visible property to True or checked (the default). Otherwise, if you want it to be hidden for example to wait for an intermediary action from the user, you can set its Visible property to False or unchecked. For example, to display a control, whether it is hidden or not, call the CWnd::ShowWindow() method and pass SW_SHOW as its argument :

```
BOOL CDialog5Dlg::OnInitDialog()
{
    CDialog::OnInitDialog();
    SetIcon(m_hIcon, TRUE);        // Set big icon
```

```
    SetIcon(m_hIcon, FALSE);    // Set small icon

    // TODO: Add extra initialization here
    m_Submit.ShowWindow(SW_SHOW);
    return TRUE; // return TRUE unless you set the focus to a control
}
```

In the same way, to hide a control, call it by passing the **SW_HIDE** constant as argument.

For the user to be able to use a control such as clicking a button, the control must allow it. This characteristic of Windows objects is controlled by the **CWnd::EnableWindow()** method. Its syntax is :

```
BOOL EnableWindow(BOOL bEnable = TRUE);
```

This method is used to enable or disable a control. The default value of the argument bEnable is set to **TRUE**, which would display a control. To disable a control, set the argument to **FALSE**. Here is an example :

```
void CDialog5Dlg::OnLetItBe()
{
    // TODO: Add your control notification handler code here
    m_Submit.EnableWindow(FALSE);
}
```

5.1.2 Buttons Messages

The most regular action users perform on a button is to click it. When a user does this, the button sends a **BN_CLICKED** message. In some but rare circumstances, you may also ask the user to double-click a button. Over all, you will take care of most message handling when the user clicks a button.

There are other messages that you can handle through a button.

To close a dialog box, you can use the Win32 API's **PostQuitMessage()** function. Its syntax is :

```
VOID PostQuitMessage(int nExitCode);
```

This function takes one argument which is an integer. The argument could be set to almost any integer value although it should be **WM_QUIT**. Here is an example :

```
void CDialog5Dlg::OnBtnClose()
{
    // TODO: Add your control notification handler code here
    PostQuitMessage(125);
}
```

Although the MFC provides enough messages associated with the various controls, in some circumstances you will need use a message that is not necessarily associated with the control. In such a case, you can call the **CWnd::SendMessage()** method. Its syntax is :

```
LRESULTSendMessage(UINTmessage WPARAM wParam = 0,LPARAM lParam= 0);
```

The first argument of this method can be a Win32 message or constant. Examples would be **WM_CLOSE** or **WM_ACTIVATE**. The wParam and lParam arguments can be additional (Win32) messages.

The WinExec() function is used to run an application. Its syntax is :

```
UINT WinExec(LPCSTR lpCmdLine, UINT uCmdShow);
```

The *lpCmdLine* argument specifies the name or path of the application you want to display. The uCmdShow specifies how the application should be displayed. It uses the same values as the **CWnd::ShowWindow()** method.

5.2 STATUS BAR PROMPTS **(MDU 2008)**

A status bar is a horizontal window, usually displayed the bottom of the parent window, in which an application can display various kinds of status information.

The status bar can provide the user with more substantial information It can provide information about what the user is doing, where he is in the document he is editing, or the mode that the application is in. This information goes beyond the Caps, Num, and Scroll lock keys that Visual C++ automatically reports on the status bar.

A status bar is divided into a number of panes pasted on the status bar. It's actually easy to add additional panes to the status bar, as well as take away the panes that are already there.

5.2.1 Adding a New Status Pane

To learn just how easy a change this is, you will add a new pane to the status bar in your application that will display the color currently in use.

Before you add a new status bar pane, you need to add a new entry to the application string table for use in the status bar pane. This string table entry will perform two functions to the status bar pane. First, it provides the object ID for the status bar pane. You will use this ID for updating the pane, as you need to update the text in the pane. The second function this string table entry performs is to size the pane. To size the pane correctly, you need to provide a caption for the string table entry that is at least as wide as the widest string that you wish to place in the status bar pane. To create a status bar pane, do the following :

1. Add a new string to your application string table. Click open the ResourceView pane of the workspace. Double click at string resource, if it is already there or

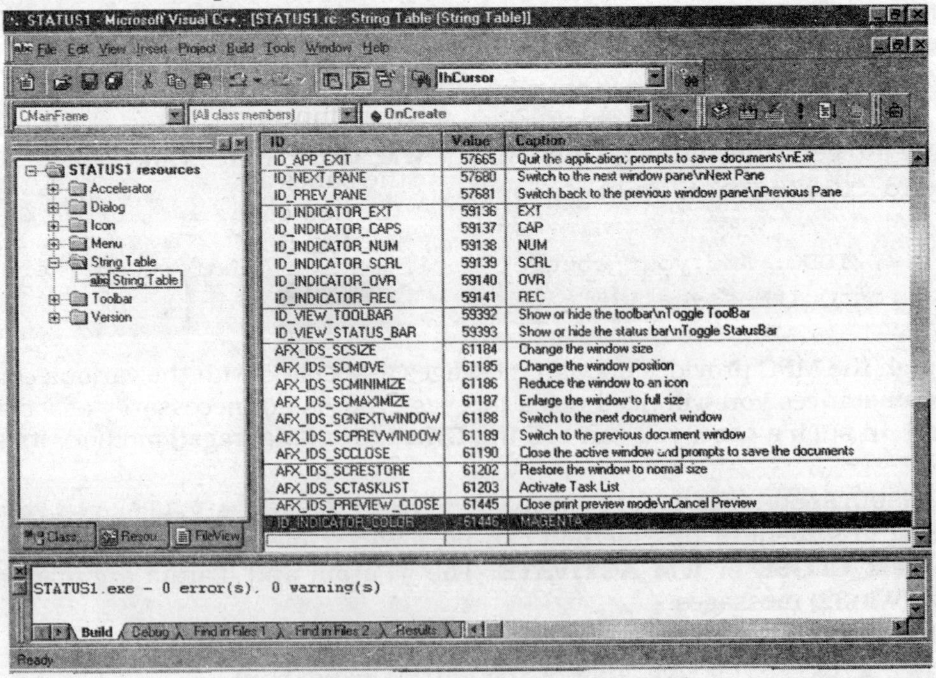

Fig. 5.1. Creating String Item in String Table

insert a new string. Specify the string ID as ID_INDICATOR_COLOR and the caption as
MAGENTA (the widest string that you will put into the status bar pane). This small table contains the object ID's of the status bar panes as table elements, in the order in which they are to appear from left to right on the status bar. Fig. 5.1.

2. A small section in the first part of the CMainFrame source code defines the status bar layout. To add the colour pane to the status bar, add the ID of the colour pane to the status bar indicator table definition, just after the message map in the source code file for the main frame. Place the colour pane ID in the table definition in the position that you want it to be on the status bar as listed below.

```
MainFrm.cpp
.
.
static UINT indicators[] =
{
    ID_SEPARATOR,        // status line indicator
    ID_INDICATOR_CAPS,
    ID_INDICATOR_NUM,
    ID_INDICATOR_SCRL,
    ID_INDICATOR_COLOR, //your status information.
};
```

If you want to drop any of the lock key indicators from the status bar, just remove them from the indicators table definition.

3. If you examine the OnCreate function, where the status bar is created (just after the toolbars are created in MainFrm.cpp), you'll see where this table is used to create the status bar with the following code :

```
If (!m_wndStatusBar.Create(this) ||
!mwndStatusBar.SetIndicators(indicators,
sizeof(indicators) / sizeof(UNIT)))
```

Once the status bar is created the Set Indicators function is called on the status bar to add the panes as they are defined in the indicator table. The strings associated with the IDs in the indicators table are used to initialise the panes and set their size. If you compile and run your application at this point, you see the new color pane on the status bar with the caption from the string table displayed within. Fig. 5.2.

Fig. 5.2. Status Bar Prompt

5.2.2 Setting a Status Bar Pane Text

Once you've added pane to the status bar, you can let the ***UPDATE_COMMAND_UI*** event do all the updating of the pane. All you need to do is add an event handler for this event on the object ID and the pane and use this event to set the pane text. Because the status bar is always visible, the UPDATE_COMMAND_UI event for the panes on the status bars triggered every time that the application is idle. This means that it is triggered after the application has finished processing just about every keystroke and mouse movement.

In the event handler, you need to create a string containing the name of the current colour (or whatever other text you want to display in the status bar pane). Next, you have to make sure that the pane is enabled. Finally, you need to set the text of the pane to the string that you have created.

To implement this in your application, you need to create an UPDATE_COMMAND_UI event handler. Once again, the ClassWizard does not know about the status bar pane, so you have to create the message handler and add it to the message map yourself. To create the message handler, add a new member function to the document class (CToolbarDoc) with a type of afx_msg void, a definition of OnUpdateIndicatorColor.(CcmdUI *pCmdUI), and an access of protected.

In this function, you followed three steps exactly: You created a string with the current colour name, made sure that the pane was enabled, and set the pane text to the string that you had created.

Now, to make sure that your new message handler is called when it is supposed to be, you need to add an ON_UPDATE_COMMAND_UI entry to the message map at the top of the document source code file.

After adding the message handler and message map entry, you should now be able to compile and run your application and see the colour status bar pane automatically updated to reflect the current drawing colour.

5.2.3 Working with Status Bars

Status bars are mostly begin objects that sit at the bottom of your application's window, doing whatever MFC instructs them to do. This consists of displaying command descriptions and the status of various keys on the keyboard, including the Caps Lock and Scroll Lock keys. In fact, status bars are so mundane from the programmer's point of view (at least they are in an AppWizard application) that they aren't even represented by a resource that you can edit like a toolbar. When you tell AppWizard to incorporate a status bar into your application, there's not much left for you to do.

Or is there? A status bar, just like a toolbar, must reflect the interface needs of your specific application. For that reason, the CStatusBar class features a set of methods with which you can customise the status bar's appearance and operation. Table 1.6 lists the methods along with brief descriptions.

Table 5.1. Methods of the CStatusBar Class

Method	*Description*
CommandToIndex()	Obtains an indicator's index, given its ID
Create()	Creates the status bar
GetItemID()	Obtains an indicator's ID, given its index
GetItemRect()	Obtains an item's display rectangle, given its index

GetPaneInfo()	Obtains information about an indicator
GetPaneStyle()	Obtains an indicator's style
GetPaneText()	Obtains an indicator's text
GetStatusBarCtrl()	Obtains a reference to the CStatusBarCtrl object represented by the CStatusBar object
SetIndicators()	Sets the indicators' Ids
SetPaneInfo()	Sets the indicators' IDs, widths, and styles
SetPaneStyle()	Sets an indicator's style
SetPaneText()	Sets an indicator's text

When you create a status bar as part of an AppWizard application, you see a window similar to that shown in Figure 1.40. (To make your own, create a project called Status1 and accept all the defaults, as you did for the Tool application.) The status bar has several parts, called panes, which display certain information about the status of the application and the system. These panes, which are marked in Figure 1.41, include indicators for the Caps Lock, Num Lock, and Scroll Lock keys, as well as a message area for showing status text and command descriptions. To see a command description, place your mouse pointer over a button on the toolbar.

5.3. DIALOG BARS **(MDU 2008)**

A dialog bar is an object of the family of Control Bars. These are objects used to complete an interface when the frame needs extra real estate to carry the desired controls of an application. One of the main justifications of control bars is that the user can display or hide them at will.

A dialog bar is a sort of intermediary between a dialog box and a toolbar. Like a dialog box or a toolbar, a dialog bar is mainly used as a container or host of other controls.

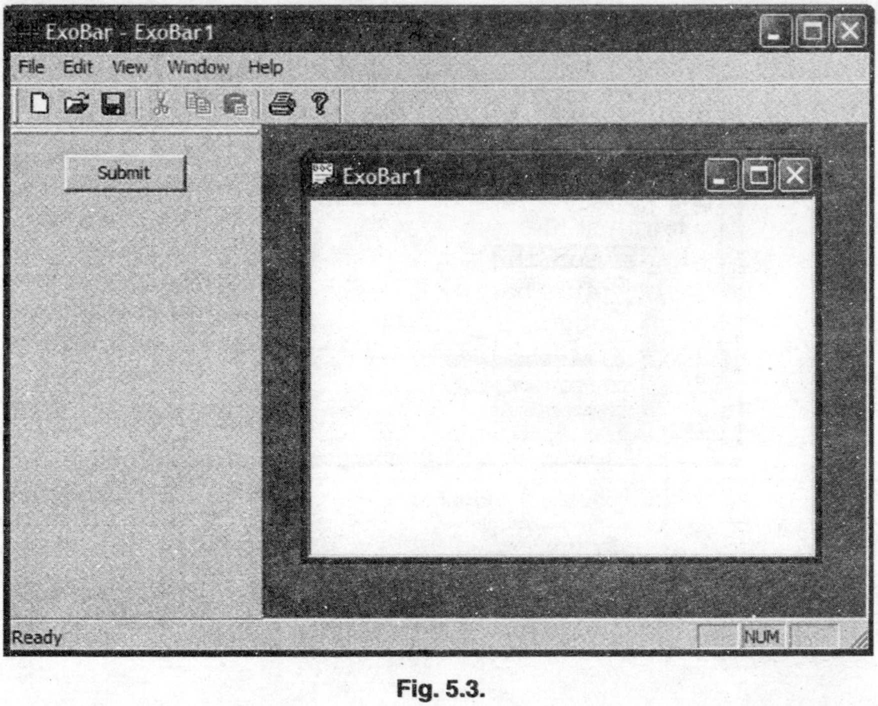

Fig. 5.3.

5.3.1. Creating a Dialog Bar

Like a dialog box, a dialog bar is created from a dialog-based resource. To create a dialog box, toolbar, you can display the Add Resource dialog box, expand the Dialog node and double-click Dialog bar. This would add a border-less dialog resource to your application.

You have two options: you can keep the dialog bar as is or you can derive a class from it. Keeping the dialog bar as created is probably easier because it is ready to be recognised by the other classes, namely the frame and the document, of your application. To add a dialog bar to your project, in the frame header class, you can declare a CDialogBar variable. To display and attach it to your interface, you can programmatically create it in the **OnCreate()** event of your frame class using one of the overloaded CDialog::Create() method :

```
virtual BOOL Create(
    CWnd* pParentWnd,
    LPCTSTR lpszTemplateName,
    UINT nStyle,
    UINT nID
);
virtual BOOL Create(
    CWnd* pParentWnd,
    UINT nIDTemplate,
    UINT nStyle,
    UINT nID
);
```

5.3.2. Introducing Dialog Bars

1. Create an MFC Application named ExoBar then click OK and Finish.
2. To add a dialog box, on the main menu, click Project -> Add Resource.
3. Expand the Dialog node and click Dialog Bar.

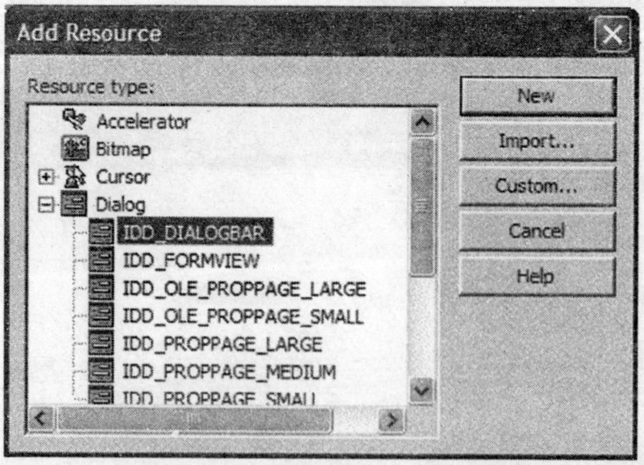

Fig. 5.4.

4. Click New.
5. Change its ID to IDD_TOOLBOX.
6. Delete the TO DO line.

7. Resize the dialog bar to have a taller height and a narrower width.
8. Add a Button to the dialog bar. Change its ID to IDC_SUBMIT_BTN and its Caption to Submit.

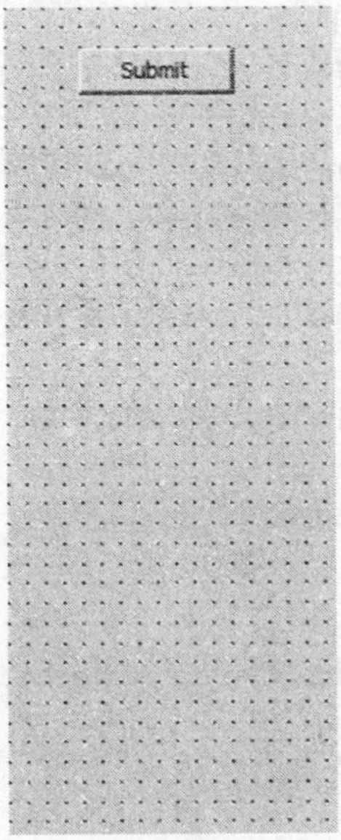

Fig. 5.5.

9. Right-click the Submit button and click Add Event Handler...
10. In the Class List, click the view class.

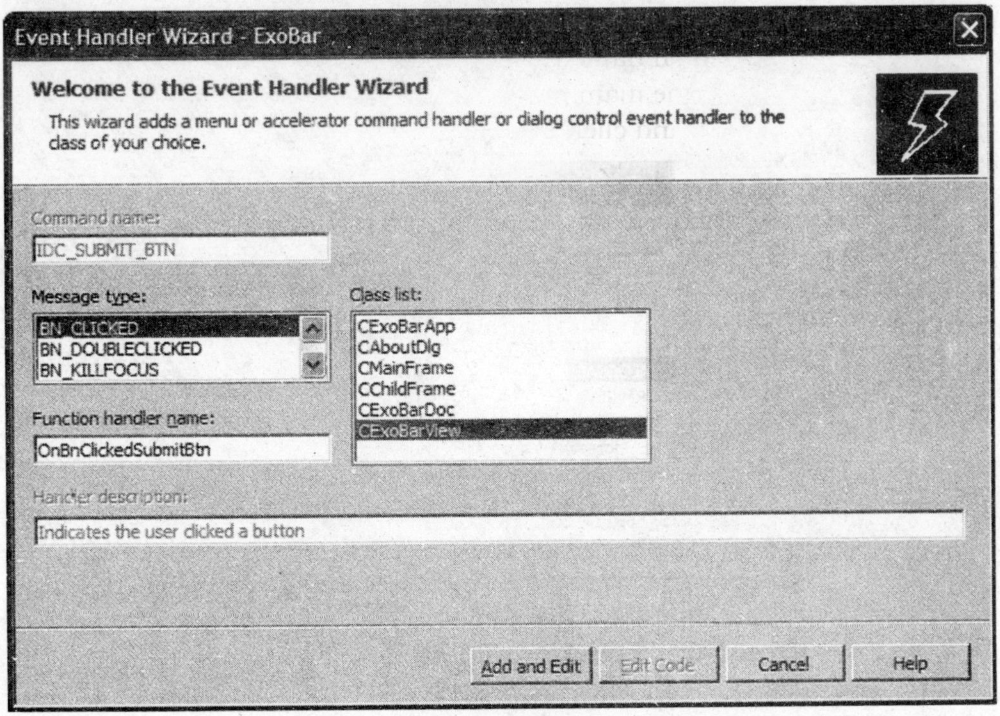

Fig. 5.6.

11. Click Add and Edit.
12. Implement the event as follows :

```
void CExoBarView::OnBnClickedSubmitBtn()
{
    // TODO: Add your control notification handler code here
    AfxMessageBox("Thank you for your consideration\n"
        "Your application has been submitted to Human Resources\n"
        "Once we have reviewed, we will contact you");
}
```

13. Access the header file of your frame and declare a CDialogBar variable named m_wndToolbox.
14. In the same section where the toolbar and the status bar are declared :

```
// MainFrm.h : interface of the CMainFrame class
//

#pragma once
class CMainFrame : public CMDIFrameWnd
{
    DECLARE_DYNAMIC(CMainFrame)
public:
    CMainFrame();
// Attributes
public:

// Operations
public:

// Overrides
public:
    virtual BOOL PreCreateWindow(CREATESTRUCT& cs);

// Implementation
public:
    virtual ~CMainFrame();
#ifdef _DEBUG
    virtual void AssertValid() const;
    virtual void Dump(CDumpContext& dc) const;
#endif

protected:  // control bar embedded members
            CStatusBar m_wndStatusBar;
            CToolBar  m_wndToolBar;
            CDialogBar  m_wndToolbox;
```

```
    // Generated message map functions
    protected:
       afx_msg int OnCreate(LPCREATESTRUCT lpCreateStruct);
       DECLARE_MESSAGE_MAP()
    };
```

15. Access the source file of the frame class and create the dialog bar as follows :

```
    int CMainFrame::OnCreate(LPCREATESTRUCT lpCreateStruct)
    {
       if (CMDIFrameWnd::OnCreate(lpCreateStruct) == -1)
          return -1;

       if  (!m_wndToolBar.CreateEx(this,  TBSTYLE_FLAT,  WS_CHILD  |
WS_VISIBLE | CBRS_TOP
       | CBRS_GRIPPER | CBRS_TOOLTIPS | CBRS_FLYBY | CBRS_SIZE_DYNAMIC)||
       !m_wndToolBar.LoadToolBar(IDR_MAINFRAME))
       {
          TRACE0("Failed to create toolbar\n");
          return -1;    // fail to create
       }

       m_wndToolBar.SetWindowText("Standard");

       // Create the dialog bar using one of its Create() methods
       if( !m_wndToolbox.Create(this,
             IDD_TOOLBOX,
             CBRS_LEFT | CBRS_GRIPPER | CBRS_TOOLTIPS | CBRS_FLYBY,
             IDD_TOOLBOX) )
       {
          TRACE0(_T("Failed to create the toolbox\n"));
             return -1;
       }

       //When the dialog bar is undocked, display a caption on its title
       bar
       m_wndToolbox.SetWindowText("Toolbox");

       if (!m_wndStatusBar.Create(this) ||
          !m_wndStatusBar.SetIndicators(indicators,
          sizeof(indicators)/sizeof(UINT)))
       {
          TRACE0("Failed to create status bar\n");
          return -1;    // fail to create
       }
```

```
    // TODO: Delete these three lines if you don't want the toolbar to
    be dockable
    m_wndToolBar.EnableDocking(CBRS_ALIGN_ANY);
    // Allow the dialog bar to be dockable on the left or the right of
    the frame
    m_wndToolbox.EnableDocking(CBRS_ALIGN_LEFT | CBRS_ALIGN_RIGHT);

    EnableDocking(CBRS_ALIGN_ANY);

    DockControlBar(&m_wndToolBar);
    DockControlBar(&m_wndToolbox);

    return 0;
}
```

16. Test the application.

5.3.3. Changing a Cursor in a Dialog Box **(V.Imp)**

Some applications, like drawing programs, use various mouse cursors depending on what is going on. Unlike most other libraries, the MFC doesn't make it easy to change a cursor. We will just view an example of doing that :

1. Start Microsoft Visual C++ .Net or MS Visual Studio .Net and create a new MFC Application named ChangeCursor.
2. Create it as Dialog Based.
3. Design or import two cursors. We will consider that one of cursors has an ID as IDC_CURSOR1 and the other has an ID as IDC_CURSOR2 .
4. Add two buttons to the dialog box. Change the ID of the first to IDC_FIRST_BTN and its Caption to First. Change the ID of the other to IDC_SECOND_BTN and its Caption to Second.
5. Add another button to the dialog box. Change its ID to IDC_DEFAULT_BTN and its Caption to Default.
6. Double-click the first button and implement its event as follows :

```
void CChangeCursorDlg::OnBnClickedFirstBtn()
{
    // TODO: Add your control notification handler code here
    SetClassLong(m_hWnd,
        GCL_HCURSOR,
        (LONG)LoadCursor(AfxGetInstanceHandle(),
            MAKEINTRESOURCE(IDC_CURSOR1)));
}
```

7. Double-click the second button and implement its event as follows :

```
void CChangeCursorDlg::OnBnClickedSecondBtn()
{
    // TODO: Add your control notification handler code here
    SetClassLong(m_hWnd,
        GCL_HCURSOR,
```

```
        (LONG)LoadCursor(AfxGetInstanceHandle(),
            MAKEINTRESOURCE(IDC_CURSOR2)));}
```

8. Double-click the Default button and implement its event as follows :

```
void CChangeCursorDlg::OnBnClickedDefaultBtn()
{
    // TODO: Add your control notification handler code here
    SetClassLong( m_hWnd,
        GCL_HCURSOR,
        (LONG)LoadCursor(NULL, IDC_ARROW) );
}
```

9. Test the application.

5.3.4. Calling One Dialog Box From Another Dialog Box

It is unusual to have a dialog-based application. This is because a dialog box is usually made to complement a frame-based application. In that case, after adding the dialog box to the application, you would also provide the ability to **call the dialog from an SDI**. In the same way, if you create an application made of various dialog boxes, at one time, you may want the user to be able to call one dialog box from another. To start, you would create the necessary dialog boxes.

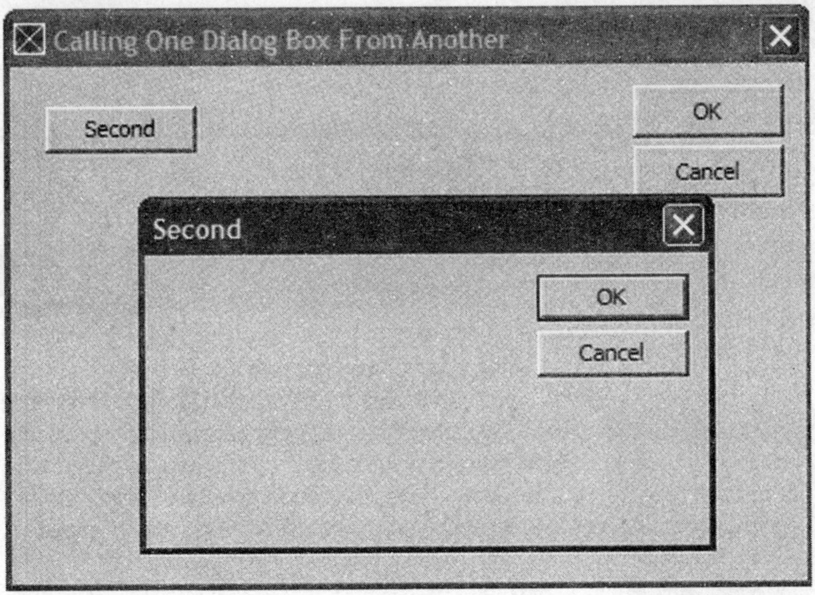

Fig. 5.7.

To call one dialog box from another, in the header file of the class of the calling dialog box, you can include the header file of the class of the called dialog box. In the section where you want to call the dialog box, you can declare a variable of that other dialog box' class and call its CDialog::DoModal() method.

5.3.4.1 Calling One Dialog Box From Another

1. Start Microsoft Visual C++ .NET or MS Visual Studio .NET.
2. Create a new MFC Application named **Primary1**.
3. Create it as Dialog Based.

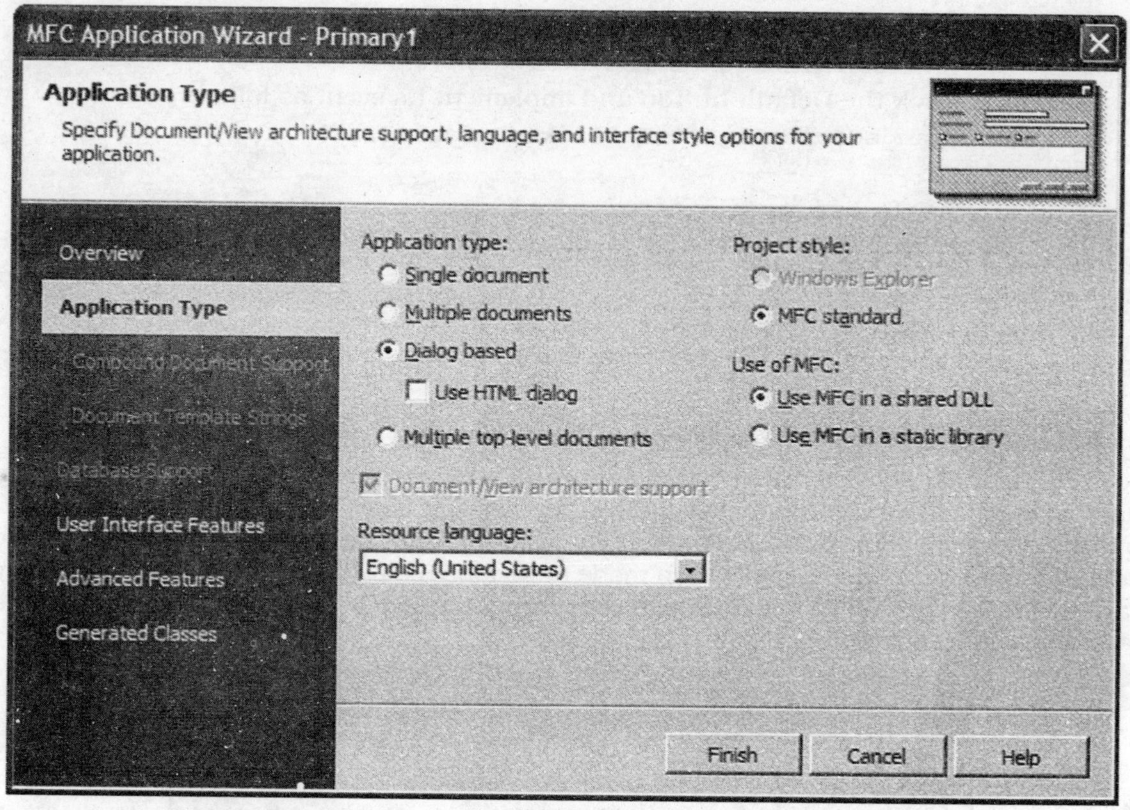

Fig. 5.8.

4. Set the Dialog Title to Calling One Dialog Box From Another.

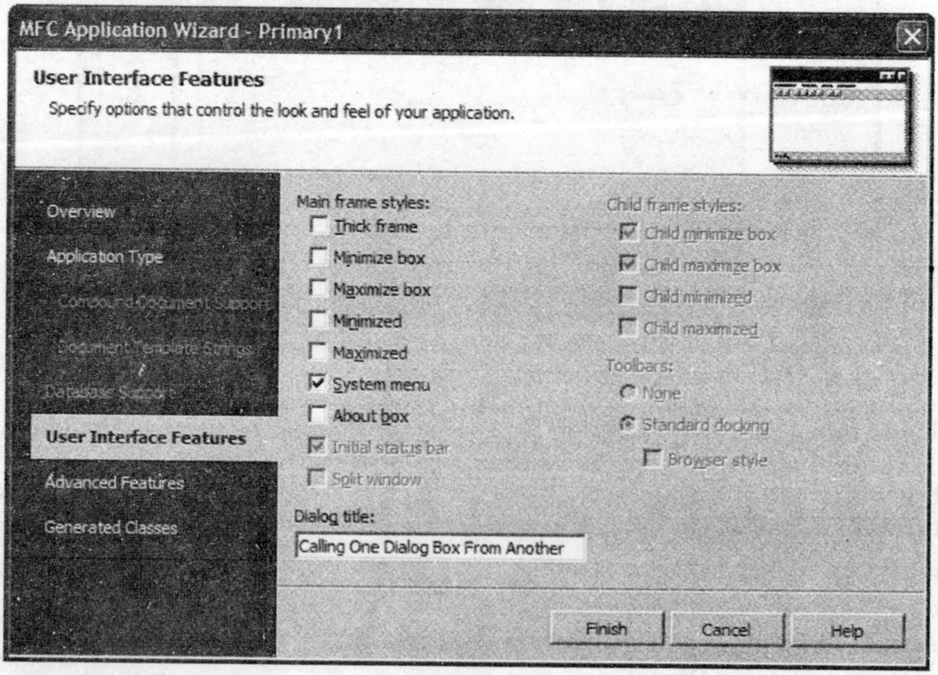

Fig. 5.9.

5. Click Finish.
6. To add another dialog box, on the main menu, click Project -> Add Resource.
7. In the Add Resource dialog box, double-click Dialog.
8. Change the ID of the new dialog to IDD_SECOND_DLG and its Caption to Second.
9. Right-click the dialog box and click Add Class....
10. Set the Class Name to CSecondDlg and base it on CDialog.
11. Click Finish.
12. From the Class View, double-click the IDD_PRIMARY1_DIALOG.
13. Using the Toolbox, add a Button to the dialog box and change its properties as follows :

    Caption : Second

    ID : IDC_SECOND_BTN
14. Double-click the new Second button and implement its event as follows :

```
// Primary1Dlg.cpp : implementation file
//

#include"stdafx.h"
#include "Primary1.h"
#include "Primary1Dlg.h"
#include ".\primary1dlg.h"
#include "SecondDlg.h"

#ifdef _DEBUG
#define new DEBUG_NEW
#endif

// CPrimary1Dlg dialog

. . . No Change

// The system calls this function to obtain
// the cursor to display while the user drags
// the minimized window.
HCURSOR CPrimary1Dlg::OnQueryDragIcon()
{
   return static_cast<HCURSOR>(m_hIcon);
}

void CPrimary1Dlg::OnBnClickedSecondBtn()
{
   // TODO: Add your control notification handler code here
   CSecondDlg Dlg;

   Dlg.DoModal();
}
```

15. Test the application

5.3.5 Calling a Dialog Box From an SDI

A Single Document Interface (SDI) is the type of application that presents a frame-based window equipped with a title bar, a menu, and thick borders. In most cases the one-frame window is enough to support the application. The advantage of creating an SDI application is the full support of the Document/View architecture.

To provide support and various options to the frame, you can add other dependent windows such as dialog boxes. When such a dialog box is needed, you can provide a menu item that the user would use to display the dialog box. For example, if you create an SDI using the wizard, it would be equipped with an About Box and the main menu would allow the user to open that dialog box. In the same way, you can add as many dialog boxes as you judge necessary to your SDI application and provide a way for the user to display them.

To display a dialog from an SDI, the easiest and most usual way consists of creating a menu item on the main menu. When implementing the event of the menu item, as normally done in C++, in the source file of the class that would call it, include the header file of the dialog box. In the event that calls the dialog box, first declare a variable based on the name of the class of the dialog box. Then call the **CDialog::DoModal()** method of the dialog box that is being called.

5.3.5.1 Calling a Dialog Box From an SDI

1. Start Microsoft Visual C++ .NET or MS Visual Studio .NET.
2. Create a new MFC Application named SDI1.
3. Create it as Single Document.
4. To add a dialog box, on the main menu, click Project -> Add Resource...
5. In the Add Resource dialog box, double-click Dialog.
6. Change the **ID** of the new dialog to **IDD_GENERIC_DLG** and its **Caption to Dependent**.
7. Right-click the dialog box and click Add Class...
8. Set the Class Name to **CDependentDlg** and base it on **CDialog**.

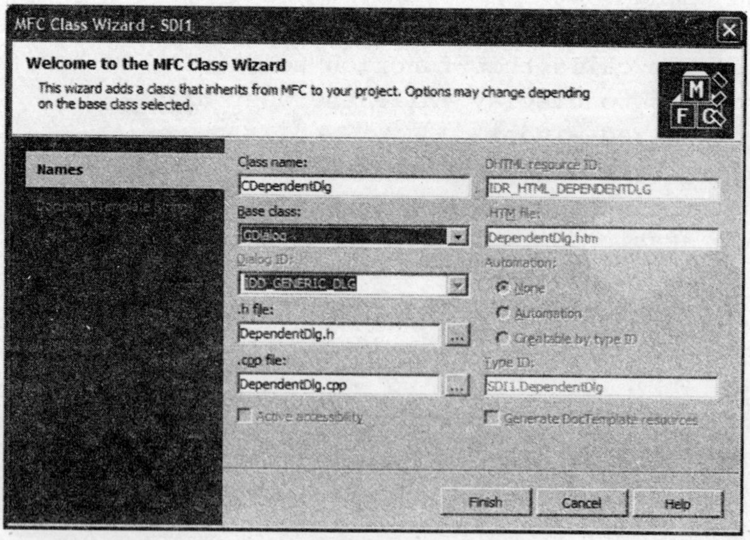

Fig. 5.10.

9. Click Finish.
10. From the Class View, double-click the IDR_MAINFRAME menu.
11. On the right side of Help, click Type Here and type &Tools.
12. Click the box under Tools and type &Dependents... and press Enter.
13. Click and drag the Tools menu category to the left to position it between View and Help .
14. Right-click the Dependents menu item and click Add Event Handler.
15. Set the Class List to the view class.

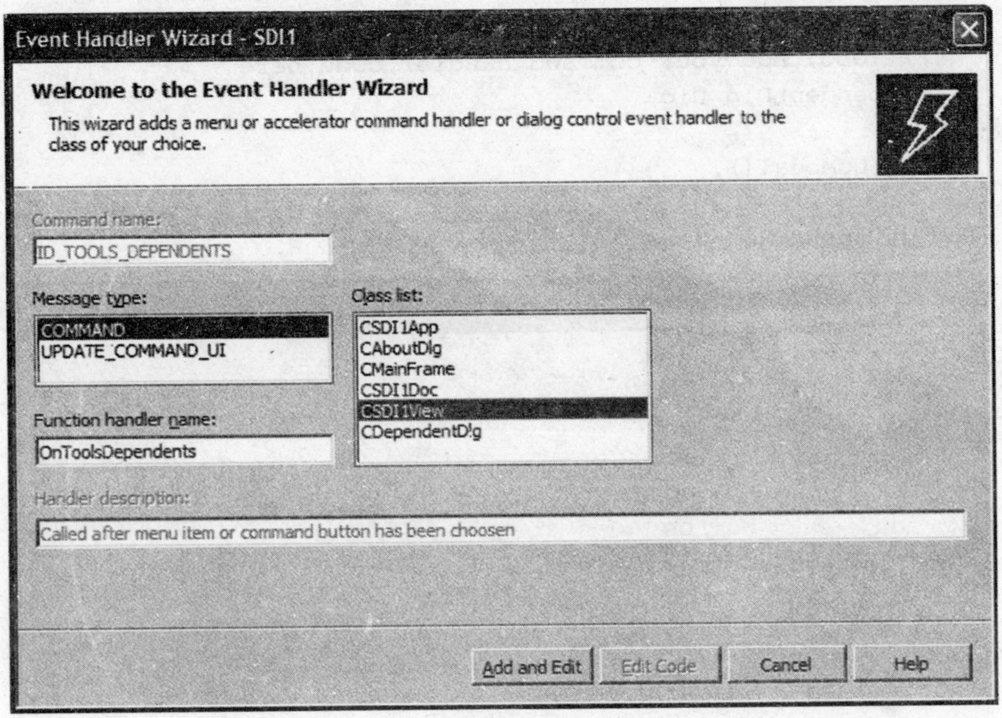

Fig. 5.11.

16. Click Add and Edit.
17. Implement the event as follows :

```
// SDI1View.cpp : implementation of the CSDI1View class
//

#include "stdafx.h"
#include "SDI1.h"

#include "SDI1Doc.h"
#include "SDI1View.h"
#include ".\sdi1view.h"
#include "DependentDlg.h"

#ifdef _DEBUG
```

```
#define new DEBUG_NEW
#endif

. . .No Change

// CSDI1View message handlers

void CSDI1View::OnToolsDependents()
{
    // TODO: Add your command handler code here
    CDependentDlg Dlg;

    Dlg.DoModal();
}
```

18. Test the application

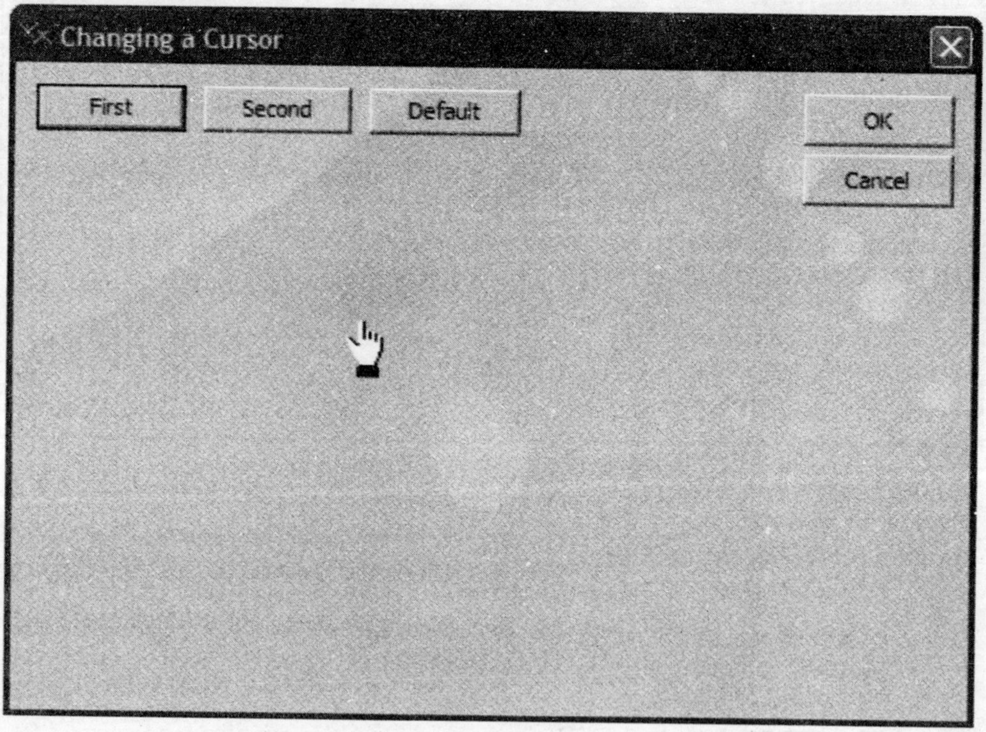

Fig. 5.12.

19. Close.

5.3.6. Controls Designs on Forms and Dialog Boxes

5.3.6.1. Selecting and Adding Controls

The controls used to provide functionality to a form or a dialog box are provided in a window called the Controls toolbox :

Visual C++ 6 Controls | Visual C++ 7 Controls

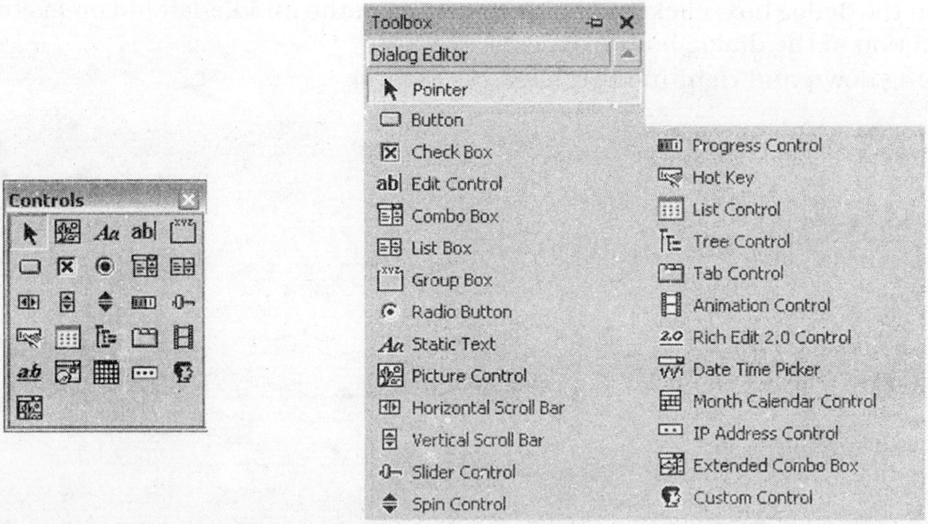

Fig. 5.13.

To make Windows controls available to your users, you will add them to their **host** object, which can be a dialog box or a form. To do this, you mostly click the desired control from the Controls toolbox and click the desired area on the host. You can keep adding controls on the dialog box as necessary.

If you want to add a control over and over again, press and hold Ctrl. Then click the control on the Controls toolbox. Then click in the desired area on the host. Every time you click, the control would be added to the form or dialog box. Once you have added enough controls, click the control again on the Controls toolbox to deselect it. You can also press Esc.You cannot select more than one control to add on a host.

5.3.6.2. Adding Controls to a Dialog Box

1. Create a new dialog-based application and name it Dialog11a.
2. Click the Toggle Grid button .
3. On the Controls window, click the button control  and click somewhere (no precision necessary) in the top left corner of the dialog box:

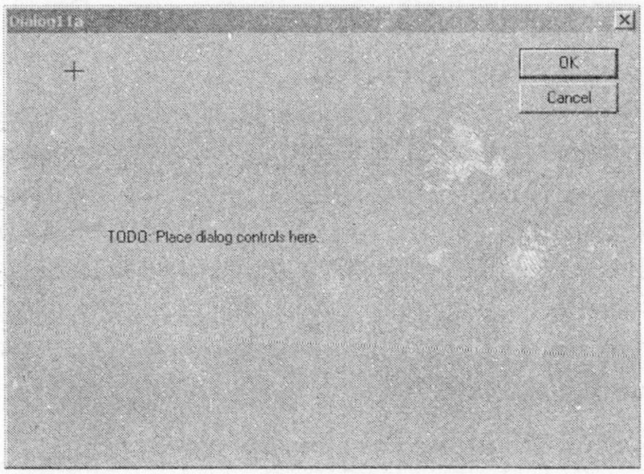

Fig. 5.14.

4. On the Controls window, click the Edit Box control ab|.
5. On the dialog box, click and hold the mouse in the middle-left (no precision needed) section of the dialog box.
6. Drag down and right (no precision necessary):

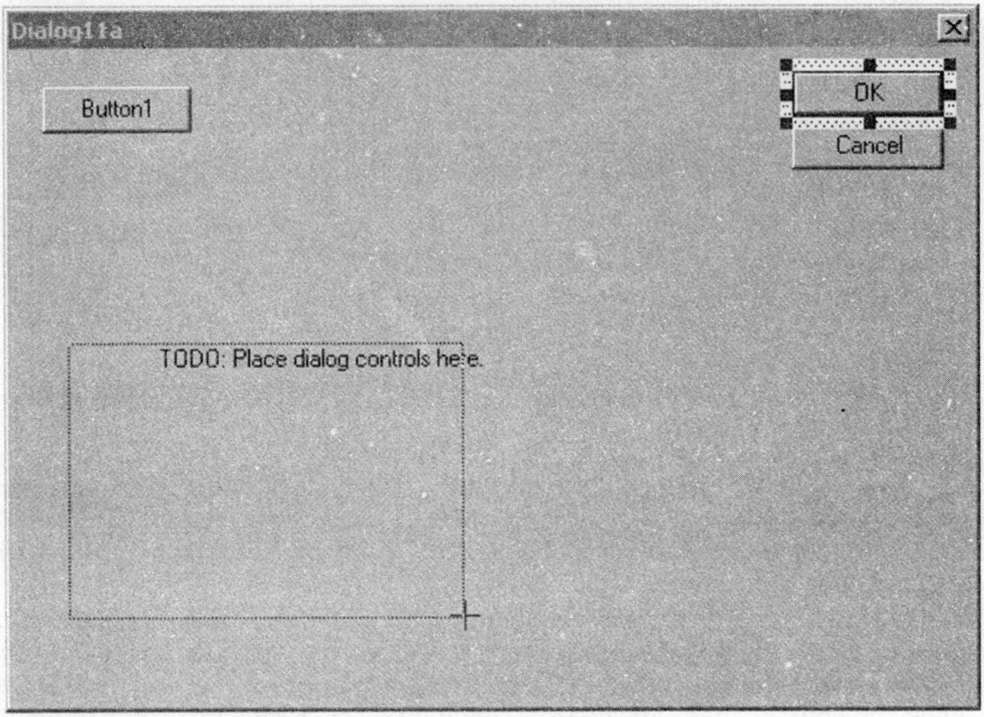

Fig. 5.15.

7. Release the mouse.
8. From the Controls window, click the Vertical Scroll Bar control and release the Ctrl key.
9. On the dialog box, click three times an unoccupied area (no precision necessary) of the dialog box.
10. To de select the Vertical Scroll Bar control , press Esc.

5.3.6.3. Selecting Controls on a Form or Dialog Box

To visually manipulate a control, you will first need to select it. To select a control, simply click it. A control that is selected is surrounded with 8 (blue) handles.

To select more than one control in the same area, click on the dialog box and draw a "fake" rectangle that encloses all of the needed controls. The first control from the selected group has 8 (blue) handles while the other control(s) from the same selected group has (have) 8 white handles (each).

To select controls at random. Click one of them. Press and hold either Shift or Ctrl. Then click each one of the needed controls.

5.3.6.4. Selecting Controls on a Dialog Box or a Form

1. Open the Form11a application.

2. If the form is not displaying, in the Resource View, expand the Form11a resources and expand the Dialog folder. Then double-click IDD_FORM11A_FORM.
   On the form, click the lower Edit box. Notice that it displays 8 handles around:

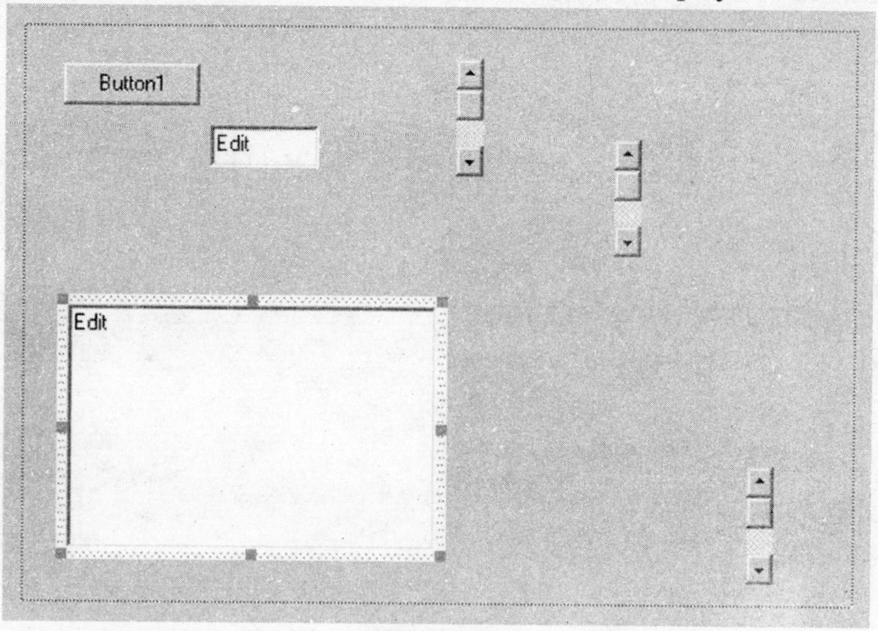

Fig. 5.16.

3 To select controls in a range, click and hold the mouse from the upper left section of the dialog box. Then drag down and to the right, drawing a rectangle that includes all three buttons in the top section of the dialog box:

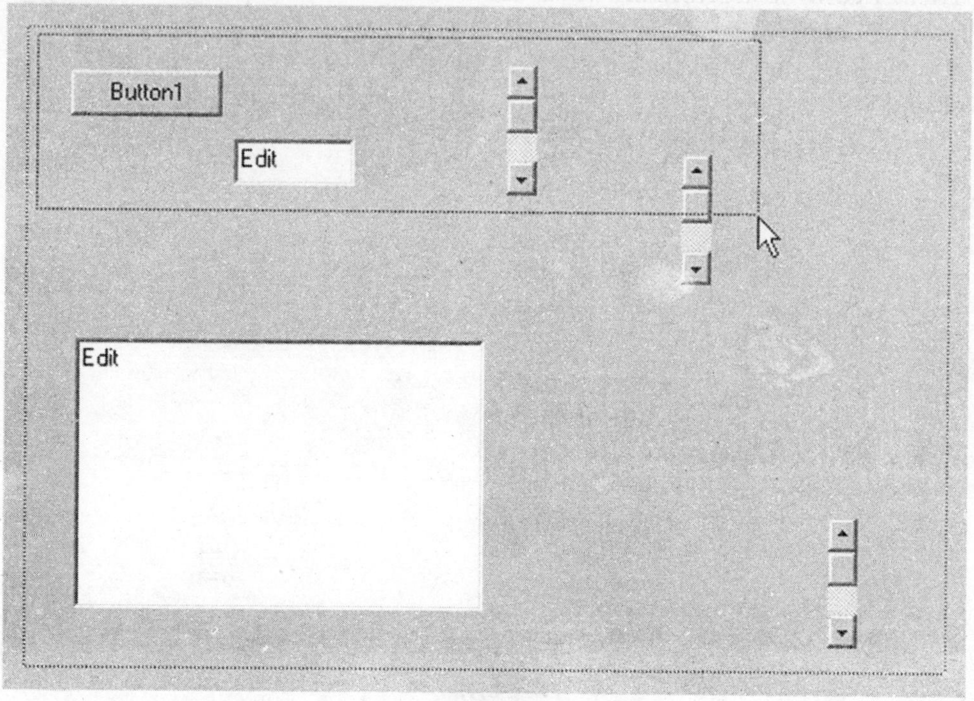

Fig. 5.17.

4. 4Notice that the first control of the selection has 8 blue handles around.

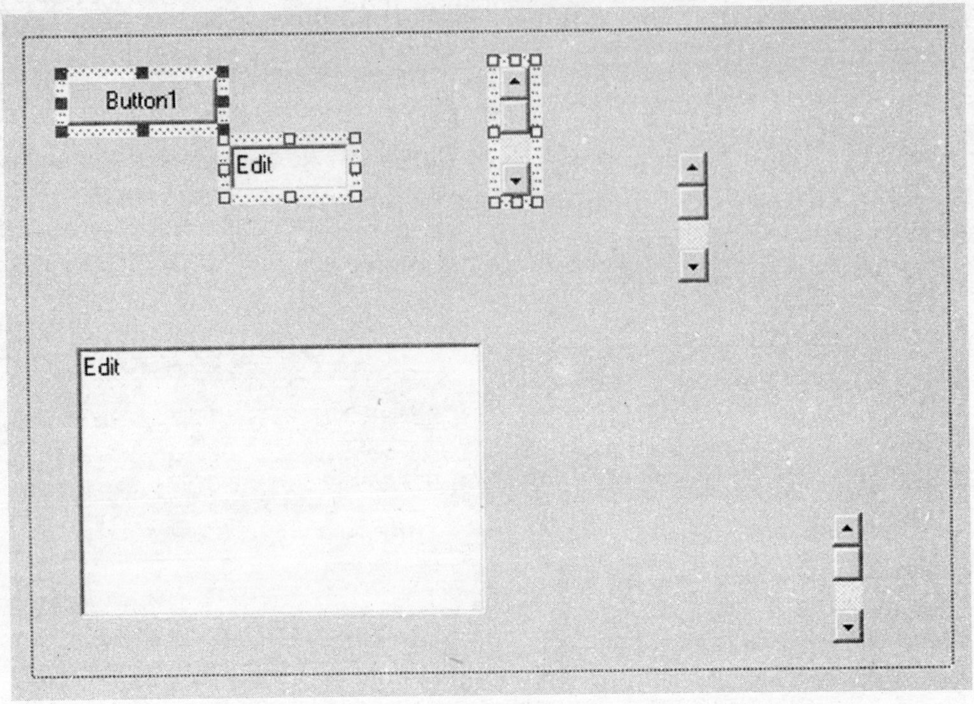

Fig. 5.18.

5. Click an unoccupied area on the dialog box to unselect everything.
6. To select controls at random, click the Button1 in the top-left section.
7. Press and hold Shift.
8. Then click the lower-right scroll bar. Click the lower left edit control.
9. Release Shift.

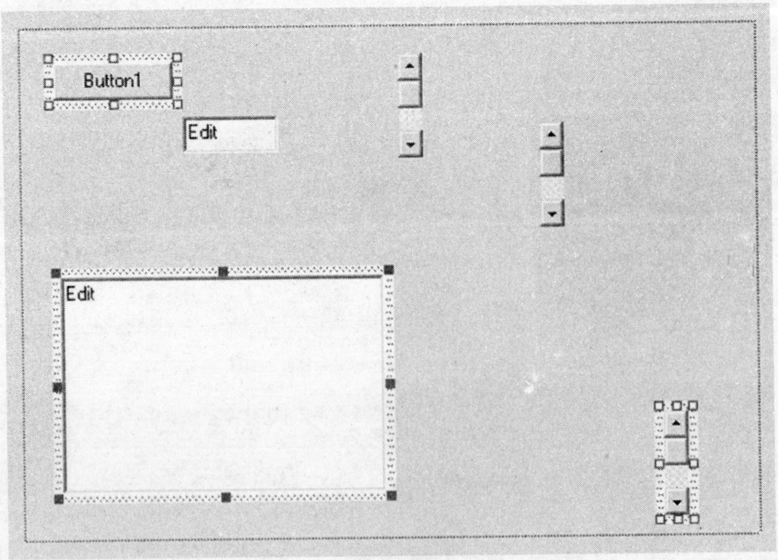

Fig. 5.19.

10. Click an empty area on the form to deselect the controls.

5.3.6.5. Controls Resizing

After you have added a control to a dialog box, it assumes either its default size or the size you drew it with. To help with the sizes of controls on the form or dialog box, Visual C++ provides a visual grid made of black points. To can display or hide the grid by clicking the Toggle Grid button . The dimensions of grid points are in dialog box units (DLUs) and are set at a default value of 5. To change this spacing, display the Guide Settings dialog box from the Layout menu and set the desired values.To change the size of a control, first select it. Then position the mouse on one of its handles. The mouse would assume a sizing cursor that indicates the possible type of resizing you can apply. The mouse cursors are :

Cursor	***Description***
	Moves the seized border in the North-West <-> South-East direction
	Shrinks or heightens the control
	Moves the seized border in the North-East <-> South-West direction
	Narrows or enlarges the control

To resize a control, that is, to give it a particular width or height, position the mouse on one of the handles and drag in the desired direction.

If the Toggle Grid button is down, in which case the dialog or form would display the grid indicators, a control can be moved or resized only on the dotted lines. If you do not want to follow the grid indicators, you have two alternatives. You can hide the grid indicators by Toggle Guides button. On the other hand, you can press and hold Alt, then click the control or one its sizing handles and drag in the desired direction.

To resize more than one control at the same time. Firs select them. Then use the following buttons from the Dialog toolbar :

Button	***Name***	***Description***
	Make Same Width will	Applies the same width to all selected controls.The width used be be that of the last control selected
	Make Same Height	Applies the same height to all selected controls.The height used will be be that of the last control selected
	Make Same Size	Applies the same width and height to all selected controls.The height used will be be that of the last control selected

5.3.6.6. Resizing Controls

1. Click the lower large Edit box to select it.
2. Position the mouse on the middle handle of the lower border.

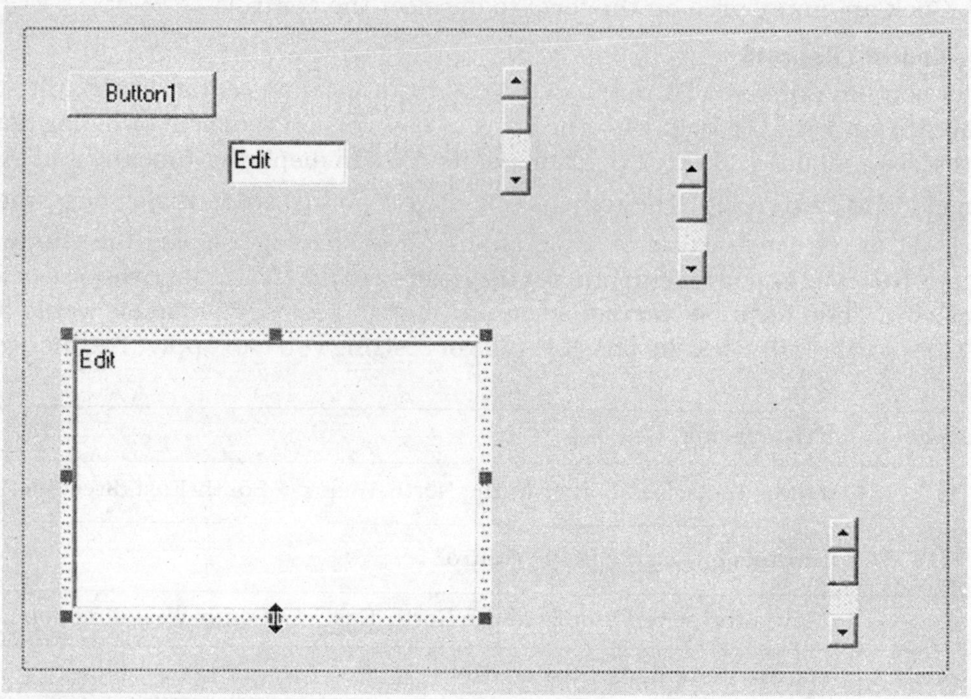

Fig. 5.20.

3. Click and drag up until the guiding horizontal line is in the middle of the control.

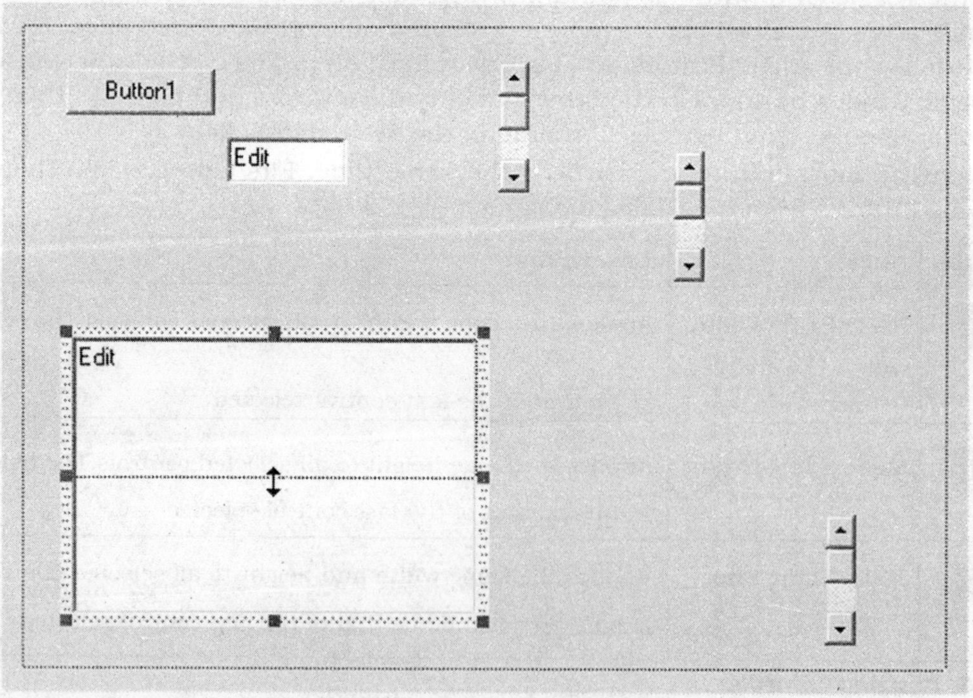

Fig. 5.21.

4. Release the mouse.
5. Open the Dialog11b application
6. Display the dialog and click the Toggle Grid button
7. Click the top Edit control. Then press and hold Ctrl
8. Click the middle Edit control and click the bottom Edit control. Release Ctrl
9. On the Dialog toolbar, click the Make Same Width button

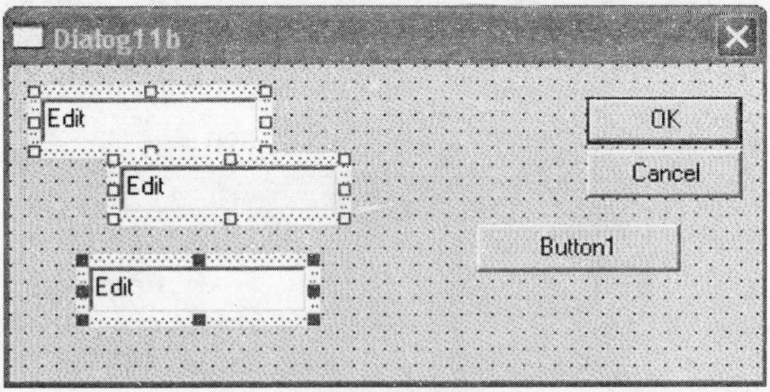

Fig. 5.22.

5.3.6.7. Controls Positions

The controls you position on a dialog box or a form assume their given place. Most of the time, these positions are not practical. You can move them around to any positions of your choice.

To move a control, click and drag it, while the mouse cursor is a cross, in the desired direction until it reaches the intended position.

To move a group of controls, first select them. Then drag the selection to the desired location.

To help with positioning the controls, Visual C++ provides the Dialog toolbar with the following buttons :

Button	*Name*	*Button*	*Name*	*Button*	*Name*
	Align Left		Align Tops		Vertical
	Align Right		Align Bottoms		Horizontal
	Across		Down		

5.3.6.8. Moving Controls

1. While the edit controls are still selected, on the Dialog toolbar, click the Down button .
2. Click an empty area on the dialog box to deselect everything.
3. Click the lower Edit control. Then press and hold Shift
4. Click the middle Edit control and click the top Edit control then release Shift

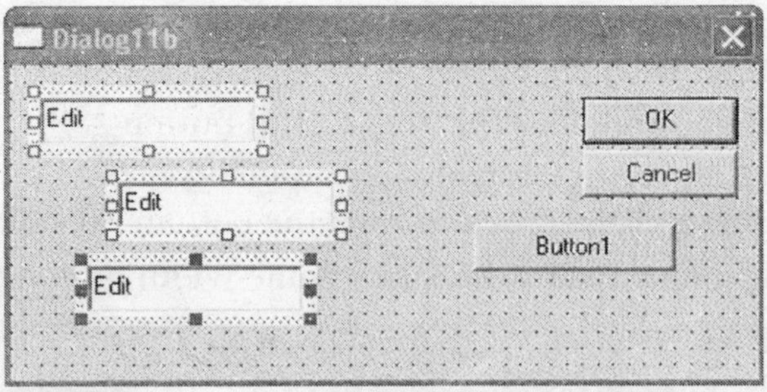

Fig. 5.23.

5. On the Dialog toolbar, click the Align Left button .

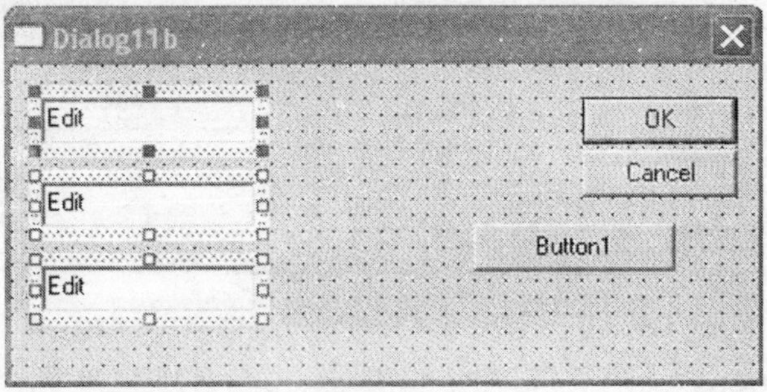

Fig. 5.24.

6. Save everything and close the application.

5.3.6.9. Tab Order

The controls you add to a form or a dialog box are positioned in a sequence that follows the order they were added. When you add a control on the host that already has other controls, regardless of the section or area you place the new control, it is sequentially positioned at the end of the existing controls. If you do not fix it, the user would have a hard time navigating the controls.

The sequence of controls navigation is called the tab order. While designing a form or a dialog box, to change the sequential order of controls, on the main menu, click Layout or Format and click Tab Order or press Ctrl + D.

5.3.6.10. Controlling Tab Order

1. On the main menu, click Layout (Visual C++ 6) or Format (Visual C++ 7) and click Tab Order.

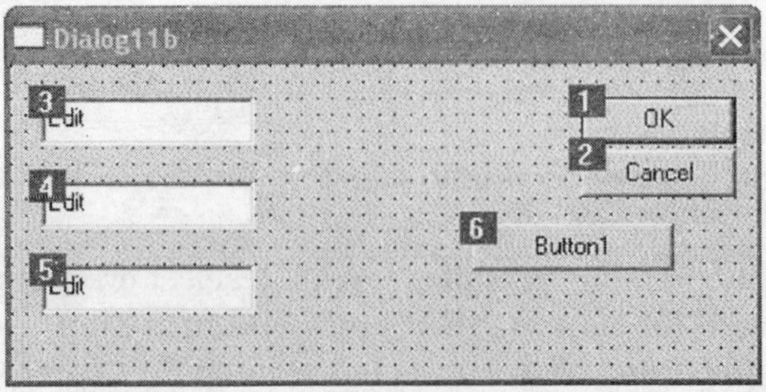

Fig. 5.25.

2. To change the order or sequence of controls, click the top Edit control. Notice that its order changes to 1.
3. Click the middle Edit control, then the lower Edit control.

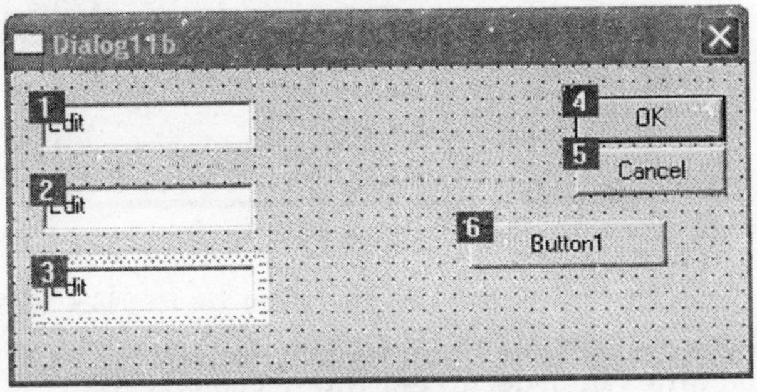

Fig. 5.26.

4. Save everything.

5.4. CHECK BOXES (MDU 2008-09)

A check box is a Windows control that allows the user to set or change the value of an item as true or false. The control appears as a small square ? When this empty square is clicked, it gets marked by a check symbol S. These two states control the check box as checked ? or unchecked ?.
Like the **radio button**, the check box doesn't indicate what it is used for. Therefore, it is usually accompanied by a label that displays an explicit and useful string. The label can be positioned on either part of the square box, depending on the programmer who implemented it.

Unlike the radio buttons that implement a mutual-exclusive choice, a check box can appear by itself or in a group. When a check box appears in a group with other similar controls, it assumes a completely independent behavior and it must be implemented as its own entity. This means that clicking a check box has no influence on the other controls.

5.4.1. Check Box Properties (MDU 2008-09)

To provide a check box to an application, on the Controls toolbox, click the Check Box button ☒ and click the desired area on the dialog box or form. If you require more than one check box, you can keep adding them as you judge necessary.Like the radio button, a check box is a button with just a different style. Therefore, it shares the same characteristics as the radio button.Like the radio button, the accompanying label of the check box can appear on the left or the right side of the square box. Also, you can control the horizontal and the vertical alignments of its label using the **Horizontal Alignment** or the **Vertical Alignment** properties.As a radio button can appear as a regular command button with 3-D borders, so can a check box. When a check box with the Push-Like property is up and then clicked, it stays down until it is clicked again, regardless of the appearance of the other controls. Remember that a check box behaves independently of the other controls even if it appears in a group.

A check box can be checked or unchecked. The check box as a control adds a third state for its appearance. Instead of definitely appearing as checked or unchecked, a check box can appear "half-checked". This is considered as a third state. To apply this behaviour to a check box, set its Tri-State property to True or add the BS_3STATE style to it :

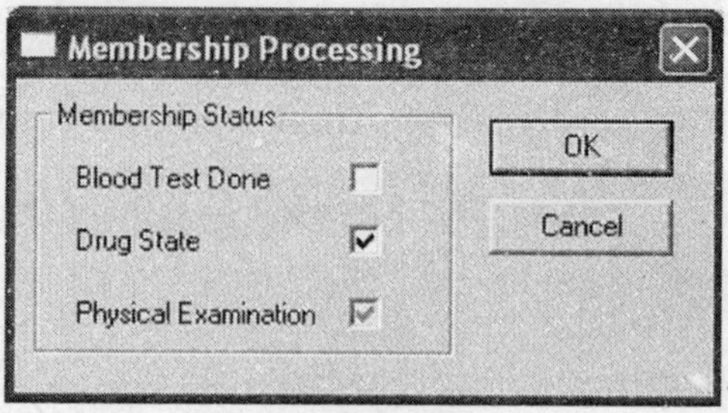

Fig. 5.27.

A check box that has the Tri-State property set or the BS_3STATE style, can appear checked, unchecked, or dimmed. When using this kind of check box, the user may have to click the control three times in order to get the necessary state.

5.4.2. Check Box Methods (V.Imp)

Like the radio button, a check box is based on the **CButton** class. Therefore, to programmatically create a check box, declare CButton variable or pointer using its constructor. Here is an example :

```
CButton *HotTempered = new CButton;
```

After declaring a variable or a pointer to **CButton**, you can call its **Create()** method to initialise the control. The syntax is the same as for the radio button. To make the button a check box, its style must include the **BS_CHECKBOX** value. If you want the control to display or hide a check mark when it gets clicked, create it with the **BS_AUTOCHECKBOX** style.

```
BOOL CCheckBoxes::OnInitDialog()
{
    CDialog::OnInitDialog();

    // TODO: Add extra initialization here
    CButton *chkDay[5];

    chkDay[0] = new CButton;
    chkDay[0]->Create("&Monday", WS_CHILD | WS_VISIBLE |
    BS_AUTOCHECKBOX | BS_LEFTTEXT,
    CRect(20, 40, 140, 55), this, 0x11);
    chkDay[1] = new CButton;
    chkDay[1]->Create("&Tuesday", WS_CHILD | WS_VISIBLE |
    BS_AUTOCHECKBOX | BS_LEFTTEXT,
    CRect(20, 60, 140, 75), this, 0x12);
    chkDay[2] = new CButton;
    chkDay[2]->Create("&Wednesday", WS_CHILD | WS_VISIBLE |
    BS_AUTOCHECKBOX | BS_LEFTTEXT,
    CRect(20, 80, 140, 95), this, 0x13);
    chkDay[3] = new CButton;
    chkDay[3]->Create("&Thursday", WS_CHILD | WS_VISIBLE |
    BS_AUTOCHECKBOX | BS_LEFTTEXT,
```

```
    CRect(20, 100, 140, 115), this, 0x14);
    chkDay[4] = new CButton;
    chkDay[4]->Create("&Friday", WS_CHILD | WS_VISIBLE |
    BS_AUTOCHECKBOX | BS_LEFTTEXT,
    CRect(20, 120, 140, 135), this, 0x15);

    return TRUE; // return TRUE unless you set the focus to a control
      // EXCEPTION: OCX Property Pages should return FALSE
}
```

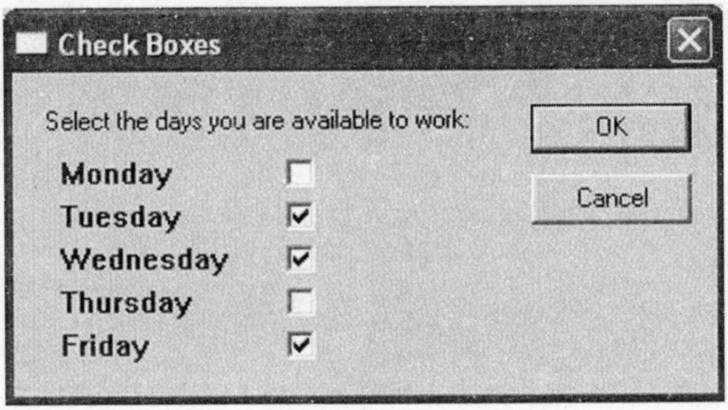

Fig. 5.28.

We mentioned already that a check box can assume one of three states: checked, unchecked, or dimmed. If you want to apply one of the states to such a check box, that is, to programmatically check, uncheck, or dim it, you can call the **CWnd::CheckDlgButton()** method whose syntax is :

```
void CheckDlgButton(int nIDButton, UINT nCheck);
```

The *nIDButton* argument is the identifier of the check box whose state you want to modify. The nCheck argument is the value to apply. It can have one of the following values :

State Value	***Description***
BST_UNCHECKED	The check mark will be completely removed
BST_CHECKED	The button will be checked
BST_INDETERMINATE	A dimmed checked mark will appear in the control's square. If the Tri-State property is not set or the BS_3STATE style is not applied to the control, this value will be ignored

Here is an example that automatically sets a check box to a dimmed appearance when the dialog box comes up :

```
BOOL CCheckBoxes::OnInitDialog()
{
    CDialog::OnInitDialog();

    // TODO: Add extra initialization here
    CheckDlgButton(IDC_CHECK_PHYSICAL, BST_INDETERMINATE);
```

```
    return TRUE; // return TRUE unless you set the focus to a control
    // EXCEPTION: OCX Property Pages should return FALSE
}
```

Alternatively, to change the state of a check box, you can call the **CButton::SetCheck()** and pass of the above state values. For a check box, to set a dimmed check mark, here is an example :

```
BOOL CCheckBoxes::OnInitDialog()
{
    CDialog::OnInitDialog();

    // TODO: Add extra initialization here
    CButton *ChkPhysical = new CButton;

    ChkPhysical=reinterpret_cast<CButton
    *>(GetDlgItem(IDC_CHECK_PHYSICAL));
    ChkPhysical->SetCheck(BST_INDETERMINATE);

    return TRUE; // return TRUE unless you set the focus to a control
    // EXCEPTION: OCX Property Pages should return FALSE
}
```

To get the current state of a check box, you can call the **CWnd::IsDlgButtonChecked()** method. If a check mark appears in the square box, this method will return 1. If the square box is empty, this method returns 0. If the check box control has the BS_3STATE style, the method can return 2 to indicate that the check mark that appears in the square box is dimmed.

5.4.3. Check Box Events

Like all other MFC buttons, the check box can natively send only the click and the double-click messages. These messages respectively are **BN_CLICKED** and **BN_DOUBLECLICKED** which originate when the user clicks or double-clicks the control. If you need to, you can use either the events of the parent window or hand code the messages yourself.

5.5. RADIO BUTTONS **(V.Imp)**

A radio button is a Windows control made of two sections: a round box O and a label. The label informs the user as to what the control is used for. The user makes his or her decision by selecting or clicking the round box. In practical usage, a radio button is accompanied by one or more other radio buttons that appear and behave as a group. The user decides which button is valid by selecting only one of them. When the user clicks one button, its round box fills with a (big) dot: ?. When one button in the group is selected, the other round buttons of the (same) group are empty. The user can select another button by clicking a different choice, which empties the previous selection. This technique of selecting is referred to as mutually-exclusive.

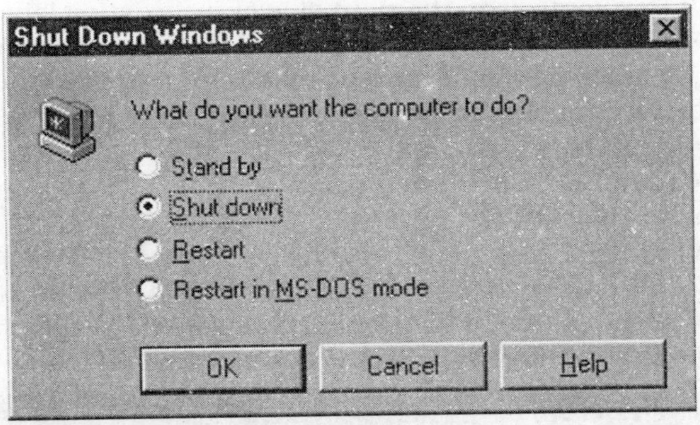

Fig. 5.29.

5.5.1. Designing Radio Buttons

Although considered a control of its own right, a radio button almost never stands by itself: radio buttons go by a group, which allows the user to select only one (from the same group). To configure radio buttons in a group, you should include them in a control that would "hold" them, as a group. The most common control to simulate this is a Group Box.

By default, a Group Box control considers the radio buttons it holds as a group. This means that, when the application is running, only one radio button in the group can be selected.

After adding the radio buttons to an MFC (or Win32) application, make sure that only one of them has its **Group** property set to true. The radio button that has the **Group** property set to true should have an easily recognisable name because the other controls will be managed from that one.

1. Start Microsoft Visual C++ (any version is fine: Microsoft Visual C++, Microsoft Visual Studio, Microsoft Visual C++.NET, Microsoft Visual Studio.NET).
2. Create a Dialog Based project called **Operations1**

   Here is the dialog box we will need:

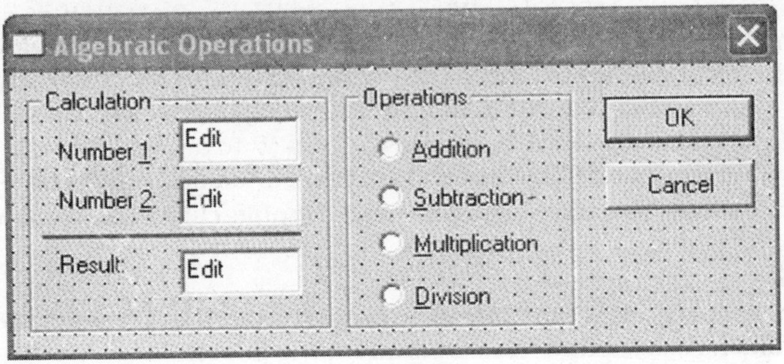

Fig. 5.30.

3. After making sure the TODO line is deleted, from the Toolbox, click the Group Box and click on the dialog box to add it. Set its caption to **Calculation**.
4. Add three Static Text controls inside the Calculation group box. Set their captions to Number &1:, **Number &2: and Result, respectively**.

5. Add three Edit Box controls to the right side of the existing labels. Set their IDs to **IDC_OPERAND1, IDC_OPERAND2**, and IDC_RESULT, respectively. Optionally, to add a line between the operands and the result, from the Toolbox, click the Picture control; on the dialog box, drag a line from the left side between Number 2 and Result and to the right.
6. Add another Group Box control to the right side of the existing one on the dialog. Set its Caption to **Operations**.
7. From the Toolbox, click the Radio Button control and click inside the Operations group box. Set the radio button's ID to IDC_ADDITION, set its Caption to and Addition, and check its Group check box (make sure it displays a check mark; in the .NET environment select True in the **Group** combo box):

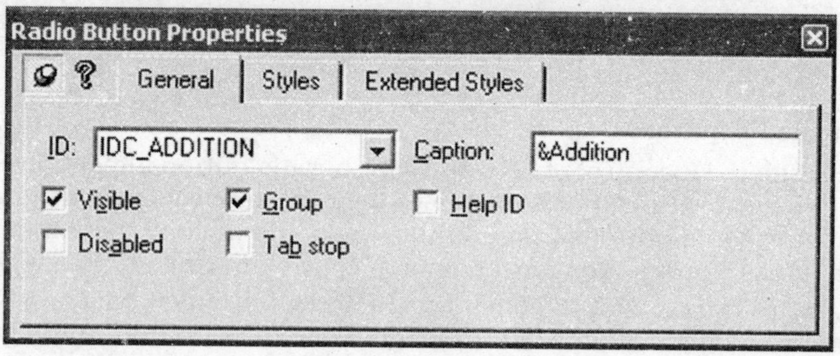

Fig. 5.31.

8. Repeat the previous step to add three more radio buttons with the IDs as **IDC_SUBTRACTION, IDC_MULTIPLICATION**, and **IDC_DIVISION** respectively. Set their captions to **&Subtraction, &Multiplication**, and **&Division** respectively. Make sure each one of these last three radio buttons has its Group check box unchecked (there must not be a check mark in it) or set to False.

5.5.2. Programming the Group of Radio Buttons

To program the radio buttons that are in a group, you only need the one that has its Group check box ON. In the MFC ClassWizard, create that radio button as an integer. The radio buttons in a group are considered an array of integers. The first radio button has a subscript of 0, the second is 1, etc. When the user clicks one of the radio buttons, you can use a condition statement (if or switch) to find out which one is selected or has focus :

1. In Microsoft Visual C++ or Studio, press Ctrl + W to access the MFC ClassWizard. In the .NET, right-click the Number 1 edit box and click Add Variable.
2. Click the Member Variables property page. Double-click each control and add the following variables:
3. Click OK.
4. On the Workspace, click the ClassView property sheet and expand the Operations1 classes folder. Also expand the COperationsDlg folder.
5. Right-click COperations1Dlg folder and click Add Member Function.
6. Set the Function Type to void and the Function Name to Calculate.
7. Click the Private radio button (since we don't intend to let this dialog communicate with an external control, we can make this function private) and click OK.

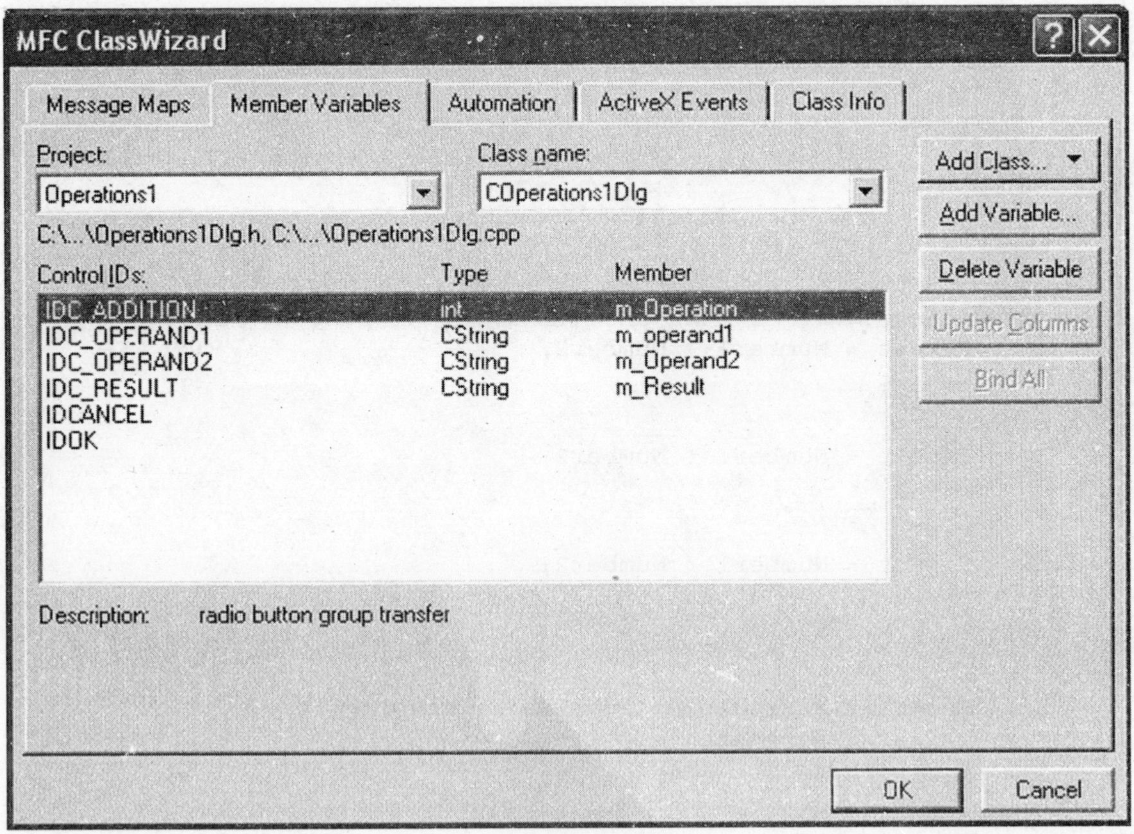

Fig. 5.32.

8. Implement the function as follows :

```
void COperations1Dlg::Calculate()
{
    double Number1, Number2, Result;

    UpdateData();

    // Evaluate the content of each operand edit box
    if( m_Operand1.IsEmpty() )
       Number1 = 0;
    else
       Number1 = atof(m_Operand1);

    if( m_Operand2.IsEmpty() )
       Number2 = 0;
    else
       Number2 = atof(m_Operand2);
    // Find out what radio button the user clicked
    // This is done considering that the radio buttons
```

```
    // are stored in an array.
    // The selected radio button can be identified by
    // its ordinal position based on the m_Operation integer
    switch(m_Operation)
    {
    case 0:
        Result = Number1 + Number2;
        break;
    case 1:
        Result = Number1 - Number2;
        break;
    case 2:
        Result = Number1 * Number2;
        break;
    case 3:
        Result = Number1 / Number2;
        break;
    }

    // Now that a result is available, display it
    // in the Result edit box
    m_Result.Format("%.2f", Result);
    UpdateData(FALSE);
}
```

9. From the ResourceView tab of the Workspace, display the dialog box and uncheck the Group check box of the Addition radio button to make it empty.
10. Press Ctrl + W to get back to the MFC ClassWizard.
11. Click the Message Maps property sheet. Click the Object ID of each radio control then, in the Messages list box, double-click BN_CLICKED, accept the function name, and click OK.
12. Click Edit Code and simply call the Calculate function every time the user clicks one of the radio buttons. Implement them as follows :

```
void COperations1Dlg::OnAddition()
{
    Calculate();
}

void COperations1Dlg::OnSubtraction()
{
    Calculate();
}

void COperations1Dlg::OnMultiplication()
{
```

```
        Calculate();
    }

    void COperations1Dlg::OnDivision()
    {
        Calculate();
    }
```

13. From the ResourceView tab of the Workspace, display the dialog box and check (put a check mark on) the Group check box of the Addition radio button.
14. Test your program.

5.6. LIST BOX

(MDU 2007)

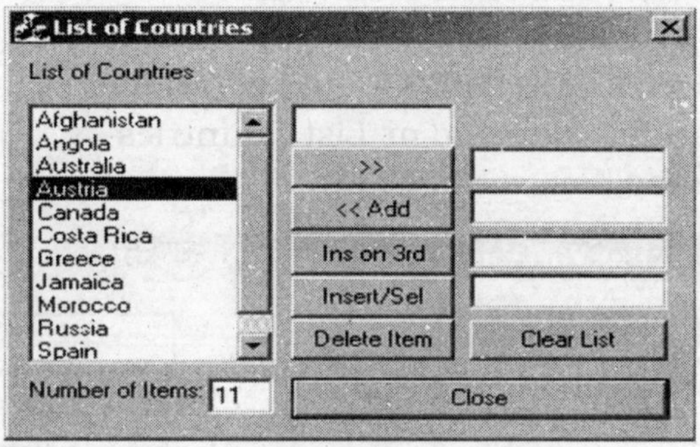

Fig. 5.33.

A list box presents a list of items to the user to choose from with each item on its own line. The user makes a selection by clicking in the list. Once clicked, an item becomes highlighted indicating that it is the current choice. A list box can be configured to allow single or multiple selections.

One of the main reasons for using a list box is to display a list of items on specific lines to the user. Sometimes the list would be very large. If the list is longer than the available space on a form, the control would provide a scroll bar that allows the user to navigate up and down and then access all items of the list. The application developer decides how many items to display on the list.

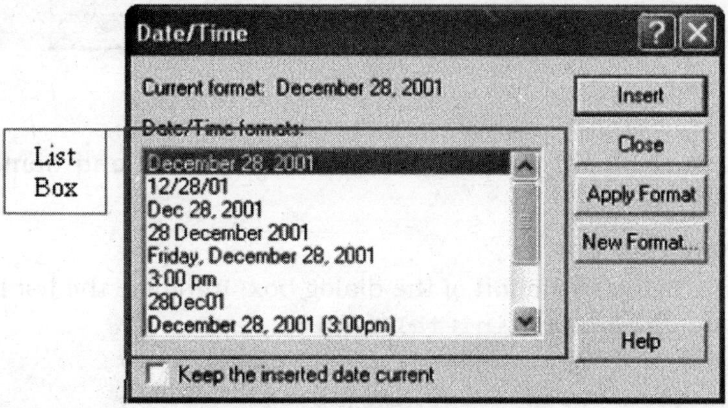

Fig. 5.34.

To create a list box, you can use the List Box button from the Controls toolbox.

5.6.1. Testing the Program as you Progress

- To test the result, anytime, press Ctrl + F5.

Creating the Application

1. Start Microsoft Visual C++
2. On the main menu, click File -> New.
3. Decide to create a Dialog Based application called List Of Countries with a title as List of Countries
4. While designing the dialog box, place a List Box control on it. Change the identifier of the list box to IDC_LIST_COUNTRIES
5. Place an Edit Box on the dialog. Call it IDC_LISTDBLCLK
6. Right-click the List Box and click ClassWizard...
7. Click the Member Variables property sheet.
8. Double-click IDC_LISTCOUNTRIES to Add a Variable
9. Set the Member Variable Name to m_List Countries
10. In the Category combo box, select Control

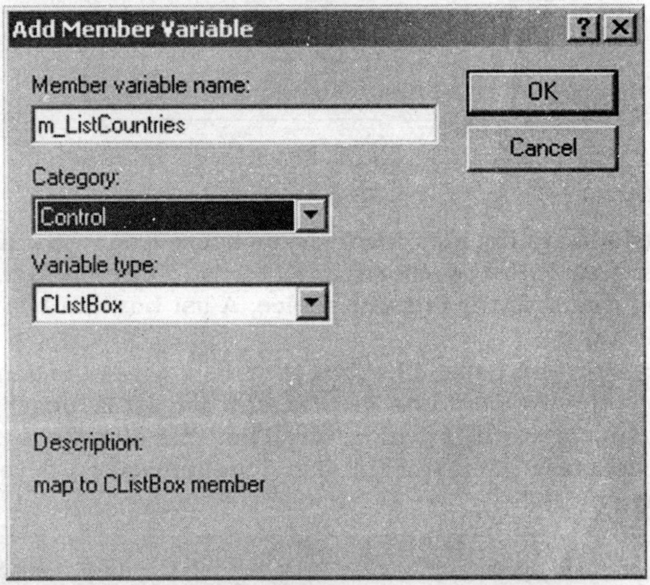

Fig. 5.35.

11. Click OK
12. Also Add a Variable for the IDC_LISTDBLCLK control and named m_ListResult based on CString.

5.6.2. Adding Items to a List Box

- In the OnInitDialog() function of the dialog box, initialise the list box as follows :

```
•    // TODO: Add extra initialisation here
•        m_ListCountries.AddString("Afghanistan");
•        m_ListCountries.AddString("Costa Rica");
•        m_ListCountries.AddString("Spain");
```

```
•        m_ListCountries.AddString("Angola");
•        m_ListCountries.AddString("Russia");
•        m_ListCountries.AddString("Canada");
•        m_ListCountries.AddString("Greece");
•        m_ListCountries.AddString("Morocco");
•        m_ListCountries.AddString("Austria");
•        m_ListCountries.AddString("Jamaica");
         m_ListCountries.AddString("Australia");
```

Select an Item Using its Position

- To set the 4th item in the list box as selected, under the last AddString() function in the OnInitDialog, add the following line.
  `m_ListCountries.SetCurSel(3);`

5.6.3. Transfer an Item to an Edit Box After Double-Clicking in the List

1. Press Ctrl + W to access the ClassWizard.
2. Click the Message Maps property sheet. In the Class Name combo box, make sure CListExoDlg is selected. In the Object IDs, click IDC_LISTCOUNTRIES
3. In the Messages list, double-click LBN_DBLCLK. Change the name of the function to OnDblclkList and click OK.
4. Click Edit Code and implement the function as follows :
5. `Void CExoListDlg::OnDblclkList()`
6. `{`
7. `m_ListCountries.GetText(m_ListCountries.GetCurSel(), m_ListResult);`
8. `UpdateData(FALSE)`
   `}`

5.6.4. Physically Transfer the Content of an Edit Box to a List Box

We have learned how to transfer a double-clicked item from the List Box to an Edit Box. The following is used if for some reason you don't want to (or can't) use the DblClick function.

1. On the dialog, add a button identified as IDC_BTN_TRANSFER
2. Set its Caption to >>
3. On its right side, add an Edit Box identified as IDC_EDIT_TRANSFER
4. Press Ctrl + W to access the ClassWizard.
5. In the Member Variables property sheet, double-click IDC_EDIT_TRANSFER
6. Set the name of the variable as m_Transfer
7. Leave its Category to Value and its Variable Type to CString.
8. Click OK.
9. In the Message Maps property sheet, double-click IDC_BTN_TRANSFER.
10. Set its function to OnTransfer and click OK.
11. While the function is still selected, click Edit Code.
12. Implement the function as follows :
13. `Void CListOfCountriesDlg::OnTransfer()`
14. `{`

15. `m_ListCountries.GetText(m_ListCountries.GetCurSel(), m_Transfer);`
16. `UpdateData(FALSE);`
    `}`

5.6.5. Add the Content of an Edit Box to a List Box

1. On the dialog, add a button identified as IDC_BTN_ADD
2. Set its Caption to << Add
3. On the right side of the previous button, add an Edit Box identified as IDC_EDIT_ADD
4. Right-click anywhere on the dialog and click ClassWizard.
5. Click the Member Variables property sheet.
6. Double-click IDC_EDIT_ADD
7. Set the name of the variable to m_Add
8. Make sure its Category is Value and its Variable Type is CString.
9. Click OK
10. On the Message Maps property sheet, click IDC_BTN_ADD.
11. In the Messages list, double-click BTN_CLICKED.
12. Set its function OnAddToList
13. Click OK and click Edit Code.
14. Implement the function as follows :
15. `Void CListOfCountriesDlg::OnAddToList()`
16. `{`
17. `UpdateData();`
18. `m_ListCountries.AddString(m_Add);`
19. `UpdateData(FALSE);`
    `}`

5.6.6. Add an Item to a Specific Position in the List

1. On the dialog, add a button named IDC_BTN_INSERT3
2. Set its Caption to Ins on 3rd
3. On the right side of that button, add an Edit Box named IDC_EDIT_INS3
4. On the main menu, click View -> ClassWizard
5. In the Member Variables property sheet, double-click IDC_EDIT_INS3
6. Set the variable's name to m_Insert3
7. Keep its Category to Value and its Variable Type to CString
8. On the Message Maps property sheet, double-click the BTN_CLICKED of the IDC_BTN_INSERT3
9. Accept the name of the function by clicking OK and click Edit Code.
10. Implement the function as follows :
11. `void CListOfCountriesDlg::OnBtnInsert3()`
12. `{`
13. `UpdateData();`
14. `m_ListCountries.InsertString(2, m_Insert3);`
15. `UpdateData(FALSE);`
    `}`

5.6.7. Insert an Item above the Selected Item on the List

1. On the dialog, add a button with a name of IDC_BTN_INSERT and a Caption of Insert/Sel
2. On the right side of that button, add an Edit Box named IDC_EDIT_INSERT
3. Press Ctrl + W.
4. On the Member Variables property sheet, double-click IDC_EDIT_INSERT
5. Set its name to m_Insert Item and click OK
6. On the Message Maps property sheet, add a BTN_CLICKED function for the IDC_BTN_INSERT button
7. Set the name of the function to OnInsertItem and click Edit Code.
8. Implement the function as follows:

```
9. void CListOfCountriesDlg::OnInsertItem()
10. {
11.         UpdateData();
12.         m_ListCountries.InsertString(m_ListCountries.GetCurSel(),
            m_InsertItem);
13.         UpdateData(FALSE);
    }
```

5.6.8. Delete an Item in the List Using its Position - Item No. 5

1. On the dialog, add a button called IDC_BTN_DELETE
2. Set its Caption to Delete Item
3. Double-click the button to access its BTN_CLICKED funtion
4. Set its name to On Delete Item and click OK.
5. Implement the function as follows :

```
6. Void CListOfCountriesDlg::OnDeleteItem()
7. {
8.      m_ListCountries.DeleteString(4);
9.      UIpdateData(FALSE);
   }
```

5.6.9. Delete the Selected Item on the List

- Change the implementation of the above function as follows :

```
• Void CListOfCountriesDlg::OnDeleteItem()
• {
•       m_ListCountries.DeleteString(m_ListCountries.GetCurSel());
•       UpdateData(FALSE);
  }
```

5.6.10. Delete/Clear the Whole List

1. On the dialog, add a button called IDC_BTN_CLEARLIST
2. Double-click the button and change its function to OnClearWholeList
3. Click OK and implement the function as follows :

```
4. Void CListOfCountriesDlg::OnClearWholeList()
5. {
6.      m_ListCountries.ResetContent();
   }
```

5.6.11. Get the Number of Items in the List

1. When deleting or inserting items in a list, you might want to know how many items are in a list box. On the dialog, under the list box, add a Static text captioned Number of Items :
2. Add an Edit Box on the right side of the previous text.
3. Set the name of the Edit Box to IDC_EDIT_COUNT
4. Press Ctrl + W to access the ClassWizard.
5. In the Member Variables property Sheet, double-click the IDC_EDIT_COUNT control.
6. Set the name of the variable to m_ListCount
7. Make sure the Category is Value. Set the Variable Type int.
8. Click OK.
9. Click the Message Maps property sheet.
10. In the Member Functions list, click OninitDialog and click Edit Code.
11. Just under the last AddString function, type
12. `m_ListCount = m_ListCountries.GetCount();`
    `UpdateData(FALSE);`
13. To update the count of items in the list box, you can use the CListBox::GetCount() function as necessary. For example, change the implementation of the OnInsertItem() function follows :
14. `void CListOfCountriesDlg::OnInsertItem()`
15. `{`
16. `    UpdateData();`
17. `    m_ListCountries.InsertString(m_ListCountries.GetCurSel(),`
    `    m_InsertItem);`
18. `    m_ListCount = m_ListCountries.GetCount();`
19. `    UpdateData(FALSE);`
    `}`

5.7. COMBO BOX

(MDU 2006)

In this exercise, you will add to the above dialog application a combo box control. You will also implement the message handler to display the item selected through MessageBox function, when an item is selected from the combo box.

1. From the Control palette, select combo box button control and place it on the dialog window. The combo box is placed with its default **ID as IDC_COMBO1**. However, again, you can change it through the property dialog box. From the property dialog box, select Droplist for type and no for ownerdraw. However, the combo box is created empty and needs to be populated with items. There are more than one ways of doing that, but for simplicity, we will populate it when the dialog box is created and initialized.
2. Create a variable m_Combo1 of type CComboBox, to be associated with this combo box using ClassWizard. To populate it, add the following code snippet into DialogDlg.cpp file as shown below :

```
BOOL CDialogDlg::OnInitDialog()
{
    CDialog::OnInitDialog();
```

```
//do not change other codes

m_Combo1.AddString("manoj");

m_Combo1.AddString("anurag");

m_Combo1.AddString("kamal");

m_Combo1.AddString("pankaj");

m_Combo1.AddString("anupam");

m_Combo1.AddString("kamini");

return TRUE; // return TRUE unless you set the focus to a control

}
```

3. When an item is selected from this combo box by clicking at the item, **CBN_SELCHANGE** message is generated. We would like our application to capture this message and process it. For that, an appropriate message handler is required to be implemented by the dialog window application. In order to do so, select the message handler as suggested by the ClassWizard for the **CBN_SELCHANGE** message. Here, handler name is OnSelchangeCombo1().
4. Add the following code into the message handler - OnSelchangeCombo1().

```
char mm[20];  //buffer to store the selected item
m_Combo1c.GetLBText(m_Combo1c.GetCurSel(),mm);
MessageBox(mm);
```

NOTE :
void GetLBText(int nIndex, CString & lpBuffer) const; returns the list item from the combo into the buffer lpBuffer (in our example mm). The item number of selected item is given by GetCurSel() function.

5. Compile the application and run. Select an item from the combo box by clicking at it. You will see the output as in Fig. 5.36.

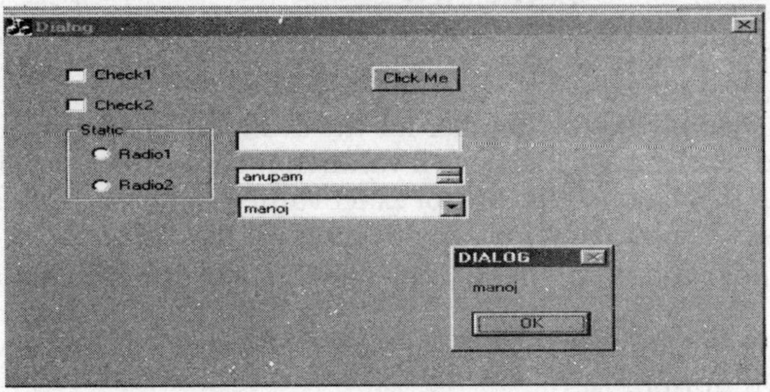

Fig. 5.36. The output window of the example application

5.8. THE SLIDER CONTROL **(MDU 2007)**

A slider is a Windows control equipped with a small bar, also called a thumb, that slides along a visible line. There are two types of sliders: horizontal and vertical :

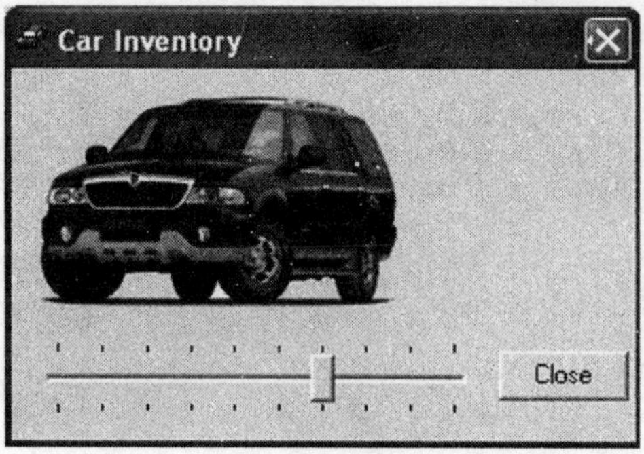

Fig. 5.37.

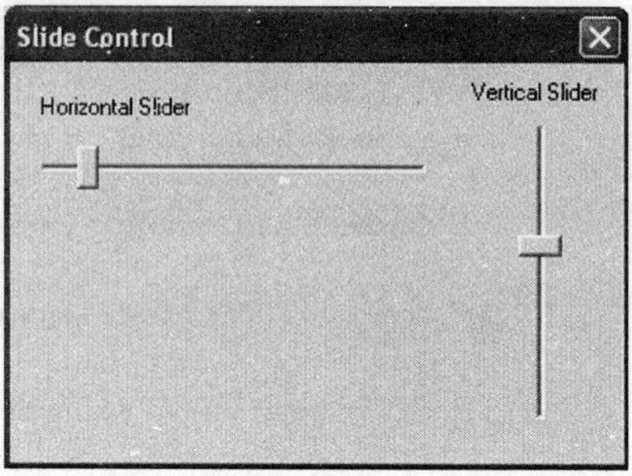

Fig. 5.38.

To use the slider control, the user can drag the thumb in one of two directions. If the slider is horizontal, the user can drag the thumb left or right. The thumb of a vertical slider can be dragged up or down. This changes the position of the thumb. The user can also click the desired position along the line to place the thumb at the desired location. Alternatively, when the slider has focus, the user can also use the arrow keys of the keyboard to move the thumb.A slider is configured with a set of values from a minimum to a maximum. Therefore, the user can make a selection included in that range. Optionally, a slider can be equipped with small indicators called ticks :

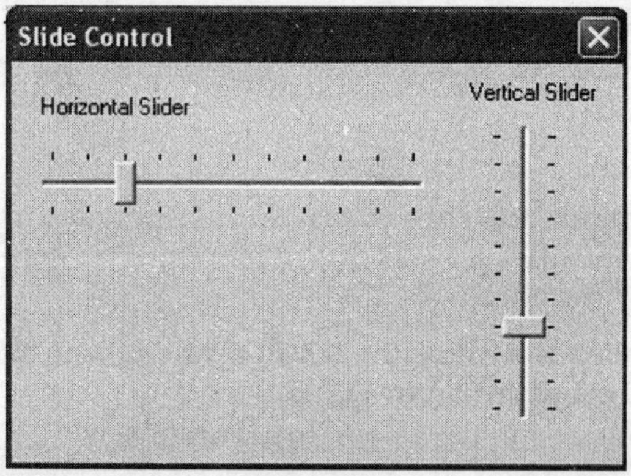

Fig. 5.39.

The ticks can visually guide the user for the available positions of the thumb mark. A slider can be used to let the user specify a value that conforms to a range. When equipped with ticks, a slider can be used to control exact values that the user can select in a range, preventing the user from setting just any desired value.

1. For this application, we will help the user display different pictures from a series of 10. If you want, you can start a new Dialog-based application named CarInventory with no About Box and set the Dialog Title to Car InventoryDelete the TODO line and the OK button.

2. Save the following pictures to your hard drive :

Fig. 5.40.

3. On the main menu, click Insert -> Resource or Project -> Add Resource...
4. On the Add Resource dialog box, click Import... and import the above pictures. If you are using MSVC 6, you may receive a message box when you import each picture, simply click OK.
5. Change the ID of the new bitmap to IDB_CIVIC, IDB_ELANTRA, IDB_ESCAPE, IDB_ESCORT, IDB_FOCUS, IDB_MARQUIS, IDB_MYSTIQUE, IDB_NAVIGATOR, IDB_SENTRA, and IDB_SEPHIA respectively.
6. Add a Picture control 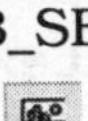 to the top side of the dialog box and change its ID to IDC_PREVIEW.
7. Type to Bitmap and set its Bitmap to IDB_CIVIC.
8. Add a Control variable for the picture control and name it m_Preview.
9. To store the values for the controls on the dialog box, in the header file of the dialog box, declare a private UINT array named CarPicture[10]

10. To initialise the array, type the following in the OnInitDialog event :

```
BOOL CSlider1Dlg::OnInitDialog()
{
    CDialog::OnInitDialog();

    SetIcon(m_hIcon, TRUE);          // Set big icon
    SetIcon(m_hIcon, FALSE);         // Set small icon

    // TODO: Add extra initialization here

    CarPicture[0] = IDB_CIVIC;
    CarPicture[1] = IDB_ELANTRA;
    CarPicture[2] = IDB_ESCAPE;
    CarPicture[3] = IDB_ESCORT;
    CarPicture[4] = IDB_MARQUIS;
    CarPicture[5] = IDB_MYSTIQUE;
    CarPicture[6] = IDB_NAVIGATOR;
    CarPicture[7] = IDB_SENTRA;
    CarPicture[8] = IDB_FOCUS;
    CarPicture[9] = IDB_SEPHIA;

    return TRUE; // return TRUE unless you set the focus to a control
}
```

11. To display default values in the dialog box, on top of its OnPaint() event, type the following :

```
void CSlider1Dlg::OnPaint()
{
    CBitmap Bmp;

    Bmp.LoadBitmap(CarPicture[0]);

    m_Preview.SetBitmap(Bmp);

    UpdateData(FALSE);

    if (IsIconic())
    {
        CPaintDC dc(this); // device context for painting

        SendMessage(WM_ICONERASEBKGND, (WPARAM) dc.GetSafeHdc(), 0);

        // Center icon in client rectangle
        int cxIcon = GetSystemMetrics(SM_CXICON);
        int cyIcon = GetSystemMetrics(SM_CYICON);
```

```
        CRect rect;
        GetClientRect(&rect);
        int x = (rect.Width() - cxIcon + 1) / 2;
        int y = (rect.Height() - cyIcon + 1) / 2;

        // Draw the icon
        dc.DrawIcon(x, y, m_hIcon);
    }
    else
    {
        CDialog::OnPaint();
    }
}
```

12. Test the application and return to MSVC.

5.8.1. Slider Creation

To provide a slider to an application, from the Controls toolbox, click the Slider button and click the desired area on the dialog box or the form.To programmatically create a slider, declare a pointer to **CSliderCtrl** using the new operator. To initialise the control, call its Create() method. Here is an example :

```
BOOL CDlgSlide::OnInitDialog()
{
    CDialog::OnInitDialog();

    // TODO: Add extra initialization here
    CSliderCtrl *TrackBar = new CSliderCtrl;

    TrackBar->Create(WS_CHILD | WS_VISIBLE,
    CRect(15, 20, 222, 50), this, 0x14);

    return TRUE; // return TRUE unless you set the focus to a control
    // EXCEPTION: OCX Property Pages should return FALSE
}
```

1. On the Controls toolbox, click the Slider button and click in lower-left section of the dialog box.
2. Save All.

5.8.2. Slider Properties

After placing a slider control on a form or other host, by default, its assumes a horizontal direction. If you want a vertical slider, change the value of the Orientation property. If you are dynamically creating the control, its default orientation would be horizontal, whose style is TBS_HORZ. If you want a vertical slider, apply the *TBS_VERT* style :

```
BOOL CDlgSlide::OnInitDialog()
{
CDialog::OnInitDialog();
```

```
    // TODO: Add extra initialization here
    CSliderCtrl *TrackBar = new CSliderCtrl;

    TrackBar->Create(WS_CHILD | WS_VISIBLE | TBS_VERT,
    CRect(20, 20, 50, 250), this, 0x14);

    return TRUE; // return TRUE unless you set the focus to a control
    // EXCEPTION: OCX Property Pages should return FALSE
}
```

The new slider appears as one line, horizontal or vertical, that guides the user with the area to slide the thumb. When sliding the thumb along the line, the user can set only the value where the thumb is positioned. Alternatively, if you want the user to be able to select a range of values instead of just a value, at design time, you can set the **Enable Selection** property to True. This is equivalent to adding the TBS_ENABLESELECTION style. A slider equipped with this style displays a 3-D "whole" in the body of the slider :

Fig. 5.41.

The selection area allows the user to select a range of values.The thumb of a slider can assume one of three shapes. By default, it appears as a rectangular box. Alternatively, you can convert one of its shorter borders to appear as an arrow. The shape of the thumb is controlled at design time by the Point property. Its default value is Both, which gives it a rectangular shape. You can get this same shape by omitting or adding the TBS_BOTH value.For a horizontal slider, you can make the thumb's arrow point to the left by changing the Point property to Top/Left. If the slider is horizontal, this Point value would orient the thumb arrow to the top :

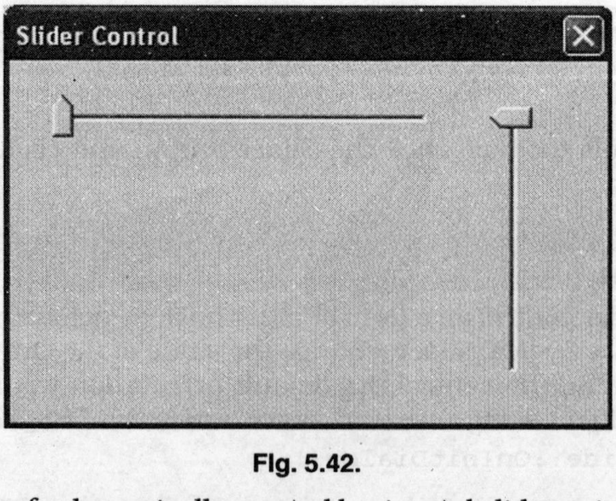

Fig. 5.42.

To make the thumb of a dynamically created horizontal slider point up, add the TBS_TOP style. If the slider is vertical, to point its thumb to the left, add the TBS_LEFT style to it.If you want the thumb to point down for a horizontal slider, set the Point property to Bottom/ Right. This same value would make the thumb of a vertical slider point n the right direction :

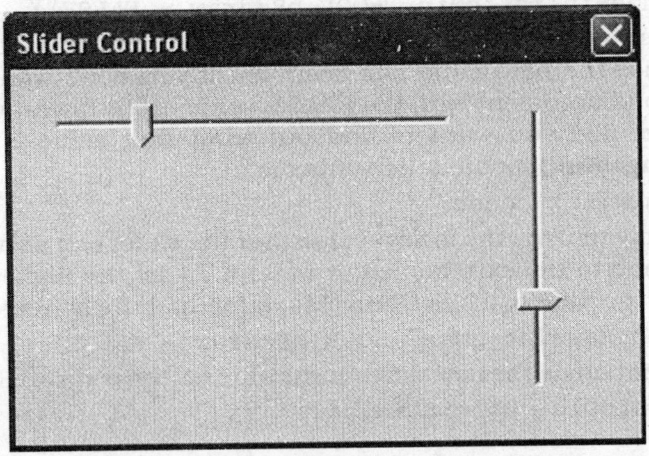

Fig. 5.43.

To point the thumb up for a horizontal slider you are programmatically creating, add the **TBS_BOTTOM**. For the thumb of a vertical slider to point right, add the **TBS_RIGHT** to it.If you want to guide the user with some ticks on the control, at design time, set the Tick Marks property to True. If you are dynamically creating the slider and you want it to display ticks, simply adding the either the **TBS_VERT** or the **TBS_HORZ** style equips the slider with ticks. If you don't want to display the ticks at all, at design time, clear the Tick Marks property or set its value to False.The ticks are positioned on the side the thumb is pointing. If the slider is created with the Both value for the Point property or the TBS_BOTH style, the ticks would appear on both sides the thumb.The thumb of a slider is used to scroll from one minimum value to another. The range of these extreme values can be divided in regular increments that can further guide the user with value selection. To display where these increments are set, at design time, set thc Auto Ticks property to True or add the **TBS_AUTOTICKS** style to a dynamic slider.

1. While the slider control is selected on the dialog box, on the Properties window, change its ID to IDC_SLIDER.
2. Check its Auto Ticks check box or set it to True.
3. Check its Tick Marks check box or set it to True.
4. Set its Point property to Top/LeftSet to True the Tooltips property of the slider control (MSVC 7).
5. Add a CSliderCtrl Control variable for the slider and name it m_CarSlider.
6. Save All.

5.8.3. Slider Methods **(V.Imp)**

As seen earlier, the slider control is based on the **CSliderCtrl** class. Therefore, we saw that we can dynamically create a slider by declaring pointer to **CSliderCtrl** and call its **Create()** method to initialise it. Once a slider has been designed and received the desired style, you can programmatically use it and provide the necessary feedback to the user.A slider is a control that provides a range of values between which the user can navigate using the control's thumb or by clicking on the slider's line. Usually, the first aspect you may need to configure on your control is to specify its limit values. To specify the minimum value of a slider control, you can call the **CSliderCtrl::SetRangeMin()** method. Its syntax is :

```
void SetRangeMin(int nMin, BOOL bRedraw = FALSE);
```

The *nMin* value is the new minimum value that the slider can assume. The control is typically redrawn once the new value has been set. If you don't want the control to be redrawn, pass a second argument with the FALSE value. If the lowest value of the control has already been set and you want to find out what that value is, you can call the **CSliderCtrl::GetRangeMin()** method. Its syntax is :

```
int GetRangeMin() const;
```

This method simply returns the lowest value that the slider can assume when the user has dragged the thumb to the extreme left or bottom.To set the highest value of a slider control, you can call the SliderCtrl::SetRangeMax() method. Its syntax is :

```
void SetRangeMax(int nMax, BOOL bRedraw = FALSE);
```

The *nMax* argument holds the new maximum value for the control. Here is an example :

```
BOOL CControlsDlg::OnInitDialog()
{
    CDialog::OnInitDialog();

    // TODO: Add extra initialization here
    m_Slider.SetRangeMin(0);
    m_Slider.SetRangeMax(50);

    return TRUE; // return TRUE unless you set the focus to a control
    // EXCEPTION: OCX Property Pages should return FALSE
}
```

If the maximum value of the control had previously been set, you can find it out by calling the **SliderCtrl::GetRangeMax()** method. Its syntax is :

```
int GetRangeMax() const;
```

This method returns the highest value that the slider control can have.To set both the minimum and the maximum values of the slider with one line of code, you can call the **CSliderCtrl::SetRange()** method. Its syntax is :

```
void SetRange(int nMin, int nMax, BOOL bRedraw = FALSE);
```

The *nMin* and the *nMax* arguments hold the lowest and the highest respective values of the control. Here is an example :

```
BOOL CControlsDlg::OnInitDialog()
{
    CDialog::OnInitDialog();

    // TODO: Add extra initialization here
    m_Slider.SetRange(0, 50);

    return TRUE; // return TRUE unless you set the focus to a control
    // EXCEPTION: OCX Property Pages should return FALSE
}
```

If the control is already functioning and you want to know its limit values, you can call the **CSliderCtrl::SetRange()** method whose syntax is :

```
void GetRange(int& nMin, int& nMax) const;
```

This method returns two values, namely the lowest value, as *nMin*, and the highest value, as *nMax*.

Once the minimum and maximum values have been set, the user can slide the thumb to select a value or a range. This value is what mostly interests you. While sliding the thumb, the value of the slider is called its position. At startup or at any time, you can set a specific position for the thumb. This can be done by calling the **CSliderCtrl::SetPos()** method. Its syntax is :

```
void SetPos(int nPos);
```

The nPos argument holds the new position of the slider. Here is an example :

```
BOOL CControlsDlg::OnInitDialog()
{
    CDialog::OnInitDialog();

    // TODO: Add extra initialization here
    m_Slider.SetRange(0, 50);
    m_Slider.SetPos(32);

    return TRUE; // return TRUE unless you set the focus to a control
    // EXCEPTION: OCX Property Pages should return FALSE
}
```

The value specified using the SetPos() method should be in the range nMax - nMin of the SetRange() method. If there is a possibility that this value is outside the valid range, you can call the CSliderCtrl::VerifyPos() method to check it. Its syntax is :

```
void VerifyPos( );
```

When the position of the thumb has change and you want to find out what it is, call the CSliderCtrl::GetPos() method whose syntax is :

```
int GetPos() const;
```

If the slider control was specified to let the user select a range, you can define your own selected range at any time by callin the CSliderCtrl::SetSelection() method. Its syntax is :

```
void SetSelection(int nMin, int nMax);
```

When calling this method, make sure you specify the nMin and the *nMax* values so that this nMin is greater than the minimum value of the slider and this *nMax* is less than the highest possible value of the slider. Furthermore, the value of this nMin must be less than that of nMax. This relationshiped can be illustrated as follows :

```
Minimum <= nMin < nMax <= Maximum
```

Here is an example :

```
BOOL CControlsDlg::OnInitDialog()
{
    CDialog::OnInitDialog();

    // TODO: Add extra initialization here
```

```
    m_Slider.SetRange(0, 50);
    m_Slider.SetPos(32);
    m_Slider.SetSelection(22, 42);

    return TRUE; // return TRUE unless you set the focus to a control
    // EXCEPTION: OCX Property Pages should return FALSE
}
```

If a selected range has been performed on the slider, if you want to get the minimum and the maximum values of the selection, you can call the CSliderCtrl::GetSelection() method whose syntax is:void GetSelection(int& nMin, int& nMax) const;This method returns two values, the minimum as nMin and the maximum as nMax.If the slider controla is configure to displays ticks, you can specify their frequncy with a call to the CSliderCtrl::SetTicFreq() method. Its syntax is :

```
void SetTicFreq(int nFreq);
```

1. In the OnInitDialog event of the dialog class, set the range of values of the slider to 0 to 9 and the frequency of its ticks to :

```
    m_CarSlider.SetRange(1, 10);
    m_CarSlider.SetTicFreq(1);

    return TRUE; // return TRUE unless you set the focus to a control
}
```

2. Test the application and return to MSVC.

5.8.4. Slider Events

On its own, the slider controls can send three notification messages :

- The NM_OUTOFMEMORY message is sent when the slider has run out of memory and could not complete a task.
- The NM_RELEASECAPTURE message is sent when the user releases the mouse on the slider.
- The NM_CUSTOMDRAW message is used if you want to draw something on the slider or you want to customise the appearance of the slider beyond what Visual C++ proposes.

For its functionality, the slider highly relies on its parent. When the user clicks the thumb or any part of the slider, which causes it to slide, a scroll event is fired. If the slider is horizontal, the CWnd::OnHScroll() event is sent. If the slider is vertical, the CWnd::OnVScroll() event is sent.

1. Change the OnPaint() event of the dialog class as follows :

```
void CCarInventoryDlg::OnPaint()
{
    int CurPos = m_CarSlider.GetPos() - 1;
    CBitmap Bmp;

    Bmp.LoadBitmap(CarPicture[CurPos]);

    m_Picture.SetBitmap(Bmp);
    UpdateData(FALSE);
```

```
    . . .

    }
```

2. Using either the ClassWizard (MSVC 6) or the Messages button , for the dialog, generate the WM_HSCROLL message and implement it as follows :

```
vvoid CCarInventoryDlg::OnHScroll(UINT nSBCode, UINT nPos,
                                   CScrollBar* pScrollBar)
{
    // TODO: Add your message handler code here and/or call default
    CRect RectPicture;

    // Get the location and dimensions of the picture area
    m_Preview.GetWindowRect(&RectPicture);

    // Convert the current coordinates from
    // the monitor (Screen) to the dialog box (Client)
    ScreenToClient(&RectPicture);
    // Repaint the picture area
    InvalidateRect(&RectPicture);

    CDialog::OnHScroll(nSBCode, nPos, pScrollBar);
}
```

3. Test the application

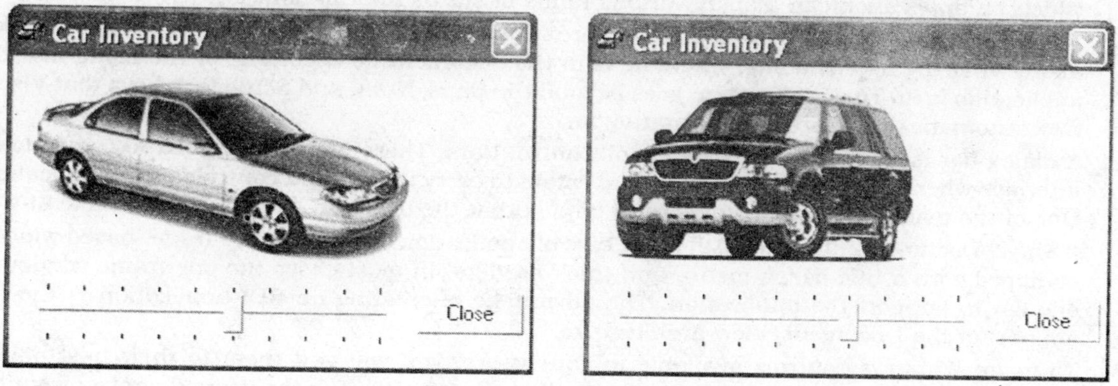

Fig. 5.45.

4. Return to MSVC

5.9. MULTIPLE DOCUMENTS

What Is an MDI Application? (MDU 2009)

As far as coding an MDI application with Visual C++, there's little difference between creating an SDI and an MDI application. However, when you get deeper into the two application styles, you'll find quite a few differences. Although an SDI application allows the user to work on only one document at a time, it also normally limits the user to working on a specific type of document. MDI applications not only enable the user to work on multiple Documents at the same time, but also MDI applications can allow the user to work on multiple types of documents.

An MDI application uses a window-in-a-window style, where there is a frame window around one or more child windows. This is a common application style with many popular software packages, including Word and Excel.

Architecturally, an MDI application is similar to an SDI application. In fact, with a sim-ple MDI application, the only difference is the addition of a second frame class to the other classes that the AppWizard creates, as shown in Figure 11.1. As you can see, the Document/View architecture is still very much the approach you use for developing MDI applications as well as SDI applications.

SUMMARY

1. There are two main ways a control is made part of your application. At design time, which is the time you are visually creating an application, you will select controls and place them on a dialog box. Another technique, referred to as run-time, consists of creating a control programmatically. In this case you must write all the code that specifies the control's appearance and behaviour.
2. A Button is a Windows control used to initiate an action. From the userâ•™s standpoint, a button is useful when clicked, in which case the user positions the mouse on it and presses one of the mouse buttons.
3. The most regular action users perform on a button is to click it. When a user does this, the button sends a BN_CLICKED message. In some but rare circumstances, you may also ask the user to double-click a button. Over all, you will take care of most message handling when the user clicks a button.
4. A status bar is a horizontal window, usually displayed the bottom of the parent window, in which an application can display various kinds of status information.
5. The status bar can provide the user with more substantial information It can provide information about what the user is doing, where he is in the document he is editing, or the mode that the application is in. This information goes beyond the Caps, Num, and Scroll lock keys that Visual C++ automatically reports on the status bar.
6. A dialog bar is an object of the family of Control Bars. These are objects used to complete an interface when the frame needs extra real estate to carry the desired controls of an application. One of the main justifications of control bars is that the user can display or hide them at will.
7. A Single Document Interface (SDI) is the type of application that presents a frame-based window equipped with a title bar, a menu, and thick borders. In most cases the one-frame window is enough to support the application. The advantage of creating an SDI application is the full support of the Document/View architecture.
8. To make Windows controls available to your users, you will add them to their host object, which can be a dialog box or a form. To do this, you mostly click the desired control from the Controls toolbox and click the desired area on the host. You can keep adding controls on the dialog box as necessary.
9. To select more than one control in the same area, click on the dialog box and draw a "fake" rectangle that encloses all of the needed controls. The first control from the selected group has 8 (blue) handles while the other control(s) from the same selected group has (have) 8 white handles (each).
10. A check box is a Windows control that allows the user to set or change the value of an item as true or false. The control appears as a small square ?. When this empty square is clicked, it gets marked by a check symbol S. These two states control the check box as checked ? or unchecked ?.

SHORT QUESTIONS WITH ANSWER

Q.1. Briefly Explain Buttons Messages. **(V.Imp)**

Ans. The most regular action users perform on a button is to click it. When a user does this, the button sends a **BN_CLICKED** message. In some but rare circumstances, you may also ask the user to double-click a button. Over all, you will take care of most message handling when the user clicks a button.

There are other messages that you can handle through a button.

To close a dialog box, you can use the Win32 API's **PostQuitMessage()** function. Its syntax is :

```
VOID PostQuitMessage(int nExitCode);
```

This function takes one argument which is an integer. The argument could be set to almost any integer value although it should be **WM_QUIT**. Here is an example :

```
void CDialog5Dlg::OnBtnClose()
{
    // TODO: Add your control notification handler code here
    PostQuitMessage(125);
}
```

Although the MFC provides enough messages associated with the various controls, in some circumstances you will need use a message that is not necessarily associated with the control. In such a case, you can call the **CWnd::SendMessage()** method. Its syntax is :

```
LRESULT SendMessage(UINT message, WPARAM wParam = 0, LPARAM lParam = 0);
```

The first argument of this method can be a Win32 message or constant. Examples would be **WM_CLOSE** or **WM_ACTIVATE**. The wParam and lParam arguments can be additional (Win32) messages.

The WinExec() function is used to run an application. Its syntax is :

```
UINT WinExec(LPCSTR lpCmdLine, UINT uCmdShow);
```

The lpCmdLine argument specifies the name or path of the application you want to display. The uCmdShow specifies how the application should be displayed. It uses the same values as the **CWnd::ShowWindow()** method.

Q.2. Briefly Explain Status Bar Prompts. **(V.Imp)**

Ans. A status bar is a horizontal window, usually displayed the bottom of the parent window, in which an application can display various kinds of status information.

The status bar can provide the user with more substantial information It can provide information about what the user is doing, where he is in the document he is editing, or the mode that the application is in. This information goes beyond the Caps, Num, and Scroll lock keys that Visual C++ automatically reports on the status bar.

A status bar is divided into a number of panes pasted on the status bar. It's actually easy to add additional panes to the status bar, as well as take away the panes that are already there.5

Q.3. Write procedure Calling One Dialog Box From Another Dialog Box.

Ans. It is unusual to have a dialog-based application. This is because a dialog box is usually made to complement a frame-based application. In that case, after adding the dialog box to the application, you would also provide the ability to **call the dialog from an SDI**. In the same way, if you create an application made of various dialog boxes, at one time, you may want the user to be able to call one dialog box from another. To start, you would create the necessary dialog boxes.

To call one dialog box from another, in the header file of the class of the calling dialog box, you can include the header file of the class of the called dialog box. In the section where you want to call the dialog box, you can declare a variable of that other dialog box' class and call its **CDialog::DoModal()** method.

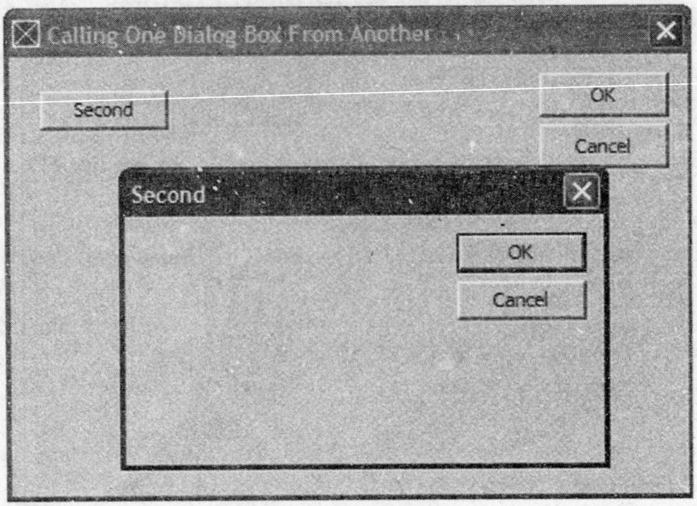

Fig. 5.46.

Q.4. Briefly Explain Check Boxes.

Ans. A check box is a Windows control that allows the user to set or change the value of an item as true or false. The control appears as a small square. When this empty square is clicked, it gets marked by a check symbol S. These two states control the check box as checked ? or unchecked ?.

Like the **radio button**, the check box doesn't indicate what it is used for. Therefore, it is usually accompanied by a label that displays an explicit and useful string. The label can be positioned on either part of the square box, depending on the programmer who implemented it.

Q.5. How to use text metrics to determine characteristics of the current font?

Ans. Characteristics of the current first using textmetrics can be determined using the caret shown in editors lextmetric is a structure type which has members as:

```
typedef struct tag TEXTMETRJC {//tm
long tmHeight;
long tmAscent;
long tm Weight;
Byte tm Italic; Byte tmCharSet;
BytetmUnderlined;
  .
  .
 }TEXTMETR1C;
```

Example:

```
void CGgView:: OnDraw (CDC ? pDc)
{
    CGgDoc • pDoc = Get Document( );
    ` ASSERT_V ALID (pDoc);
    if (Icrtcre)
    {
    TEXTMETRICtext;
    pDC -> Get Text Metrics (& txt);
    //txt.tmAveCharWidth -» To get the character width
    //txt.tmHeight -» To get the character height
```

```
        :
    //Likewise, other attributes can be found, from the
    //textmetric object.
}
```

Q.6. Write the steps to create a multiple document visual basic program and explain application object.

Ans. MDI frame windows can display multiple child windows inside them. In fact, VB IDE itself is an MDI frame window.

For e.g., if you have a program based on a single form, form 1 and you want to make that into an MDI child window inside an MDI frame, follow these steps:

1. Add a new MD! form to the project using project/Add MDI form item.
2. Set MDI child property of the form you want to use as the MDI child form to true.
3. Run the program, the form you have made into the MDI child form appears in MDI form. Arranging MOI Child Windows:

So, you have made MDI program, you could arrange MDI child forms with their left, top, width and height properties you can use MDI form method arrange( ).

For e.g., If you add a menu item to an MDI form named say "Arrange Air", you can use the average ( ) method to average all the window is the form in a cascade.

```
Private sub arrange All_clutch( )
Mc. Arrange Vb cAscade
End sub.
```

Ways of Arranging MD! Child Window:

Constant	Value	Does this
Vb Cascade	0	Cascades all non-minimized MDI child windows
Vb Tile Horizontal	1	Tele all non-minimized child form horizontally
Vb Tile Vertical	2	Tiles are non-minimized MDI child forms vertically.
VbArrange Icon	3	Arrange icon for minimized MDI child forms.

Q.7. Explain various parts of VC++ program with example.

Ans. These are four major parts of a visual **C++** app. wizard programs:

1. Application object
2. Main window object
3. Document object
4. View object

1. Application Object: The application object is what window actually runs first. When this object is started, it places the main window on the screen windows calls the win main( ) method in the application object to get our program started.

The application object has mainly three following tasks:

(*a*) To start the program

(*b*) To launch air main window.

(*c*) To pass windows messages on from windows to our mainwindow.

Windows Messages: These are special codes along with a minimal amount of data sent between various objects in windows to communicate.

e.g., when the user quits our program, windows sends the application object the WIN-QUITmessage. When the user resizes our program's window, we get the WM-SIZE message.

2. Main Window Object: The main window object displays the program itsell and this is where we find the menu bar, the tittle bar of the window and the total bar. The main window is responsible for everything that surrounds the area where the action the drawing text and other graphics goes on in our window. That area is called the client area in a window.

3. View Object: The view object handles the client area that is where we will format, display the data in our program, such as the text we are editing, e.g., we are creating a word processing program. The data we display in the view object is stored in the document object.

4. Document Object: In the document object, we store the data for our program. The data may be stored in the view object but if we have large amount of data to be displayed then we store it in the document object and then handle the display of the data that will fit into the & client area in the view object.

LONG QUESTIONS WITH ANSWER

Q.1. How would you Change a Cursor in a Dialog Box?

Ans. Some applications, like drawing programs, use various mouse cursors depending on what is going on. Unlike most other libraries, the MFC doesn't make it easy to change a cursor. We will just view an example of doing that.

1. Start Microsoft Visual C++ .Net or MS Visual Studio .Net and create a new MFC Application named ChangeCursor.
2. Create it as Dialog Based.
3. Design or import two cursors. We will consider that one of cursors has an ID as IDC_CURSOR1 and the other has an ID as IDC_CURSOR2.
4. Add two buttons to the dialog box. Change the ID of the first to IDC_FIRST_BTN and its Caption to First. Change the ID of the other to IDC_SECOND_BTN and its Caption to Second.
5. Add another button to the dialog box. Change its ID to IDC_DEFAULT_BTN and its Caption to Default.
6. Double-click the first button and implement its event as follows :

```
void CChangeCursorDlg::OnBnClickedFirstBtn()
{
    // TODO: Add your control notification handler code here
    SetClassLong(m_hWnd,
        GCL_HCURSOR,
        (LONG)LoadCursor(AfxGetInstanceHandle(),
            MAKEINTRESOURCE(IDC_CURSOR1)));}
```

7. Double-click the second button and implement its event as follows :

```
void CChangeCursorDlg::OnBnClickedSecondBtn()
{
    // TODO: Add your control notification handler code here
    SetClassLong(m_hWnd,
        GCL_HCURSOR,
        (LONG)LoadCursor(AfxGetInstanceHandle(),
            MAKEINTRESOURCE(IDC_CURSOR2)));
}
```

8. Double-click the Default button and implement its event as follows :

```
void CChangeCursorDlg::OnBnClickedDefaultBtn()
{
    // TODO: Add your control notification handler code here
    SetClassLong( m_hWnd,
        GCL_HCURSOR,
        (LONG)LoadCursor(NULL, IDC_ARROW) );
}
```

9. Test the application.

Q.2. Write Procedure for Calling a Dialog Box From an SDI.

Ans. A Single Document Interface (SDI) is the type of application that presents a frame-based window equipped with a title bar, a menu, and thick borders. In most cases the one-frame window is enough to support the application. The advantage of creating an SDI application is the full support of the Document/View architecture.

To provide support and various options to the frame, you can add other dependent windows such as dialog boxes. When such a dialog box is needed, you can provide a menu item that the user would use to display the dialog box. For example, if you create an SDI using the wizard, it would be equipped with an About Box and the main menu would allow the user to open that dialog box. In the same way, you can add as many dialog boxes as you judge necessary to your SDI application and provide a way for the user to display them.

To display a dialog from an SDI, the easiest and most usual way consists of creating a menu item on the main menu. When implementing the event of the menu item, as normally done in C++, in the source file of the class that would call it, include the header file of the dialog box. In the event that calls the dialog box, first declare a variable based on the name of the class of the dialog box. Then call the CDialog::DoModal() method of the dialog box that is being called.

Calling a Dialog Box From an SDI

1. Start Microsoft Visual C++ .NET or MS Visual Studio .NET.
2. Create a new MFC Application named **SDI1**.
3. Create it as Single Document.
4. To add a dialog box, on the main menu, click Project -> Add Resource.
5. In the Add Resource dialog box, double-click Dialog.
6. Change the **ID** of the new dialog to **IDD_GENERIC_DLG** and its Caption to Dependent.
7. Right-click the dialog box and click Add Class....
8. Set the Class Name to **CDependentDlg** and base it on **CDialog**.
9. Click Finish.
10. From the Class View, double-click the IDR_MAINFRAME menu.
11. On the right side of Help, click Type Here and type **&Tools**.
12. Click the box under Tools and type **&Dependents**... and press Enter.
13. Click and drag the Tools menu category to the left to position it between View and Help.
14. Right-click the Dependents menu item and click Add Event Handler...
15. Set the Class List o the view class.
16. Click Add and Edit.
17. Implement the event as follows :

```
// SDI1View.cpp : implementation of the CSDI1View class
//

#include "stdafx.h"
#include "SDI1.h"

#include "SDI1Doc.h"
#include "SDI1View.h"
#include ".\sdi1view.h"
#include "DependentDlg.h"

#ifdef _DEBUG
#define new DEBUG_NEW
#endif

. . . No Change
```

```
// CSDI1View message handlers

void CSDI1View::OnToolsDependents()
{
    // TODO: Add your command handler code here
    CDependentDlg Dlg;

    Dlg.DoModal();
}
```

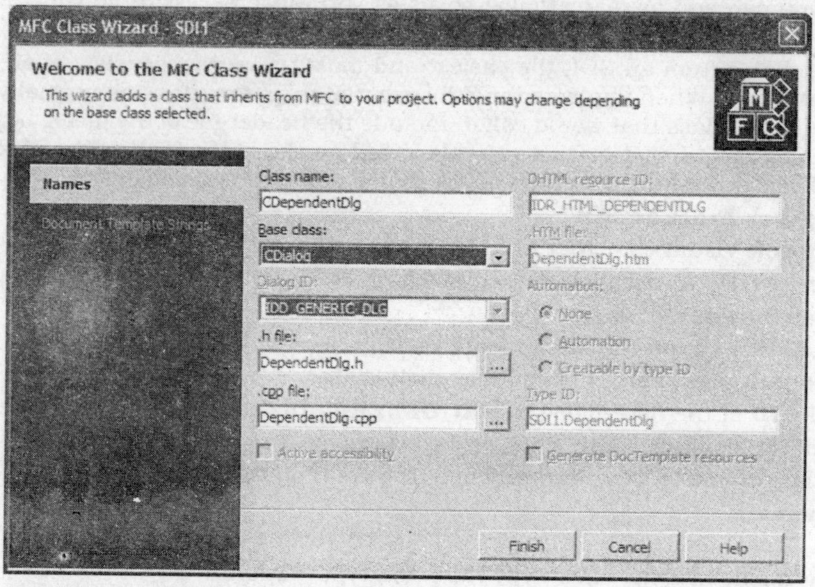

Fig. 5.47.

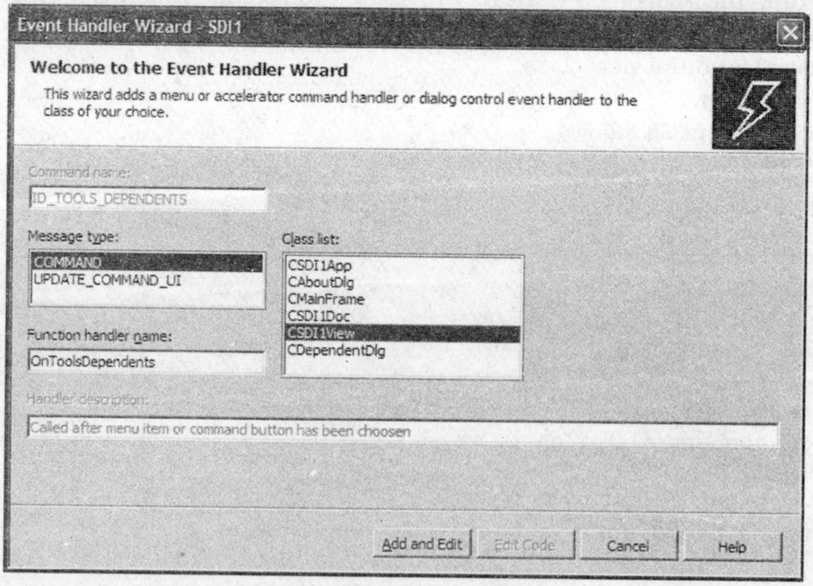

Fig. 5.48.

18. Test the application.

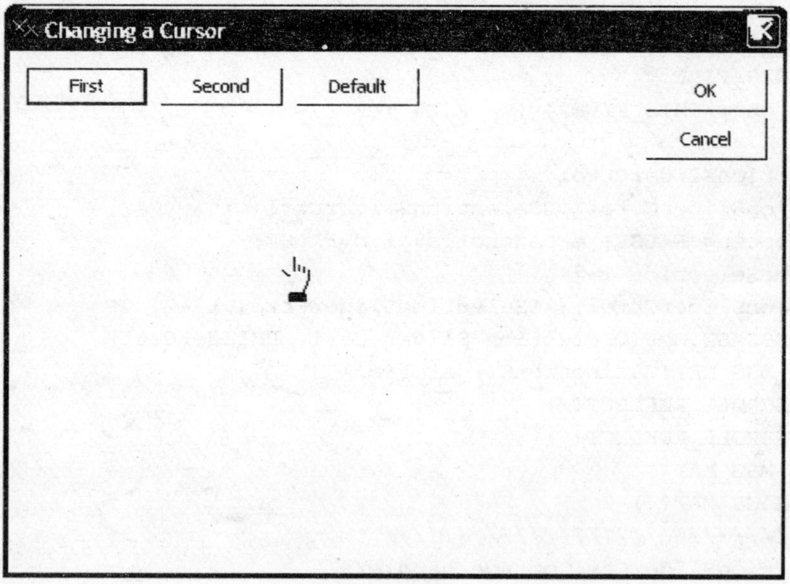

Fig. 5.49.

19. Close.

Q.3. What is a slider ? Write a program to create a slider and also initialize its range.

Ans. Slider: This derivation of CSliderCtrol allows the user to use the mouse or the keyboard to select range in the slider control. The program can then use CSliderCtrl:: GetSelection ( ) to retrieve the range select by the user. This is the way I thought that CSliderCtrol worked in the first place, only to discover that selection was just a static representation of the values set by CSliderCtrol::SetSelection ( ). This class correct this shortcoming of the slider control.

```
//CSelection SliderCtrol Window
Class CSelectionsilder Ctrol: public CSliderCtrl
{
// Construction
public: CSelection Slider Ctrl ( );
// Attributes
public:
//Operations
public:
public: virtual~CSelectionSliderCtrl ( );
protected:
BOOL m_bSelecting;
BOOL m_bAnchorlsValid;
int m_iAnchorSelection
protected:
//{{AFX_MSG (CSelection Slider Ctrl)
afx_msg void HScroll (UINT n TBCode, UINT nPos);
afx_msg void VScroll (UINT n TBCode, UINT nPos);
//}} AFX_MSG
DECLARE_MESSAGE_MAP ( )};
//Selection SliderCtrol.opp: implementation file//
```

```
#include "StdAFX.h"
#ifdef_DEBUG
#define new DEBUG_NEW
#undefTHIS_FILE.
#static char THIS_FILE [ ] =_FILE_;
#endif
//CSelectionSliderCtrol
CSelectionSliderCtrol::CSelectionSliderCtrl( )
{m_bSelecting=FALSE; m-bAnchorlsValid=FALSE;
m-iAnchorSelection=1;}
CSelectionSliderCtrol::~CSelectionSliderCtrl( ) { }
BEGIN_MESSAGE_MAP(CSelection Slider Ctrl, CSliderCtrl)
//{{AFX_MSG_MAP(CSelectionSliderCtrol)
ON_WM_HSCROLL_REFLECT( )
ON_WM_VSCROLL_REFLECT( )
//}}AFX_MSG_MAP
END_MESSAGE_MAP( )
///////////////////////////////////
//CSelectionSliderCtrl messge handlers
void CSelectionSlider Ctrol:: HScroll (UINT nTBCode, UINT/*nPos*/)
{   if (GetStyle( ) & TBS_ENABLESELRANGE)
{   BOOL bEndScroll=FALSE;
    int iPos = GetPos( );
    int iprevPos = iPos;
    switch (nTBCode)
    case TB_PAGEUP: iPrevpos=iPos + GetpageSize ( );
    break;
    case TB_PAGEDOWN: iPrevPos=iPos - GetpageSize ( ); break;
    case TB_LINEUP: iPrevpos=iPos + GetLineSize ( );
    break;
    case TB_LINEDOWN: iPrevPos=iPos - GetLineSize ( ); break;
}
    switch (nTBCode)
{
    case TB_ENDTRACK:
    bEndScroll=True;
    break;
    case TB_PAGEUP:
    case TB_PAGEDOWN:
    case TB_LINEUP:
    case TB_IINEDOWN:
    case TB_THUMBPOSITION:
    case TB_THUMBTRACK:
{
    const BOOL bShiftIspressed = (::Getkeys State (VK_SHIFT)<
    if (!M_bSelecting&&lbShiftIs Pressed)
{   m_iAnchorSelection=-1;
    m-bAnchorIsValid=FALSE;    }
```

```
        if(lm_bSelecting&&bShiftIs Pressed)
      { m_bSelecting + TRUE;
        if(!m_bAnchorlsValid)
      { m_iAnchorSelection = iprevpos;
        m_bAnchorlsValid = TRUE; }}
        if(m_bSelecting) {
Set Selection (min(m_iAnchorSelection, ipos), max(m_iAnchorSelection, ipos));
        Invalidate (FALSE); break; }
        if(bEndScroll)
      { if(lm_bSelecting)::GetkeyState (VK_SHIFT) > 0)
      { M_bAnchorlsValid = FALSE;
        m_iAnchorSelection =-1;    }
        m_bSelecting = FALSE; }}}
        void CSelection Slider Ctrol::VScroll (UINT nTBCode, UINT nPos)
      { HScroll (nTBCode, nPos);   }
```

Q.4. Define multiple document. Write program that handle multiple documents.

Ans. Multiple document Interface: Graphical computer applications with a **Multiple Document Interface** (MDI) are those whose windows reside under a single parent window (usually with the exception of modal windows), as opposed to all windows being separate from each other (single document interface). The initialism MDI is usually not expanded. In the usability community, there has been much debate over which interface type is preferable. Generally SDI is seen as more useful in cases where users work with more than one application. Companies have used both interfaces with mixed responses. For example, Microsoft has changed its Office applications from SDI to MDI mode and then back to SDI, although the degree of implementation varies from one component to another. The disadvantage of MDI usually cited is the lack of information about the currently opened windows. In order to view a list of windows open in MDI applications, the user typically has to select a specific menu ("Window list" or something similar), if this option is available at all. With an SDI application, the window manager's task bar or task manager displays the currently opened windows. In recent years, applications have increasingly added "task-bars" and "tabs" to show the currently opened windows in an MDI application, which has made this criticism somewhat obsolete. Some people use a different name for this interface, "tabbed document interface" (TDI). When tabs are used to manage windows, individual ones can usually not be resized.

First of all you should create MDI application, generate IDispatch wrapper class CP GAX for the control and add member variable to the view class. Import type library to access all ActiveXinterfaces:

```
#import "prographer.tlb" no_namespace,named_guids
View class should be like the following:
classCMDITView: publicCView
{ DECLARE_DYNAMIC(CMDITtestView)
public:
CMDITestView( );
virtual~CMDITestView( )
protected:
afx_msg int OnCreate (LPCREATESTRUCT ipCreateStract);
afx_msg void OnDestroy( );
afx_msg void OnSize (UNIT nType, int cx, int cy);
DECLARE_MESSAGE_MAP( )
private:
  template<classT>class CComplannerObject: public
  CComObjectNoLock <T>
{ STDMETHOD_(ULONG,Release)( )
```

```
{   return Internal Release( );          }   };
    class CDocEventSink:
    public CComObjectRootEx,
    public ID isp EventImpl<1, CDocEventSink,
    & DIID_IDocumentEvent,&LIBID_ProGrapherFoundation, 1,0>
{   public:
    CDocEventSink( ) { }
    private:
        BEGIN_COM_MAP (CDocEventSink)
        END_COM_MAP( )
        BEGIN_SINK_MAP (CDocEventSink)
        SINK_ENTRY_EX(1,DIID_IDocumentEvents,1,OnOpen)
        END_SINK_MAP( )
        void stdcall OnOpen( ); };
        class CWndEventSink:
        public CComObjectRootEx,
    public IDispEventLmp !<1, CWndEventSink, &DIID_IWindowEvents,
    &LIBID_ProGrapherFoundation, 1, 0>
{   public:
    CWndEventSink( ) { }
  private:
    BEGIN_COM_MAP (CWndEventSink)
    END_COM_MAP( )
    BEGIN_SINK_MAP (CWndEventSink)
    SINK_ENTRY_EX(1,DIID_IWindowEvents,1,OnMouseMove)
    END_SINK_MAP( )
    void stdcall OnMouseMove (long Flags, double X, double Y);   };
public:
    CComInnerObject<CDocEventSink>m_DocEventSink;
    CComInnerObject<CWndEventSink>m_WndEventSink;
    CPGAX  m_PGAX;
    IDocumentPtrm_spDocument;
    IWindowPtr m_sp Window; };
```

Q.5. How a combo box is different from list box ? Write a program to create a combo box and fill it with items.

Ans. Combo Box: A combo box is a component with a drop-down arrow that users click 160 display an associated list of choices. If the list is too long to display fully, a vertical scrollbar appears. The current selection appears in an editable or non-editable text field bnext to the drop-down arrow. The user displays the list by clicking or dragging the drop-down arrow. The currently selected item appears in the pulled-down combo box. As a user moves the pointer over the list, each option under the pointer is highlighed. If the user selects an option from the list, that option replaces the current selection. In the following figure, the currently selected item is Vanilla, and the Guanabana option will replace Vanilla when the mouse button is lifted or the space bar is pressed.

Users can close editable or non-editable combo boxes by clicking the drop-down arrow in the combo box again, selecting an item from the list, or clicking anywhere outside the combo box.

List Box: A list box present a list of items to the user to choose from with each item on items own line. The users makes a selection by clicking in the list. Once clicked, an item becomes highlighted indication that it is the current choice. A list box can be configured to allow single or multiple selections, One of the main reasons for using at list box is to display a list of items on specific lines to the user. Sometimes the list would be very large. If the list is longer than the available space on a form, the control would provide a scroll bar that allows the user to navigate up and down and then access all items of the list. The application developer decides how many items to display on the list. A combo box is based on the **CCombo Box** class. Therefore, if you want to dynamically create this control declare a variable or a pointer to CComboBox using its default constructor. To initialize the control, call its Create ( ) method. Here is an example:

```
BOLL CExerciseDig::::OnInitDialog ( )
{ CDialog::::OnInitDialog ( );
Selection(m_hlcon, TRUE); //Set big icon
Selection (m_hlocon, FALSE); // Set SMALL ICON
//TODO ; Add extra initialization here
CComboBox *Majors = new CComboBox;
Majors->Create(WS_CHILD|WS_VISIBLE)
WS_VSCROSS|CBS_DROPDOWNLIST,
CRect (10,50,100,150), this, ox1448);
return TRUE; //return TRUE unless you set the focus to a control}
```

Q.6. What is MDI ? What is different in an MDI program ? Explain giving suitable example.

Ans. MDI: Graphical computer application with a **Multiple Document Interface (MDI)** are those whose windows reside under a single parent window (usually with the exception of modal windows), as opposed to all windows being separate from each other (single document interface). The initialism MDI is usually not expanded. In the usability community, there has been much debate over which interface type is preferable. Generally SDI is seen as more useful in cases where users work with more than one application. Companies have used both interfaces with mixed responses. For example, Microsoft has changed its Office applications from SDI to MDI mode and then back to SDI, although the degree of implementation varies from one component to another. The disadvantage of MDI usually cited is the lack of information about the currently opened windows: In order to view a list of windows open in MDI applications, the user typically has to select a specific menu ("Window list" or something similar), it this option is available at all. With as SDI application, the window manager's task bar or task manager displays the currently opened windows. In recent years, applications have increasingly added "task-bars" and "tabs" to show the currently opened windows in an MDI application, which has made this criticism somewhat obsolete. Some people use a different name for this interface, "tabbed document interface" (TDI). When tabs are used to manage windows, individual ones usually cannot be resized.

Advantages:

- With MDI (and also TDI), a single menu bar and/or toolbar is shared between all child windows, reducing clutter and increasing efficient use of screen space.
- An application's child windows can be hidden/shown/minimized/maximized as a whole.
- Features such as "Tile" and "Cascade" can be implemented for the child windows.
- Possibly faster and more memory efficient, since the application is shared, and only the document changes, the speed of switching between the internal windows is usually faster than having the OS switch between external windows.
- Some application have keyboard shortcuts to quickly jump to the functionality you need (faster navigating), and this doesn't need the OS or windows manager support, since it happens inside the application.

Disadvantages:

- Can be tricky to implement on desktops using monitors as the parent window may need to span both monitors.
- Virtual desktops cannot be spanned by children of the MDI. However, in some cases, this is solvable by initiating another parent window; this is the case on Opere, for example, which allows trabs/child windows to be dragged outside of the parent window to start their own parent window (on Windows).

- MDI can make it more difficult to work with several; application at once, by restricting the ways in which windows from multiple application can be arranged together.
- Without an MDI frame window, floating toolbars from one application can clutter the workspace of other applications, potentially confusing users with the jumble of interfaces.
- The shared menu changes, which may cause confusion of interfaces.
- MDI child windows behave differently rom those in single document interface applications, requiring users to learn two subtly different windowing concepts. Similarly, the MDI parent window behaves like the desktop in many respects, but has enough differences to confuse some users.
- Many window manages have built-in-support for manipulating groups of separate windows, which is typically more flexible than MD in that windows can be grouped and ungrouped arbitrarily. A typical policy is to group automatically windows that belong to the same application. This arguably makes MDI redundant by providing a solution to the same problem.

Q.7. What do you mean by serialization ? Write all the steps to show how a class can be made serialized.

Ans. Serialization is the method used by MFC to store an object to a file or another type of storage. The MFC serialization architecture uses three classes. The CFile class models a Win32 file 1/0 object. Some differences exist between Win32 file objects and MFC CFile file objects.

- CObjects provides a common framework for objects to serialize themselves—a property known as persistence.
- CArchieve handles the serialization of a CObject instance to a CFile object.
- CFile is used as a base class for all MFC file classes. CFile is a general file class, and the other classes are used for more specialized types of files:
- CStdioFile wraps a standard C stream file handle like the standard output handle with a CFilederived class.
- CMemFile manages a file that is hosted in memory rather than on disk.
- CSharedFile allows for files to be shared in memory.
- COleStreamFile provides CFile access to Component Object Model (COM) compound files.
- CSockedFile wraps a Windows socked with a CFile-derived class.

Opening Files: There are two ways to open a file using CFile.

- The simplest method is to pass the name of the file to be opened as a parameter to the constructor.
- Alternatively, you can create a CFile instance using the default constructor and explicitly open a file using the Open member function.

If you open the file during construction, you must use C++ exception handling to handle error conditions. If you don't handle exceptions with this type of construction and an error occurs, your program will be terminated due to the unhandled exception.

The standard pattern for opening a CFile during construction and handling any exceptions looks like this try.

```
{    CFile myFile ("C:\"MyFile", CFile::modeRead);
     II use the here
}    catch(CFileException*pe)
{    char szMsg(256);
     pe->GetErrorMessages(szMsg, size of(szMsg));
     MessageBox(NULL, szMsg; 3grv(O), MB_ICONHAND);
     pe->Delete ( );}
```

The second parameter passed to the CFile constructor is used to hold one or more flags that indicate how the file be opened. Multiple flags are combined using the C++ or symbol, I. The most commonly used values follows:

- mode Read opens the file for reading only.
- mode Write opens the file for writing only.
- mode ReadWrite opens the file for both reading and writing.
- modeCreates the file if it doesn't already exist. If the file exists, the current file is erased, unless mode NoTruncate is used.
- shareDenyNone gives other processes read and write access to the file.
- shareDenyWrite prevents other processes from writing to the file.
- shareDenyRead prevents other processes from reading the file.
- shareExclusive prevents other processes from writing to or reading the file.

Other mode flags are used to control inheritance of the underlying file handle and to prevent an existing file from being truncated. Check the documentation for CFile for details.

To open a file explicitly, use the CFile::Open member function, passing the filename and file open flags as parameters:

```
CFile file;
BOOL bOpened = file.open ("C:\MyFile", CFile::modeRead);
if(bOpened ==FALSE)
{   II Handle error
}
```

You should use the CFile constructor that accepts a filename only when you are fairly certain that the file can be opened. If you aren't sure whether the file will be opened successfully, it usually is easier to handle errors by using CFile::Open, which doesn't raise exceptions. The CFile::Close member function is used to explicitly close an open CFile. When an instance of CFile is destroyed, the CFile destructor checks and closes any open files.

Reading and Writing from a CFile Object: Reading and writing to MFC CFile instances is much simpler than using the Win32 alternatives. To read from a CFile instance, the Read member function is used:

```
try
    {   BYTE buff [4096];
        CFile myFile ("C:\MyFile", CFile::modeRead);
        DWORD dwRead = myFile.Read(&buff, sizeof(buff";}
        catch(CFileException*pe)
 {//Handle exception...
 pe->DeleteO; II Free exception instance}
```

Read takes two parameters:

- The address of a buffer to be filled
- The maximum number of bytes that can be stored in the buffer.

The Read member function returns the number of bytes read from the CFile. If an error occurs, an exception is raised.

To write to a CFile instance, use the Write member function:

```
try
{      CFile myFile("C:\MyFile", CFile::modeCreateICFile::modeWrite);
fileDest.Write(&buff, dwBytesTowrite); }
catch(CFileException*pe)
{    II Handle exception ...
pe->DeleteO; II Free exception instance
}
```

Write takes two parameters:

- The address of a buffer to be written to the CFile instance.
- The number of bytes to be written to the file.

Write has no return value; if an error occurs, an exception is raised.

Q.8. How a menu item can be added to an existing menu using menu editor? Also explain how short-cut keys and status bar prompts can be added.

Ans. Adding Menu item to existing Menu editor; You can use the Menu Editor to create new menus and menu bars, add new commands to existing menus, replace existing menu commands with your own commands, and change and delete existing menus and menu bars.

To display the Menu Editor:

- From the Tools menu, choose Menu Editor,

or

Click the Menu Editor button on the tool bar.

The Menu Editor: While most menu control properties can be set using the Menu Editor, all menu properties are available in the Properties window. The two most important properties for menu controls are:

- **Name:** This is the name you use to reference the menu control from code.
- **Caption:** This is the text that appears on the control.

Other properties in the Menu Editor, including Index, Checked, and NegotiatePosition, are described later in this chapter.

Using the list box in the menu editor: The menu control list box (the lower portion of the Menu Editor) lists all the menu controls for the current form. When you type a menu item in the Caption text box, that item also appears in the menu control list box. Selecting an existing menu control from the list box allows you to edit the properties for that control. For example, Figure shows the menu controls for a File menu in a typical application. The position of the menu control in the menu control list box determines whether the control is a menu title, menu item, submenu title, or submenu item:

- A menu control that appears flush left in the list box is displayed on the menu bar as a menu title.
- A menu control that is indented once in the list box is displayed on the menu when the user clicks the preceding menu title.
- An indented menu control followed by menu controls that are further indented becomes a submenu title. Menu controls indented below the submenu title become items of that submenu.
- A menu control with a hyphen (-) as its Caption property setting appears as a separator bar. A / separator bar divides menu items into logical groups.

To create menu controls in the Menu Editor:

1. Select the form.
2. From the Tools menu, choose Menu Editor.

   or

   Click the Menu Editor button on the toolbar.
3. In the Caption text box, type the text for the first menu title that you want to appear on the menu bar. Also, place an ampersand (&) before the letter you want to be the access key for that menu item. This letter will automatically be underlined in the menu.

   The menu title text is displayed in the menu control list box.
4. In the Name text box, type the name that you will use to refer to the menu control in code. See "Menu Title and Naming Guidelines" later in this chapter.
5. Click the left arrow or right arrow buttons to change the indentation level of the control.
6. Set other properties for the control, if you choose. You can do this in the Menu ditor or later, in the Properties window.
7. Choose Next to create another menu control, or

   Click Insert to add a menu control between existing controls.

   You can also click the up arrow and down arrow buttons to move the control among the existing menu controls.

8. Choose OK to close the Menu Editor when you have created all the menu controls for that form.

The menu titles you create are displayed on the form. At design time, click a menu title to drop down its corresponding menu items.

Separating menu items: A separator bar is displayed as a horizontal line between items on a menu. On a menu with many items, you can use a separator bar to divide items into logical groups. For example, the File menu in Visual Basic uses separator bars to divide its menu items into three groups, as shown in figure.

To create a separator bar in the menu editor:

1. If you are adding a separator bar to an existing menu, choose Insert to insert a menu control between the menu items you want to separate.
2. If necessary, click the right arrow button to indent the new menu item to the same level as the msnu items it will separate.
3. Type a hyphen (-) in the Caption text box.
4. Set the Name property.
5. Choose OK to close the Menu Editor.

Assigning access keys and shortcut keys: You can improve keyboard access to menu commands by defining access keys and shortcut keys. Access keys:

Access keys allow the user to open a menu by pressing the AL T key and typing a designated letter. Once a menu is open, the user can choose a control by pressing the letter (the access key) assigned to it. For example, ALT+ E might open the Edit menu, and P might select the Paste menu item. An access-key assignment appears as an underlined letter in the menu control's caption.

To assign an access key to a menu control in the menu editor:

1. Select the menu item to which you want to assign an access key.
2. In the Caption box, type an ampersand (&) immediately in front of the letter you want to be the access key.

For example, if the Edit menu shown in Figure 6.9 is open, the following Caption property settings respond to the corresponding key.

Menu control caption	Caption property	Access keys
Cut	Cu&t	*t*
Copy	C&opy	*o*
Paste	&Paste	*p*
Delete	De&lete	*l*
Select All	Select &A11	*a*
Time/Date	Time/&Date	*d*

Shortcut keys: Shortcut keys run a menu item immediately when pressed. Frequently used menu items may be assigned a keyboard shortcut, which provides a single-step method of keyboard access, rather than a three-step method of pressing ALT, a menu title access character, and then a menu item access character. Shortcut key assignments include function key and control key combinations, such as CTRL +F1 or CTRL +A. They appear on the menu to the right of the cot responding menu item, as shown in figure.

To assign a shortcut key to a menu item:

1. Open the Menu Editor.
2. Select the menu item.
3. Select a function key or key combination in the Shortcut combo box.

To remove a shortcut key assignment, choose "(none)" from the top of the list.

Q.9. How to create, how and hide carets in a window?

Ans. Let us assume file name is CGgview. Then we have,

```
CGgview
-> ClassCGgView: public CView
{
      protected:
      CGgViewQ;
      DECLARE_DYNCREATE (CGgView);
      intx,y;
      Cpoint crtpos;
      brol crtcre;
      public:
      CGgDoc * GetDocument();
      }
CGgView
->CGgView()
-> CGgView:: CGgView()
      {
      x = 0;
      y = 0
      crtcre = false;
}
void CGg View:: OnDraw (CDC * pDC)
{
      CGgDoc * pDoc = GetDocument ( );
      ASSERT_VALID (pDoc); if(!crtcre)
      {
         TEXT METRIC txt;
         pDC -> GetText Metrics (& txt);
         Create Solid Caret (txttm A veChar Width/8, txt.tmHeight),
         Crtpos.x = crtpos.y = 0;
         SetCaret pos (Crtpos);
         Showcaret();
         crtcre « true;
      }
      pDC -» Text Out (0,0, pDoc -* st);
      Hidecaret(); //To hide the caret
      CSize size = pDc -* Get Text Extent (pDoc -> st);
      crtpos.x = x + size, ex;
      //Crtpos. y - y;    //For moving the cursor to position
                              //mouse clicked
      Set caret prs (crt pos);
      show caret( );
         }.
```

EXERCISES

1. Add another pane to the status bar to display the current width selected.
2. Remove the Num Lock status indicator from the status bar.

3. Set the text in the first section of the status bar other than using menu and toolbar prompts.
4. The above server application can handle one client connection at a time. If a second client tries to open a connection to the server while the server is busy communicating with the first client, the application will crash. Modify, by adding one more socket object, to reject additional client connections until the first client closes the connection.
5. Modify the above application to handle two clients at a time.
6. What function is called to determine that a connection has been established?
7. A server application may be a client application as well. How?

□□□

6

Serialisation and File Handling

6.0. INTRODUCTION

Chapter Outline

- Introduction
- Serialising the class
- Introduction to File Processing
- Archiving
- File Handling
- Debugging and Profiling COM components

In the previous section you learnt how to store and retrieve individual data items. In object-oriented programming paradigm, however, the data items are not usually individual but are members of some object(instance of a class) existing in the memory. Ever wondered how to store the member data items of an object into a file and later on recreate the object from the member data items stored in the file!

Before the data members of an object may be saved in a file, the object must be serialised. ***Serialisation*** is a process by which an object writes its state(the current values of its member variables) onto a file in such a manner that enables it to recreate itself back into the state it was by dserialisation of the stored data.

Serialization is an important concept in Windows Programming because most of the data of an application exists in object form. It enables an application to open and save documents alongwith its data. Merely saving the data of a document would not enable the application to recreate the document from the stored data.

MFC supplies built-in support for serialisation in the class CObject and all classes derived from CObject can take advantage of its serialisation service. MFC uses an object of the CArchive class(a derivative of Cobject class as an intermediary between the object to be serialised and the storage media. This object is always associated with a Cfile object, from which it obtains the necessary information for sereliasation including the filename and whether the requested operation is a read or a write.

As data is serialised to an archive, the archive accumulates the data until its buffer is full. Then the archive writes its buffer to the Cfile object pointed to by the Carchive object. Similarly, as you read data from an archive, it reads data from the file to its buffer and then from the buffer to your de-serialised object.This buffering reduces the number of times a storage device (e.g., hard disk) is physically read, thus improving your application's performance.

6.1. SERIALISING THE CLASS (MDU 2007-08)

After you have your custom class with all the variables defined and initialised, you need to make the class serialisable. Making your class serialisable involves three steps.

1. The first step is adding the Serialise function to the class. This function writes the variable values to, and reads them back from, the CArchive object using C++ streams.
2. Second step involves adding of the DECLARE_SERIAL macro.
3. The Third step consists of adding IMPLEMENT_SERIAL macro.

After these elements are added, your custom class will be serialisable and ready for your application.

To add the *Serialise* function to your custom class, add a member function through the Class View tab in the workspace pane. Specify the function as type void, the function declaration as *Serialize(CArchive &ar)*, and the access as public and check the Virtual check box. This should add the Serialise function and place you in the editor.

In the *Serialise* function, the first you want to do is to first call the ancestor's *Serialize function.* When you do this, any foundation information that has been saved is restored first, providing the necessary support for your class before the variables in your class are restored. Once you call the ancestor function, you need to determine whether you need to read or write the class variables. You can do this by calling the method *IsStoring* of the class *CArchive.* This function returns TRUE if the archive is being written to and FALSE if the archive id being read from. If the *IsStoring* function returns TRUE, C++/IO streams can be used to write all the class variables to the archive. If the function returns false, C++ streams can be used to read from the archive. It must be sure, in both the cases, to order the variables in the same order for both reading and writing.

Example 3

An example of a Serialise function is shown in the Listing.

```
void Cperson::Serialize(CArchive &ar)
{
    CObject::Serialize(ar);

    // Calling the ancestor function.

    If(ar.IsStoring())

    {

        // Writing all the variables in the proper order.
```

```
        ar << m_sname << m_iAge << m_iMaritalStatus << m_bEmployed;
        else

        // Reading all the variables in the order.

        ar >> m_sName >> m_iAge >> m_iMaritalStatus >> m_bEmployed;
```

Once you place the Serialise function in the order, you need to add macros to your custom class. The firstmacro, DECLARE_SERIAL, needs to go in the class header and is passed the class name as its only argument.

6.2. INTRODUCTION TO FILE PROCESSING

File processing is the ability to create, store, and/or retrieve the contents of a file from a medium such as a hard drive, a floppy disk, a CD-ROM, or a DVD-ROM, etc. One of the areas in which the MFC excels is file processing. MFC makes it particularly easy to perform most file related operations.

6.2.1. What to Process, When, and Where

Before creating a file, you should first plan your program since applications deal with different types of files. An application that manipulate pictures may need to save only one value for the whole file. Another application used for word processing may also need to save only one value but of a completely different kind. If an application is list-based, it may need to save various values that, when put together, constitute one file.

After deciding what type of file the user would be processing, you would also specify how and/or who would decide when to process a file. For example, if you create a certain application that allows a user to change the internal settings and store them in a file, you can process the file automatically and store it somewhere without asking the user to decide when, where, or how to save the file (some applications use the Registry to take care of this). For example, in tthe bits of data that are arranged in a particular way to produce a usable document. For easy storage, location, and management, the bits are stored on a medium such as a hard disc, a floppy disc, a compact disc, or any valid and support type of storage. When these bits belong to a single but common entity, the group is referred to as a file. For even greater management, files can be stored in a parent object called a directory or a folder. Since a file is a unit of storage and it stores information, it has a size which is the number of bits it contains. To manage it, a file also has a location also called a path that specifies where and/or how the file can be retrieved. Also, for better management, a file has attributes that indicate what can be done on a file or that provide specific information that the programmer or the operating system can use when dealing with the file.

For example, in Jasc Paint Shop Pro, users are allowed to create new customised gradients that were not installed with the application. Once a user clicks OK, the new gradient is saved :

Another way you can deal with this is to let a user decide when to save the file and where to save it. This is usually taken care of by providing the File Dialog to the user.

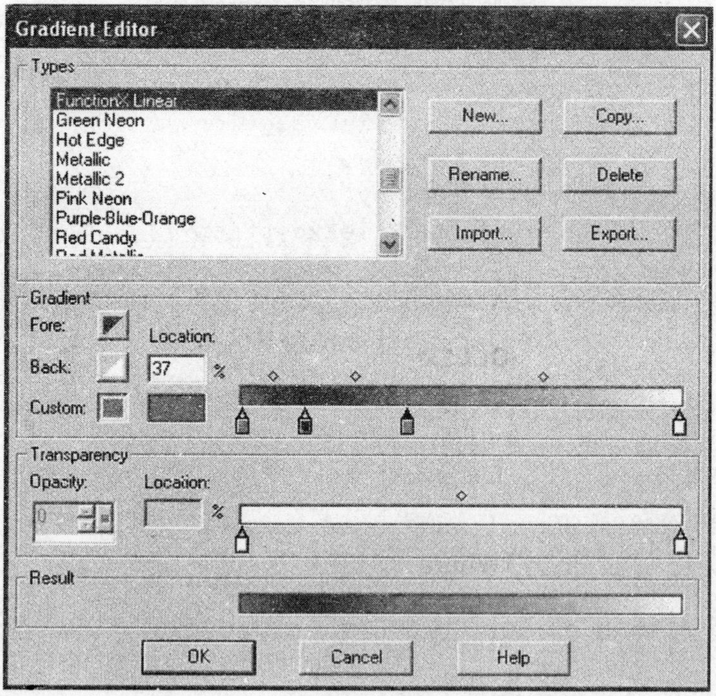

Fig. 6.1.

6.3. ARCHIVING **(V.Imp)**

Most of the file processing in an MFC application is performed in conjunction with a class called **CArchive**. The **CArchive** class serves as a relay between the application and the medium used to either store data or make it available. The **CArchive** class is an accessory and not an end to itself. As such, it provides its services to the MFC class that need to perform file transfer.

Although used by MFC, the **CArchive** class is semi-independent. That is, it is not derived from CObject. It only provides its services to the MFC objects that need archiving. To provide this support, the parent class of MFC objects, **CObject**, is equipped with a method called Serialize(). The CObject::Serialize() method takes one argument, which is a **CArchive** reference variable. The syntax of the CObject::Serialise() method is :

```
virtual void Serialize(CArchive& ar);
```

This means that the MFC library already provides file processing at a very high level. This also implies that almost every MFC class is ready to save its value(s) from a medium or to retrieve existing (an) value(s) from a medium. This convenience is given at a very low price, as file processing is one of MFC strengths. Based on this, to provide file processing for an object in your application, you can simply override the Serialise() method.

6.3.1. The Process of Archiving

One of the most regular ways you will use the **CArchive** class consists of storing or retrieving values through their **Serialize()** method. To do this directly and easily in the class, the **CArchive** class overloads two operations: << and >>.

To store one or more values, the **CArchive** class provides the << operator. The values you can store are of MFC, **BYTE, short, WORD, int, unsigned int, long, LONG, DWORD,**

float, or **double** types. The class also provides an exception handler for each kind. Therefore, the syntaxes for this operator are :

```
friend CArchive& operator <<( CArchive& ar, const CObject* pOb );
throw( CArchiveException, CFileException );

CArchive& operator <<( BYTE by );
throw( CArchiveException, CFileException );

CArchive& operator <<( WORD w );
throw( CArchiveException, CFileException );

CArchive& operator <<( int i );
throw( CArchiveException, CFileException );

CArchive& operator <<( LONG l );
throw( CArchiveException, CFileException );

CArchive& operator <<( DWORD dw );
throw( CArchiveException, CFileException );

CArchive& operator <<( float f );
throw( CArchiveException, CFileException );

CArchive& operator <<( double d );
throw( CArchiveException, CFileException );
```

As you may have realised, there is only one version of the Serialise() method. That is, this member function is called once for both storing and retrieving. Therefore, when using it, you must specify what operation you want to perform. This checking process can be performed by calling the **CArchive::IsStoring()** Boolean method. Its syntax is :

```
BOOL IsStoring() const;
```

This method returns TRUE if you are storing data. If the answer is then TRUE, you can use the << operator to store each necessary value.

As opposed to storing value(s), to retrieve a value, the CArchive class provides the >> operator. It comes in the following syntaxes :

```
friend CArchive& operator >>( CArchive& ar, CObject *& pOb );
throw( CArchiveException, CFileException, CMemoryException );

friend CArchive& operator >>( CArchive& ar, const CObject *& pOb );
throw( CArchiveException, CFileException, CMemoryException );

CArchive& operator >>( BYTE& by );
throw( CArchiveException, CFileException );

CArchive& operator >>( WORD& w );
```

```
throw( CArchiveException, CFileException );

CArchive& operator >>( int& i );
throw( CArchiveException, CFileException );

CArchive& operator >>( LONG& l );
throw( CArchiveException, CFileException );

CArchive& operator >>( DWORD& dw );
throw( CArchiveException, CFileException );

CArchive& operator >>( float& f );
throw( CArchiveException, CFileException );

CArchive& operator >>( double& d );
throw( CArchiveException, CFileException );
```

This indicates that you can retrieve an MFC object, a **BYTE**, a **short**, a **WORD**, an int, an unsigned int, a **long**, a **LONG**, a **DWORD**, a **float**, or a **double** value.

As mentioned above, the Serialise() method is used for both operations. Therefore, when using this method, you should first check what operation is being performed. This can be done by calling the IsLoading() method. Its syntax is :

```
BOOL IsLoading() const;
```

This method returns TRUE if you are retrieving a value or some values. If it is true, you can then use the >> operator to retrieve the value(s).

6.3.2. Introducing File Processing

1. If you want to follow this exercise, create a new application using MFC AppWizard (exe) or MFC Application.
2. Name it **Palace Ice Cream.**
3. Change the design of the IDR_MAINFRAME icon as follows :

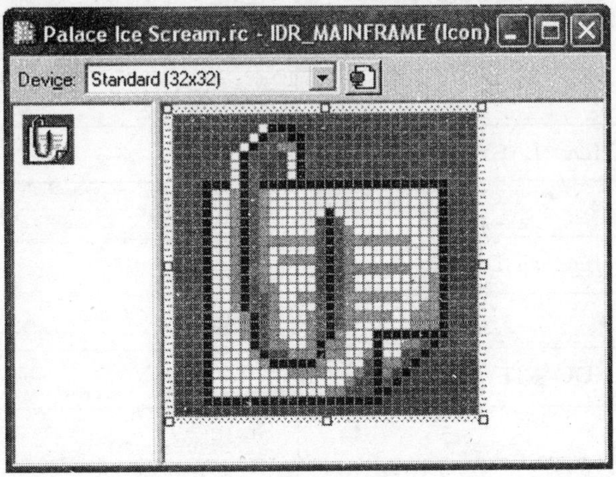

Fig. 6.2.

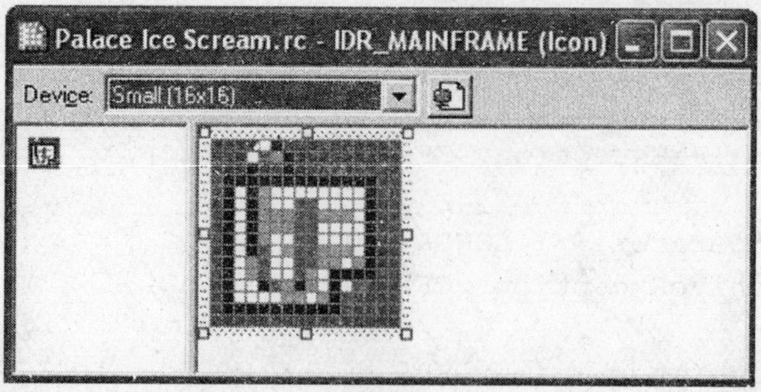

Fig. 6.3.

4. Design the dialog box as follows :

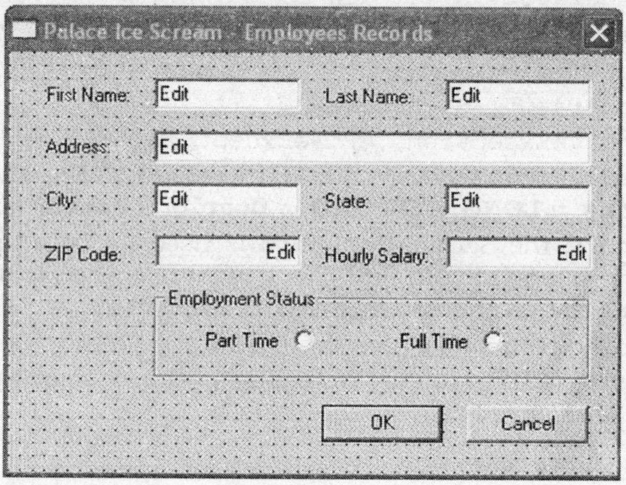

Fig. 6.4.

Control		***ID***	***Caption***	***Additional Properties***
Static Text	Aa		First Name:	
Edit Box	ab|	IDC_FIRST_NAME		
Static Text	Aa		Last Name:	
Edit Box	ab|	IDC_LAST_NAME		
Static Text	Aa		Address:	
Edit Box	ab|	IDC_ADDRESS		
Static Text	Aa		City:	
Edit Box	ab|	IDC_CITY		
Static Text	Aa		State:	
Edit Box	ab|	IDC_STATE		
Static Text	Aa		ZIP Code:	

Edit Box		IDC_ZIP_CODE		
Static Text			Hourly Salary:	
Edit Box		IDC_HOURLY_SALARY		
Group Box			Employment Status	
Radio Button		IDC_PART_TIME	Part Time	Group: True Left Text: True
Radio Button		IDC_FULL_TIME	Full Time	Left Text: True
Button		IDOK		
Button		IDCANCEL		

5. Using either the ClassWizard or the Add Member Variable Wizard, add the following member variables to the controls :

Identifier	***Value Variable***	
IDC_FIRST_NAME	CString	m_FirstName
IDC_LAST_NAME	CString	m_LastName
IDC_ADDRESS	CString	m_Address
IDC_CITY	CString	m_City
IDC_STATE	CString	m_State
IDC_ZIP_CODE	long	m_ZIPCode
IDC_HOURLY_SALARY	double	m_HourlySalary
IDC_PART_TIME	int	m_EmploymentStatus

6. Save everything

6.3.3. MFC Support of File Processing

MFC supports file processing at different levels. There is so much file processing that one of your biggest difficulties would consist of deciding what file processing system you want to use. MFC supports C, C++, Win32, and MFC file processing. Even if you select only one of these categories, some of them have other variances. For example, the C++ language provides more than one way to perform file processing. MFC itself is tremendously wide on this issue.

One of the classes used to perform file processing on MFC applications is **CFile**. Using this class, to create a file, you can first declare a variable of type CFile using one of its three constructors. They are :

```
CFile( );
CFile( int hFile );
CFile( LPCTSTR lpszFileName, UINT nOpenFlags );
```

The default constructor is used if you are not ready to specify the information needed to create the file. The second constructor takes an argument that is a handle to a file created using the Win32's CreateFile() function. The third constructor is used to create a file by specifying the file name and the flags that specify the operation to perform when saving or opening the file.

1. Assuming that the user would have typed information about an employee, to automatically save the file, double-click the OK button.
2. Accept the suggested name of the event and implement it as follows :

```
void CPalaceIceCreamDlg::OnOK()
{
    // TODO: Add extra validation here
    // TODO: Add your control notification handler code here
    this->UpdateData();

    CFile flEmployees;

    flEmployees.Open("employees.mpl", CFile::modeCreate |
    CFile::modeWrite);
    CArchive ar(&flEmployees, CArchive::store);

    ar << m_FirstName << m_LastName << m_Address << m_City << m_State
    << m_ZIPCode << m_HourlySalary << m_EmploymentStatus;

    ar.Close();
    flEmployees.Close();

    CDialog::OnOK();
}
```

3. Execute the application and fill the dialog with some information.

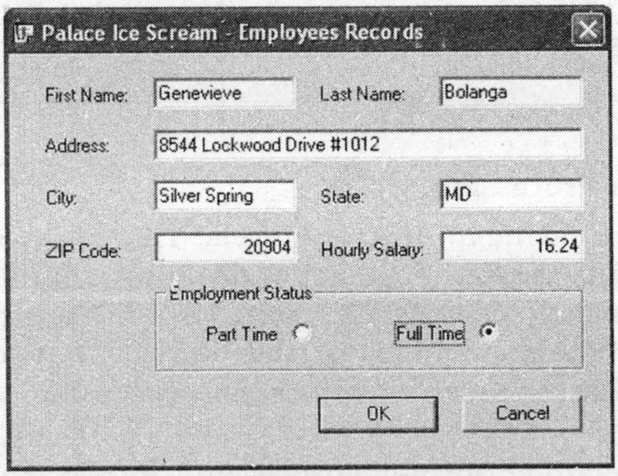

Fig. 6.5.

4. Click OK to close the dialog box.
5. To automatically load the saved file when the dialog box comes up, generate the WM_CREATE message for the dialog box and implement its OnCreate event as follows :

```
int CPalaceIceCreamDlg::OnCreate(LPCREATESTRUCT lpCreateStruct)
{
    if (CDialog::OnCreate(lpCreateStruct) == -1)
        return -1;
```

```
    // TODO: Add your specialised creation code here
    CFile flEmployees;

    flEmployees.Open("employees.mpl", CFile::modeRead);
    CArchive ar(&flEmployees, CArchive::load);

    ar >> m_FirstName >> m_LastName >> m_Address >> m_City >> m_State
    >> m_ZIPCode >> m_HourlySalary >> m_EmploymentStatus;

    ar.Close();
    flEmployees.Close();

    return 0;
}
```

6. Execute the application.
7. Close it .

Object serialisation consists of saving the values that are part of an object, mostly the value gotten from declaring a variable of a class. At the current standard, C++ doesn't inherently support object serialisation. To perform this type of operation, you can use a technique known as binary serialisation.

When you decide to save a value to a medium, the **fstream** class provides the option to save the value in binary format. This consists of saving each byte to the medium by aligning bytes in a contiguous manner, the same way the variables are stored in **binary numbers**.

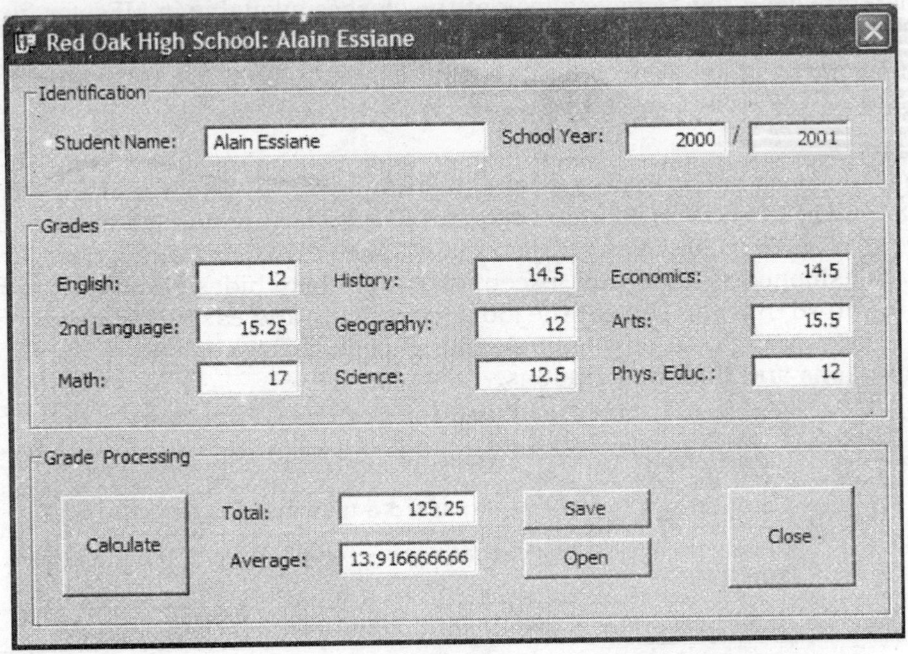

Fig. 6.6.

6.4. FILE HANDLING

(MDU 2008-09)

File basics

Every piece of information(whether program or data) is stored on the secondary storage devices(floppy disc, hard disc, CD ROM etc.) in the form of a unit called ***file***. There are a number of operations which can be and must be performed on files. Some of these file related operations are - file reading(reading data from a file stored on a secondary storage device), file writing(writing data into a file existing on the secondary storage device), file copying, file deleting, file moving, file re-naming etc.

For each of these file operations, the operating system of the computer has corresponding services in the form of functions and an application program may use these functions by calling them. The nature of these functions is complex and operating system dependent. Thus, the way one operating system(say Unix) handles the file operations/functions is generally different from that of the other operating system(say Windows NT). It implies that each operating system has its own(generally unique) file systems. This is the reason why a file saved on the secondary device by one operating system may not be accessible by another operating system.

Since file handling is largely system dependent, applications written for one operating system may not run on the other. This affects the portability of the application program. In order to make these operations system independent, programming languages generally offer functions implemented for the underlying operating system grouped in library files(API's - Application Program Interfaces). All the programmer has to know is the services offered by such API functions, their syntax and the input required as well as the output produced.

In the context of Windows programming, a number of vendors have provided commercially available file related API's. Microsoft has made available different file handling API's for different hardware and operating system platforms. Probably the most widely used API for this purpose is the MFC(Microsoft Foundation Classes) framework.

It is therefore desirable to have a look at the classes available in MFC for file handling operations.

The CFile class and its derived classes can be used to handle files in almost the same manner as in non-Windows programs.

The CFile Class

MFC's CFile class encapsulates all the functions you need to handle any type of file. Whether you want to perform common sequential data saving and loading or construct a random access file, the CFile class gets you there. Using the CFile class is a lot like handling files the old-fashioned C-style way, except that the class hides some of the busy-work details from you so that you can get the job done quickly and easily. For example, you can create a file for reading with only a single line of code. Table 6.1 shows the CFile class's member functions and their descriptions.

Table 6.1. Member Functions of the CFile Class

Function	***Description***
Cfile	Creates the CFile object. If passed a filename, it opens the file.
Destructor	Cleans up a CFile object that's going out of scope. If the file is open, it closes that file.
Abort()	Immediately closes the file with no regard for errors.
Close()	Closes the file.

Duplicate()	Creates a duplicate file object.
Flush()	Flushes data from the stream.
GetFileName()	Gets the file's filename.
GetFilePath()	Gets the file's full path.
GetFileTitle()	Gets the file's title (the filename without the extension).
GetLength()	Gets the file's length.
GetPosition()	Gets the current position within the file.
GetStatus()	Gets the file's status.
LockRange()	Locks a portion of the file.
Open()	Opens the file.
Read()	Reads data from the file.
Remove()	Deletes a file.
Rename()	Renames the file.
Seek()	Sets the position within the file.
SeekToBegin()	Sets the position to the beginning of the file.
SeekToEnd()	Sets the position to the end of the file.
SetFilePath()	Sets the file's path.
SetLength()	Sets the file's length.
SetStatus()	Sets the file's status.
UnlockRange()	Unlocks a portion of the file.
Write()	Writes data to the file.

Apart from the Cfile ckass, MFC also offers a few of its derivatives such as CstdioFile and CmemFile. While these classes suffice the programming needs of an application, a programmer can always derive own custom classes from the above mentioned classes.(The beauty of object oriented programming!).

Using files in Windows applications

As you can see from Table 6.1, the CFile class offers plenty of file-handling power. This section demonstrates how to call a few of the CFile class's member functions. However, most of the other functions are just as easy to use.

Writing into File

Here's a sample snippet of code that creates and opens a file, writes a string to it, and then gathers some information about the file :

```
// Create the file.
CFile file("TESTDATAFILE.TXT", CFile::modeCreate | CFile::modeWrite);
// Create the string to be stored in the file.
CString message("Hello file!");
// Compute the number of characters of the string in variable
length.
int length = message.GetLength();
```

```
// Write the content of the string variable message into the file.
file.Write((LPCTSTR)message, length);
// Obtain path information of the data file.
CString filePath = file.GetFilePath();
// Obtain the size(number of bytes) of the data file.
Int fileLength = file.GetLength();
```

Notice that you don't have to explicitly open the file when you pass a filename to the constructor, whose arguments are the name of the file and the file access mode flags. You can use several flags at a time simply by ORing their values together, as in the little snippet above. These flags, which describe how to open the file and which specify the types of valid operations, are defined as part of the CFile class and are described in Table 6.2.

Table 6.2. The File Mode Flags

Flag	***Description***
CFile::modeCreate	Creates a new file or truncates an existing file to length 0
CFile::modeNoInherit	Disallows inheritance by a child process
CFile::modeNoTruncate	When creating the file, doesn't truncate the file if it already exists
CFile::modeRead	Allows read operations only
CFile::modeReadWrite	Allows both read and write operations
CFile::modeWrite	Allows write operations only
CFile::shareCompat	Allows other processes to open the file
CFile::shareDenyNone	Allows other processes read or write operations on the file
CFile::shareDenyRead	Disallows read operations by other processes
CFile::shareDenyWrite	Disallows write operations by other processes
CFile::shareExclusive	Denies all access to other processes
CFile::typeBinary	Sets binary mode for the file
CFile::typeText	Sets text mode for the file

CFile::Write() takes a pointer to the buffer containing the data to write and the number of bytes to write. Notice the LPCTSTR casting operator in the call to Write(). This operator is defined by the CString class and extracts the string from the class.

One other thing about the code snippet: There is no call to Close()--the CFile destructor closes the file automatically when file goes out of scope.

Reading from file

Reading data from a file isn't much different from writing to one :

```
// Opening the file.
CFile file("TESTDATAFILE.TXT", CFile::modeRead);
// Creating variable s to store the data read from the file.
char s[81];
// Reading data from the file and storing it in the variable s.
int bytesRead = file.Read(s, 80);
// Terminating the value read by null value.
s[bytesRead] = 0;
```

```
// Storing the value into variable message.
CString message = s;
```

This time the file is opened by the CFile::modeRead flag, which opens the file for read operations only, after which the code creates a character buffer and calls the file object's Read() member function to read data into the buffer. The Read() function's two arguments are the buffer's address and the number of bytes to read. The function returns the number of bytes actually read, which in this case is almost always less than the 80 requested. By using the number of bytes read, the program can add a 0 to the end of the character data, thus creating a standard C-style string that can be used to set a CString variable.

6.5. DEBUGGING AND PROFILING COM COMPONENTS (MDU 2008)

Debugging is the process of correcting or modifying the code in your project so that your project can build, run smoothly, act as you expected, and be easy to maintain later.

To this end, Visual C++ provides a variety of tools to help with the varied tasks of tracking down errors in the code and program components.

The debugger interface provides special menus, windows, dialog boxes, and spreadsheet fields. Drag-and-drop functionality is available for moving debug information between components. Occasionally the debugger is paused in break mode, meaning the debugger is waiting for user input after completing a debugging command (like break at breakpoint, step into/over/out/to cursor, break at exception, break after Break command or Restart).

To start Debugging

1. Click **Start Debug** on the **Build** menu.
2. Click **Go, Step Into**, or **Run to Cursor**.

To halt Debugging

1. Click **Break** on the **Debug** menu.
2. Control returns to Microsoft Visual C++.

Use the **Breakpoints** dialog box (accessed by the **Breakpoints** command on the **Edit** menu), to set, remove, disable, enable, or view breakpoints. The breakpoints you set will be saved as a part of your project.

> **Note :**
> You must have a project open before you can set a breakpoint. With no project open, the Breakpoints command does not appear on the Edit menu.

You can set breakpoints on a source-code line or in the Disassembly or Call Stack window.

If you set more than one location breakpoint on a line, and some breakpoints are disabled while others are enabled, a gray dot appears in the left margin in the **Source** window, **Disassembly** window, and/or **Call Stack** window (not the **Breakpoints** dialog). The first time you click the **Enable/Disable Breakpoint** toolbar button, all breakpoints on the line become disabled, and the gray dot changes to a hollow circle. If you click the **Enable/Disable Breakpoint** button again, all breakpoints on the line become enabled, and the hollow circle changes to a red dot.

An asterisk (*) in the breakpoint check box indicates that the breakpoint is not supported on the current platform.

To run the program and execute the next statement (Step Into) :

1. While the program is paused in break mode (program is waiting for user input after completing a debugging command), click Step Into from the Debug menu.

The debugger executes the next statement, then pauses execution in break mode. If the next statement is a function call, the debugger steps into that function, then pauses execution at the beginning of the function.

2. Repeat step 1 to continue executing the program one statement at a time.

   If you step into a nested function call, the debugger steps into the most deeply nested function. For example, on the line of code Fun(Fun2); the debugger steps into the function Fun2, then pauses.

To Step into a Specific Function

1. Set a breakpoint just before the nested function call.

   \- or -

   Use the **Step Into, Step Over,** or **Run To Cursor** command to advance the program execution to that point.
2. In a source window, select the function that you want to step into.
3. On the **Debug** menu, or on the source window shortcut menu, click Step Into Name, where *Name* is the name of the selected function.

   The debugger executes the function call and pauses execution at the beginning of the selected function.

   Step Into Specific Function works for any number of nesting levels. You can also sometimes use **Step Into Specific Function** to step into a member function; for example, MemberFn in the line of C++ code CMyClass::MemberFn();.

To Step Over a Function

1. Open a source file, and start debugging.
2. Execute the program to a function call.
3. On the **Debug** menu, click **Step Over**.

   The debugger executes the next function, but pauses after the function returns.
4. Repeat step three to continue executing the program, one statement at a time.

To Step Out of a Function

1. Start debugging, and execute the program to some point inside the function.
2. On the **Debug** menu, click **Step Out**.

   The debugger continues until it has completed execution of the return from the function, then pauses.

Caution In general, to avoid very slow execution, you should not step out of a function containing a loop. Instead, you should set a breakpoint at the end of the function, and then choose **Go** from the **Debug** menu to execute to the end of the function. Then choose **Step Out**.

With Edit and Continue (a new feature in Visual C++ 6.0), you can make changes to your source code while the program is being debugged (with some limitations). You can apply code changes while the program is running or halted under the debugger.

To apply code changes to a program you are debugging :

From the **Debug** menu, click **Apply Code Changes**. Some code changes cause the point of execution to move to a new location when Edit and Continue applies the changes. Edit and continue places the point of execution as accurately as possible, but the results may not be correct in all cases. A dialog box informs you when the point of execution changes. You should verify that the location is correct before you continue debugging and use Set Next Statement if necessary.

Edit and Continue can also apply changes automatically when you select a **Go** or **Step** command for a program that is halted. You can turn off this automatic Edit and Continue, if you prefer. (If you turn the automatic feature off, you can still apply code changes manually using **Apply Code Changes**.)

To enable/disable Automatic Edit and Continue :

1. From the **Tools** Menu, click **Options**.
2. In the **Options** dialog box, select the **Debug** tab.
3. On the **Debug** tab, select or clear the **Debug commands invoke Edit and Continue** check box, as appropriate.

Because **Debug commands invoke Edit and Continue** is a tools option rather than a project option, altering this setting affects all projects you work on. You do not need to rebuild your application after changing this setting. You can change the setting even while debugging.

Edit and Continue uses the Program Database to store information needed to apply code changes. When you create a new project, Visual C++ chooses the appropriate option (**Program Database for Edit and Continue**) in the **Project Settings** dialog box. If you change this option (in order to use C7-compatible debug information, for example), Edit and Continue is disabled.

To enable/disable Edit and Continue for the current project :

1. From the **Project** menu, click **Settings**.
2. In the **Project Settings** dialog box, select the **C/C++** tab and the **General** category.
3. On the **C/C++** tab, select the **Debug Info** list box and select **Program Database for Edit and Continue** to enable Edit and Continue or select any other option to disable it.
4. Click **OK** to close the dialog box.
5. Rebuild your application with the new setting.

Choosing **Program Database for Edit and Continue** selects the /ZI compiler option.

Note :
Do not enable Edit and Continue if you are debugging optimised code. Edit and Continue is incompatible with optimizations. The result is a compiler error.

To view the value of a variable or expression or contents of a register :

1. Wait for the debugger to stop at a breakpoint.

   - or -

   Click **Break** on the **Debug** menu to halt the debugger.
2. On the **Debug** menu, click **QuickWatch**.
3. Type or paste the variable or register name or an expression into the Expression text box, and click **Recalculate**.
4. Click **Close**.

To view the value of a variable using QuickWatch :

1. When the debugger is stopped at a breakpoint, switch to a source window, and click the right mouse button on a variable (Var, for example).
2. On the shortcut menu, click **QuickWatch** *Var*.
3. Click **Close**.

To view the value of a variable or expression or the contents of a register **in the Watch** window :

1. Start debugging, and pause the debugger in break mode (program is waiting for user input after completing a debugging command).
2. On the View menu, click Debug Windows, then click Watch.
3. Select a tab for the variable or expression.
4. Type, paste, or drag the variable or register name or an expression into the Name column on the tab.

   Press ENTER (if typing).

   The Watch window evaluates the variable or expression immediately and displays the value or an error message.

   If you add an array or object variable to the Watch window, plus sign (+) or minus sign (-) boxes appear in the Name column. Use these boxes to expand or collapse your view of the variable described in Spreadsheet Fields.

   Change the Watch window display format (to display Unicode characters, for example) using formatting symbols, described in Symbols for Watch Variables.

To view type information for a variable in the Watch window :

1. In the Watch window, select the line containing the variable whose type you want to see.
2. On the **View** menu, click **Properties**.

To view a variable in the Variables window :

1. Start debugging, and pause the debugger in break mode (program is waiting for user input after completing a debugging command).
2. On the **View** menu, click **Debug Windows**, then click **Variables**.
3. Click the **Auto** tab, **Locals** tab, or **This** tab, according to the type of variables you want to see.

To view type information for a variable in the Variables window :

1. In the Variables window, click the **Auto tab**, **Locals** tab, or **This** tab.
2. Select the line containing the variable whose type you want to see.

   On the **View** menu, click **Properties**.

To display meaningful values for a custom datatype:

1. The file autoexp.dat controls how the debugger displays data types in windows, Quickwatch, and DataTips. You can modify this file to add meaningful displays for your own datatypes. For details, see the autoexp.dat file itself (located in the /bin directory).
2. By default, the Disassembly window displays disassembled code with source-code annotations and symbols. Use options on the **Debug** tab in the **Options** dialog box on the **Tools** menu to change the display.

To view specific source code in the Disassembly window :

Start debugging, and pause the debugger in break mode (program is waiting for user input after completing a debugging command).

On the **View** menu, click **Debug Windows**, then click **Disassembly**.

To view the call stack for a function :

1. Place the insertion point in the function.
2. On the **Debug** menu, click **Run To Cursor** to execute your program to the location of the insertion point.

   The **Locals** tab of the Variables window is updated automatically to display the local variables for the function or procedure.

3. On the **View** menu, click **Debug Windows**, then click **Call Stack**.

The calls are listed in the calling order, with the current function (the most deeply nested) at the top.

To run the program to the return address, select the function in the Call Stack window, and choose **Run To Cursor** from the **Start Debug** menu.

To set or remove a breakpoint at a function return address, select the function in the Call Stack window, and click the **Insert/Remove Breakpoint** toolbar button.

To change the call stack display :

1. On the **Tools** menu, click **Options**.
2. Click the **Debug** tab.
3. Under Call stack window, select the check boxes for **Parameter Values or Parameter. Types**, according to the information you want to display.

The Visual C++ debugger uses CodeView-format debug information. If CodeView-format information is not available, Visual C++ can use two additional types of information :

- COFF-format debugging information
- DLL export tables

COFF is a format used by some other debuggers. You can find COFF-format information in the MSVBVM50.DBG file shipped provided with Visual Basic 5.01, for example.

Reading symbolic information from DLL export tables can be useful if you are working with Windows 95 system DLLs, Windows messages, Windows procedures (**WindowProcs**), COM objects, or marshalling.

Reading COFF and DLL export information entails some overhead, so this capability is turned off by default.

To use COFF-format information or DLL exports :

1. On the **Tools** menu, click **Options**.
2. The **Options** dialog appears. Select the **Debug** tab.
3. Select the **Load COFF and Exports** check box.

   The calls are listed in the calling order, with the current function (the most deeply nested) at the top.

   To see whether a DBG file contains CodeView- or COFF-format information, use dumpbin /headers. Look in the Debug Directories section of the output file for COFF (coff) or CodeView (cv) entries.

   To see what symbols are available in the export table of a DLL, use dumpbin / exports. Symbols are available for any 32-bit system DLLs. By reading the dumpbin /exports output, you can see the exact function name, including non-alphanumeric characters. This is useful for setting a breakpoint on a function. Function names from DLL export tables may appear truncated elsewhere in the debugger.

   When the program is paused at a breakpoint or between steps, you can change the value of any **non-const** variable or contents of any register (including Intel MMX registers).

The Profiler is an analysis tool used to examine the run-time behaviour of the programs. By using profiler information, it can be easily determined which sections of code are working efficiently, or the areas of code that are not being executed, or that are taking a long time to run/execute.

Since profiling is a tuning process, you should use the profiler to make your programs run better and not to find bugs. The profiler can analyse the execution of your code in two ways, *viz.* Line Profiling and **Function Profiling**.

Line Profiling shows how many times each line is executed in response to certain input data and can be useful for checking the validity of an algorithm. There are two line-profiling options *viz.* Line Counting and **Line Coverage**.

Line Counting shows how many times each line was executed.

Line Coverage shows you which lines were executed at least once.

Function Profiling is used for detecting inefficient code. The profiler provides four ways to profile by function: **Function Timing, Function Counting, Function Coverage** and **Function Attribution**.

To enable profiler, check out the profiler option in Project->Settings->Link menu.

SUMMARY

1. Serialisation is an important concept in Windows Programming because most of the data of an application exists in object form. It enables an application to open and save documents along-with its data. Merely saving the data of a document would not enable the application to recreate the document from the stored data.
2. To add the Serialise function to your custom class, add a member function through the Class View tab in the workspace pane. Specify the function as type void, the function declaration as Serialise(CArchive &ar), and the access as public and check the Virtual check box. This should add the Serialise function and place you in the editor.
3. In the Serialize function, the first you want to do is to first call the ancestor's Serialise function. When you do this, any foundation information that has been saved is restored first, providing the necessary support for your class before the variables in your class are restored. Once you call the ancestor function, you need to determine whether you need to read or write the class variables. You can do this by calling the method IsStoring of the class CArchive. This function returns TRUE if the archive is being written to and FALSE if the archive id being read from. If the IsStoring function returns TRUE, C++/IO streams can be used to write all the class variables to the archive.
4. File processing is the ability to create, store, and/or retrieve the contents of a file from a medium such as a hard drive, a floppy disk, a CD-ROM, or a DVD-ROM, etc. One of the areas in which the MFC excels is file processing. MFC makes it particularly easy to perform most file related operations.
5. Most of the file processing in an MFC application is performed in conjunction with a class called CArchive. The CArchive class serves as a relay between the application and the medium used to either store data or make it available. The CArchive class is an accessory and not an end to itself. As such, it provides its services to the MFC class that need to perform file transfer.
6. Object serialisation consists of saving the values that are part of an object, mostly the value gotten from declaring a variable of a class. At the current standard, C++ doesn't inherently support object serialisation. To perform this type of operation, you can use a technique known as binary serialisation.
7. Since file handling is largely system dependent, applications written for one operating system may not run on the other. This affects the portability of the application program. In order to make these operations system independent, programming languages generally offer functions implemented for the underlying operating system grouped in library files(API's - Application Program Interfaces).
8. The debugger interface provides special menus, windows, dialog boxes, and spreadsheet fields. Drag-and-drop functionality is available for moving debug information between components.

9. In general, to avoid very slow execution, you should not step out of a function containing a loop. Instead, you should set a breakpoint at the end of the function, and then choose Go from the Debug menu to execute to the end of the function.
10. The Profiler is an analysis tool used to examine the run-time behaviour of the programs. By using profiler information, it can be easily determined which sections of code are working efficiently, or the areas of code that are not being executed, or that are taking a long time to run/execute.

SHORT QUESTIONS WITH ANSWER

Q.1. What is Archiving?

Ans. Most of the file processing in an MFC application is performed in conjunction with a class called **CArchive**. The **CArchive** class serves as a relay between the application and the medium used to either store data or make it available. The **CArchive** class is an accessory and not an end to itself. As such, it provides its services to the MFC class that need to perform file transfer.

Although used by MFC, the **CArchive** class is semi-independent. That is, it is not derived from **CObject**. It only provides its services to the MFC objects that need archiving. To provide this support, the parent class of MFC objects, **CObject**, is equipped with a method called Serialise(). The CObject::Serialise() method takes one argument, which is a **CArchive** reference variable. The syntax of the CObject::Serialise() method is :

```
virtual void Serialize(CArchive& ar);
```

This means that the MFC library already provides file processing at a very high level. This also implies that almost every MFC class is ready to save its value(s) from a medium or to retrieve existing (an) value(s) from a medium. This convenience is given at a very low price, as file processing is one of MFC strengths. Based on this, to provide file processing for an object in your application, you can simply override the **Serialise()** method.

Q.2. Define serializing. Write program to make a class serializable.

Ans. Serializing::::serialization is the process of saving an object onto a storage medium (such as a file or a memory buffer) or to transmit it across a network connection link in binary form. The series of bytes or the formation can be used to re-create an object that is identical in its internal state to the original object (actually, a clone). This process of serializing an object is also called deflating or marshalling an object. The opposite operation, extracting a data structure from a series of bytes, is deserialization (which is also called inflating or unmarshalling). The Serialize member function, which is defined in the CObject class, is responsible for actually serializing the data necessary to capture an object's current stage. The Serialize function has a Carchive argument that it uses to read and write the object data. The CArchieve object has a member function, isStoring, which indicates whether Serialize is storing (writing data) or loading (reading data). Using the results of Is Storing as a guide, you either insert your object's data in the CArchieve object with the insertion operator (<<) or extract data with the extraction operator (>>). Consider a class that is derived from CObject and has two new member variables, of types CString and WORD. The following class declaration fragment shows the new member variables and the declaration for the overridden Serialize member function:

```
class CPerson: public Cobject {
public:: DECLARE_SERIAL (CPerson)
CPerson ( ); virtual_CPerson ( );
CString m_name; WORD m_number; void Serialize (CArchive & archive);};
```

Q.3. What do you mean by connectivity to a database ? Differentiate between ADIO and RDO connectivity methods.

Ans. Making connectivity to database means that connecting the front end application with beckoned and storing the data. There are different methods to make the connectivity in different languages. The connectivity methods provides the ease to store the data in the database.

Difference between ADO and DAO: DAO can access most local databases (actually through ODBCDirect which uses RDO). There are performance issues with connecting remotely though, RDO will connect remotely to most any data source, but it doesn't handle local databases well, performance wise. It is an easy interface to the MS ODBC API (Application Programming Interface) ADO is the newest technology, it is simpler and more efficient in resources that the above. It will access local and remote databases, as well as files, and email programs like Exchange... It is also used in ASP. It is what Microsoft recommend using, now.

Using ADO Objects: ADO has three objects, the Connection, the Recordest, and the Command object if (IpOleDraw ! = (FARPROC)NULL)

```
{  (*iLoIEdRAW) (PUnkown, dwAspect, hdcDraw, IpreBouds) ; }
FreeLibrary (hOle2D||); }
```

The procedure for explicit run-time linking is the same in any language that supports pointers to functions, since it depends on the Windows API rather than language constructs.

Q.4. What are the advantages of using an ActiveX control?

Ans. Advantages of ActiveX control: There are many advantages to ActiveX controls:

Good performance: ActiveX controls always run in process. However, there is additional overhead involved in using an ActiveX control that does not occur with an ActiveX DLL server. This will be discussed further in Part 3.

- Controls are compatible with many containers, including Microsoft Office applications and Internet browsers.
- Controls offer seamless integration into the VB environment.
- Property pages allow design time user interface, as well as runtime interface within Visual Basic.
- Controls have the ability to persist design time properties in most containers.

Disadvantages are:

- Controls are considerably foster than ActiveX EXE servers, but somewhat slower than ActiveX DLL servers due to the ActiveX overhead.
- There is some complexity involved in creating good quality controls.
- Controls increase the complexity of deploying an application.
- Registration, version checking, and component verification are required for safe distribution.

LONG QUESTIONS WITH ANSWER

Q.1.Write down the procedure for Serialising the class ?

Ans. After you have your custom class with all the variables defined and initialized, you need to make the class serialisable. Making your class serialisable involves three steps.

1. The first step is adding the Serialise function to the class. This function writes the variable values to, and reads them back from, the CArchive object using C++ streams.
2. Second step involves adding of the DECLARE_SERIAL macro.
3. The Third step consists of adding IMPLEMENT_SERIAL macro.

After these elements are added, your custom class will be serialisable and ready for your application.

To add the *Serialise* function to your custom class, add a member function through the Class View tab in the workspace pane. Specify the function as type void, the function declaration as *Serialise(CArchive &ar)*, and the access as public and check the Virtual check box. This should add the Serialise function and place you in the editor.

In the *Serialise* function, the first you want to do is to first call the ancestor's *Serialize* function. When you do this, any foundation information that has been saved is restored first, providing the necessary support for your class before the variables in your class are restored. Once you call the ancestor function, you need to determine whether you need to read or write the class variables. You can do this by calling the method *IsStoring* of the class *CArchive*. This function returns TRUE if the archive is being written to and FALSE if the archive id being read from. If the *IsStoring* function returns TRUE, C++/IO streams can be used to write all the class variables to the archive. If the function returns false, C++ streams can be used to read from the archive. It must be sure, in both the cases, to order the variables in the same order for both reading and writing.

Example 3

An example of a Serialise function is shown in the Listing.

```
void Cperson::Serialise(CArchive &ar)
{

    CObject::Serialise(ar);
    // Calling the ancestor function.

    If(ar.IsStoring())

    {
        // Writing all the variables in the proper order.

    ar << m_sname << m_iAge << m_iMaritalStatus << m_bEmployed;
    else

    // Reading all the variables in the order.

    ar >> m_sName >> m_iAge >> m_iMaritalStatus >> m_bEmployed;
```

Once you place the *Serialise* function in the order, you need to add macros to your custom class. The firstmacro, DECLARE_SERIAL, needs to go in the class header and is passed the class name as its only argument.

Q.2. In File Processing, explain What to Process, When, and Where?

Ans. Before creating a file, you should first plan your program since applications deal with different types of files. An application that manipulate pictures may need to save only one value for the whole file. Another application used for word processing may also need to save only one value but of a completely different kind. If an application is list-based, it may need to save various values that, when put together, constitute one file.

After deciding what type of file the user would be processing, you would also specify how and/ or who would decide when to process a file. For example, if you create a certain application that allows a user to change the internal settings and store them in a file, you can process the file automatically and store it somewhere without asking the user to decide when, where, or how to save the file (some applications use the Registry to take care of this). For example, in tthe bits of data that are arranged in a particular way to produce a usable document. For easy storage, location, and management, the bits are stored on a medium such as a hard disc, a floppy disc, a compact disc, or any valid and support type of storage. When these bits belong to a single but common entity, the group is referred to as a file. For even greater management, files can be stored in a parent object called a directory or a folder. Since a file is a unit of storage and it stores information, it has a size which is the number of bits it contains. To manage it, a file also has a location also called a path that specifies where and/or how the file can be retrieved. Also, for better management, a file has attributes that indicate what can be done on a file or that provide specific information that the programmer or the operating system can use when dealing with the file.

For example, in Jasc Paint Shop Pro, users are allowed to create new customised gradients that were not installed with the application. Once a user clicks OK, the new gradient is saved :

Another way you can deal with this is to let a user decide when to save the file and where to save it. This is usually taken care of by providing the File Dialog to the user.

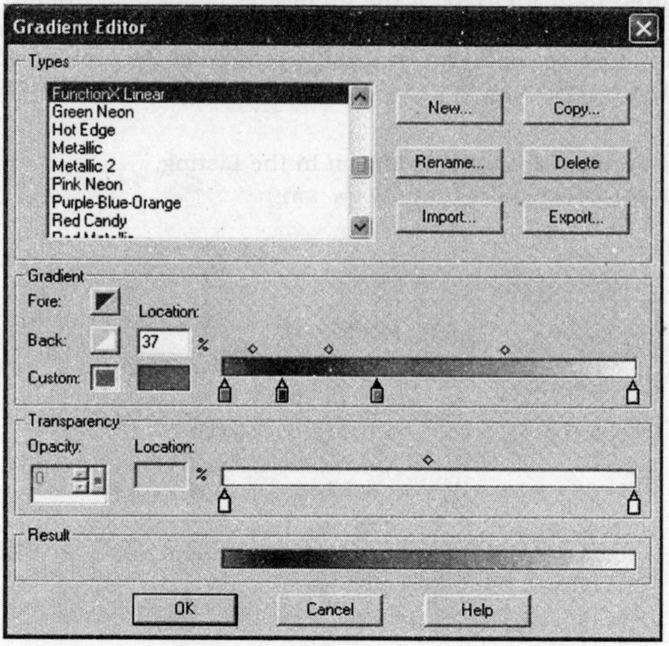

Fig. 6.7.

Q.3. What is The Process of Archiving?

Ans. One of the most regular ways you will use the **CArchive** class consists of storing or retrieving values through their **Serialise()** method. To do this directly and easily in the class, the **CArchive** class overloads two operations: << and >>.

To store one or more values, the **CArchive** class provides the << operator. The values you can store are of MFC, **BYTE, short, WORD, int, unsigned int, long, LONG, DWORD, float,** or **double** types. The class also provides an exception handler for each kind. Therefore, the syntaxes for this operator are :

```
friend CArchive& operator <<( CArchive& ar, const CObject* pOb );
throw( CArchiveException, CFileException );

CArchive& operator <<( BYTE by );
throw( CArchiveException, CFileException );

CArchive& operator <<( WORD w );
throw( CArchiveException, CFileException );

CArchive& operator <<( int i );
throw( CArchiveException, CFileException );

CArchive& operator <<( LONG l );
throw( CArchiveException, CFileException );

CArchive& operator <<( DWORD dw );
throw( CArchiveException, CFileException );
```

```
CArchive& operator <<( float f );
throw( CArchiveException, CFileException );

CArchive& operator <<( double d );
throw( CArchiveException, CFileException );
```

As you may have realised, there is only one version of the Serialize() method. That is, this member function is called once for both storing and retrieving. Therefore, when using it, you must specify what operation you want to perform. This checking process can be performed by calling the **CArchive::IsStoring()** Boolean method. Its syntax is :

```
BOOL IsStoring() const;
```

This method returns TRUE if you are storing data. If the answer is then TRUE, you can use the << operator to store each necessary value.

As opposed to storing value(s), to retrieve a value, the CArchive class provides the >> operator. It comes in the following syntaxes :

```
friend CArchive& operator >>( CArchive& ar, CObject *& pOb );
throw( CArchiveException, CFileException, CMemoryException );

friend CArchive& operator >>( CArchive& ar, const CObject *& pOb );
throw( CArchiveException, CFileException, CMemoryException );

CArchive& operator >>( BYTE& by );
throw( CArchiveException, CFileException );

CArchive& operator >>( WORD& w );
throw( CArchiveException, CFileException );

CArchive& operator >>( int& i );
throw( CArchiveException, CFileException );

CArchive& operator >>( LONG& l );
throw( CArchiveException, CFileException );

CArchive& operator >>( DWORD& dw );
throw( CArchiveException, CFileException );

CArchive& operator >>( float& f );
throw( CArchiveException, CFileException );

CArchive& operator >>( double& d );
throw( CArchiveException, CFileException );
```

This indicates that you can retrieve an MFC object, a **BYTE**, a **short**, a **WORD**, an int, an unsigned int, a long, a **LONG**, a **DWORD**, a **float**, or a **double** value.

As mentioned above, the Serialise() method is used for both operations. Therefore, when using this method, you should first check what operation is being performed. This can be done by calling the IsLoading() method. Its syntax is :

```
BOOL IsLoading() const;
```

This method returns TRUE if you are retrieving a value or some values. If it is true, you can then use the >> operator to retrieve the value(s).

Q.4. What do you mean by File Handling?

Ans. Writing into File: Here's a sample snippet of code that creates and opens a file, writes a string to it, and then gathers some information about the file :

```
// Create the file.
CFile file("TESTDATAFILE.TXT", CFile::modeCreate | CFile::modeWrite);
// Create the string to be stored in the file.
CString message("Hello file!");
// Compute the number of characters of the string in variable length.
int length = message.GetLength();
// Write the content of the string variable message into the file.
file.Write((LPCTSTR)message, length);
// Obtain path information of the data file.
CString filePath = file.GetFilePath();
// Obtain the size(number of bytes) of the data file.
Int fileLength = file.GetLength();
```

Notice that you don't have to explicitly open the file when you pass a filename to the constructor, whose arguments are the name of the file and the file access mode flags. You can use several flags at a time simply by ORing their values together, as in the little snippet above. These flags, which describe how to open the file and which specify the types of valid operations, are defined as part of the CFile class and are described in Table 6.3.

Table 6.3. The File Mode Flags

Flag	***Description***
CFile::modeCreate	Creates a new file or truncates an existing file to length 0
CFile::modeNoInherit	Disallows inheritance by a child process
CFile::modeNoTruncate	When creating the file, doesn't truncate the file if it already exists
CFile::modeRead	Allows read operations only
CFile::modeReadWrite	Allows both read and write operations
CFile::modeWrite	Allows write operations only
CFile::shareCompat	Allows other processes to open the file
CFile::shareDenyNone	Allows other processes read or write operations on the file
CFile::shareDenyRead	Disallows read operations by other processes
CFile::shareDenyWrite	Disallows write operations by other processes
CFile::shareExclusive	Denies all access to other processes
CFile::typeBinary	Sets binary mode for the file
CFile::typeText	Sets text mode for the file

CFile::Write() takes a pointer to the buffer containing the data to write and the number of bytes to write. Notice the LPCTSTR casting operator in the call to Write(). This operator is defined by the CString class and extracts the string from the class.

One other thing about the code snippet: There is no call to Close()--the CFile destructor closes the file automatically when file goes out of scope.

Reading from file: Reading data from a file isn't much different from writing to one :

```
// Opening the file.
CFile file("TESTDATAFILE.TXT", CFile::modeRead);
```

```
// Creating variable s to store the data read from the file.
char s[81];
// Reading data from the file and storing it in the variable s.
int bytesRead = file.Read(s, 80);
// Terminating the value read by null value.
s[bytesRead] = 0;
// Storing the value into variable message.
CString message = s;
```

This time the file is opened by the CFile::modeRead flag, which opens the file for read operations only, after which the code creates a character buffer and calls the file object's Read() member function to read data into the buffer. The Read() function's two arguments are the buffer's address and the number of bytes to read. The function returns the number of bytes actually read, which in this case is almost always less than the 80 requested. By using the number of bytes read, the program can add a 0 to the end of the character data, thus creating a standard C-style string that can be used to set a CString variable.

Q.5. Briefly Explain Debugging and Profiling COM components.

Ans. Debugging is the process of correcting or modifying the code in your project so that your project can build, run smoothly, act as you expected, and be easy to maintain later.

To this end, Visual C++ provides a variety of tools to help with the varied tasks of tracking down errors in the code and program components.

The debugger interface provides special menus, windows, dialog boxes, and spreadsheet fields. Drag-and-drop functionality is available for moving debug information between components. Occasionally the debugger is paused in break mode, meaning the debugger is waiting for user input after completing a debugging command (like break at breakpoint, step into/over/out/to cursor, break at exception, break after Break command or Restart).

To start debugging :

1. Click **Start Debug** on the **Build** menu.
2. Click **Go, Step Into**, or **Run to Cursor**.

To halt debugging :

1. Click **Break** on the **Debug** menu.
2. Control returns to Microsoft Visual C++.

Use the **Breakpoints** dialog box (accessed by the **Breakpoints** command on the **Edit** menu), to set, remove, disable, enable, or view breakpoints. The breakpoints you set will be saved as a part of your project.

The Profiler is an analysis tool used to examine the run-time behaviour of the programs. By using profiler information, it can be easily determined which sections of code are working efficiently, or the areas of code that are not being executed, or that are taking a long time to run/execute.

Since profiling is a tuning process, you should use the profiler to make your programs run better and not to find bugs. The profiler can analyse the execution of your code in two ways, *viz.* Line Profiling and **Function Profiling**.

Line Profiling shows how many times each line is executed in response to certain input data and can be useful for checking the validity of an algorithm. There are two line-profiling options *viz.* Line Counting and **Line Coverage**.

Line Counting shows how many times each line was executed.

Line Coverage shows you which lines were executed at least once.

Function Profiling is used for detecting inefficient code. The profiler provides four ways to profile by function: **Function Timing, Function Counting, Function Coverage** and **Function Attribution**.

To enable profiler, check out the profiler option in Project->Settings->Link menu.

Q.6. Define debugging. Write steps to set and remove break points for debugging.

Ans. Debugging: Debugging is the process of location add fixing errors (Known as bugs), in a computer program, or hardware device, To debug a program or hardware device, you start with a known problem, isolate the source of the problem, they imply that they have fixed the program, so that the bugs no longer exist in it. Debugging is a necessary process in almost any new software, or hardware development process, whether a commercial product, an enterprise, or personal application program. For complex, products, debugging is done periodically throughout the development, and again during the customer beta test stages. Because most computer programs and many programmed hardware devices contain thousands of lines of code, almost any new product is likely to contain a few bugs. Invariably, the bugs in the functions that get the most use, are found and fixed first. An early version of a program that has lots of bugs is referred to as "" buggy". "Debugging tools help identify coding errors at various stages of development. Some programming language packages include a facility coding errors at various stages of development. Some programming language packages include a facility for checking the code for errors as it is being written. To use he debugger in the Visual C++ programming environment you must have compiled and linked your program using the "Debug" configuration (not in "Release" configuration). The configuration is set using the menu options **Build|Set Active Configuration**. This is the default configuration, so you don't have to set it unless you previously changed the configuration to "Release". Usually it is not helpful to step through the statements in a program beginning with the test statement in **main ( )**. It is better to have the program run until it gets to that part of your code where you think the bug resides. There are two ways to tell the debugger to run to this point. You can locate the cursor at that place in your code and press **Ctrl-F10** (or use **Build | Start Debug | Run to Cursor)**. The program will be run by the debugger and if all goes well it will stop at the place in your code that you selected, and the debugger windows will be opened. Of course, it is possible that the program may terminate before your intended stopping place is reached. In which case, choose a place earlier in the program. The second way to tell the debugger to run to the place in your code that you wish to inspect is to set a "breakpoint" there. To set a break point, position the cursor at the desired place and press F9 (or use **Right-mouse-button |Insert/Remove Breakpoint)**. Once you have set a breakpoint then press **F5** (or use **Build | Start Debug | Go)** to cause the debugger to run the program and halt at the breakpoint, whereupon the debugger windows are opened (if they are not already open). The windows at the bottom left and bottom right of the screen hold variable names and values. Those in the left window include variables local to the function which is being executed. Those in the right window are user-specified variables. To specify a variable click on the left part of the first empty line then type the name of the variable. After pressing Enter the value of that variable (if it has a value) will be displayed. You can then view how the displayed variables change in value as execution of the program is continued. Pressing **F1** brings up context-sensitive Help. For example, it tells us the difference between "Auto" variables and "Locals" variables. The Auto tab displays information about variables used in the current. Statement and the previous statement. The Locals tab displays information about variables that are local to the current function.

To view the value of variable without having to type in the name, position the cursor somewhat within the variable name and press **Shift -F9** (or **Debug |Quick Watch**). You can continue program execution in several ways. This simplest is to press **F10** (or **Debug | Step Over)** to execute the next statement (which is indicated y to small yellow arrow in the left margin of the code window). Pressing **F10** repeatedly steps through successive program statements. Attend to the information given in the button windows. For example, if you are at a statement.

if (j = atoi (buffer)) > 2000)

(which converts the string in buffer to an integer, places the value in variable j and compress this to 2000) then after pressing **F10** you will be informed as to the value returned by ratio ( ). If you are at a function call then **F10** causes that function to be executed and when the debugger returns control you are at the next statement. At a function call you can press **F11** (or **Debug | Step Into**) to enter that function (if te code for that function was compiled in "Debug" mode). You can then use **F10** to step through the statements in that function. When you are within a function you can press **Shift-F11** (or **Debug Out**) to run through the rest of the function and return to the place where the function was called. When the debugger starts it opens a window for the program's output, so you can switch between the debugger and this window to see what the program is sending to the screen. In addition to **F10, F11** and **Shift-F11** you can use **Ctrl-F10** or (**Debug | Run to Cursor**) to run to the cursor position. You can also set further breakpoints and use **F5** (or **Debug | Go**) to run to each one in turn. You can have several breakpoints active, and you can disable or remove them at will by positioning the cursor at them and pressing **F9**.

If you want to see the definition of a variable place the cursor within the variable name and use **Right-mouse-button | Go to Definition**. It may happen that the debugger will tell you that it can't do this until it has created the Browser database, in which case tell it do so. The Browsers a useful tool which you can use to get the definition of any variable in your program and the places in your program where the variable is used (the "variable references"). You can invoke it by using **Tools | Source Browser**. If your program takes command line arguments then during debugging you can set these using **project | Setting | Debug | Program arguments.** To restart the program from the beginning press **Ctrl-shift-F5** (or **Debug | Restart**). To stop the debugger and return to edit mode use **Shift-F5** (or **Debug | Stop Debugging**).

Q.7. What do you mean by connectivity to a database ? Differentiate between DAO and ADO connectivity methods.

Ans. Make connectivity to database means that connecting the front end application with backed and storing the data. There are different methods to make the connectivity in different languages. The connectivity methods provides the ease to store the data in the data base.

Difference between ADO and DAO: DAO can access most local databases, and remote databases (actually through ODBCDirect which uses RDO). There are performance issues with connecting remotely though. RDO will connect remotely to most any data source, but it doesn't handle local databases well, performance wise. It is an easy interface to the MS ODBC API (Application programming Interface) ADO is the newest technology, it is simpler and more efficient in resources than the above. It will access local and remote databases, as well as files, and email programs like Exchanges. It is also used in SP. It is what Microsoft recommend using, now.

Using ADO objects: ADO has three objects, the Connection, the Recordest, and the Command object. Command is not normally needed. To proceed from there, look up "ADO Connection" then lookup "ADO Recordest", to connect and retrieve records from a data source.

You can also use the ADOData control and bound controls, to make simple apps without code. There is the Data Environment, too, which basically tries to do the same as the ADO Data control i.e., making simple database apps with little or no code.

Q.8. Using uppwizard WAP in VC++ to add a caret of given dimension to a window.

Ans.

```
#include "MainFrame.h"
BEGIN_MESSAGE_MAP(CMainFrame, CFrameWnd)
// message calls this function
ON-WM_PAINTO II void OnPaint ( )
ON-WM_CHARO II void OnChar( .. )
ON-WM_LBUTTONDOWNO II void OnLButtonDown( .. )
ON-WM_SETFOCUSO //void OnSetFoucs
(CWnd*pOldWnd)
ON-WM_KILLFOCUSO II void OnKillFocus
(CWnd*pNewWnd)
END_MESSAGE_MAPO
CMainFrame::CMainFrameO
{--Same s in Ex04f -
};
void CMainFrame::OnPaint( ){
CPaintDC dc(this); //removes WM_PAINT msg
```

```
HideCaretO; mLength. Display( ); mWidth.Display( );
mArea.DisplayO; AdjustCaretPositionO;// see details
ShowCaret( ); }
void CMainFrame::OnChar (UINT nChar, UNIT nRep, UNIT nFlags)
{   --same as Ex04f --
}   void CMainFrame::OnLButtonDown (UINT nFlags, CPoint pt)
{   --same s Ex04f -
}   void CMainFrame::AdjustCaretPosition( ){
    CPoint pt;switch (mActiveStr)
{   case 'L':
    pt = mLength.GetTextEndPt( );
    SetCaretpos (pt);
    break;
    case 'W':
    pi = mWidth.GetTextEndPt( );
    SetCaretpos (pt);
    break;
    case 'A':
    pt = mArea.GetTextEndPt( );
    SetCaretpos (pt);
    break;
    default;
    break; }}
    //MainFrame.h
    #include <afxwin.h>
    #include "TextBox.h"
    class CMainFrame: public CFrame Wnd
    {
        private:
        CTextBox mLength;
        CTextBox mWidth;
        CTextBox mArea;
        char mActiveStr; /I'L' = mstrLen, etc.
        public:
        CMainFrameO;
        // data manipulation function
        void AdjustCaretPositionO;
        II window msg handling functions below
        afx_msg void OnPaintO;
        afx_msg void OnChar (UINT nChar,
        UINT nRep,
        UINT nFlags);
        afx_msg void OnLButtonDown (UINT nFlags,
        CPoint pt);
        afx_msg void OnSetFocus (CWnd* pOldWnd);
        afx_msg void OnKillFocus (CWnd*pNewWncl);
        DECLARE_MESSAGE_MAP9);
    };
```

```
void CMainFrame::OnSelfFocus (CWnd*pOldWnd)
{
    CreateSolidCaret (2, 16); II width = 2, ht = 16 pixels Show CaretO;
}
void CMainFrame::OnKillFoucs (CWnd* pNew Wnd)
{
    HideCaret ( );
    ::DestroyCaretO; II A Win32 function
}
```

Q.9. What are different file system controls available in VB?

Ans. File system controls in VB: The file system is what makes up everything that Windows sees on a disk drive. This includes files and folders along with these file and folders properties. Using Visual Basic commands we can do many if not almost anything you couid want with the file system. Such as copying files, changing file properties, deleting, moving, etc, etc. Along with this you can use the VB6 file commands to do things such as checking creation dates, last modified dates, file size, and many other useful things. Over all you should be able to do almost anything you could do at a DOS prompt in the VB6 environment. VB provides a set of statements that perform DOS-like (or Windows command line-like) commands. These commands are listed and described in the table below:

VB Statement Equivalent DOS/Windows:

	Command Line Command	**Description**	**Syntax**
MkDir	MKDIR or MD	Creates a directory	MkDir pathname
RmDir	RMDIR or RD	Removes a directory	RmDir pathname
ChDrive	drive: (like typing A:	Changes the current drive	ChDrive drive
ChDir	CHDIR or CD	Changes current directory	ChDir drive
Kill	ERASE or DEL	Deletes a file	Kill filespec
Name	RENAME or REN	Renames a file	Nameoldfilespec As newfilespec
File Copy	COPY	Copies a file	FiieCopy oldfilespec, newfilespec

In the syntax above, pathname is a string expression specifying a valid path (it may optionally include the drive); drive is a string expression specifying a drive letter; and filespec, oldfilespec, and newfilespec are string expressions that specify a file (they may optionally include the drive and path). Following is a set of functions that can be used with files (all are functions except SetAttr, which is a statement):

Function	**Description**	**Syntax**
File Date Time	Returns a date/time value indicating the date and time that a file was created or last modified.	FileDateTime(filespec)
GetAttr	Returns an integer representing the attributes of a file, directory, or folder	GetAttr(filespec)
SetAttr	Statement that lets you specify the attributes for a file.	SetAttr(filespec, attributes)
CurDir$(or CurDir)	Returns a string that indicates the current path for a specified disk drive. In the syntax on the right, drivename is a string expression that specifies a valid disk drive designation.	CurDir$(drivename)

DirS(orDIr)	Returns a string that indicates a file or directory matching specified conditions.	Dir$(filespec [.attributes])
FileLen	Returns a Long specifying the length of a file in bytes. If the specified file is open when the FileLen function is called, the value returned represents the size of the file-immediately before it was opened.	FileLen(pathnanie)
LOF	Returns a Long representing the size, in bytes, of a file opened using the Open statement.	FileLen(filenumber)

Q.10. Define serializing? Write program to make a class serializable.

Ans. Serializing: Serializing is the process of writing or reading an object of some type to a "persistent storage medium" which means a disk. When we place code in the document's serialize( ) function to serialize our data to and from disk, we will automatically support program's file-hand I ijig items in its file menu.

Program to make a class serializable: Create a SDI program named serializer and go to File → New → C/C++ header file named CData.h. Open it and place the following code:

```
class CData: public Cobject
{
   private:
   CString data;
   DECLARE_SERIAL (CData);
   public: CData()
   data = CString(" ");
}
void AddText (CString text)
{
   data t = text;
   void DrawText(CDC*pDC)
{
   pDC → TextOut (0,0, data);
}
void clear Text()
{
   data "?";
}
void serialize (Archieve & archieve);
}
```

Then include new file in the document's header, seriaHzer.

```
//Serial izerDoc.h
        :
include "CData.h"
   :
//Attributes
public:
CData DataObject;
void CSerializer View:: OnCharflJFNT nchar, UINTnKepCut, UINT nFlags)
{
```

```
    CSerializer Doc * pDcc ~ Get Document);
    ASSERT VALID (pDoc);
    pDoc → DataObjectAdd Text(CSlring(nehaf));
    Invalidated):
    CView:; OnChar(nChar. nRepCut, nF!ags);
}
//To draw the text in the Data Object
void CSerializer View:: OnDraw (CDC*pDC)
{
    CSeriaiizer Doc ? pDoc = GetDocument( );
    ASSERT_VALID(pDoc);
    pDoc → DataObject.DrawText(pDc);
}
```

Now, add a new file to the project CData.cpp and define Serialize();

```
#include "stdafx.h"
#include "serializer Doc.h"
void CData:: Serializer(CArchieve & archieve)
{
    CObject:: Serialize (archieve);
    if (archieve.Is strlng())
    {
        archieve << data;
    }
    else
    {
        archieve >> data;
    }
}
lMPLEMENT_SERIAL(CData, CObject, 0);
//Implement_Serial is a macro which adds the
//Additional methods that VC++ needs for serialization
void CSerial Doc:: Serialize (CArchieve & ar)
{
    DataObject.Serialize(ar);
}
```

Q.11. Explain the concept of file handling with suitable example.

Ans. Concept of file handling includes, how to perform read/write operation. Consider the example of a dialog based application.

Create a dialog based application, name it "Filer".

```
    //filer Dlg.h     //Initilize variable to
      :                 //edit box 1 named m_text 1
    protected:          //and m_text 2 to the edit box 2
    HlCON m_hlCon;
    char OutString [6] [20];
    charInString[20];
    }
}
```

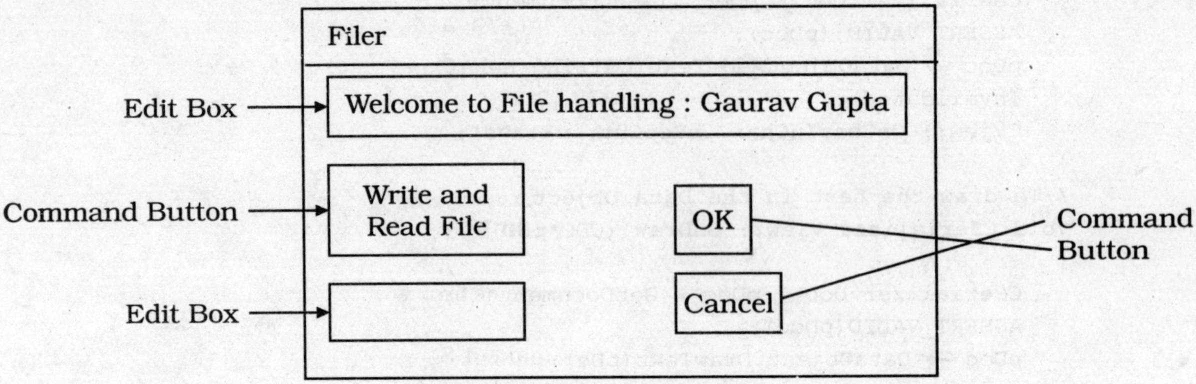

Fig. 6.8.

Initializing filer's data: Add OnInitDialog to CFiler Dig

```
class through class wizard, edit code:
BOOL CFiler Dig:: Onlnit Dialog( )
{
   CDialog:: Onlnit Dialog();
   strcpy (OutString [10], "welcome'¹);
   strcpy (OutString [1], "to");
   strcpy (OutString [2], "file");
   strcpy (OutString [3], "Handling:)";
   strcpy (OutString [4], "Gourav:)";
   strcpy (OutString [5], "Gupta:)";
      :
   m_text 1 = CString (OutString [0]) + CString (Outstrmg [ 1 ])+
   CString (OutString [2]) + CString (OutString [3]);
   UpdateData (false);
   }
```

Writing a File: OnButton 1 is the name given to command button:

```
   Void CFiler Dig:: OnButton 1()
   {
      CFile OutFile ("Data.dat", CFile:: mode Create: CFile:: mode
   Write);
      for(int loop_index = 0; loop_index < 4; loop_jndex + +)
   {
   OutFile.Write (OutString [loop-index], 20);
   }
   OutFile.Close();       //End of writing a file
                           //Reading a File
 CFile inFile ("data.dat", CFile:: modeRead);
 for (loop-index = 0; loop-index < 4; loop-index + +)
   {
      InFile.Seek(20* loop-index, CFile:: begin);
      //Seek() positions the file pointer at the beginning of each
      //succesive string.
         int Number characters = InFile.Read (InString, 20);
```

```
//Returns the number of characters actually read and store
//that value in integer number characters,
m-text 2 + = CString (InSTring);
}
pdate Data (False);
lnfile.ClQse();
```

Q.12. Explain various debugging techniques used in VC++ and VB.

Ans. In this, we have to debug our program:

1. In this program, you have to find the average of first five position numbers 1 -5 and we expect the output is 3.
2. Create the program using APP wizard and make it SDI with name "Buffy."
3. Store the first five integers is an away data( ) and add in document header, // buggy Doc.h public: int data (s);
4. CBugger Doc:: (CBuggy Doc( )

```
{
date[0] = 1;
data[1] = 2;
date[2] = 3;
date[3] = 4;
date[4] = 5;
```

5. Now using menu editor, add a new menu to the program, calculate (place this menu between file and edit) with one item in it. "Calculate Average."
6. Now use CW to add event handler to the menu item, *i.e.*, on Calculate Calculate Average( ).
7. In thus method, we will perform the calculation and display results on the screen. We need a front to the document.

```
Void (Buffer new:: On Calculate calculate Average
{
  CBuffer Doc * pDoc « Get Document ();
  ASSERT-VALUE (pDoc);
  float sum; float average;
  for (int i = 0, i < 5; i++)
  {
     sum = pDOC ® date [i];  }
     Average = Sum / (float) 5.0;
     Output string format ('Average: %3f, Average);
     Invalidate ( );
  }
```

8. Now set up output string object in view's header //Bufiy view.h protected: (String Output String;}
9. We place a prompt in that string for inverting the user to select the calculate average item in calculate menu in view's constructor like

```
(Buffer View:: (Buffer View( )
{
Output String: "Select Calculate Average item";
}
```

10. Finally, display output string in OnDraw();

```
Void (Buffer View:: OnDraw (CDC * pDC)
{
```

```
(Buffer Doc * pDoc » Get Document();
ASSERT-VALID(pDOC);
pDC → Text Out (0,0, Output String);
```

11. When we run the program and click calculate average menu item them average will be shown and , then output is 2818446765003.000.
12. But output is not correct. So, we have to debug our program. For debugging, set a Backpair. We start debugging where we suspect a problem.
13. To set a support, place the insertion point court at the beginning of that line and from, e.g., key or clutch the button in the toolbar with an unpaised hand icon.
14. Now we can single step through our code one line at a time by pressing F10 key; pressing that key once moves us to the next line of code.
15. Now run the program with debugger select Build → Start → Debug → Go.
16. Examine variable as program runs. To examine the value of sum variable which hold first integer, just set the mouse cursor over that variable.
    A tool tip appears with variable values which is 1.40817e + 0.19,cv
    i.e., garbage value.
17. Look at code, you forget to initiate the sum to 0.
    Float sum = 0;
18. Now run again, you will get the average is 2.800 as output but it is closer to 3.00 but not exact.
19. So, start debugging by selecting build → start → Debug → go again and run the program till Breakport. Press F10 to start the for loop.
20. Now look at the value that we are adding to sum, *i.e.*, pDoc → data (;).
21. So, now change the value of i to 0 line for (int i = 0; i < 5; i +)
22. Now run the program and get the output.
23. To remove be appear, place the count one the breakport code line and from F9 to off the breakport the a now.

Q.13. What do you mean by serialization? How a document can be made serialize. Explain with example.

Ans. Serializing in the process of writing or reading an object of some type to a "persistent storage medium." When we place code in the document's serialize 1. Function to serialize our data to and from disk, we will automatically support program's file handling items in to file menu. Now take an example "water" and it will write a string of text that the user types in to disk and read it back in again on demand.

Creating Program: Use appwizard, create program using SDI.

1. Add the code to lit the water to accept and display keystrokes. We will store the characters the user types in a (string object string Date.

```
Class (Writer Doc:: public CDocument
{
    (String string Data;
}
```

2. Initialize that object to the empty string in document's constructor.
   String Date = " ";
3. Use CW, connect WM_CHAR message to the method On Char( ) and add code to store the character the user has just typed in

```
String Data Object
Void (Writer View:: On Char (–)
{
    –
```

```
    -
    pDoc → String Data +m = nChar;
    Invalidate ( );
    pDoc → Set Modified Flag()
}
```

4. To display the new string of text in On Draw( )

```
Void (Writer View:: OnDraw (CDC.pDC(
{
pDc → TextOut(0,0pDoc → String Data);
```

5. Now program can read and display characters that the user types and stores in String Date.

Serializing the String Data Object:

6. Serialize ( ) method in document class

```
void (Writer Doc:: Serialize ((archieve & or)
(
if(or.IsString())
{
    ar << String Date;
}
else
{
    ar >> String Date;}}
```

EXERCISES

1. Write a code block creating a green pen three pixels wide.
2. Write a code block that gets the handle to the previously selected pen.
3. Create an application in VC++ that generates a message box on the left click of the mouse. The message should display the text "Left mouse button clicked", and the title of the application window should be "Left Click".
4. Write a VC++ application program that creates the menu resource named 'Color'. Clicking on this menu, three menu items should appear – namely Red, Green and blue. Clicking on these menu items should fill the application window with the corresponding colors(i.e. clicking on red menu item should fill the application window with the red color.
5. Write an application code to implement a Serializable class.
6. Write the VC++ program that responds to the double-click. On double-clicking the right mouse button, a message box should appear with the text as "Right mouse button clicked." and with the title as "Right-Click". Left-clicking the left mouse button should display "Left mouse button clicked", with the title in the message box window as "Left-Click". (Single) Clicking on the middle button should quit the application.
7. Write an application program that illustrates a device context, including some drawing function and some more other functions.
8. Write a program to draw a thick line using the class CPen.
9. Create an application that initially has the default cursor shape as an arrow. After right-clicking the mouse button, the shape of the cursor should change to hourglass.

❑❑❑

7

Win32 Static DLL

7.0. INTRODUCTION

Chapter Outline

- Introduction
- Creating a Project DLL
- Coding the DLL
- Selecting the DLL
- Using the DLL in a Console Application
- Using the DLL in an MFC Application
- OLE object Technologies
- Creating Internet Programs - Including a Web Browser
- Creating ActiveX Controls

A Dynamic Link Library (DLL) is a program that holds one or more functions or some functionality that other programs can use (that's the shortest definition I could come up with...). As such, you can create one DLL and use it in many programs or even distribute it to other programmers (because a DLL is usually treated as an "executable" program, you cannot e-mail it to somebody as an attachment; most e-mail programs would not carry it). Most of the time, a DLL is just a file that has the extension .dll but it could also have another extension.

The DLL we are going to create can be used in any program that uses MFC. Unfortunately, and based on my experience, you should not distribute it to people who are using other compilers such as Borland C++ Builder or Dev-C++. In reality you can. Because they will not be able to simply "plug" it in their programs, apparently there are some things they can do to make it work. I cannot cover that. In fact, the only reason I am writing this is because I can't believe how hard it is to find this information, whether in books or on the Internet; people who write books either assume that everybody knows how to create a DLL or they simply don't know, which is incredible considering, as you will see soon, how easy this thing is (I myself looked for it in many places, to no avail).

7.1. CREATING A PROJECT DLL

To start, we need a project that would carry the DLL :

1. Start Microsoft Visual C++. On the main menu, click File -> New -> Project...
2. In the New Project dialog box, in the Project Types list, click the Visual C++.

3. In the Templates list, click Win32 Project Projects tab and click Win32 Dynamic Link Library.
4. In the Project Name, type a name such as ExoDLL and specify your desired folder in the Location box.

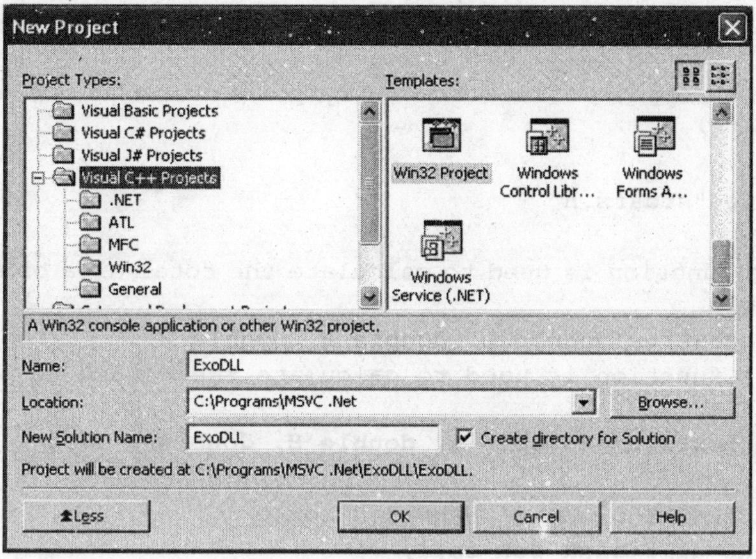

Fig. 7.1.

5. Click OK.
6. In the **Win32 Application Wizard**, click **Application Settings**.
7. In the Application Type section, click the DLL radio button.

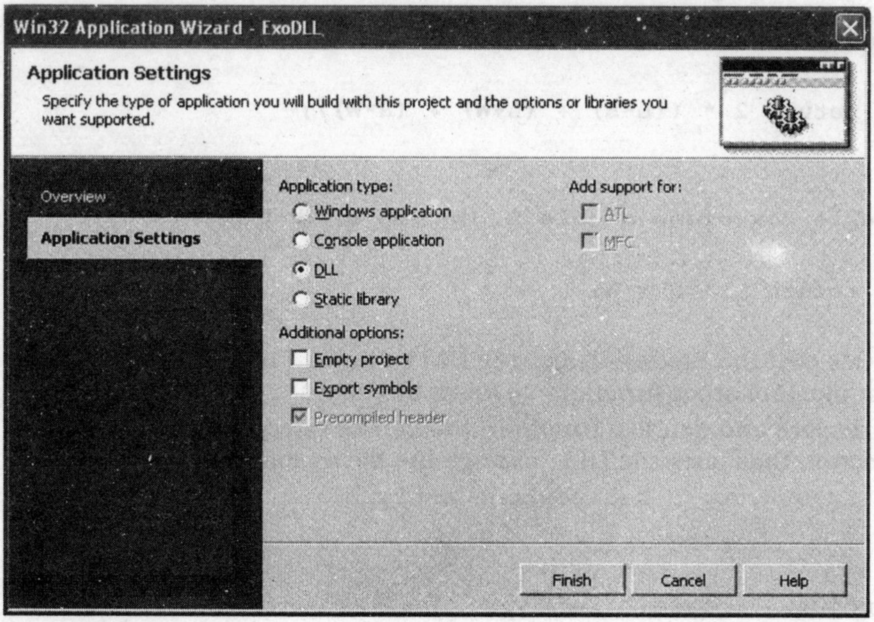

Fig. 7.2.

8. Click Finish.

7.2. CODING THE DLL

To make this project a little useful for our learning process, we will use two types of functions: those that are used internally by the DLL only, and those that can be used by external functions, that is, the other programs that will need this DLL.

1. To create functions that are used internally by the DLL, change the program as follows :

```
// ExoDLL.cpp : Defines the entry point for the DLL application.
//

#include "stdafx.h"

// This function is used to calculate the total area of a parallelepid
rectangle
double BoxArea(double L, double H, double W);
// This function is used to calculate the volume of a parallelepid
rectangle
double BoxVolume(double L, double H, double W);

BOOL APIENTRY DllMain( HANDLE hModule,
    DWORD ul_reason_for_call,
    LPVOID lpReserved)
{
    return TRUE;
}

double BoxArea(double L, double H, double W)
{
    return 2 * ((L*H) + (L*W) + (H*W));
}

double BoxVolume(double L, double H, double W)
{
    return L * H * W;
}
```

2. Notice that the BoxArea() and the BoxVolume() functions are declared and defined just like any other function we know already.

   To declare and define a function that can be used outside of the DLL, that is, by a function that uses the DLL, change the file as follows :

```
// ExoDLL.cpp : Defines the entry point for the DLL application.
//

#include "stdafx.h"

// This function is used to calculate the total area of a parallelepiped
rectangle
```

```
double BoxArea(double L, double H, double W);
// This function is used to calculate the volume of a parallelepiped
rectangle
double BoxVolume(double L, double H, double W);
// This function is used to get the dimensions of a rectangular
parallelepiped
// calculate the area and volume, then pass the calculated values to
the
// function that called it.
extern "C" __declspec(dllexport)void BoxProperties(double Length,
double Height, double Width, double& Area, double& Volume);

BOOL APIENTRY DllMain( HANDLE hModule,
    DWORD ul_reason_for_call,
    LPVOID lpReserved)
{
    return TRUE;
}

double BoxArea(double L, double H, double W)
{
    return 2 * ((L*H) + (L*W) + (H*W));
}

double BoxVolume(double L, double H, double W)
{
    return L * H * W;
}

void BoxProperties(double L, double H, double W, double& A, double&
V)
{
    A = BoxArea(L, H, W);
    V = BoxVolume(L, H, W);}
```

3. To save the project, on the main menu, click File -> Save All.
4. To create the DLL, on the main menu, click Build -> Build ExoDLL.dll. At the end, you should receive Build: 1 succeeded, 0 failed, 0 skipped.

7.3. SELECTING THE DLL

At this time, the DLL is ready. Probably the easiest way to use a DLL you have created is to create a program in the same folder and test it. To make it more fun, we will use our DLL in a separate program on another folder. the technique we will use is the same you would use if you want to create a program in the same folder the DLL project was created.

1. To see the DLL, open Windows Explorer or My Computer.
2. Locate the folder where the DLL project was created.

3. Inside the folder, display the content of the Debug folder. Notice two files: one has the extension .dll and another has .lib.
4. Click the ExoDLL.dll file to select it.
5. Press and hold Ctrl. then click ExoDLL.lib file to select both.
6. Release Ctrl.

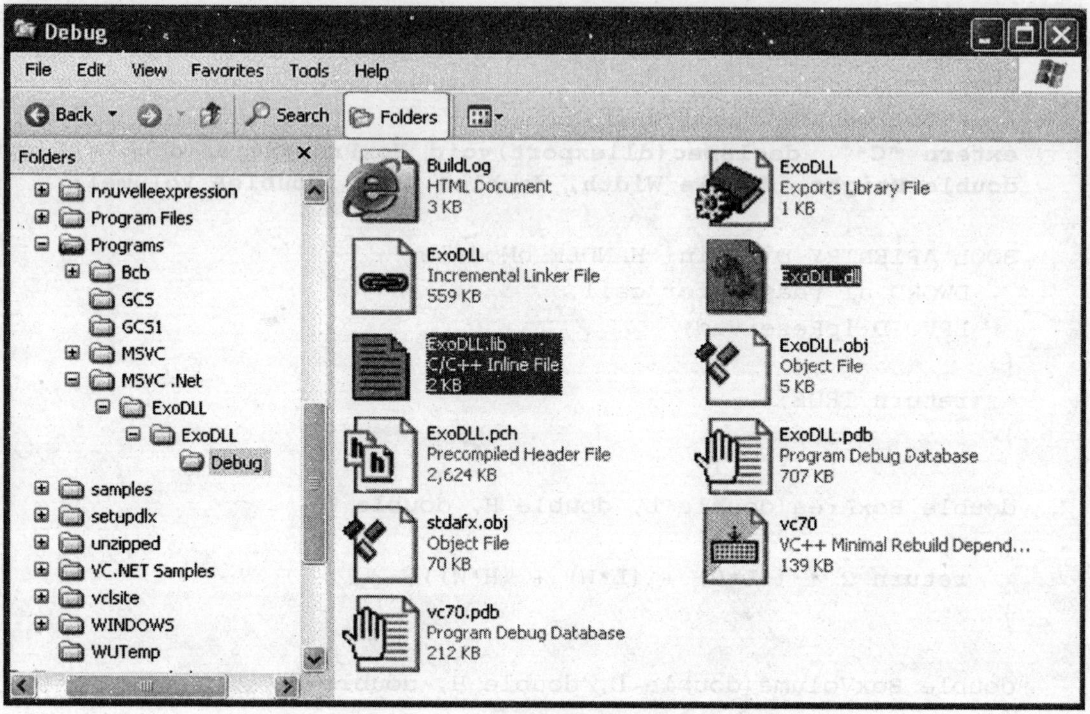

Fig. 7.3.

7. On the main menu of Windows Explorer or My Computer, click Edit -> Copy.

7.4. Using the DLL in a Console Application

The DLL we have created can be used in a Console application or an MFC program. We will start with a console application.

1. To start a new C++ program, on the main menu of MSVC .Net, click File -> New -> Project.
2. In the Project Types list, click Visual C++ Projects.
3. In the Templates list, click Win32 Project.
4. In the Project Name, type a name, such as ExoDLLTest.
5. Click OK.
6. In the Win32 Application Wizard, click Application Settings.
7. In the Application Type section, click Console Application.
8. In the Additional Options section, click the Empty Project check box and click Finish./
9. Go back to Windows Explorer or My Computer.
10. Locate the folder that has the current exercise. For me, that would be the ExoDLLTest folder.
11. Right-click in the folder and click Paste.The previously copied files should be pasted

now in the new project folder.

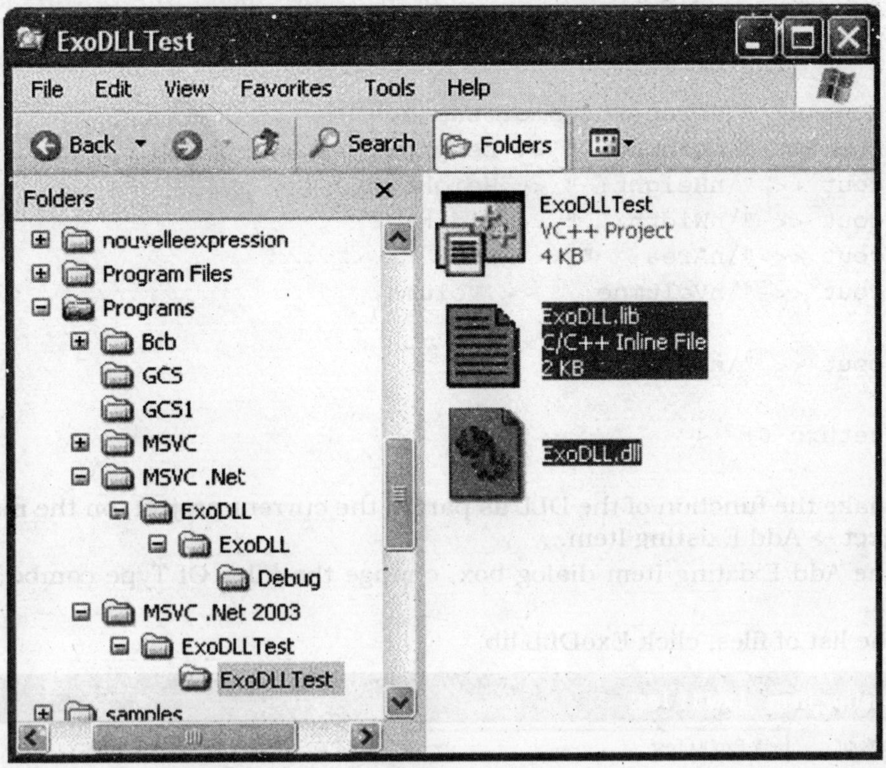

Fig. 7.4.

12. To create a new source file, on the main menu of MSVC, click Project -> Add New Item.
13. In the Add New Items dialog box, int the Templates section, click C++ File (.cpp).
14. In the Name edit box, type a name such as Exercise and click Open.
15. To use the DLL, in the empty file, type the following :

```
#include <iostream.h>
extern "C"_declspec(dllimport)void BoxProperties(double L,double H,
     double W, double& A, double& V);

int main()
{
   double Length, Height, Width, Area, Volume;

   cout << "Enter the dimensions of the box\n";
   cout << "Length: ";
   cin >> Length;
   cout << "Height: ";
   cin >> Height;
   cout << "Width: ";
```

```
    cin >> Width;

    BoxProperties(Length, Height, Width, Area, Volume);

    cout << "\nProperties of the box";
    cout << "\nLength: " << Length;
    cout << "\nHeight: " << Height;
    cout << "\nWidth:  " << Width;
    cout << "\nArea:   " << Area;
    cout << "\nVolumne: " << Volume;

    cout << "\n\n";

    return 0;
}
```

16. To make the function of the DLL as part of the current project, on the main menu, Project -> Add Existing Item...
17. In the Add Existing Item dialog box, change the Files Of Type combo box to All Files.
18. In the list of files, click ExoDLL.lib.

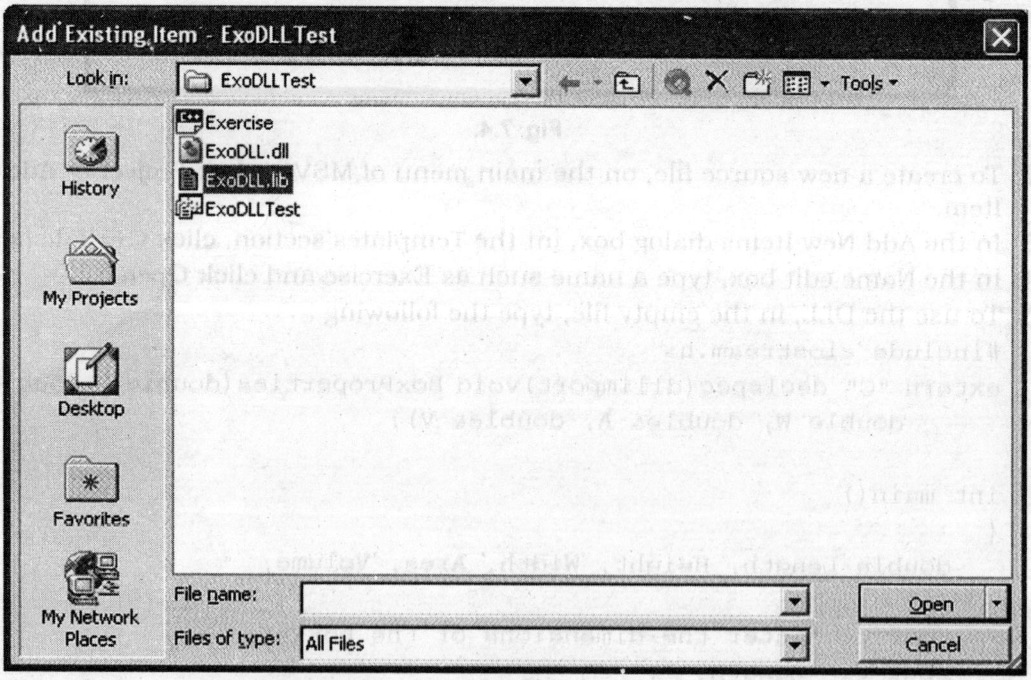

Fig. 7.5.

19. Click Open.
20. To execute the program, press Ctrl + F5. Here is an example :

```
Enter the dimensions of the box
```

```
Length: 16.52
Height: 12.74
Width: 8.46

Properties of the box
Length: 16.52
Height: 12.74
Width:  8.46
Area:  916.009
Volumne: 1780.53

Press any key to continue
```

21. Close the DOS window and return to MSVC.

7.5. USING THE DLL IN AN MFC APPLICATION

We will proceed with the same ease to use our DLL in an MFC application.

1. To create a new project, on the main menu, click File -> New -> Project...
2. In the Project Types list, click Visual C++ Projects.
3. In the Templates list, click MFC Application.
4. In the Name edit box, type a name such as ExoDLLTest2.
5. Click OK.
6. In the MFC Application Wizard, click Application Type.
7. In the Application Type section, click Dialog Based and click Finish.
8. Open Windows explorer or My Computer. Locate the folder where the ExoDLL project was created and copy the ExoDLL.dll and ExoDLL.lib files.
9. Paste them in the ExoDLLTest2 folder.
10. Return to MSVC.
11. To make the function of the DLL as part of the current project, on the main menu, Project -> Add Existing Item.
12. In the Add Existing Item dialog box, change the Files Of Type combo box to All Files.
13. In the list of files, click ExoDLL.lib and click Open.
14. Design the dialog box as follows :

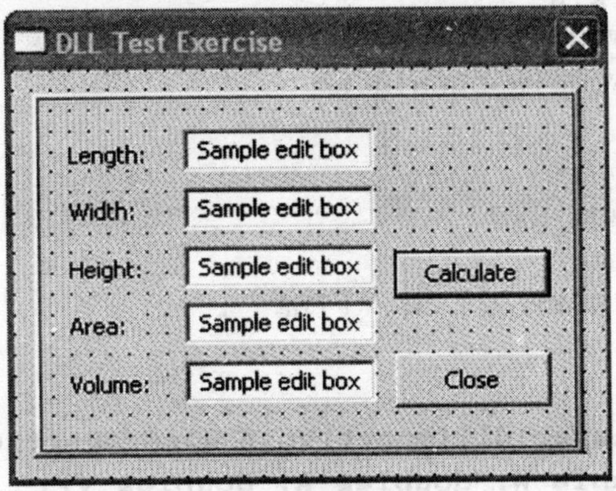

Fig. 7.6.

Control	ID	Caption	Others
Static Text		Length:	
Edit Control	IDC_LENGTH		Align Text: Right
Static Text		Width:	
Edit Control	IDC_WIDTH		Align Text: Right
Static Tex		Height	
Edit Control	IDC_HEIGHT		Align Text: Right
Button	IDC_CALCULATE_BTN	C&alculate	Default Button: True
Static Text		Area:	
Edit Control	IDC_AREA		Align Text: Right
Static Text		Volume:	
Edit Control	IDC_VOLUME		Align Text: Right
Button	IDCANCEL	Close	

15. To prepare the dialog for coding, right-click each Edit Control and click Add Variable.
16. In the Category combo box, select Value and click the Name edit box. Create the control variable as follows :

ID	Type	Name
IDC_LENGTH	CString	m_Length
IDC_WIDTH	CString	m_Width
IDC_HEIGHT	CString	m_Height
IDC_AREA	CString	m_Area
IDC_VOLUME	CString	m_Volume

17. In the dialog box, double-click the Calculate button and change the file as follows :

```
// ExoMFCUsingDLLDlg.cpp : implementation file
//

#include "stdafx.h"
#include "ExoMFCUsingDLL.h"
#include "ExoMFCUsingDLLDlg.h"

#ifdef _DEBUG
#define new DEBUG_NEW
#undef THIS_FILE
static char THIS_FILE[] = __FILE__;
#endif

extern"C"__declspec(dllimport)void BoxProperties(double L,double H,
        double W, double& A, double& V);
```

```
////////////////////////////////////////////////////////////////////////////
```

18. To make the library part of the current project, on the main menu, click Project -> Add To Project -> Files.
19. On the **Insert Files Into Project** dialog box, change the Files Of Type combo box to Library Files (*.lib).
20. In the list of files, click ExoDLL.lib and click OK.
21. Implement the OnCalculate function as follows :

```
void CExoMFCUsingDLLDlg::OnCalculate()
{
    // TODO: Add your control notification handler code here
    double Length, Height, Width, Area, Volume;

    UpdateData(TRUE);
    Length = atof(m_Length);
    Height = atof(m_Height);
    Width = atof(m_Width);

    BoxProperties(Length, Height, Width, Area, Volume);

    m_Area.Format("%.2f", Area);
    m_Volume.Format("%.2f", Volume);

    UpdateData(FALSE);
}
```

22. To test your program, press Ctrl + F5.

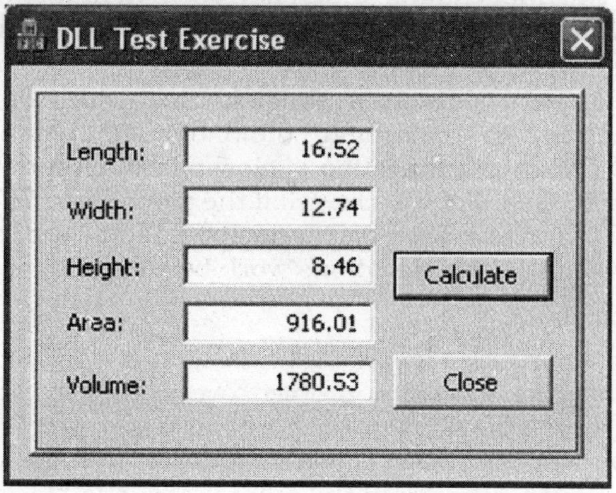

Fig. 7.7.

7.6. OLE OBJECT TECHNOLOGIES

ActiveX is part of an evolutionary process that began with OLE. To understand ActiveX is to understand OLE, and *vice versa*. Since ActiveX begins with OLE, we will begin there too. In its original conception, OLE was designed to allow one application to link or embed information created by another application. When this is done, a compound document is created. A compound document is also referred to as an OLE document. While this original purpose of OLE is still supported, it is important to understand that OLE has gone through a rather extensive evolution. OLE version 1.0 was devised in 1991, but it was a seldom-used system. With the advent of OLE version 2.0, the capabilities and applicability of OLE took a quantum leap forward. Perhaps the single most important new feature of OLE 2.0 was the definition of the Component Object Model (COM).

It is this feature that ultimately led to the ActiveX technology. In response to the Internet, Microsoft defined ActiveX in 1996. Based on OLE, it added support for Web-based applications. However, since it supports all of OLE, the term ActiveX has nearly supplanted the term OLE.

7.7. CREATING INTERNET PROGRAMS - INCLUDING A WEB BROWSER (MDU 2007-08)

Most applications that communicate over a network, whether it's the Internet or a small office network, use the same principles and functionality to perform their communication. One application sits on a computer, waiting for another application to open a communication connection. This application is "listening" for this connection request, much like you listen for the phone to ring if you are expecting someone to call.

Meanwhile, another application, most likely on another computer (but not necessarily), tries to connect to the first application. This attempt to open a connection is similar to calling someone on the telephone. You dial the number and hope that the other person listening for the phone on the other end. As the person making the call, you have to know the phone number of the person you are calling. If you don't know the phone number, you can look it up using the person's name. Likewise, the application trying to connect to the first application has to know the network location, or address, of the first application.

Once the connection is made between the two applications, messages can pass back and forth between the two applications, much like you can talk to the person on the other end of the phone. This connection is a two-way communications channel, with both sides sending information.

Finally, once one or both sides have finished their sides of the conversation, the connection is closed, much like you hang up the phone after you have finished talking to the person you called. Once the connection is closed from either side, the other side can detect it and close its side, just like you can tell if the person on the other end of the phone has hung up on you or if you have been disconnected by some other means. This is a basic explanation of how network communications work between two or more applications.

Note :

This is a basic description of how network communications work with the TCP/IP network protocol, which is the primary network protocol over the Internet. Many other network protocols use a subtle variation on this description. Other protocols, such as the UDP protocol, are more like radio broadcasts, where there is no connection between the two applications; one sends messages, and the other is responsible for making sure that it receives all of the messages. These protocols are more involved than we have the luxury to discuss today. If you want to learn more about network protocols and how they work, many books cover this one topic and look at the various Internet applications and how they communicate over the connections they establish.

Sockets, Ports, and Addresses

(MDU 2008)

The basic object used by applications to perform most network communications is called a socket. Sockets were first developed on UNIX at the University of California at Berkley. Sockets were designed so that most network communications between applications could be performed in the same way that these same applications would read and write files. Sockets have progressed quite a bit since then, but the basics of how they work are still the same.

During the days of Windows 3.x, before networking was built into the Windows operating system, you could buy the network protocols required for network communications from numerous different companies. Each of these companies had a slightly different way that an application performed network communications. As a result, any applications that performed network communications had a list of the different networking software that the application would work with. Many application developers were not happy with this situation. As a result, all the networking companies, including Microsoft, got together and developed the Winsock (Windows Sockets) API. This provided all application developers with a consistent API to perform all network communications, regardless of the networking software used.

When you want to read or write a file, you must use a file object to point to the file. A socket is an object used to read and write messages that travel between applications. Making a socket connection to another application does require a different set of information than opening a file. To open a file, you need to know the file's name and location, To open a socket connection, you need to know the computer on which the other application is running and the port on which it's listening. A port is like a phone extension, and the computer address is like the phone number. If you call someone at a large office building, you may dial the main office number, but then you need to specify the extension number. Likewise, ports are used to route network communications. As with the phone number, there are means of looking up the port number, if you don't already know what it is, but this requires your computer to be configured with the information about which port the connecting application is listening on. If you specify the wrong computer address or port number, you may get a connection to a different application; with making the phone call, someone other than the person you called may answer the phone call. You also may not get an answer at all if there is no application listening at the other end.

> **Note :**
> Only one application may be listening on any specific port on a single computer. Although numerous applications may listen for connection requests, on a single computer at the same time, each of these applications must on a different port.

Creating a Socket

When you build applications with Visual C++, you can use the MFC Winsock classes to add network communications capabilities with relative ease. The base class, CAsyncSocket, provides complete, event-driven socket communications. You can create your own descendent Socket class that captures and responds to these events.

To create a socket that you can use in your application, the first thing you need to do is declare a variable of CAsyncSocket (or your descendent class) as a class member for one of the main application classes :

```
class CMyDig : public CDialog
{
private:
```

```
    CAsyncSocket m_sMySocket;
  };
```

Before you can begin using the socket object, you must call its Create method. This actually creates the socket and prepares it for use. How you call the Create method depends on how you will be using the socket. If you will be using the socket to connect to another application, as the one placing the call (the client), then you do not need to pass any parameters to the Create method :

```
if (m_sMySocket .Create ())
{
// Continue on
}
else
// Perform error handling here
```

However, if the socket is going to be listening for another application to connect to it, waiting for the call (the server), then you need to pass at least the port number on which the socket should be listening :

```
if (m_sMySocket .Create (4000))
{
// Continue on
}
else
// Perform error handling here
```

You can include other parameters in the Create method call, such as the type of socket to create, the events that the socket should respond to, and the address that the socket should listen on (in case the computer has more than one network card).

Making a Connection

Once you create a socket, you are ready to open a connection with it. Three steps go alongwith opening a single connection. Two of these steps take place on the server, the application listing for the connection, and the third step lakes place on the client, the one making the call.

For the client, opening the connection is a simple matter of calling the Connect method. The client has to pass two parameters to the Connect method: the computer name, or network address, and the port of the application to connect to. The Connect method could be used in the following two ways :

```
if (m_sMySocket.Connect("computerOne", 4000))
{
//Continue on
}
else
// Perform error handling here
```

The second form is

```
if (m_sMy Socket. Connect ("178. 1.25.82", 4000))
{
// Continue on
}
```

```
else
// Perform error handling here
```

Once the connection is made, an event is triggered to let your application know that it is connected or that there were problems and the connection couldn't be made (to be discussed later). For the server, or listening side of the connection, the application first must tell the socket to listen for incoming connections by calling the Listen method. The Listen method takes only a single argument, which you do not need to supply. This parameter specifies the number of pending connections that can be queued, waiting for the connection to be completed. By default this value is 5, which is the maximum. The Listen method can be called as follows :

```
if (m_sMySocket .Listen ())
{
// Continue on
}
else
// Perform error handling here
```

Whenever another application is trying to connect to the listening application, an event is triggered to let the application know that the connection request is there. The listening application must accept the connection request by calling the Accept method. This method requires the use of a second CAsyncSocket variable, which is connected to the other application. Once a socket is placed into listen mode, it stays in listen mode.

Whenever connection requests are received, the listening socket creates another socket, which is connected to the other application. This second socket should not have the Create method called for it because the Accept method creates the socket. You call the Accept method as follows :

```
if (m_sMySocket.Accept (m_sMySecondSocket) )
{
// Continue on
}
else
// Perform error handling here
```

At this point, the connecting application is connected to the second socket on the listening application.

Sending and Receiving Messages

Because you can use sockets to send any kind of data, and they don't care what the data is, the functions to send and receive data expect to be passed a pointer to a generic buffer. For sending data, this buffer should contain the data to be sent. For receiving data, this buffer will have the received data copied into it. As long as you are sending and receiving strings and text, you can use fairly simple conversions to and from CStrings with these buffers.

To send a message through a socket connection, you use the Send method. This method requires two parameters and has a third, optional parameter that can be used to control how the message is sent. The first parameter is a pointer to the buffer that contains the data to be sent. If your message is in a CString variable, you can use the LPCTSTR operator to pass the CString variable as the buffer. The second parameter is the length of the buffer. The method returns the amount of data that was sent to the other application. If error occurs, the Send function returns SOCKET_ERROR. You can use the Send method as follows :

```
CString strMyMessage;
int iLen:
int iAmtSent:

iLen = strMyMessage.GetLength();
iAmtSent = m_sMySocket.Send(LPCTSTR(strMyMessage), iLen);
if (iAmtSent == SOCKET_ERBOR)
{
// Do some error handling here
}
else
{
// Everything's -fine
}
```

When data is available to be received from the other application, an event is triggered the receiving application. This lets your application know that it can receive and process the message. To get the message, the Receive method must be called. This method takes the same parameters as the Send method with a slight difference. The first parameter is a pointer to a buffer into which the message may be copied. The second parameter is the size of the buffer. This tells the socket how much data to copy (in case more is received than will fit into the buffer). Like the Send method, the Receive method will return the amount that was copied into the buffer. If an error occurs, the Receive method also returns SOCKET_ERROR. If the message your application is receiving is a text message, it can be copied directly into a CString variable. This allows you to use the Receive method as follows :

```
Char *pBuf = new char[1025];
int iBuffSize = 1024;
int iRcvd;
CString strRecvd;
iRcvd = m_sMySocket. Receive (pBuf, iBuffSize);
if (iRcvd == SOCKET_ERROR)
{
// Do some error handling here
}
else
{
pBuf [iRcvd) = NULL;
strRecvd = pBuf;
// Continue processing the message
}
```

When receiving text messages, it's always a good idea to place a NULL in the buffer position just after the last character received, as in the preceding example. There may be garbage characters in the buffer that your application might interpret as part of the message if you don't add the NULL to truncate the string.

Closing the Connection

Once your application has finished all of its communications with the other application, it can close the connection by calling the Close method. The Close method doesn't take any parameters, and you use it as follows :

```
m_sMySocket.Close();
```

The Close function is one of the few CAsyncSocket methods that does not return any status code. For all the previous member functions that we have examined, you can capture the return value to determine if an error has occurred.

Socket Events

The primary reason that you create your own descendent class of CAsyncSocket is that you want to capture the events that are triggered when messages are received, connections are completed, and so on. The CAsyncSocket class has a series of functions that are called for each of these various events. These functions all use the same definition, the function name is the only difference, and they are intended to be overridden in descendent classes. All of these functions are declared as protected members of the CAsyncSocket class and probably should be declared as protected in your descendent classes. The functions all have a single integer parameter, which is an error code that should be checked to make sure that no error has occurred.

Table 7.1. Lists these event, functions and the events they signal.

Table 7.1. CAsyncSocket Overridable Event-Notification Functions

Function	*Event Description*
OnAcceot	This function is called on a listening socket to signal that a connection request from another application is waiting to be accepted.
OnClose	This function is called on a socket to signal that the application on the other end of the connection has closed its socket or that the connection has been lost. This should be followed by closing the socket that received this notification.
OnConnect	This function is called on a socket to signal that the connection with another application has been completed and that the application can now send and receive messages through the socket.
OnReceive	This function is called to signal that data has been received through the socket connection and that the data is ready to be retrieved by calling the Receive function.
OnSend	This function is called to signal that the socket is ready and available for sending data.This function is called right after the connection has been completed. Usually, the other time that this function is called is when your application has passed the Send function more data than can be sent in a single packet. In this case, this is a signal that all of the data has been sent, and the application can send the next buffer-full of data.

Detecting Errors

Whenever any of the CAsyncSocket member functions return an error, either FALSE for most functions or SOCKET_ERROR on the Send and Receive functions, you can call the GetLastError method to get the error code. This function returns only error codes, and you have to look up the translation yourself. All the Winsock error codes are defined with constants, so you can use the constants in your code to determine the error message to display for the user, if any. You can use the GetLastError function as follows :

```
int iErrCode;
iErrCode = m_sMySocket. GetLastError ();
switch (iErrCode)
{
case WASNOTINITIALISED:
.
.
.
}
```

Example : Internet Application

For the sample application that you will now build, you'll create a simple dialog application that can function as either the client or server in a Winsock connection. This will allow you to run two copies of the sample application, one for each end of the connection, on the same computer or to copy the application to another computer so that you can run the two copies on separate computers and see how you can pass messages across a network. Once the application has established a connection with another application, you will be able to enter text messages and send them to the other application. When the message has been sent, it will be added to a list of messages sent. Each message that is received will be copied into another list of all messages received. This will allow you to see the complete list of what is sent and received. It will also allow you to compare which one copy of the application has sent and what the other has received. (The two lists should be the same.).

7.7.1. Creating the Application Shell

Create a new MFC dialog-based application named Sock using AppWizard. On the first step of the AppWizard, specify that the application will be a dialog-based application. On the second step of the AppWizard specify that the application should include support for Windows Sockets (Fig. 7.8.). You can accept the default settings for the rest of the options in the AppWizard.

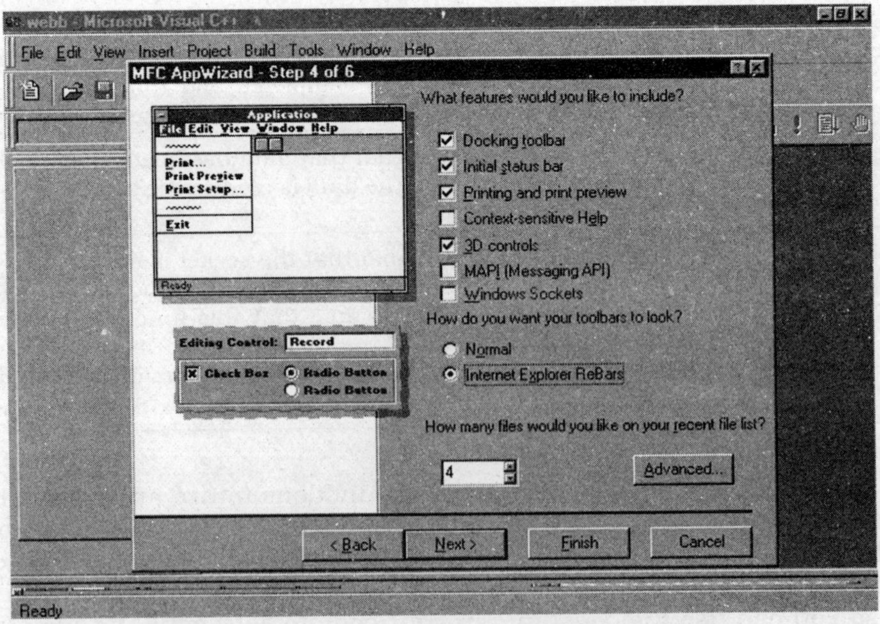

Fig. 7.8. Create a new MFC Dialog-based Application

Window Layout and Startup Functionality

Once you create your application shell, you can layout the main dialog for your application. On this dialog, you'll need a set of radio buttons to specify whether the application is running as the client or server. You'll also need a couple of edit boxes for the computer name and port that the server will be listening on. Next, you'll need a command button to start the application listening on the socket or opening the connection to the server, and a button to close the connection. You'll also need an edit box for entering the message to be sent to the other application and a button to send the message. Finally, you'll need a couple of list boxes into which you can add each of the messages sent and received. Place all these controls on the dialog (Fig. 1.62), setting all the control properties as specified in table 7.2.

Table 7.2.

Object	*Property*	*Setting*
Group Box	ID	IDC_STATICTYPE
	Caption	Socket Type
Radio Button	ID	IDC_RCLIENT
	Caption	&Client
	Group	Checked
Radio Button	ID	IDC_RSERVER
	Caption	&Server
Static Text	ID	IDC_STATICNAME
	Caption	Server &Name :
Edit Box	ID	IDC_ESERVNAME
Static Text	ID	IDC_STATICPORT
	Caption	Server &Port :
Edit Box	ID	IDC_ESERVPORT
Command Button	ID	IDC_BCONNECT
	Caption	C&onnect
Command Button	ID	IDC_BCLOSE
	Caption	C&lose
	Disabled	Checked
Static Text	ID	IDC_STATICMSG
	Caption	&Message :
	Disabled	Checked
Edit Box	ID	IDC_EMSG
	Disabled	Checked
Command Button	ID	IDC_BSEND
	Caption	S&end

	Disabled	Checked
Static Text	ID	IDC_STATIC
	Caption	Sent :
List Box	ID	IDC_LSENT
	Tab Stop	Unchecked
	Sort	Unchecked
	Selection	None
Static Text	ID	IDC_STATIC
	Caption	Received:
List Box	ID	IDC_LRECVD
	Tab Stop	Unchecked
	Sort	Unchecked
	Selection	None

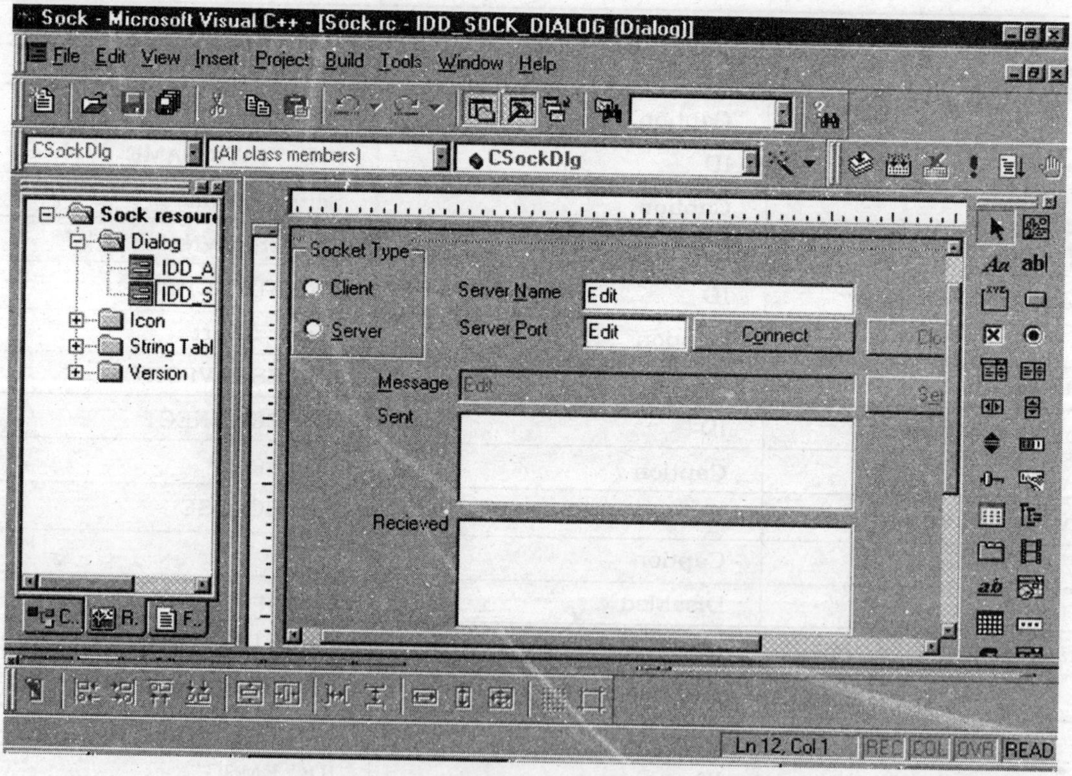

Fig. 7.9.

Once you have the dialog designed, open the Class Wizard to attach variables to the controls on the dialog, as specified in Table 7.3.

Table 7.3. Control Variable

Object	*Name*	*Category*	*Type*
IDC_BCONNECT	m_ctlConnect	Control	CButton
IDC_EMSG	m_strMessage	Value	CString
IDC_ESERVNAME	m_strName	Value	CString
IDC_ESERVPORT	m_iPort	Value	int
IDC_LRECVD	m_ctlRecvd	Control	CListBox
IDC_LSENT	m_ctlSent	Control	CListBox
IDC_RCLIENT	m_iType	Value	int

So that you can reuse the Connect button to place the server application into listen mode you'll add a function to the clicked event message for both radio buttons, changing the text on the command button depending on which of the two is currently selected. To add this functionality to your application, add a function to the BN_CLICKED event message for the IDC_RCLIENT control ID, naming the function OnRType. Add the same function to the BN_CLICKED event message for the IDC_RSERVER control ID. Edit this function adding the code as :

```
Void CSockDlg::OnRType()
{
UpdateData(TRUE);
if (m_iType == 0)
m_ctlConnect.SetWindowText("C&onnect");
else
m_ctlConnect.SetWindowText("&Listen");
}
```

Now, if you compile and run your application, you should be able to select one and then the other of these two radio buttons, and the text on the command button should change to reflect the part the application will play as in Fig. 7.10.

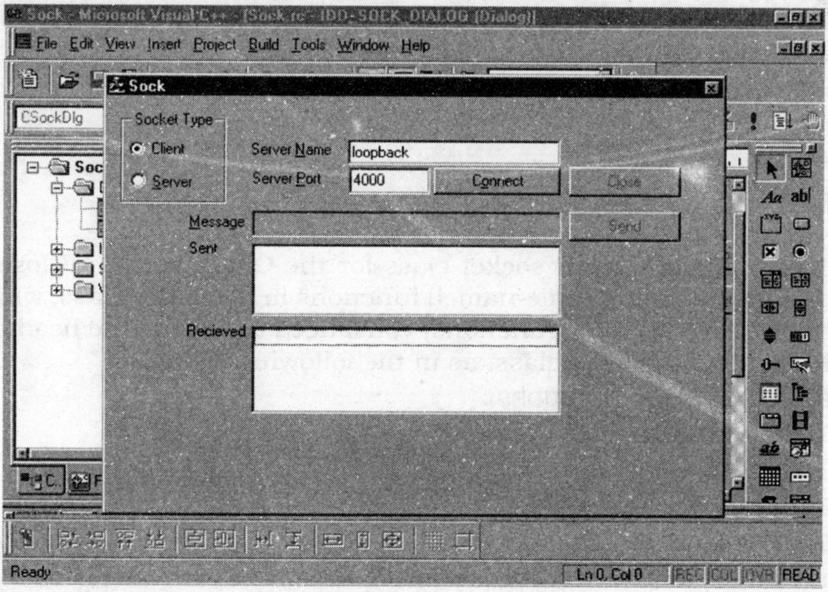

Fig. 7.10. Client Server at work

Inheriting from the CAsyncSocket Class

So that you will be able to capture and respond to the socket events, you'll create your own descendent class from CAsyncSocket. This class will need its own versions of the event functions, as well as a means of passing this event to the dialog that the object will be a member of. So that you can pass each of these events to the dialog-class level, you'll add a pointer to the parent dialog class as a member variable of your socket class. You'll use this pointer to call event functions for each of the socket events that are member functions of the dialog, after checking to make sure that no errors have occurred (of course).

To create this class in your application, select Insert, New Class from the menu. In the New Class dialog, leave the class type with the default value of MFC Class. Enter a name for your class, such as CMySocket, and select CAsyncSocket from the list of available base classes. This is all that you can specify on the New Class dialog, so click the OK button to add this new class to your application. Once you have created the socket class, add a member variable to the class to serve as a pointer to the parent dialog window. Specify the variable type as CDialog*, the variable name as m_pWnd, and the access as private. You also need to add a method to the class to set the pointer, so add a member function to your new socket class. Specify the function type as void, the declaration as SetParent (CDialog* pWnd), and the access as public. Edit this new function, setting the pointer passed as a parameter to the member variable pointer, as in following listing :

```
// THE CMySocket SetParent FUNCTION.
void CMySocket: :SetParent(CDialog *pWnd)
{
m_pWnd = pWnd;
}
```

The only other thing that you need to do to your socket class is add the event functions, which you'll use to call similarly named functions on the dialog class. To add a function for the OnAccept event function, add a member function to your socket class. Specify the function type as void, the function declaration as OnAccept (int nErrorCode), and the access as protected and check the virtual check box. Edit this function, adding the code :

```
//The CMySocket OnAccept FUNCTION.
void CMySocket: :OnAccept(int nErrorCode)
{
if (nErrorCode == 0)
((CSockDlg*)m_pWnd)->OnAccept();
}
```

Add similar functions to your socket class for the OnConnect, OnClose, OnReceive, and OnSend functions, calling same-named functions in the dialog class, which you'll add later. After you've added all these functions, you'll need to include the header file for your application dialog in your socket class, as in the following listing :

```
// THE CMySocket INCLUDES.
#include "stdafx.h"
#include "Sock.h"
#include "MySocket.h"
#include "SockDIg.h"
```

Once you've added all the necessary event functions to your socket class, you'll add a variable of your socket class to the dialog class. For the server functionality, you'll need two variables in the dialog class, one to listen for connection requests and the other to be connected to the other application. Because you will need two socket objects, add two member variables to the dialog class (CSockDIg). Specify the type of both variables as your socket class (CMySocket) and the access for both as private. Name one variable m_sListenSocket, to be used for listening for connection requests, and the other m_sConnectSocket, to be used for sending messages back and forth.

Once you've added the socket variables, you'll add the initialisation code for all the variables. As a default, set the application type to client, the server name as loopback, and the port to 4000. Alongwith these variables, you'll set the parent dialog pointers in your two socket objects so that they point to the dialog class. You can do this by adding the following code to the OnInitDialog function in the dialog class.

```
//THF CRonknin OnlnitDtalon FUNCTION-
BOOL CSockDlg::OnlnitDialog()
{
CDialog::OnlnitDialoa();
Setlcon(m_hlcon, FALSE);
m_iTyp3 = 0;
m_strNa.ne = "loopback";
m Pori = 4000;
UpdateData (FALSE);
in_sConnectSocket. SefParent (this ) ;
in_sListenSocket.SetParent (this) ;
return TRUE;
```

Note :

The computer name loopback is a special name used in the TCP/IP network protocol to indicate the computer you are working on. It's an internal computer name that is resolved to the network address 127.0.0.1. This is a computer name and address that is commonly used by applications that need to connect to other applications running on the same computer.

Connecting the Application

When the user clicks the Connect button, you'll disable all the top controls on the dialog. At this point, you don't want the user to think that she is able to change the settings of the computer that she's connecting to or change how the application is listening. You'll call the Create function on the appropriate socket variable, depending on whether the application is running as the client or server. Finally, you'll call either the Connect or Listen function to initiate the application's side of the connection. To add this functionality to your application, open the Class Wizard and add a function to the BN_CUCKED event message for the Connect button (ID IDC_BCONNECT). Edit this function, adding the code :

```
//The CSockDIg OnBconnect FUNCTION.
void CSockDlg::OnBconnect()
{
UpdateData(TRUE);
GetDlgltem(IDC_BCONNECT) ->EnableWindow(FALSE);
GetDlgltem(IDC_ESERVNAME) ->EnableWindow(FALSE);
```

```
GetDlgltem(IDC_E§ERVPORT) ->EnableWindow(FALSE);
GetDlgltem(IDC_STATICNAME) ->Enable.Window(FALSE);
GetDlgltem(IDC_STATICPORT) ->EnableWindow(FALSE);
GetDlgltem(IDC_RCLIENT) ->EnableWindow(FALSE);
GetDlgltem(IDC_RSERVER) ->EnableWindow( FALSE);
GetDlgltem(IDC_STATICTYPE) ->EnableWindow(FALSE);
if (m iType == 0)
{
m_sConnectSocket. Create ();
m_sConnectSocket. Connect (m_strName, m iPort);
}
else
{
in_sListenSocket.Create(m_iPort);
m_sListenSocket. Listen ( ) ;
}
}
```

Next, to complete the connection, you'll add the socket event function to the dialog class for the OnAccept and OnConnect event functions. These are the functions that your socket class is calling. They don't require any parameters, and they don't need to return any result code. For the OnAccept function, which is called for the listening socket when another application is trying to connect to it, you'll call the socket object's Accept function, passing in the connection socket variable. Once you've accepted the connection, you can enable the prompt and edit box for entering and sending messages to the other application.

To add this function to your application, add a member function to the dialog class(CSockDIg). Specify the function type as void, the declaration as OnAccept, and the access as public. Edit the function, adding the code :

```
//THE CSockDIg OnAccept FUNCTION.
void CSockDIg:: OnAccept ()
{
m_sListenSocket.Accept(iii_sConnectSocket);
GetDlgltem(IDC_EMSG) ->EnableWindow(TRUE);
GetDlgltem(IDC_BSEND) ->EnableWindow(TRUE);
GetDlgltem(IDC_STATICMSG) ->EnableWindow(TRUE);
}
```

For the client side, there's nothing to do once the connection has been completed except enable the controls for entering and sending messages. You'll also enable the Close button so that the connection can be closed from the client side (but not the server side). To add this functionality to your application, add another member function to the dialog class (CSockDIg). Specify the function type as void, the function declaration as OnConnect, and the access as public. Edit the function, adding the code :

```
//THE CSockDIg OnConnect FUNCTION.
void CSockDIg:: OnConnect 0
{
GetDlgltem(IDC_EMSG) ->EnableWindow(TRUE);
```

```
GetDlgltem(IDC_BSEND)->EnableWindow(TRUE);
GetDlgltem(IDC_STATICMSG) ->EnableWindow(TRUE);
GetDlgltem(IDC_BCLOSE) ->EnableWindow(TRUE);
}
```

If you could compile and run your application now, you could start two copies, put one into listen mode, and then connect to it with the other. Unfortunately, you probably can't even compile your application right now because your socket class is looking for several functions in your dialog class that you haven't added yet. Add three member functions to the dialog class (CsockDlg). Specify all of them as void functions with public access. Specify the first function's declaration as OnSend, the second as OnReceive, and the third as OnClose. You should now be able to compile your application.

Once you have compiled your application, start two copies of the application, side-by-side. Specify that one of these two should be the server, and click the Listen button to put it into listening mode. Leave the other as the client and click the Connect button.

NOTE :

Be sure that server listens before client attempts to connect, the request, otherwise, will be rejected. The application will not be able to detect that the connection was rejected because you have not added any error handling to detect this event.

Also this application will not run if TCP/IP network protocol is not available to your network.

Sending and Receiving

Now that you are able to connect the two running applications, you'll need to add functionality to send and receive messages. Once the connection is established between the two applications, the user can enter text messages in the edit box in the middle of the dialog window and then click the Send button to send the message to the other application. Once the message is sent, it will be added to the list box of sent messages. To provide this functionality, when the Send button is clicked, your application needs to check whether there is a message to be sent, get the length of the message, send the message, and then add the message to the list box. To add this functionality to your application, use the Class Wizard to add a function to the clicked event of the Send (IDC_BSEND) button. Edit this function, adding the code :

```
//The CSockDIg OnBsend FUNCTION.
void CSockDlg::OnBsend()
{
int iLen;
int iSent;
UpdateData (TRUE);
if (m_strMessage != "")
   {
   iLen = m_strMessage.GetLength();
   iSent = m_sConnectSocket.Send(LPCTSTR(m_strMessage), iLen);
   if (iSent == SOCKET_ERROR)
      {
      }
   else
```

```
            {
            m_ctlSent.AddString(m_strMessage);
            UpdateData(FALSE);
        }
    }
}
```

When the OnReceive event function is triggered, indicating that a message has arrived, you'll retrieve the message from the socket using the Receive function. Once you've retrieved the message, you'll convert it into a CString and add it to the message received list box. You can add this functionality by editing the OnReceive function of the dialog class, adding the code :

```
//THE CSockDIa OnReceive FUNCTION.
void CSockDlg::OnReceive()
{
    char *pBuf = new char[ 1025];
    int iBufSize = 1024;
    int iRcvd;
    CString strRecvd;
    iRcvd = in_sConnectSocket.Receive(pBuf, iBufSize);
    if (iRcvd == SOCKET_ERROR)
    {
    }
    else
    {
    pBuf [iRcvd} = NULL;
    strRecvd =p.Buf;
    m_ctlRecvd.AddSt ring (strRecvd);
    UpdateData( FALSE );
    }
}
```

At this point, you should be able to compile and run two copies of your application, connecting them as you did earlier. Once you've got the connection established, you can enter a message in one application and send it to the other application.

Ending the Connection

To close the connection between these two applications, the client application user can click the Close button to end the connection. The server application will then receive the OnClose socket event. The same thing needs to happen in both cases. The connected socket needs to be closed, and the message sending controls need to be disabled. On the client, the connection controls can be enabled because the client could change some of this information and open a connection to another server application. Meanwhile, the server application continues to listen on the port that it was configured to listen to. To add all this functionality to your application, edit the OnClose function, adding the code :

```
//THE CSockDIg OnClose FUNCTION.
void CSockDlg::OnClose()
{
```

```
    m_sConnectSocket. Close ();
    GetDlgltem(IDC_EMSG) ->EnableWindow(FALSE);
    GetDlgltem(IDC_BSEND) ->EnableWindow(FALSE);
    GetDlgltem(IDC_STATICMSG) ->EnableWindow(FALSE);
    GetDlgltem(IDC_BCLOSE) ->EnableWindow(FALSE);
    if (m_iType == 0)
       {
       GetDlgltem(IDC_BCONNECT) ->EnableWindow(TRUE) ;
       GetDlgltem(IDC_ESERVNAME) ->EnableWindow(TBUE);
       GetDlgltem(IDC_ESERVPOBT) ->EnableWindow(TRUE);
       GetDlgltem(IDC_STATICNAME) ->EnableWindow(TRUE);
       GetDlgltem(IDC_STATICPORT) ->EnableWindow(TRUE);
       GetDlgltem(IDC_RCLIENT) ->EnableWindow(TRUE) ;
       GetDlgltem(IDC_RSERVER) ->EnableWindow(TRUE);
       GetDlgltem(IDC STATICTYPE) ->EnableWindow(TRUE);
       }
}
```

Finally, for the Close button, call the OnClose function. To add this functionality to your application, use the Class Wizard to add a function to the clicked event for the Close button (lDC_BCLOSE). Edit the function to call the OnClose function, as :

```
//THE CSockDIa OnBclose FUNCTION.
void CSockDlg::OnBclose()
{
   OnCloseO;
}
```

If you compile and run your application, you can connect the client application to the server, send some messages back and forth, and then disconnect the client by clicking the Close button. You'll see the message sending controls disable themselves in both applications. You can reconnect the client to the server by clicking the Connect button again and then pass some more messages between the two, as if they had never been connected in the first place.

7.7.2. Building a Web-Browsing Application

For an example of how you can integrate the Internet Explorer Web browser component into your own applications, you will build a simple Web browser application. You will create an SDI application using the CHtmlView class as the base for your own view class. You'll add a menu with functions for the back and forward navigation options. You'll also add a dialog for getting from the user a URL that you will use to navigate the browser to the specified Web page.

Creating the Application Shell

To create a Web browser application, you can create a standard SDI or MDI application shell. The only other thing that you need to ensure is that Internet Explorer is installed on the computer where your application will run. For your development computer, this is not a problem because the Visual C++ installation probably required you to install the latest version of Internet Explorer. On any computers where you run your application, however, you might need to make sure that Internet Explorer is installed or install it yourself.

To create the shell of the application that you will build now, start a new project using the MFC AppWizard to create the application shell. Give the project a suitable name, such as WebBrowse, and click the OK button to start the AppWizard.

In the AppWizard you can just as easily create an MDI Web browsing application as you can create an SDI application. For the purposes of the sample application that you are building, go ahead and specify that the application is a Single document (SDI) application. You can accept the default settings for the rest of the AppWizard; for this example, however, choose the Internet Explorer ReBars for the toolbar appearance on the fourth AppWizard step. (See Fig. 7.11.).

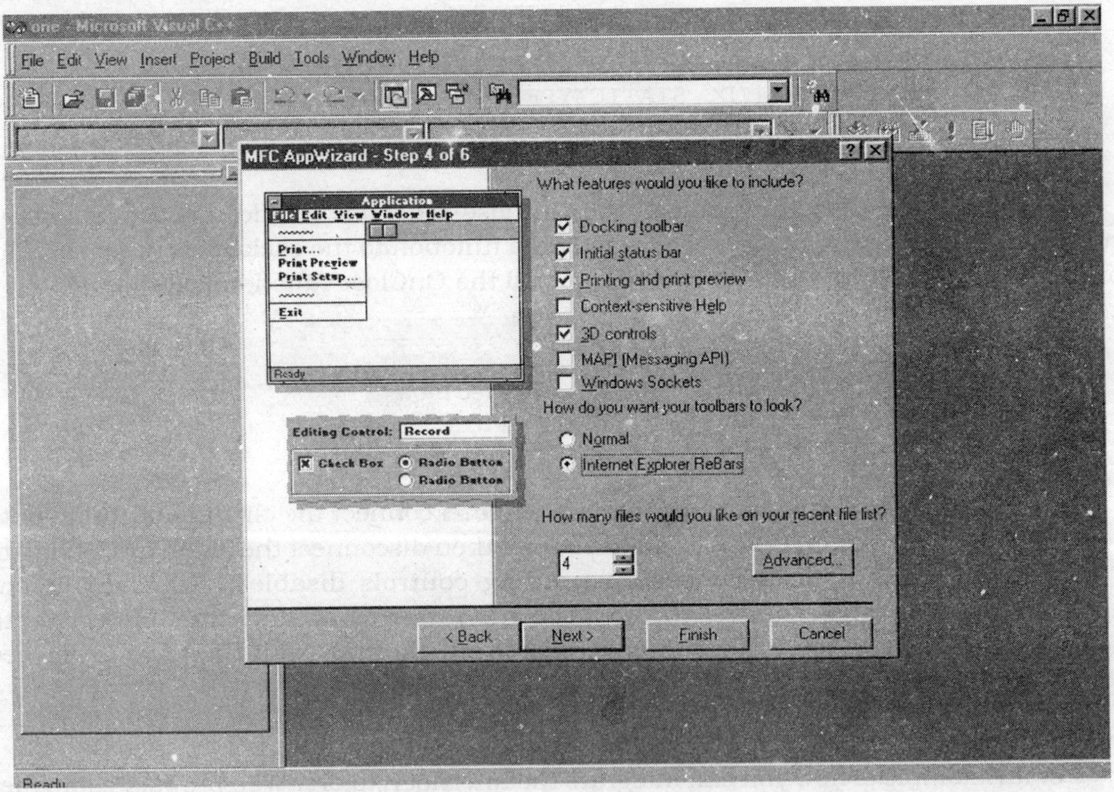

Fig. 7.11. The Internet Explorer ReBars for the Toolbar Appearance on the Fourth AppWizard Step

Finally, on the sixth step, specify the CHtmlView class as the base class for your view class. This causes your application to be created using the Internet Explorer Web browser control as the main application view.

Once you Finish generating the shell for your application, if you compile and run it while connected to the Internet, you'll find that you already have a working Web browser. However, you do not have the ability to specify where your browser will take you, other than clicking links in the Web pages displayed as shown in Fig 7.12.

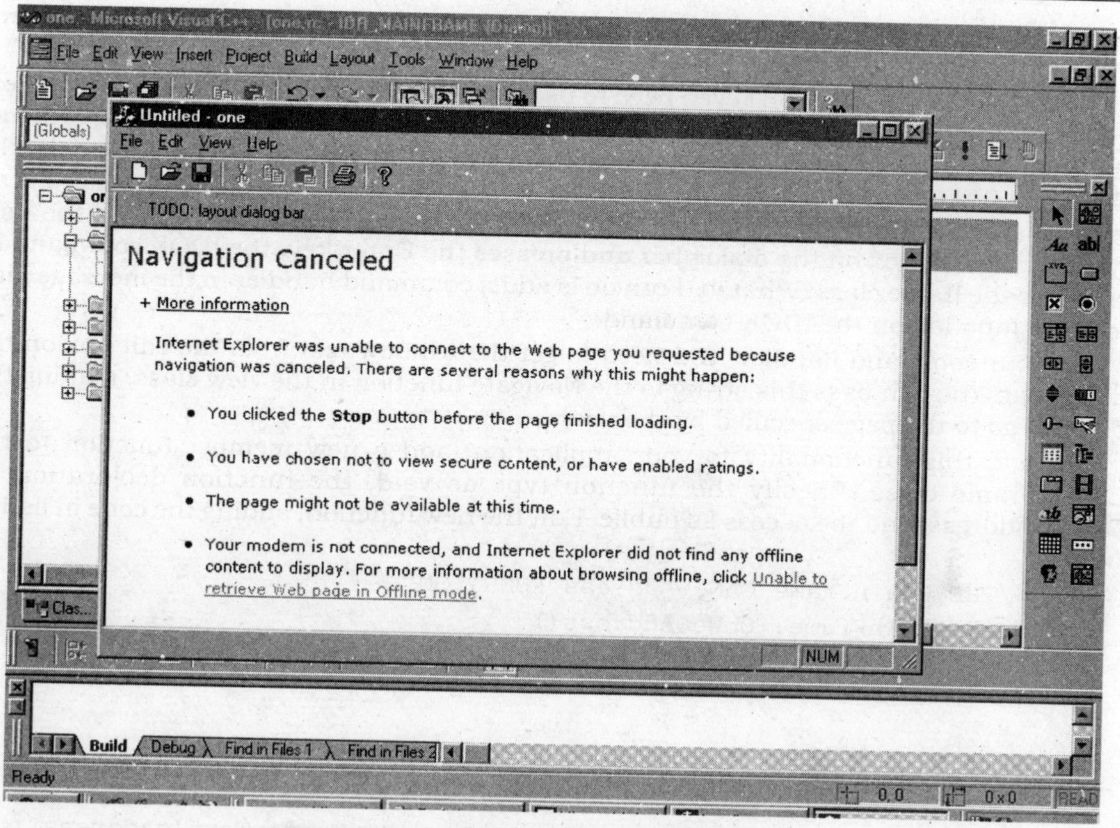

Fig. 7.12. The Display of the Web Browser Screen

Adding Navigation Functionality

Now that you've got a working Web browser, it would be nice if you could control where it takes you. What you need to add is an edit control where the user can enter a URL. Looking at the toolbar of the running application, you notice there's a place to put this control.

Specifying a URL

You probably noticed when you ran your application that the second toolbar had some static text telling you to lay out the dialog bar. The dialog bar is different from what you have worked with before. It's a toolbar with dialog controls on it. You even designed the bar in the dialog layout designer. When you look for this dialog bar in the resource tab, you won't find it in the toolbar folder; it's in the dialogs folder.

If you open the dialog's folder and double-click the IDR_MAINFRAME dialog to open it in the dialog designer, you'll see that it's the second toolbar in your application. You can place edit boxes, buttons, combo boxes, and check boxes on this toolbar. You can place any control that you can use on a dialog window on this toolbar.

For your Web browser, modify the static text control already on the dialog bar and add an edit box. Specify an ID for the edit box; for this example, use the ID IDC_EADDRESS.

Before you open the Class Wizard to begin adding variables and event functions to the dialog bar, be aware that the dialog bar will automatically send its events to the Mainframe class in your application. When you open the Class Wizard, it assumes that you need to associate the dialog bar with a class and prompts you to create a new class. This association is not necessary because you can map all its events through the frame and from there feed them to the view or document classes.

For this example, you don't even need to use the Class Wizard to add any event handlers to the dialog bar. You need to trigger an action when the user finishes entering a URL into the edit box. The closest event available to you through the Class Wizard is the EN_CHANGED event, which will trigger for each letter the user types. What you need is an event that will trigger when the user presses the Enter key. Fortunately, when the user types in the edit box on the dialog bar and presses the Enter key, the IDOK command ID is sent to the frame class. What you can do is add a command handler in the message map to call a function on the IDOK command.

In your command handler, you need to get the window text from the edit box on the dialog bar. You can pass this string to the Navigate function in the view class, making the browser go to the page specified by the user.

To add this functionality to your application, add a new member function to the CMainFrame class. Specify the function type as void, the function declaration as OnNewAddress, and the access as public. Edit the new function, adding the code in listing below.

```
//THE CMainFrame OnNewAddress FUNCTION.
void CMainFrame::OnNewAddress()

{


CString sAddress;

m_wndDlgBar.GetDlgltem(IDC_EADDRESS) ->GetWindowText(sAddress);

( (CWebBrowseView*) GetActiveView() ) -> Navigate(sAddress) ;

}
```

The dialog bar was declared in the CMainFrame class as the m_wndDlgBar variable, so you were able to use the GetDIglte function on the dialog bar variable to get a pointer to the edit box.

In line 5, you cast the return pointer from the GetActiveView function as a pointer to the CWebBrowseView class. This allowed you to call the Navigate function on the view class, passing it the URL that was entered into the edit box.

Now that you are able to take the URL that the user entered and tell the browser component to go to that Web page, how do you trigger this function? You have to add the message-map entry by hand because this is one that the Class Wizard isn't able to add. In the message-map, after the closing marker of the AFX_MSG_MAP (the section maintained by the Class Wizard), add the ON_COMMAND macro, specifying the IDOK command and your new function as the handler to be called, as in listing below. You can also add this entry before the Class Wizard section as long as it's on either side and not inside the section maintained by the Class Wizard.

```
//THE CMainPrame MFSSAGE MAP.
BEGIN_MESSAGE_MAP(CMainFrame, CFrameWnd)
```

```
    ON_WM_CREATE()

    ON_COMMAND(IDOK, OnNewAddress)

END_MESSAGE_MAP()
```

If you compile and run your application, you can enter a URL into the edit box on the toolbar and press the Enter key, and your application should browse to the Web page you specified.

Displaying the Current URL **(MDU 2008-09)**

When surfing the Web, you often follow links on Web pages that take you to other Web sites. When you do this, you wouldn't know what Web site you accessed if your browser didn't place the URL into the address box, indicating the site where you are and providing the opportunity to copy or modify the URL to find another page on the same site.

Getting the current URL from the browser is a simple matter of calling the GetLocationURL function and passing the result to the dialog bar. The problem is when to get the URL. It turns out that some event functions in the CHtmlView class can be overridden in your class. These functions will be triggered on various events that are triggered by the browser control. There are event functions for starting the navigation, beginning a download, monitoring a download's progress, and, most important for our needs, indicating a download has finished. None of these event handler functions can be added to your view class through the Class Wizard. You have to add them all.

To add the download-complete event handler to your application, add a new member function to the view class of your application. Specify the function type as void, the function declaration as OnDocumentComplete(LPCTSTR IpszUrl), and the access as public. Edit the function, adding the code in listing below.

```
//THE GWebBrowseView OnDocumentGomplete FUNCTION.
void CWebBrowseView::OnDocuiTientComplete(LPCTSTR lpszURL)

{

((CMainFrame*->GetParentFrame( )) ->SetAddress(IpszURL);

}
```

You'll notice in this function that you didn't need to call the GetLocationURL function after all. The URL that is downloaded is passed as an argument to this function. This allows you to pass the URL along to the frame, where you'll add another function to populate the edit box on the dialog bar with the URL. To add the function to populate the dialog bar with the new URL, add a member function to the main frame class, CMainFrame. Specify the function type as void, the function declaration as SetAddress(LPCTSTR IpszURL), and the access as public. Edit the function, adding the code in listing below.

```
//THE CMainFrame SetAddress FUNCTION.
void CMainFrame::SetAddr-ess(LPCTSTR IpszURL)

{
```

```
    m_wndDlgBar.GetDlgltem(IDC_EADDRESS) ->SetWindowText(lpszURL);

    }
```

In this function, you took the opposite path from the one you used to get the text from the edit box. You used the SetWindowText to change the text in the edit box to the URL that you are passing in. When you run your application, you should be able to see the URL address on the dialog bar change to reflect the Web page that you are viewing.

Back and Forth

Now that you can enter a URL into the dialog bar and have your application go to that Web site, and you can see the address of any Web sites that you view, it would be nice if you could back up from where you might have gone. This is a simple matter of calling the GoBack and GoForward functions on the view class in your application. You can call these functions from menu entries, which also allows you to attach toolbar buttons to perform the same calls.

To add this functionality, open the main menu in the Menu Designer. You can delete the Edit menu from the bar and all of the entries below it because they are of no use in the application that you are building. Grab the blank menu entry on the bar, and drag it to the left of the Help menu. Open the properties dialog for this menu entry and give it a caption of &Go. This is the menu where all navigation functions will be located.

To provide the back-and-forth functionality, you need to add two menu entries, one for the GoBack function and one for the GoForward function. Specify the properties for these two menu entries as shown in Table 7.4. below.

Table 7.4. Menu Property Settings

Object	*Property*	*Setting*
Menu Entry	ID	IDM_GO_BACK
	Caption	&Back\tCtrl + B
	Prompt	Back to the previous page\nBack
Menu Entry	ID	IDM_GO_NEXT
	Caption	&Next\tCtrl + N
	Prompt	Go forward to the next page\nNext

Once you add the menu entries, you can use the Class Wizard to add functions to the view class on both of these menu events. For the IDM_GO_BACK menu ID, add an event function on the COMMAND event message. Edit the function, adding the code in listing ;

```
//THE CWebBrowseView OnGoBack FUNCTION.
void CWebBrowseView::OnGoBack()
{
I/Goto the previous page
GoBackO;
}
```

Open the Class Wizard again, and add an event-handler function for the IDM_GO_NEXT object ID on the COMMAND event message. Edit this function with the code in listing :

```
//THE GWebBrowseView OnGoNext FUNCTION.

void CWebBrowseView::OnGoNext()
```

```
{

// TODO: Add your command handler code here

GoForwardOd;

}
```

Now you can run your application and use the menus to back up to the previous Web pages from wherever you surfed to and then trace your steps forward again. However, it's somewhat difficult using the menus, so what you need to do is add an accelerator for?

If you open the accelerator table in the resources tree, you see a bunch of accelerators tied to menu IDs. Each of these accelerators consists of an ID and a key combination. If you right-click anywhere in the accelerator table, you see the option of adding a new accelerator to the table. Choosing this option presents you a dialog to enter the accelerator information. First, you need to specify the menu ID that the accelerator will be tied to. (As with toolbar buttons, accelerators are tied to menu entries.) Below that, you can enter the key that will trigger the accelerator, or you can select a key from the drop-down list.

On the right side of the dialog, you can select the modifiers for the key. Modifiers are the other keys that must be pressed in combination with the key that you've already specified for the accelerator to be triggered. Once you've entered all the necessary information for the accelerator, close the dialog and the information you specified is added to the table.

It's recommended that you use either the Ctrl or Alt key as one of the modifier keys on all accelerators using standard keys. If you don't use one of these two keys as part of the accelerator, your application might get confused about when the user is typing information into your application and when the user is triggering an accelerator.

To add accelerators to the back and forward menus in your application, delete the accelerator for the ID_FILE_OPEN menu ID because you won't use it in this application. Add a new accelerator and specify the ID as IDM_GO_BACK and the key as B and select the Ctrl modifier. Add a second accelerator, specifying the ID as IDM_GO_NEXT and the key as N and select the Ctrl modifier. When you run your application, you can use the Ctrl+B key combination to back up to the previous page and the Ctrl+N key combination to go forward.

To really make your application work like most available Web browsers, you would also add toolbar buttons for these two menu entries with arrows pointing to the left for back and to the right for forward.

Controlling the Browser

Often when browsing, you come across a Web page that you don't want to wait to download. You'll want to stop the transfer part-way through. Maybe you entered the wrong URL or maybe the download is taking too long. It doesn't matter why you want to stop the download; it's enough that you want to stop it. This is why the CHtmlView class has the Stop function. It cancels the download currently in progress. To add this functionality to your application, add a new menu entry to the View menu in the Menu Designer. Specify the menu entry properties in Table 7.5.

Table 7.5. Menu Property Settings

Object	Property	Setting
Menu Entry	ID	IDM_VIEW_STOP
	Caption	Sto&p
	Prompt	Stop the current transfer\nStop

Using the Class Wizard, add an event-handler function to the view class for this menu ID on the COMMAND event message. Edit the function with the code in listing below.

```
//THE GWebBrowseView OnViewStop FUNCTION.

void CWebBrowseView::OnViewStop()

{
    Stop();
}
```

If you run your application, you can use this menu entry to stop any download of a Web page that you don't want to wait on. It would be more convenient if you added a toolbar button for this menu ID.

7.8. CREATING ACTIVEX CONTROLS (MDU 2008-09)

What is an ActiveX control?

An ActiveX control, like other Windows controls you have used so far, is a self-contained object packed with related functionalities. Similar to the Windows controls, it can also be included within a container object (such as a dialog window). However, it cannot run on its own.

Using ActiveX technique, you can create your own customised controls with desired set of properties, methods and events.

Properties

Properties are attributes or controls that are visible to, and often modifiable by, the container in which it is placed. There are four basic types of properties of ActiveX controls.

Ambient Properties are provided by the container application of the control - such as background color etc. - so that the control looks like a part of the container application.

Extended Properties are those properties of the control that are provided and implemented by the container application such as tab order.

Stock Properties are implemented by the ActiveX control development kit, such as control font or control background colour.

Custom Properties are the properties that you would like your control to have specific to its functionality.

You can specify any properties you need in your control using the Automation tab on the ClassWizard. You can select various aspects of properties of your control to ClassWizard, which in turn adds all the necessary codes and variables to your control project.

Methods

Methods of the ActiveX control are functions that can be called by its container application or object. These functions are made available to other applications through IDispatch interface. For detailed treatment of how this interface works, see Unit II.

You can add a new method to your control through ClassWizard's Automation window, where you specify particulars of a method and ClassWizard does the necessary coding for implementing IDispatch interface.

Events

Events are notification messages that are sent form the control to the container. They notify the container application that an event has happened in relation to the control, and if desired, the application may take an appropriate action in response.

Similar to properties and methods, events can also be added to the ActiveX control project through ClassWizard's Automation window.

SUMMARY

1. A Dynamic Link Library (DLL) is a program that holds one or more functions or some functionality that other programs can use (that's the shortest definition I could come up with...). As such, you can create one DLL and use it in many programs or even distribute it to other programmers (because a DLL is usually treated as an "executable" program, you cannot e-mail it to somebody as an attachment; most e-mail programs would not carry it).
2. ActiveX is part of an evolutionary process that began with OLE. To understand ActiveX is to understand OLE, and *vice versa*. Since ActiveX begins with OLE, we will begin there too. In its original conception, OLE was designed to allow one application to link or embed information created by another application. When this is done, a compound document is created.
3. Most applications that communicate over a network, whether it's the Internet or a small office network, use the same principles and functionality to perform their communication. One application sits on a computer, waiting for another application to open a communication connection.
4. Once the connection is made between the two applications, messages can pass back and forth between the two applications, much like you can talk to the person on the other end of the phone. This connection is a two-way communications channel, with both sides sending information.
5. When you build applications with Visual C++, you can use the MFC Winsock classes to add network communications capabilities with relative ease. The base class, CAsyncSocket, provides complete, event-driven socket communications.
6. Once you create a socket, you are ready to open a connection with it. Three steps go alongwith opening a single connection. Two of these steps take place on the server, the application listing for the connection, and the third step lakes place on the client, the one making the call.
7. To send a message through a socket connection, you use the Send method. This method requires two parameters and has a third, optional parameter that can be used to control how the message is sent. The first parameter is a pointer to the buffer that contains the data to be sent.
8. An ActiveX control, like other Windows controls you have used so far, is a self-contained object packed with related functionalities. Similar to the Windows controls, it can also be included within a container object (such as a dialog window). However, it cannot run on its own
9. Properties are attributes or controls that are visible to, and often modifiable by, the container in which it is placed. There are four basic types of properties of ActiveX controls.
10. Methods of the ActiveX control are functions that can be called by its container application or object. These functions are made available to other applications through IDispatch interface. For detailed treatment of how this interface works.

SHORT QUESTIONS WITH ANSWER

Q.1. What is DLL?

Ans. A Dynamic Link Library (DLL) is a program that holds one or more functions or some functionality that other programs can use (that's the shortest definition I could come up with...). As such, you can create one DLL and use it in many programs or even distribute it to other programmers (because a DLL is usually treated as an "executable" program, you cannot e-mail it to somebody as an attachment; most e-mail programs would not carry it). Most of the time, a DLL is just a file that has the extension .dll but it could also have another extension.

The DLL we are going to create can be used in any program that uses MFC. Unfortunately, and based on my experience, you should not distribute it to people who are using other compilers such as Borland C++ Builder or Dev-C++. In reality you can. Because they will not be able to simply "plug" it in their programs, apparently there are some things they can do to make it work. I cannot cover that. In fact, the only reason I am writing this is because I can't believe how hard it is to find this information, whether in books or on the Internet; people who write books either assume that everybody knows how to create a DLL or they simply don't know, which is incredible considering, as you will see soon, how easy this thing is (I myself looked for it in many places, to no avail).

Q.2. Briefly Explain OLE object Technologies. **(V.Imp)**

Ans. ActiveX is part of an evolutionary process that began with OLE. To understand ActiveX is to understand OLE, and *vice versa*. Since ActiveX begins with OLE, we will begin there too. In its original conception, OLE was designed to allow one application to link or embed information created by another application. When this is done, a compound document is created. A compound document is also referred to as an OLE document. While this original purpose of OLE is still supported, it is important to understand that OLE has gone through a rather extensive evolution. OLE version 1.0 was devised in 1991, but it was a seldom-used system. With the advent of OLE version 2.0, the capabilities and applicability of OLE took a quantum leap forward. Perhaps the single most important new feature of OLE 2.0 was the definition of the Component Object Model (COM).

It is this feature that ultimately led to the ActiveX technology. In response to the Internet, Microsoft defined ActiveX in 1996. Based on OLE, it added support for Web-based applications. However, since it supports all of OLE, the term ActiveX has nearly supplanted the term OLE.

Q.3. Write the concept of Creating Internet Programs.

Ans. Most applications that communicate over a network, whether it's the Internet or a small office network, use the same principles and functionality to perform their communication. One application sits on a computer, waiting for another application to open a communication connection. This application is "listening" for this connection request, much like you listen for the phone to ring if you are expecting someone to call.

Meanwhile, another application, most likely on another computer (but not necessarily), tries to connect to the first application. This attempt to open a connection is similar to calling someone on the telephone. You dial the number and hope that the other person listening for the phone on the other end. As the person making the call, you have to know the phone number of the person you are calling. If you don't know the phone number, you can look it up using the person's name. Likewise, the application trying to connect to the first application has to know the network location, or address, of the first application.

Once the connection is made between the two applications, messages can pass back and forth between the two applications, much like you can talk to the person on the other end of the phone. This connection is a two-way communications channel, with both sides sending information.

Finally, once one or both sides have finished their sides of the conversation, the connection is closed, much like you hang up the phone after you have finished talking to the person you called. Once the connection is closed from either side, the other side can detect it and close its side, just like you can tell if the person on the other end of the phone has hung up on you or if you have been disconnected by some other means. This is a basic explanation of how network communications work between two or more applications.

Q.4. What is an ActiveX control?

Ans. An ActiveX control, like other Windows controls you have used so far, is a self-contained object packed with related functionalities. Similar to the Windows controls, it can also be included within a container object (such as a dialog window). However, it cannot run on its own.

Using ActiveX technique, you can create your own customised controls with desired set of properties, methods and events.

Properties: Properties are attributes or controls that are visible to, and often modifiable by, the container in which it is placed. There are four basic types of properties of ActiveX controls.

Ambient Properties are provided by the container application of the control - such as background color etc. - so that the control looks like a part of the container application.

Extended Properties are those properties of the control that are provided and implemented by the container application such as tab order.

Stock Properties are implemented by the ActiveX control development kit, such as control font or control background colour.

Custom Properties are the properties that you would like your control to have specific to its functionality.

You can specify any properties you need in your control using the Automation tab on the ClassWizard. You can select various aspects of properties of your control to ClassWizard, which in turn adds all the necessary codes and variables to your control project.

Methods: Methods of the ActiveX control are functions that can be called by its container application or object. These functions are made available to other applications through IDispatch interface. For detailed treatment of how this interface works, see Unit II.

You can add a new method to your control through ClassWizard's Automation window, where you specify particulars of a method and ClassWizard does the necessary coding for implementing IDispatch interface.

Events: Events are notification messages that are sent form the control to the container. They notify the container application that an event has happened in relation to the control, and if desired, the application may take an appropriate action in response.

Similar to properties and methods, events can also be added to the ActiveX control project through ClassWizard's Automation window.

Q.5. Bring out the similarities and difference between DLL and EXE.

Ans. DLLvs.EXE: DLL stands for dynamic-link library and is Microsoft's implementation of a shared library. This means that many different programs can use this library to do their tasks making it easier on the programmers so that they do not have to keep reinventing the wheel each time they write software. In simple terms a .DLL file will contain logic that other programs will use.

EXE stands for executable and denotes that a program is executable This just means that if you double click on the file a program will run, normally with some kind of interface for a user to interact with. The file formats for EXE and DLL actually the same.

Advantages of ActiveX DII: I. An in-process component shares its client's address space, so property and method calls don't have to be marshaled. This results in much faster performance.

Disadvantages of ActiveX DII:

I. If an unhandled error occurs it will cause the client process to stop operating. Advantages of ActiveX Exe.
    1. The component can run as a standalone desktop application, like Microsoft Excel or Microsoft Wont, in addition to providing objects,
    2. The component can process requests on an independent thread of execution* notifying the client o task completion using events or asynchronous call-backs. This frees the client to respond to the user.
    3. If an error occurs the client processes can continue to operate. Disadvantages of ActiveX Exe.

I. Generally slower than an ActiveX dll alternative.

Q.6. What do you mean by connectivity to database? Differentiate between ADO and DAO connectivity methods.

Ans. Database connectivity: Database connectivity provides a standard software API method for using database management systems (DBMS). The designer of ODBC (Open Database Connectivity) aimed to make it independent of programming language, database systems and operating systems.

Difference between ADO and DAO connectivity methods:

→ DAO, Data Access Objects, introduced in Access 1,0 back in 1992, is a technology built around the JET engine capable of accessing microsoft access Databases, external ISAM databases including dBase, Foxpro and ODBC data sources Visual Basic 3.0 provided support for DAO and the microsoft JET database engine and for a while, it was the most popular method of communicating with Access and ISAM data sources.

→ ADO (Active X Data Objects), is a wrapper around OLE DS that was first released in 1996 to hide the complicated syntax of OLE DB. It has similar features of DAO and RDO and although, it is designed for minimal network traffic, it has good performance and easy to learn. ADO provides all the capabilities of OLE DB, which means we can access data stores from various different vendors. This model was designed towards a tightly coupled connected architecture, while still allowing for a disconnected setup. Disconnected means data is kept in memory without any active connection to the database to allow work to be done on the data without having to rely on a network that might be slow or not be available all the time. It also allows greater scalability in that server resources are freed up, hence allowing more connections. Tightly coupled means that the components or objects of a system must be connected to each other, as opposed to loosely coupled where the components can work independently of each other and only communicate when needed.

LONG QUESTIONS WITH ANSWER

Q.1. How would you Create a Project DLL?

Ans. To start, we need a project that would carry the DLL :

1. Start Microsoft Visual C++. On the main menu, click File -> New -> Project.
2. In the New Project dialog box, in the Project Types list, click the Visual C++.
3. In the Templates list, click Win32 Project Projects tab and click Win32 Dynamic Link Library.
4. In the Project Name, type a name such as ExoDLL and specify your desired folder in the Location box.
5. Click OK.
6. In the **Win32 Application Wizard**, click **Application Settings**.
7. In the Application Type section, click the DLL radio button.
8. Click Finish.

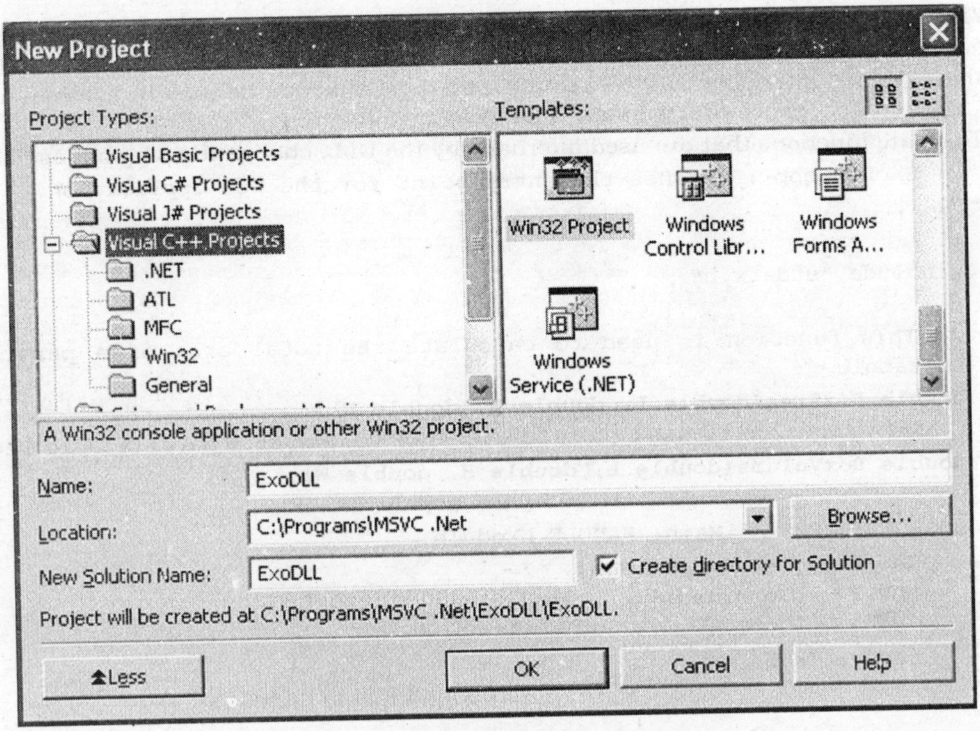

Fig. 7.13.

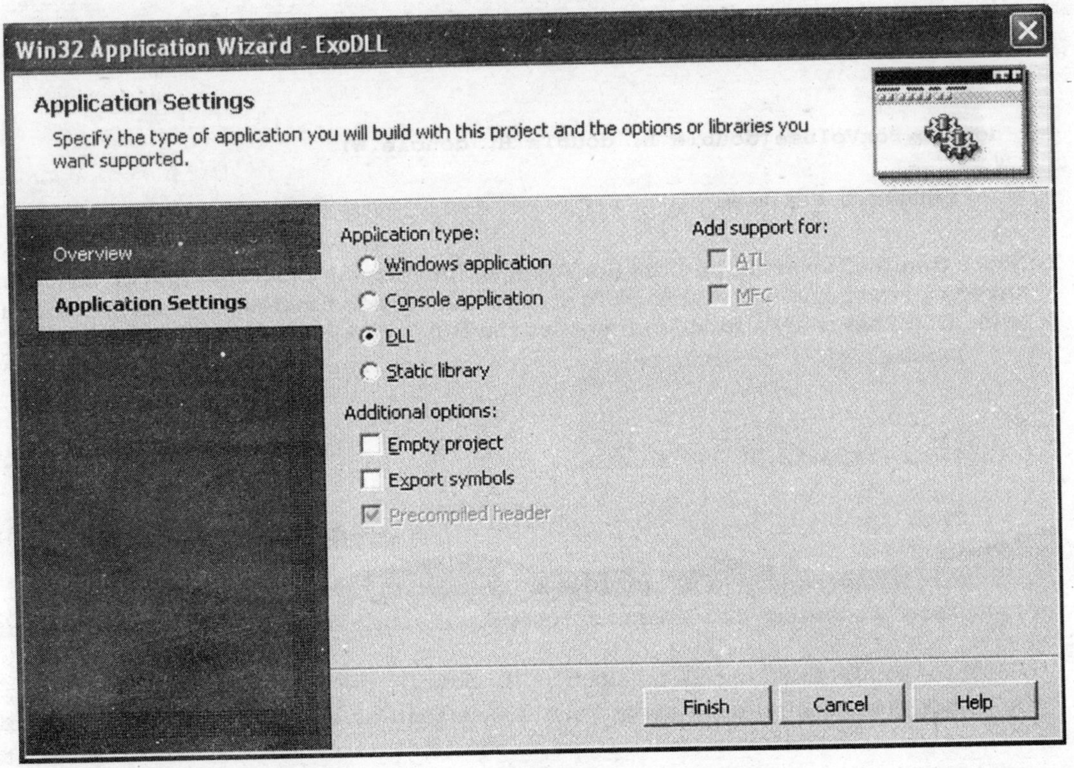

Fig. 7.14.

Q.2. What is the procedure for Coding the DLL?

Ans. To make this project a little useful for our learning process, we will use two types of functions: those that are used internally by the DLL only, and those that can be used by external functions, that is, the other programs that will need this DLL.

1. To create functions that are used internally by the DLL, change the program as follows :

```
// ExoDLL.cpp : Defines the entry point for the DLL application.
//

#include "stdafx.h"

// This function is used to calculate the total area of a parallelepid
rectangle
double BoxArea(double L, double H, double W);
// This function is used to calculate the volume of a parallelepid rectangle
double BoxVolume(double L, double H, double W);

BOOL APIENTRY DllMain( HANDLE hModule,
    DWORD ul_reason_for_call,
    LPVOID lpReserved)
{
    return TRUE;
}

double BoxArea(double L, double H, double W)
{
    return 2 * ((L*H) + (L*W) + (H*W));
}

double BoxVolume(double L, double H, double W)
{
    return L * H * W;
}
```

2. Notice that the BoxArea() and the BoxVolume() functions are declared and defined just like any other function we know already.To declare and define a function that can be used outside of the DLL, that is, by a function that uses the DLL, change the file as follows :

```
// ExoDLL.cpp : Defines the entry point for the DLL application.
//

#include "stdafx.h"

// This function is used to calculate the total area of a parallelepiped
rectangle
double BoxArea(double L, double H, double W);
// This function is used to calculate the volume of a parallelepiped
rectangle
double BoxVolume(double L, double H, double W);
// This function is used to get the dimensions of a rectangular parallelepiped
// calculate the area and volume, then pass the calculated values to the
// function that called it.
```

```
extern "C" __declspec(dllexport)void BoxProperties(double Length, double
Height,
            double Width, double& Area, double& Volume);

BOOL APIENTRY DllMain( HANDLE hModule,
    DWORD ul_reason_for_call,
    LPVOID lpReserved)
{
    return TRUE;
}

double BoxArea(double L, double H, double W)
{
    return 2 * ((L*H) + (L*W) + (H*W));
}

double BoxVolume(double L, double H, double W)
{
    return L * H * W;
}

void BoxProperties(double L, double H, double W, double& A, double& V)
{
    A = BoxArea(L, H, W);
    V = BoxVolume(L, H, W);
}
```

3. To save the project, on the main menu, click File -> Save All.
4. To create the DLL, on the main menu, click Build -> Build ExoDLL.dll. At the end, you should receive Build: 1 succeeded, 0 failed, 0 skipped.

Q.3. What is Sockets, Ports, and Addresses? **(V.Imp)**

Ans. The basic object used by applications to perform most network communications is called a socket. Sockets were first developed on UNIX at the University of California at Berkley. Sockets were designed so that most network communications between applications could be performed in the same way that these same applications would read and write files. Sockets have progressed quite a bit since then, but the basics of how they work are still the same.

During the days of Windows 3.x, before networking was built into the Windows operating system, you could buy the network protocols required for network communications from numerous different companies. Each of these companies had a slightly different way that an application performed network communications. As a result, any applications that performed network communications had a list of the different networking software that the application would work with. Many application developers were not happy with this situation. As a result, all the networking companies, including Microsoft, got together and developed the Winsock (Windows Sockets) API. This provided all application developers with a consistent API to perform all network communications, regardless of the networking software used.

When you want to read or write a file, you must use a file object to point to the file. A socket is an object used to read and write messages that travel between applications. Making a socket connection to another application does require a different set of information than opening a file. To open a file, you need to know the file's name and location, To open a socket connection, you need to know the computer on which the other application is running and the port on which it's listening. A port is like a phone extension, and the computer address is like the phone number. If you call someone at a large office building, you may dial the main office number, but then you need to specify the extension number. Likewise, ports are used to route network communications. As with the phone number, there are means of looking up the port number, if you don't already know what it is, but this requires your computer to be configured with the information about which port the connecting application is listening on. If you specify the wrong computer address or port number, you may get a connection to a different application; with making the phone call, someone other than the person you called may answer the phone call. You also may not get an answer at all if there is no application listening at the other end.

✍ Note :

Only one application may be listening on any specific port on a single computer. Although numerous applications may listen for connection requests, on a single computer at the same time, each of these applications must on a different port.

Q.4. Write the procedure for Creating a Socket.

Ans. When you build applications with Visual C++, you can use the MFC Winsock classes to add network communications capabilities with relative ease. The base class, CAsyncSocket, provides complete, event-driven socket communications. You can create your own descendent Socket class that captures and responds to these events.

To create a socket that you can use in your application, the first thing you need to do is declare a variable of CAsyncSocket (or your descendent class) as a class member for one of the main application classes :

```
class CMyDig : public CDialog
{
private:
    CAsyncSocket m_sMySocket;
};
```

Before you can begin using the socket object, you must call its Create method. This actually creates the socket and prepares it for use. How you call the Create method depends on how you will be using the socket. If you will be using the socket to connect to another application, as the one placing the call (the client), then you do not need to pass any parameters to the Create method :

```
if (m_sMySocket .Create ())
{
// Continue on
}
else
// Perform error handling here
```

However, if the socket is going to be listening for another application to connect to it, waiting for the call (the server), then you need to pass at least the port number on which the socket should be listening :

```
if (m_sMySocket .Create (4000))
{
// Continue on
}
else
// Perform error handling here
```

You can include other parameters in the Create method call, such as the type of socket to create, the events that the socket should respond to, and the address that the socket should listen on (in case the computer has more than one network card).

Q.5. Write procedure for Making a Connection? **(MDU 2008-09)**

Ans. Once you create a socket, you are ready to open a connection with it. Three steps go alongwith opening a single connection. Two of these steps take place on the server, the application listing for the connection, and the third step lakes place on the client, the one making the call.

For the client, opening the connection is a simple matter of calling the Connect method. The client has to pass two parameters to the Connect method: the computer name, or network address, and the port of the application to connect to. The Connect method could be used in the following two ways :

```
if (m_sMySocket.Connect("computerOne", 4000))
{
// Continue on
}
else
// Perform error handling here
```

The second form is

```
if (m_sMy Socket. Connect ("178. 1.25.82", 4000))
{
// Continue on
}
else
// Perform error handling here
```

Once the connection is made, an event is triggered to let your application know that it is connected or that there were problems and the connection couldn't be made (to be discussed later). For the server, or listening side of the connection, the application first must tell the socket to listen for incoming connections by calling the Listen method. The Listen method takes only a single argument, which you do not need to supply. This parameter specifies the number of pending connections that can be queued, waiting for the connection to be completed. By default this value is 5, which is the maximum. The Listen method can be called as follows :

```
if (m_sMySocket .Listen ())
{
// Continue on
}
else
// Perform error handling here
```

Whenever another application is trying to connect to the listening application, an event is triggered to let the application know that the connection request is there. The listening application must accept the connection request by calling the Accept method. This method requires the use of a second CAsyncSocket variable, which is connected to the other application. Once a socket is placed into listen mode, it stays in listen mode.

Whenever connection requests are received, the listening socket creates another socket, which is connected to the other application. This second socket should not have the Create method called for it because the Accept method creates the socket. You call the Accept method as follows :

```
if (m_sMySocket.Accept (m_sMySecondSocket) )
{
// Continue on
}
else
// Perform error handling here
```

At this point, the connecting application is connected to the second socket on the listening application.

Q.6. What are Socket Events? Briefly Explain.

Ans. The primary reason that you create your own descendent class of CAsyncSocket is that you want to capture the events that are triggered when messages are received, connections are completed, and so on. The CAsyncSocket class has a series of functions that are called for each of these various events. These functions all use the same definition, the function name is the only difference, and they are intended to be overridden in descendent classes. All of these functions are declared as protected members of the CAsyncSocket class and probably should be declared as protected in your descendent classes. The functions all have a single integer parameter, which is an error code that should be checked to make sure that no error has occurred.

Table 7.6. Lists these event, functions and the events they signal.

Table 7.6. CAsyncSocket Overridable Event-Notification Functions

Function	*Event Description*
OnAcceot	This function is called on a listening socket to signal that a connection request from another application is waiting to be accepted.
OnClose	This function is called on a socket to signal that the application on the other end of the connection has closed its socket or that the connection has been lost. This should be followed by closing the socket that received this notification.
OnConnect	This function is called on a socket to signal that the connection with another application has been completed and that the application can now send and receive messages through the socket.
OnReceive	This function is called to signal that data has been received through the socket connection and that the data is ready to be retrieved by calling the Receive function.
OnSend	This function is called to signal that the socket is ready and available for sending data.This function is called right after the connection has been completed. Usually, the other time that this function is called is when your application has passed the Send function more data than can be sent in a single packet. In this case, this is a signal that all of the data has been sent, and the application can send the next buffer-full of data.

Q.7. What is an active X control ? Write a program to add an event handler to an Active X control ?

Ans. Active X Controls: An **Active X control** is a reusable component which implements the Unknown interface. Such components do no amount to an entire application, rather they provide a small building-lock that can be shared by different software. The fact that command buttons look the same in almost any program on a platform is an example of component reusability that is not just limited to Active X controls. Active X controls can be compared in some sense to a Jave applet, since both technologies strive to act as an abstraction layer between the developer and the operating system. From there, obvious differences surface. For instance, Java applets can run on nearly any platform, while Active X components are only compatible with a Microsoft web browser and a Microsoft operating system. Active X controls re also granted a much higher level of control over Windows than Java applies, making them both more powerful add dangerous. Malware, such as computer viruses and spyware are often distributed as Active X applets in webpages. Active X controls can be written in MFC, ATL. C++, Borland Delphi and Visual Basic. Common examples of Active X controls are the command button, list box, dialog boxes and even the Internet Explorer browser.

Program for Event Handler:

Handling Events: In the declaration (b) file: Declare the message handler manually, outside of the AFX_MSG section. e.g.,

```
// Message maps
// {{AFX_MSG (CSubSectionCtrol)
afx_msg int OnCreate (LOPCREATESTRUCT lpzCreateStruct);
    // }}AFX_MSG
    afx_msg void OnExampleButtonClicked ( );
    DECLANRE_MESSAGE_MAP ( )
    COM INTERFACES
```

The afx_msg void OnExampleButtonClicked ( ) line is the one we added.

In the definition (.cpp) file: Define the message map manually, outside of the AFX_MSG_MAP section, *e.g.*,

```
BEGIN_MESSAGE_MAP (CSubSectionCtrol, COleControl)
// {{AFX_MSG (CSubSectionCtrol)
ON_WM_CREATE ( )
    // }}AFX_MSG_MAP
    ON_BN_CLICKED(IDR_EXAMPLEBUTTON, OnExampleButtonClicked)
    ON_OLEVERB(AFX_IDS_VERB_PROPERTIES, OnProperties)
    END_MESSAGE_MAP ( )
    The ON_BN_CLICKED ... line is the one we added.
```

Then implement the message handler method.

e.g.,

```
void On ExampleButtonClicked ( ) {...}
```

Q.8. What is MFC ? Differentiate between API and MFC by giving suitable programming example.

Ans. MFC: The Microsoft Foundation Class Library (also Microsoft Foundation Classes or MFC) is a library that wraps portions of the Windows API in C++ classes, including functionality that enables them to use a default application framework. Classes are defined for many of the handle-managed. Windows objects and also for predefined windows and common controls. When MFC was introduced, Microsoft extended the C++ syntac with a series of macros for management of Windows messages (via Message Maps), exceptions, custom design run time type identification, serialization and dynamic class installations (many of these features-exceptions or run time type identification, for example — were not built into Microsoft's compiler at the time, but would appear in future versions). The macros for Windows messages were intended to reduce memory required by avoiding gratuitous Virtual table use and provide a more concrete structure for various Visual C++ supplied tools to edit and manipulate code without parsing the full language. The message-handling macros replaced the virtual function mechanism provided by C++. The macros which implemented serialization, exception support, and dynamic runtime types predated availability of standards-based language extensions by a number of years. 32-bit versions of MFC, for windows NT 3.1 and later Windows operating systems, used compilers that implemented the language features and updated the macros to simply wrap the language are nothing else but standard C preprocessor constructs. Using one of these macros, one can add unicode support to MFC unicode.

MFC is basically a library of OO wrapper classes that wrap the Win32api, and provide objects for the basic window components (windows, buttons, checkboxes etc..). Essentially it is the win32api objectified. Also MFC provides some classes that resemble classes found in the STL. As MFC was made before STL was fully standardized. But that is the basic idea. The Windows API (win 32) use a C interface to access windows functionality. It requires that you write all of your own code to manage windows and message handling. In contrast, MFC, which stands or Microsoft Foundation Classes, are a collection of C++ classes which encapsulate the Win32 API. MFC has some wizards which make the initial creation of a program quick and easy, but I have to admit the learning curve of MFC can sometimes be frustrating as Microsoft seems to have done everything in a way that can at times seen counter-intuitive.

Win32 and the MFC class library approach applications input and output in different ways:

- When using win32 file I/O functions, you create a file handle and manipulate the file handle with functions such as ReadFile and WriteFile. If there is an error in a Win32 file I/O function, these functions return FALSE, and you must call the Win32 GetLastError function to determine the specific cause of the error.
- When using MFC file I/O input and output are performed in a more object-oriented manner using the CFile and Carchive classes. If an error occurs, a C++ exception is raised that contains detailed error information.

Q.9. Write short notes on:

(a) Hungarian Notation

(b) Resource editor

(c) Window Registry

(d) Tooltip

Ans. (a) Hungarian Notation: Hungarian notation is a naming convention in computer programming, in which the name of a variable indicates its type or intended use. There are two types of Hungarian notation: Systems Hungarian notation and Apps Hungarian notation. Hungarian notation was designed to be language-Independent, and found its first major use with the BCPL programming language. Because BCPL has no data types other than the machine word, notation aims to remedy this by providing the programmer with explicit knowledge of each variable's data type. In Hungarian notation, a variable name starts with one or more lower-case letters which are mnemonics for the type or purpose of that variable. Followed by whatever the name the programmer has chosen; this last part is sometimes distinguished as the given name. The first character of the given name can be capitalized to separate it from the type indicators. Otherwise the case of this character denotes scopes.

(b) Resource Editor: NET programming tool (WinCV.exe) for finding and displaying name space and class information within a NET assembly.

(c) Window Registry: The **Window registry** is a directory which stores settings and options for the operating system for Microsoft Windows 32-bit versions, 64-bit versions, and Windows Mobile. It contains information and settings for all the hardware, operating system software, most non-operating system software, users preferences of the PC, etc. Whenever a user makes changes to Control Panel settings, file associations, systems policies, or most installed software, the changes are reflected and stored in the registry. The registry also provides a window into the operation of the kernel, exposing run-time information such as performance counters and currently active hardware. This use of registry mechanism is conceptually similar to the easy that System and protectors expose runtime information through the file system (traditionally viewed as a place for permanent storage), though the information made available by each of them differs tremendously. The windows registry was introduced to tidy up the profusion of per-program INI files that had previously been used to store configuration settings for Windows program. These files tended to be scattered all over the system which made them difficult to track.

(d) Tooltip: The **tooltip** is a common graphical user interface element. It is used to conjunction with a cursor, usually a mouse pointer. The user hovers the cursor over an item, without clicking it, and a small "hover box" appears with supplementary information regarding the item being hovered over.

Q.10. What are active X controls ? How it can be used in a VCC program ?

Ans. Active X controls: Active X is a component object model (COM) developed by Microsoft for Windows platforms. By using the COM runtime, developers can create software components that perform a particular function or a set of functions. A software can then compose one or more components in order to provide the functionality it intends to. Many Microsoft Windows applications — including many of those from Microsoft such as Internet Explorer, Microsoft Officer, Microsoft Visual Studio, Windows Media Player, etc. - use Active X controls to build their feature set as well as encapsulate their functionality as Active X controls so that the functionality can be embedded in other applications. Internet Explorer also allows the Active X controls to be embedded inside web pages. Active X controls are small program building blocks that can be used to create distributed applications that work over the Internet through web browsers. Examples include customized applications for gathering data, viewing certain kinds of files, and displaying animation.

Active X controls can be compared in some sense to a Java applet, since both are designed to be downloaded and executed from web browsers. However, there are few differences. Java applets can run on nearly any platform, while Active X components are only compatible with Microsoft Internet Explorer web browser and the Microsoft Windows operating system. Active X controls are also granted a much higher level of control over Windows than Java applets, making them both more powerful and dangerous. Malware, such as computer viruses and spvware, is often distributed as Active X applets in webpages.

1. Start Visual C++, and follow File->New.
2. In the dialogue box New, choose Projects Tab.
3. Select MFC AppWizard (exe), Fill in the name "Ultima" in project name and hit OK.
4. Select Dialogue-based for "What type of application would you like to create ?", then high Finish.
5. Hit OK to create the project files.
6. If you see a dialogue box template lab led as Ultima on your screen, you can proceed to Step 11).
7. From Workspace pane, select Resource.
8. Click on the + sign in front of Ultima resources and you should see the folder lab led as Dialog.
9. Click on the + sign in front of the folder Dialog, and you should see IDD_ULTIMA_DIALOG.
10. Double-click on IDD_ULTIMA_DIALOG and you will see the dialogue box template labelled as Ultima on your screen.
11. Move the cursor to the dialog box template Ultima, and click the right button on your mouse, and follow insert Active X control.
12. In several seconds, you will see a dialog box labelled as Insert Active X control.
13. Double click on DQChart Control in the list (If you don't see this control int he list, you need to install Ultima Serial.)
14. Repeat 11-13) to insert UltimaSerial Control.
15. Remove the OK button and the "TODO Place dialog controls here" text from the dialog box. Also, resizeIre-arrange the objects on the dialog box if you like.
16. Follow View->Class Wizard ... to bring up the MFC Class Wizard.
17. From the Member Variables tab of the MFC Class Wizard, Double-click on IDC_DQCHARTCTRL1, and the Wizard will ask if it is OK to generate a C++ wrapper class for it.
18. Select OK to create the wrapper class for DQ Chart Control. When the Add Member Variable dialog box comes up, set the member variable name to m_dqchart.
19. Repeat 16-18) to add a member variable m_ultima for IDC_ULTIMASERIALCTRL 1.
20. Hit OK to exit MFC Class Wizard.
21. On the dialog box template Ultima, right-click on UltimaSerial control and follow Events.
22. You will see a dialog box labelled as New Windows Message and Event Handlers for CUltimaDig.
23. Double-click on NewData under New Windows messages/events list, then hit OK to use the default name.
24. Hit OK to exit the dialog box New Windows Message and Event Handlers for class CUltimaDig.
25. From the Workspace pane, follow File view tab, and open UltimaDig.cpp.
26. Add the follow codes to BOOL CUltimaDig::OnInitDialogO, right after the comment/1 TODO: Add extra initialization here (If you decide to type in the codes, you will notice the properties and methods will be displayed in a drop-down menu once you typed in m_ultima).

```
m_ultima.SetDevice(194);
m_ultima.SetCommPort(1);
m_ultima.SetEventLevel(2);
m_ultima.SetSampleRate(240);
m_ultima.SetChannelCount(1);
m_ultima.Start( );
```

27. Add the follow codes OnNewDataUltimaserialctrl1(short Count)

```
m_dqchart.Chart(m_ultima.GetData( ));
```

Q.11. What are break points ? How they help in debugging of program ?

Ans. Breakpoints: A breakpoint is a signal that tells the debugger to temporarily suspend execution of your program at a certain point. When execution is suspended at a breakpoint, your program is said to be in break mode. Entering break mode does not terminate or end the execution of your program. Execution can be resumed at any time. You can think of break mode as being like a time out. All the elements remain, functions, variables, and objects remain in memory, for example, but their movements and activities are suspended. During break mode, you can examine their positions and states to look for violations or bugs. You can make adjustments to the program while in break mode. You can change the value of a variable, for example. You can move the execution point, which changes the statement that will be executed next when execution resumes. In C++, C#, and Visual Basic, you can even make a changes to the code itself while in break mode using a powerful feature called Edit and Continue. Breakpoints provide a powerful tool that enables you to suspend execution where and when you need to. Rather than stepping through your code line-by-line or instruction-by-instruction, you can allows your program to run until it hits a breakpoint, then start to debug. This speeds up the debugging process enormously. Without this ability, it would be virtually impossible to debug a large program. Many programming languages have statements or constructs that suspend execution and put your program into break mode. Visual Basic, for example, has the Stop statement. Breakpoint differ from these statements because breakpoints are not actual source code that has to be added to your program. You do not type a breakpoint statement into a source window. You request a breakpoint through the debugger interface and the debugger sets it for you. To insert a line breakpoint, click in the gray margin next to the line where you want to set the breakpoint. More complex breakpoints can be handled through a full-featured Breakpoints window. Breakpoints have many advantages over debugging constructs such as the Visual Basic Stop statement. Breakpoints can be deleted or changed without having to change your program's source code. Because breakpoints are not statements, they never produce any extra code when you build a release version of your program. If you use Stop statements in your program, you need to manually remove the Stop statements before building the release version, or use conditionals.

To provide even greater flexibility, the Visual Studio debugger enables you to set properties that modify the behaviour of a breakpoint.

- Hit Count enables you to determine how many times the breakpoint is hit before the debugger breaks execution. By default, the debugger breaks execution every time the breakpoint is hit. You can set a hit count to tell the debugger to break every two times the breakpoint is hit, or every 10 times, or every 512 times, any number you choose. Hit counts can be useful because some bugs do not appear the first time your program executes a loop, calls a function, or accesses a variable. Sometimes, the bug might not show up until the 100th or 1000th iteration. To debug such a problem, you can set a breakpoint with a hit count of 100 or 1000.
- Condition is an expression that determines whether the breakpoint is hit or skipped. When the debugger reaches the breakpoint, it will evaluate the condition. The breakpoint will be hit only if the condition is satisfied. You can use a condition with a location breakpoint to stop at a specified location only when a certain condition is true. Suppose you are debugging a banking program, for example, where the account balance is never allowed to go below zero. You might set breakpoints at certain locations in the code and attach condition like balance < a to each one. When you run the program, execution will break at those locations, only when the balance is less than zero. You can examine variables and program state at

the first breakpoint location, then continue execution to the second breakpoint location, and so on.

- Action specifies something that should occur when the breakpoint is hit. By default, the debugger breaks execution, but you can choose to print a message or run a Visual Studio macro instead. If you choose to print a message instead of breaking, the breakpoint has an effect very similar to Trace statement. This method of using breakpoints is called tracepoints.
- Filter provides a way to specify a process of thread for the breakpoint.

Q.12. What features are available in VC++ for program debugging?

Ans. Using debugger in VC++: To use the debugger in the Visual C++ programming environment you must have compiled and linked your program using the "Debug" configuration (not the "Release" configuration). The configuration is set using the menu options Build I Set Active Configuration. This is the default configuration, so you don't have to set it unless you previously changed the configuration to "Release". Usually it is not helpful to step through the statements in a program beginning with the first statement in main(). It is better to have the program run until it gets to that part of your code where you think the bug resides. There are two ways to tell the debugger to run to this point. You can locate the cursor at that place in your code and press Ctrl-F 10 (or use Build | Start Debug j Run to Cursor). The program will be run by the debugger and if all goes well it will stop at the place in your code that you selected, and the debugger windows will be opened. Of course, it is possible that ihe program may terminate before your intended stopping place is reached. In which case, choose a place earlier in the program. The second way to tell the debugger to run to the place in your code that you wish to inspect is fo set a "breakpoint" there. To set a break point, position the cursor at the desired place and press F9 (or use Right-mouse-button | Insert/Remove Breakpoint). If a breakpoint has been set then you can remove it by placing the cursor next to it and pressing F9. (Breakpoints can also be disabled by using F9 or by Rightmouse-button | Disable Breakpoint.) Once you have set a breakpoint then press F5 (or use Build | Start Debug | Go) to cause the debugger to run the program and halt at the breakpoint, whereupon the debugger windows are opened (if they are not already open). The windows at the bottom left and bottom right of the screen hold variable names and values. Those in the left window include variables local to the function which is being executed. Those in the right window are user-specified variables. To specify a variable click on the left part of the first empty line then type the name of the variable. After pressing Enter the value of that variable (if it has a value) will be displayed. You can then view how the displayed variables change in value as execution of the program is continued. Pressing Fl brings up context-sensitive Help. For example, it tells us the difference between "Auto" variables and "Locals" variables:

The Auto tab displays information about variables used in the current statement and the previous statement. The Locals tab displays information about variables that are local to the current function.

To view the value of a variable without having to type in the name, position the cursor somewhere within the variable name and press Shift-F9 (or Debug / Quick Watch). You can continue program execution in several ways. The simplest is to press F10 (or Debug | Step Over) to execute the next statement (which is indicated by the small yellow arrow in the left margin of the code window). Prcssing F10 repeatedly steps through successive program statements. Attend to the information given in the bottom windows. For example, if you are at a statement

```
if ((j = atoi(buffer) ) >= 2000)
```

(which converts the string in buffer to an integer, places the value in variable j and compares this to 2000) then after pressing F10 you will be informed as to the value returned by atoi( ). If you are at a function call then F10 causes that function to be executed and when the debugger returns control you are at the next statement. At a function call you can press Fl 1 (or Debug | Step Into) to enter that function (if the code for that function was compiled in "Debug" mode). You can then use F10 to step through the statements in that function. When you are within a function you can press Shift-Fl I (or Debug j Step Out) to run through the rest of the function and return to thej^e where the function was called. When the debugger starts it opens a window for the program's outpuf,~sb you can switch between the debugger and this window to see what the program is sending to the screen. In addition to F10, F11 and Shift-F 11 you can use Ctrl-F 10 (or Debug J Run to Cursor) to run to the cursor position. You can also set further breakpoints and use F5 (or Debug j Go) to run to each one in turn. You can haveiSeveral breakpoints active, and you can disable or remove them at will by positioning the cursor at them and pressing F9. If you want to see the definition of a variable place the cursor within the variable name and use Right-mouse-button | Go to Definition. It may happen that the debugger will tell you that it can't do this until it has created the Browser database, in which case tell it to do so. The Browser is a useful tool which you can use to get the definition of any variable in your program and the places in your program where that variable is used (the "variable references"). You can invoke it by using Tools I Source Browser. If your program takes command line arguments then during debugging you can set these using Project j Settings | Debug / Program arguments. To restart the program from the beginning press Ctrl-Shift-F5 (or Debug | Restart). To stop the debugger and return to edit mode use Shift-F5 (or Debug | Stop Debugging).

Q.13. Explain the concept of MDI with suitable example.

Ans. With multiple view, the user can move around in the same document, working on different sections simultaneously. Each of these views will get its own window.

An MDI program handler multiple documents, and connects view and document classes to document emplate as well as the CChild Frame Class which supports **MDI** child windows, the type of windows that appear inside the main window.

Example: Create on MDI workspace.

Here IDR MAIN FRAME is the menu system (resource) that will appear if there are no documents open.

IDR_MULTITYPE is the menu system that appears if one or more documents are open.

Reading keystrokes in multiview:

```
//MulfiViewDoc.h
  :
public:
CString string Data;
  :
//Initilize the string
{
CMultiview Doc:: CMultiview Doc()
String Data =* " ";
}
//To store our document data on disk, use the
//Serialize() method:
void CMulti ViewDoc:; Serialize (CArchieve & ar)
{
    if(ar.IsString ())
    ar << String Data;
}
else
{
    ar >> StringData;
}
}
    //To display data add code:
    void CMultiview:: OnDraw (CDC * pDC
{
    CMuIti View Doc *pDoc = GetDocument();
```

```
    ASSERT_VALID (pDoc);
    pDC → TextOut (0,0, pDoc -> StringData):
}
void CMuIti View View:: OnChar (UINT nChar, UINT nRepcut, UINTnFlags)
{
    CMultiview Doc * pDoc = GetDocument();
    ASSERT-VALID (pDoc);
    pDoc → String Data + = n char;
    Invalidate ();
    pDoc → Update All Views (this, OL, NULL);
    //This is a pointer to the view i.e., calling the method
    pDoc → SetModified FIag();    //If the user tries to close
    // the current document without saving it, the
    //program will ask if they want to save it before the data
       in it is lost.
    CView:: OnChar CnChar, nRepCut, nFlags);
}
```

Now run the program.

Q.14. Write a VC++ program to create Active × Control. Also discuss the steps to create active X control.

Ans. ActiveX controls, are the controls that are specially designed to be embedded in programs and especially in web pages that browser can display.

Creating ActiveX control mean we will be able to place those controls in other program.

Boxer ActiveX Control: In this, small rectangle divided into four sections when we click one of them, we will shade that section in blow.

Start VC++ to create boxer program. Select MFC ActiveX control wizard and then accept all the defaults by pressing the finish button and allow contrbl wizard to create our new ActiveX control.

Drawing the ActiveX Control:

1. Out ActiveX control's on Draw( ) method looks like.

```
Void (Boxer Ctrl: = On Draw (CDC * pdc, const (Ret & rC Bonds, Const
Put & xlnvalid)
{
    pdc → Fill Rect (xBounds, CBrush:: From Header
    (HBRUSH) Get Stock Object (WHITE-BRUSH)));
    pdc → Ellipse (x Bonds);
}
```

2. Now, name those rectangles and declare then.

```
Class (Boxer ctrl: public cole control
}
protected:
CBoxerCtrII( )
→ (Rect box 1;
(Rect box 2; box 3; box 4;
before fill 1, fill 2, fill 3, fill 4;          }
```

3. Now write code for OnDraw( )

```
Void (Boxer Ctrl:: On Draw (CDC -> pdc, Count CRet & rCBonds, Cont
CRect & rCInvalid)
(rCBounds.left; xBounds.top, xBonds, right/2, xBounds.botton/2);
```

```
box 2 = (Rect (xBonds.left, rCBounds, bottom/2,
xBounds.right/2, xBounds.button);
box 3 = (Rect (xBounds.right/2 xBonds, top, xBonds.right, xBonds
bottom/2);
box 4 = (Rect (xBounds.right/2, xBonds.bottom/2, xBounds.right
xBounds.botton);
pdc → Rect angle (& box 1);
pdc → Rectangle (& box 2);
pdc → Rectangle (& box 3);
pdc → Rectangle (& box 4);
if (fill 1) pdc ~> Fill select Rect (& box 1, RGB (0,0,0));
if (fill 2) pdc -> Fill solid rent (& box 2, RGB (0,0,0));
if (fill 3) pdc ~> Fill solid rect (& box 3, RGB (0,0,0));
if (fill 4) pdc -» Fill solid rect (& box 4, RGB (0,0/0));
}
```

Add and Event Handler to an ActiveX Control:

1. Use class wizard to connect WM-UButton Down message to our control so that we can handle mouse clicks.
2. Code for On Button Down( ) method:

```
Void CBoxer Ctrl:: On CButton Down (UINT, nplays, CPoint point)
{
    fill I = box 1 .Ptln Rect (point);
    fill 2 = box 2. Ptlnrect (point);
    fill 3 = box 3.PtInRect (point);
    fill 4 = box 4.Pt InRect (point);
    invalidate();
}
```

After setting the boolean flags to indicate which rectangle the user checked, we call lnvalidate( ) to reduce the view. This calls on Draw( ) and there all we need to do check the four boolean flags and fill the corresponding rectangle using the CDC method Fill Solid Rect( ).

Testing an ActiveX Control:

1. To test, start by selecting the Build Boxer. OCx menu item in the Build Menu to create boxer. OCx and register that control with windows.
2. Next, select ActiveX control Test container item in Tools Menu, which brings up the very useful test container.
3. Now select the Insert New Control item in the test container's Edit Menu and double click the Boxer control in the Insert control box that appears.
4. You can click one of the boxes to the control box now. Click another rectangle shades that rectangle instead.

Q.15. Discuss database connectivity in Visual Basic.

Ans. There are number of steps which are given below.

1. Start visual basic and create a new project and select standard exe project.
2. Go to she project hence and select the component option and then components window will appear.
3. From this component window, select microsoft remote data control 6.0 option and click apply and then OK button.
4. A small iron of remote data control will appear in the toolbox with MSRDC as tooltip text.
5. Select this and drop it on the form.
6. Right click the MSRDC on the form and select properties.

7. Then property page dialog box will appear.
8. Select date source name from the combo box option and click OK button after giving set query in sql text box.
9. Now create GUI with three labels and three textbox and also add button.
10. Select Text Box 1 and right click it and choose properties and then click on Data source and choose MSRDC 1 and then click one data field and choose option name for text box 1.
11. Similarly, for text box 2 and text box 3.

Code:

```
Private sub first-click ()
MSRDC! ResultSet.Move First
End Sub.
Private sub Next-click()
MSRDC1 .Resultset. Move Next
If (MSRDC 1. Result Set.EOF = True) then
MSRDC 1.Resultset. Move hart
Message Box ("This is last record")
Endif
End sub.
```

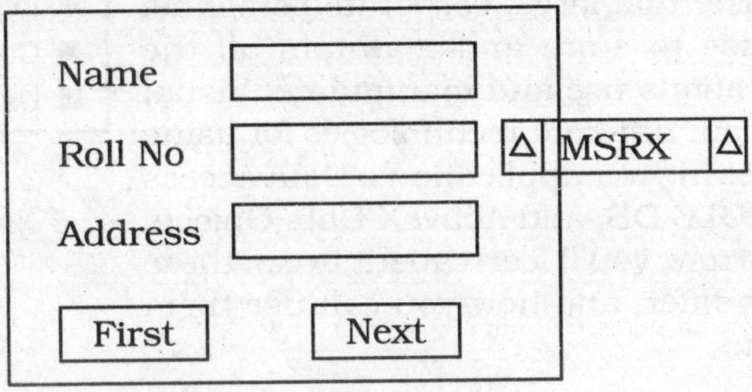

Fig. 7.15.

EXERCISES

1. What are the possible return codes that your application might receive from the MessageBox function call when you specify the MB_RETRYCANCEL button combina-tion?
2. What are the common dialogs that are built into the Windows operating systems that are defined as MFC classes?
3. What is the difference between a modal dialog and a modeless dialog?
4. How can you display a File Save dialog for the user instead of the File Open dia-log that you did have in your application?
5. Why did you not need to create any functions and add any code to your custom dialog?
6. Modify your application so that it includes the directory with the filename in the application. (Hint: The GetFileName function returns the path and filename that was selected in the File Open dialog.)

8

Database Connectivity

8.0. INTRODUCTION

A large number of applications use a database. Everything from a personal organiser to a large, corporate personnel system uses a database to store and maintain all the records that the applications use and manipulate. Visual C++ provides you with four different technologies for using and accessing databases in your applications, Data Access Objects (DAO), ODBC, OLE DB, and ActiveX Data Objects (ADO). Today and tomorrow, you'll learn about two of these technologies, how they differ, and how you can use them in your own applications.

Chapter Outline

- Introduction
- Connecting To Data Base
- Using Visual Base

8.1. CONNECTING TO DATA BASE **(V.Imp)**

8.1.1. Creating a Table

In order to use this lesson, you should be familiar with Microsoft SQL Server as we have covered it already. We will use SQL to create our table for data entry.

1. Start Microsoft SQL Server and open the SQL Query Analyser.
2. In the empty query window, type and execute the following command (you execute the command by pressing F5) :
   ```
   CREATE DATABASE LiquorStore
   ```
3. After database has been created and you receive confirmation, delete the previous line.
4. Type and execute the following instructions that create a table and enters four records :
   ```
   USE LiquorStore
   CREATE TABLE Employees
   (
      EmployeeID  INT IDENTITY(1, 1) NOT NULL,
   ```

```
    EmployeeNo  CHAR(6)   NULL,
    DateHired   VARCHAR(40) NULL,
    FirstName   VARCHAR(20) NULL,
    MI     CHAR(1)   NULL,
    LastName   VARCHAR(20) NULL,
    Address    VARCHAR(40) NULL,
    City     VARCHAR(32) NULL,
    State     CHAR(2)    NULL,
    ZIPCode    VARCHAR(12) NULL,
    Country    VARCHAR(30) NULL,
    Salary    MONEY    NULL,
    HomePhone    VARCHAR(16) NULL,
    EmailAddress VARCHAR(40) NULL,
    MaritalStatus BIT     NULL,
    Notes  Text   NULL
)
/* Data Entry */
-- First Record--
INSERT INTO Employees(EmployeeNo, DateHired, FirstName, MI, LastName,
Address, City, State, ZIPCode, Country, Salary, HomePhone,
EmailAddress, MaritalStatus)
   VALUES('GT-882','10/05/1995','Geraldine','F','Thomas','802 Epsilon
   Ave',
      'Silver Spring','MD','20904','USA',12.55,'(301) 524-
      7822','gthomas@hotmail.com',1)
-- Second Record --
INSERT Employees(EmployeeNo, DateHired, FirstName, MI, LastName,
Address, City, State, ZIPCode, Country, Salary, HomePhone,
EmailAddress, MaritalStatus)
   VALUES('EB-405','8-22-1996', 'Ernest','','Bilong','1060 Calisto
   Rd #D12',
      'Alexandria','VA','22231-1244','USA',8.72,'(703) 276-
      8676','ebilong@yahoo.com',0)
-- Third Record --
INSERT Employees(EmployeeNo, DateHired, FirstName, MI, LastName,
Address, City, State, ZIPCode, Country, Salary, HomePhone,
EmailAddress, MaritalStatus)
   V A L U E S ( ' D J - 6 1 4 ' , 8 / 2 2 /
   1996,'James','D','Datts','','','DC','','USA',10.18,
      '','dattsj@netscape.com',1)
-- Fourth Record --
INSERT Employees(EmployeeNo, DateHired, FirstName, MI, LastName,
Address, City, State, ZIPCode, Country, Salary, HomePhone,
EmailAddress, MaritalStatus)
```

```
VALUES('BC-200',4/15/1998,'Catherine','','Bollack','12442 Lockwood
Drive',
'Rockville','MD','','USA',10.44,'','bollackc1288@csumd.edu',1)
```

5. Close SQL Query Analyzer.
6. When asked whether you want to save the text, click Yes.
7. Change the name of the file to LiquorStore and make sure that a desired folder, such as My Documents, is selected in the Save In combo box. Click Save.

8.1.2. Creating an ODBC Data Source **(V.Imp)**

Here are the steps to create the needed data source :

1. Start or open **Control Panel** and, in Control Panel, double-click **Administrative Tools**.
2. In the Administrative Tools window, double-click **Data Sources (ODBC)**. Alternatively, from the Taskbar, you could have clicked Start -> Programs -> Administrative Tools -> Data Sources (ODBC).
3. In the ODBC Data Source Administrator property sheet, click the **Add** button :

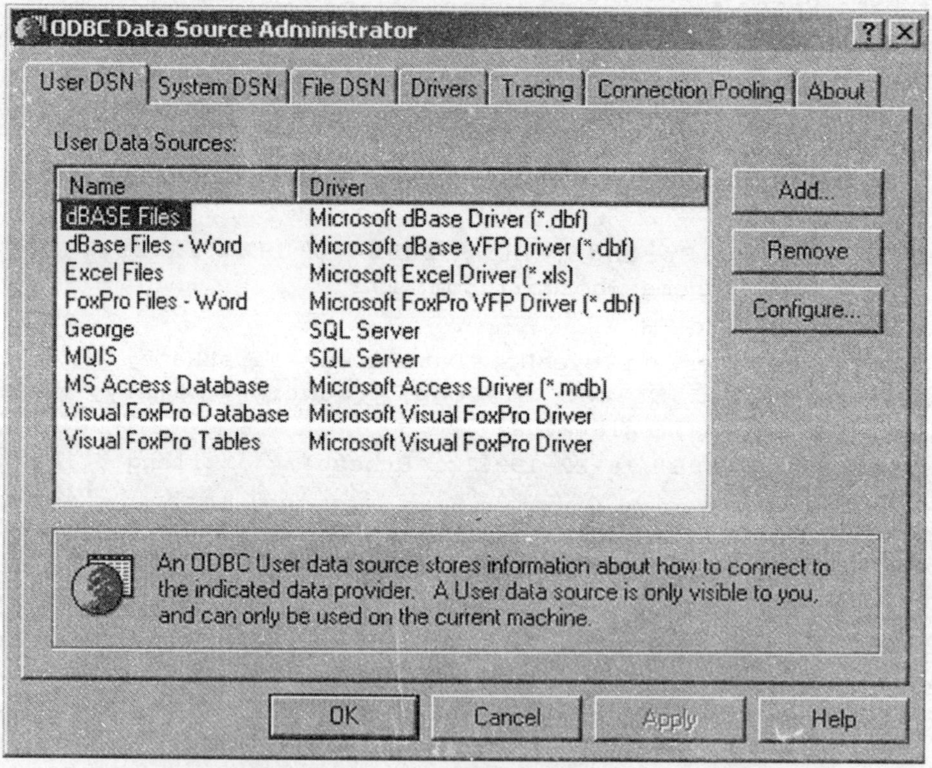

Fig. 8.1.

4. In the **Create New Data Source** wizard, scroll down in the list box and click **SQL Server** :

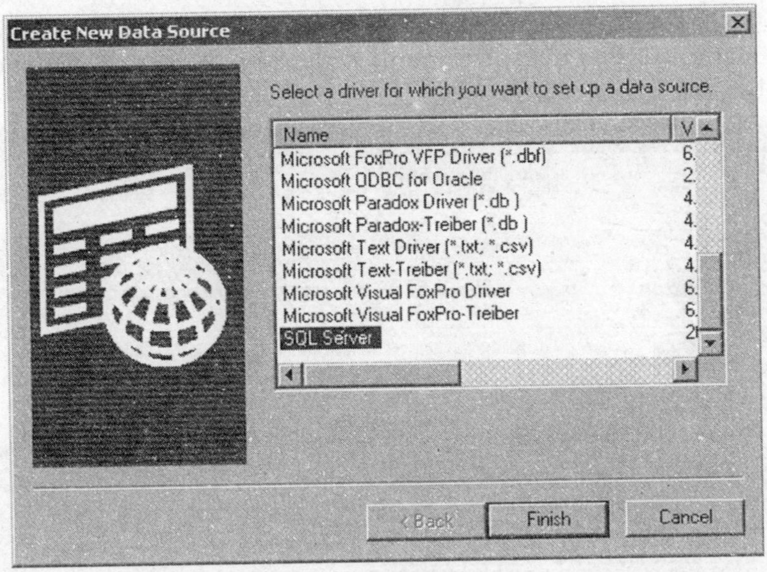

Fig. 8.2.

5. Click Finish.
6. In the Name edit box of the Create A New Data Source To SQL Server wizard, type LiqStore and press Tab.
7. In the Description edit box, type The Liquor Store Application.
8. Click the arrow of the Server combo box and select the server where the above (LiquorStore) database resides :

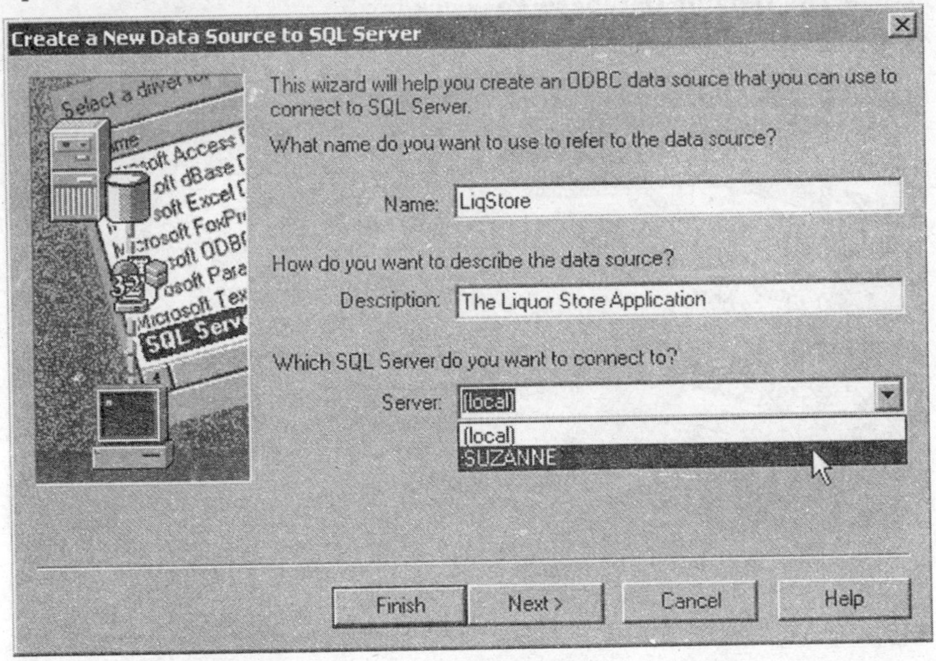

Fig. 8.3.

9. Click Next.

10. Select the desired authentication. For this example, I accept the Windows NT Authentication. I also accept to connect to SQL Server with Default Settings by leaving the check box untouched :

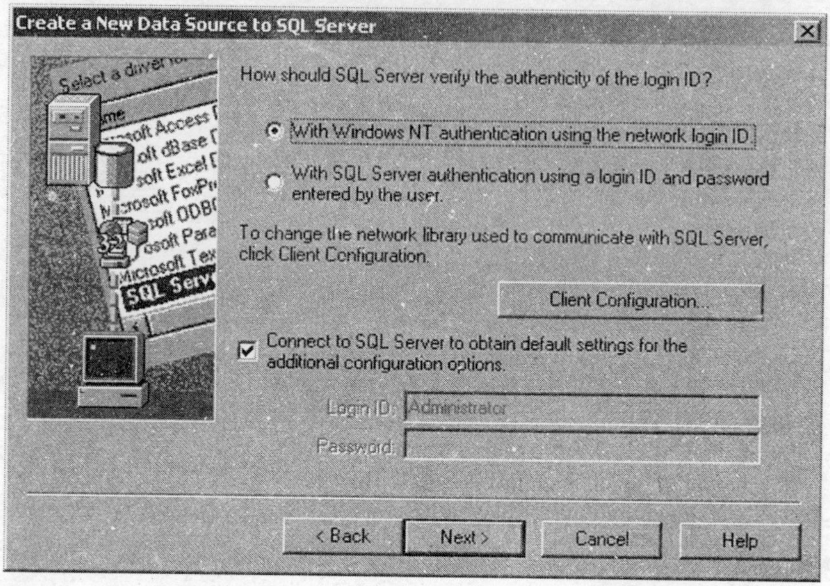

Fig. 8.4.

11. Click Next.
12. In the new page of the Create A New Data Source To SQL Server wizard, click the **Change The Default Database To** check box.
13. Once the combo box is enabled, click its arrow and select **LiquorStore** :

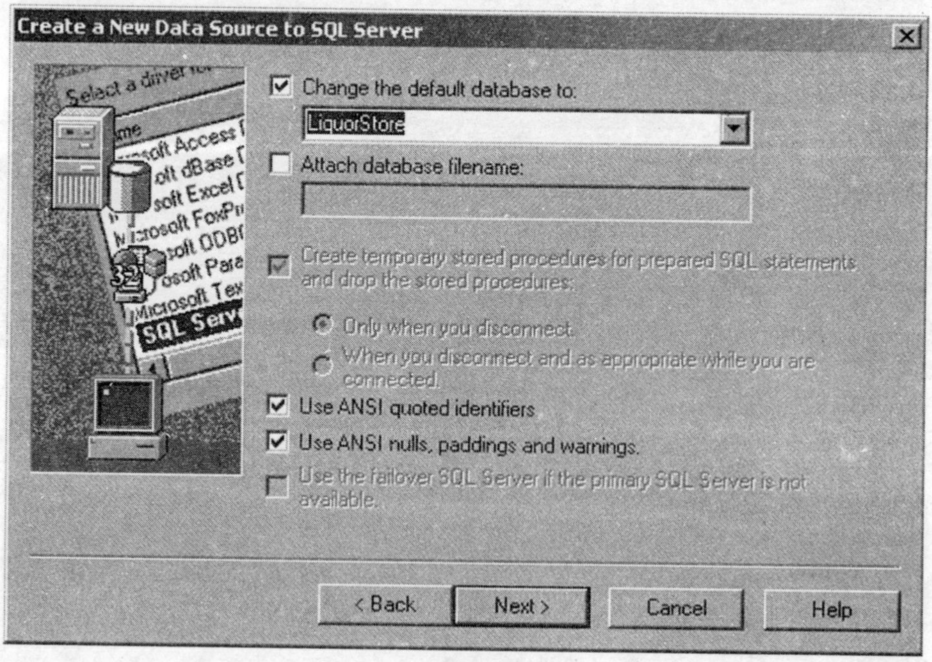

Fig. 8.5.

14. For this example, leave the Attach Database FileName check box unchecked and accept the other defaults. Click Next.
15. Unless you have any other reason, accept the default (language, encryption, etc) settings of the new page :

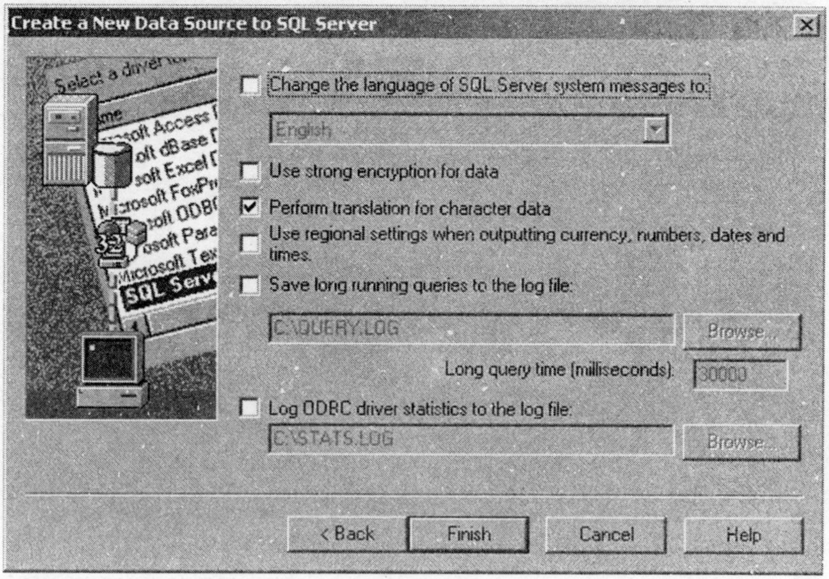

Fig. 8.6.

16. Click Finish.
17. On the ODBC Microsoft SQL Server Setup dialog box, click the Test Data Source button :

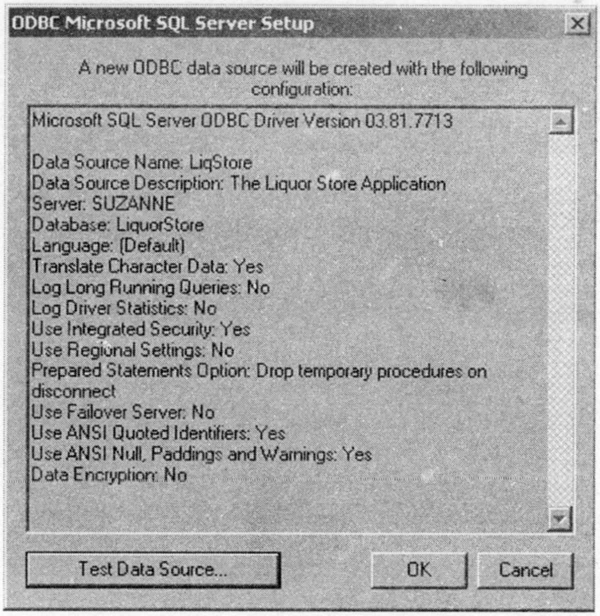

Fig. 8.7.

18. When the SQL Server ODBC Data Source Test dialog box confirms that the TESTS COMPLETED SUCCESSFULLY!, click OK :

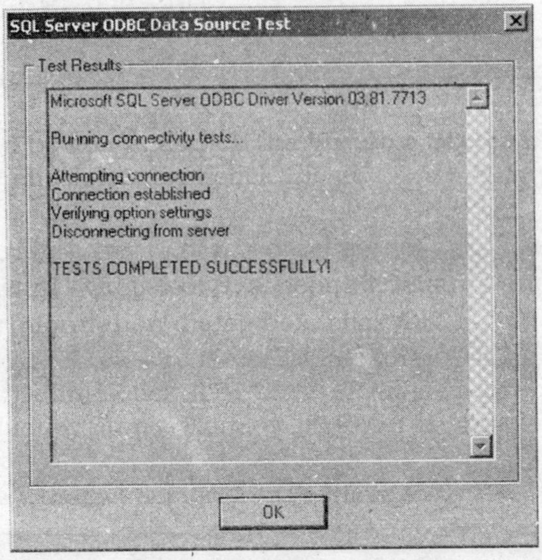

Fig. 8.8.

19. On the ODBC Microsoft SQL Server Setup dialog box, click OK.
20. In the User DSN property page of the ODBC Data Source Administrator dialog box, make sure that the LiqStore data store is listed in the User Data Sources list box :

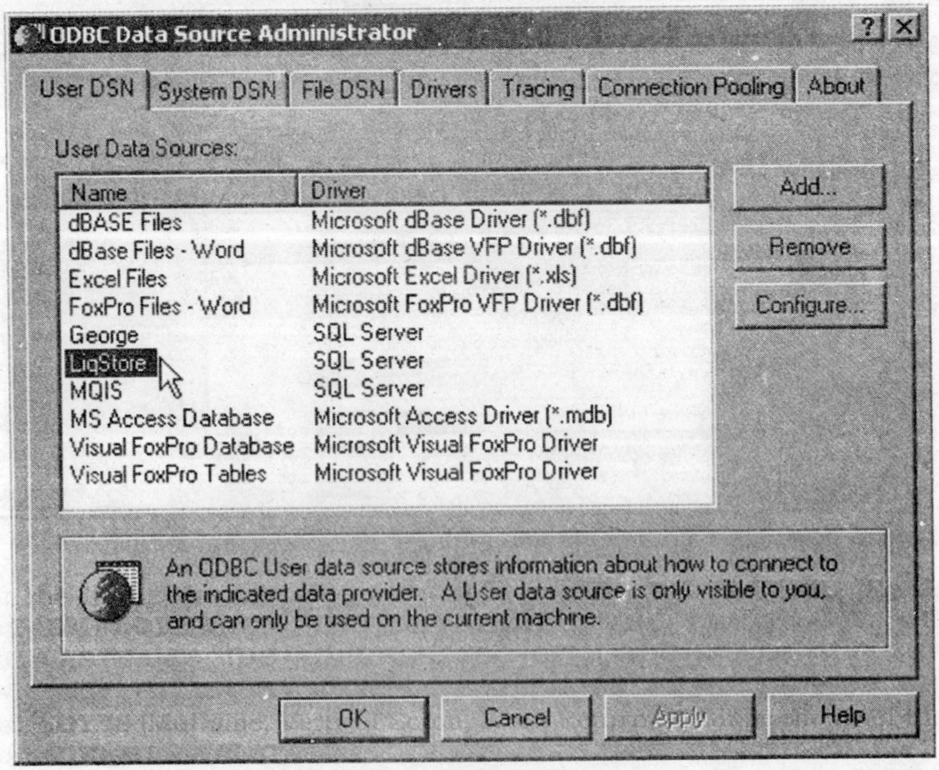

Fig. 8.9.

21. Click OK.

8.1.3. Creating the Application

1. Start Microsoft Visual C++.
2. On the main menu, click File -> New.
3. In the New dialog box, click Projects and click **MFC AppWizard (exe)**.
4. In the Location edit box, select or type the desired folder for the application. In the Project Name edit box, type **LiqStoreDB** and click OK.
5. In the MFC AppWizard - Step 1 dialog box, click the **Single Document** radio button and click **Next.**
6. In the MFC AppWizard - Step 2 of 6, click the **Database View Without File Support** (because we will not need and will not use file support for this exercise).
7. Click the **Data Source** button.
8. In the Database Options dialog box, in the **Datasource** section, make sure the **ODBC** radio button is selected and click the arrow of its combo box. Select **LiqStore.**
9. In the **Recordset Type** section, click the **Dynaset** radio button.

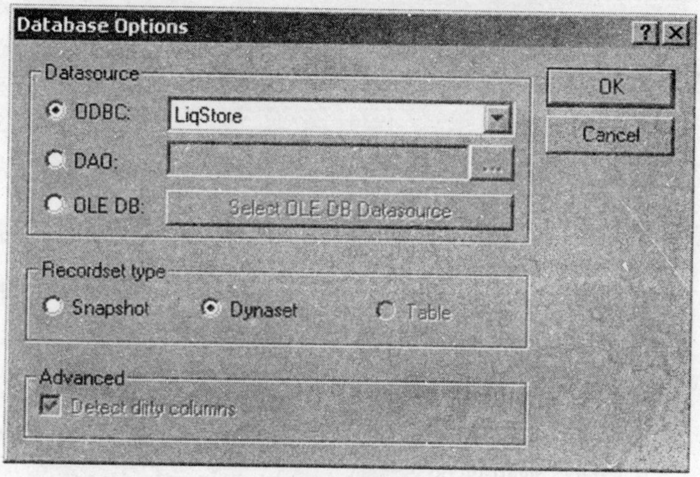

Fig. 8.10.

10. Click OK.

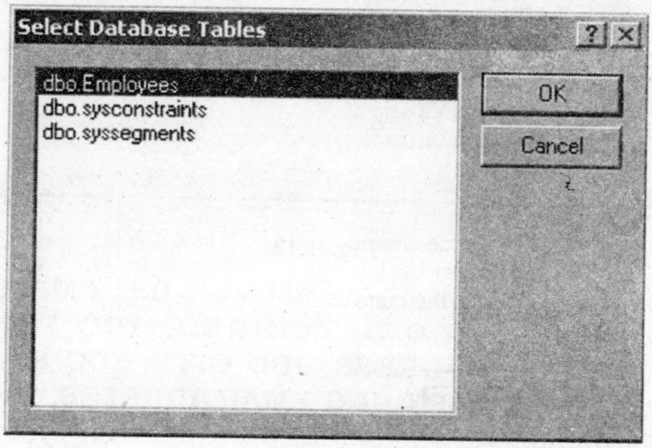

Fig. 8.11.

11. In the Select Database Tables, click dbo.Employees and click OK.
12. In the MFC AppWizard - Step 2 of 6, click Next.
13. In the MFC AppWizard - Step 3 of 6, accept all defaults and click Next.
14. In the MFC AppWizard - Step 4 of 6, uncheck the **Printing And Print Preview** check box.
15. Click the **Advanced** button and click the **Window Styles** property page. Uncheck the **Maximize Box** check box and click Close.
16. Click Next.
17. In the MFC AppWizard - Step 5 of 6, accept all defaults and click Next.
18. In the MFC AppWizard - Step 6 of 6, accept all defaults and click Finish.
19. In the New Project Information dialog box, click OK.
20. In the Workspace, click the ResourceView tab, expand the String Table, and double-click String Table. Double-click IDR_MAINFRAME and replace its entire Caption with Four-Corner Liquor Store - Employees Records and close the floating window. Also close the String Table window.
21. Using the Controls window, design form as follows :

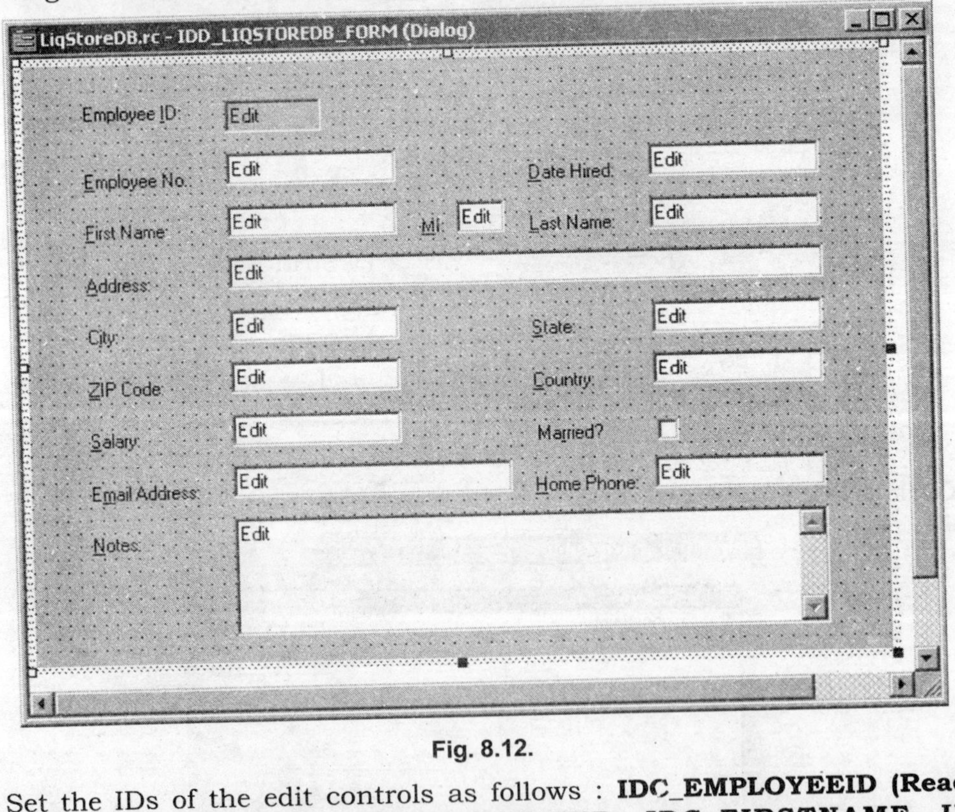

Fig. 8.12.

22. Set the IDs of the edit controls as follows : **IDC_EMPLOYEEID (Read-Only), IDC_EMPLOYEENBR, IDC_DATEHIRED, IDC_FIRSTNAME, IDC_MI, IDC_LASTNAME, IDC_ADDRESS, IDC_CITY, IDC_STATE, IDC_ZIPCODE, IDC_COUNTRY, IDC_SALARY, IDC_EMAILADDRESS, IDC_HOMEPHONE,** and **IDC_NOTES** (Multiline, Vertical Scroll, Want Return). Change the ID of the check box to **IDC_MARITALSTATUS**.
23. Save everything.

24. In the Workspace, click ClassView tab. Expand the LiqStoreDB classes node and double-click CLiqStoreDBSet. Notice that the CRecordset class already has variables declared for each control of our form.
25. To associate each control with the appropriate variable, on the main menu, click View -> ClassWizard...
26. In the MFC ClassWizard dialog box, click the Member Variables property sheet. In the Class Name combo box, select CLiqStoreDBView.
27. In the list box, double-click IDC_ADDRESS.
28. In the Add Member Variable dialog box, click the arrow of the **Member Variable Name** combo box and select m_pSet->m_Address :

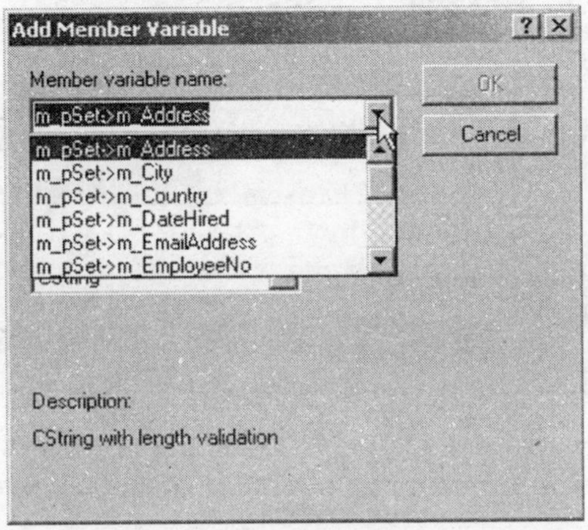

Fig. 8.13.

29. Make sure that the **Variable Type** is set to CString and click OK.
30. Do the same for the other controls :

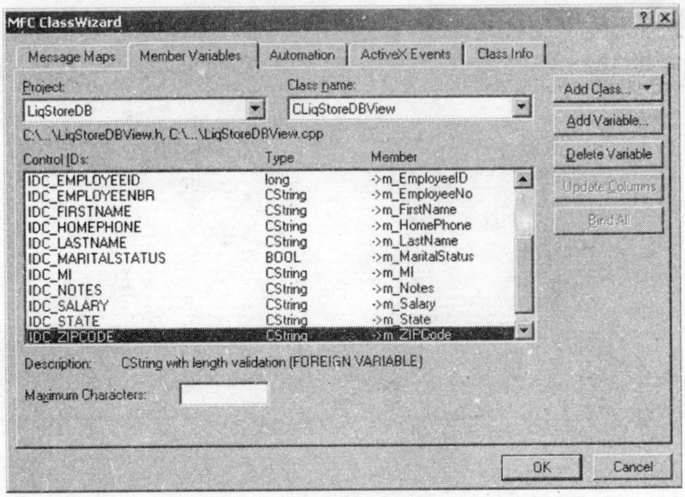

Fig. 8.14.

31. On the MFC ClassWizard, click OK.
32. Test the application.

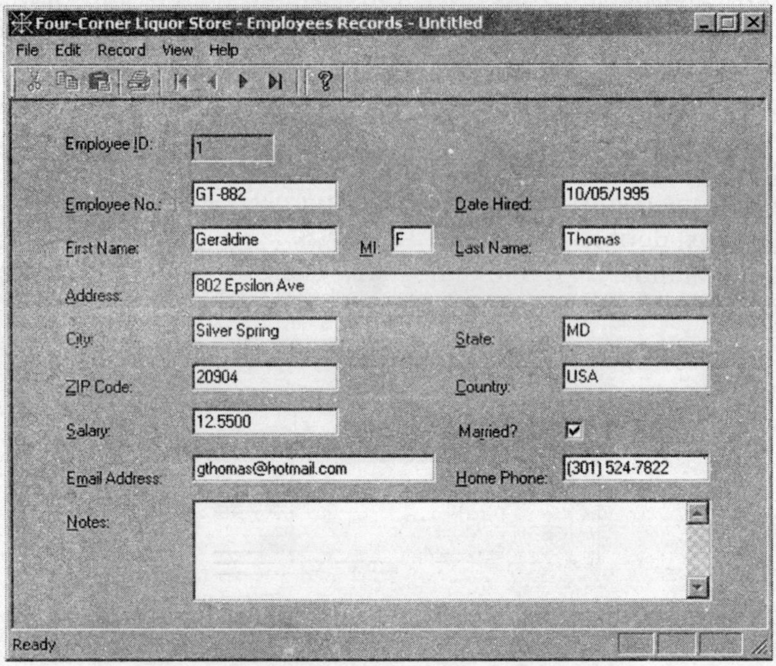

Fig. 8.15.

33. Close the application as you can see, this database was created without a single line of code.

8.1.4. Adding a Record

1. In the Workspace, click the ResourceView tab. Expand the LiqStoreDB Resource tree and expand the Menu node.
2. Double-click IDR_MAINFRAME and add an Add menu item :

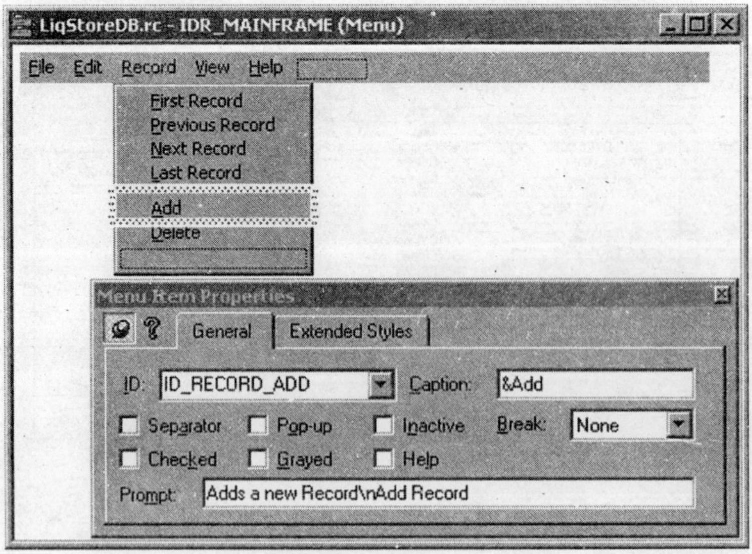

Fig. 8.16.

3. Add a button on the toolbar and associate to it the same identifier as the Add menu item :

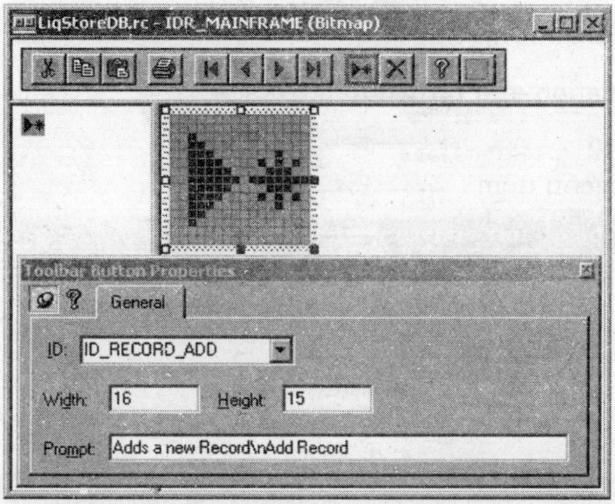

Fig. 8.17.

4. Press Ctrl + W to access the ClassWizard.
5. In the Class Name combo box, select CLiqStoreDBView and, in the Object IDs list box, select ID_RECORD_ADD.
6. In the Messages list box, double-click COMMAND.
7. Accept the suggested name of the function and click Edit Code
8. Implement the function as follows :

```
void CLiqStoreDBView::OnRecordAdd()
{
    SetDlgItemText(IDC_EMPLOYEEID, "");
    SetDlgItemText(IDC_EMPLOYEENBR, "");
    SetDlgItemText(IDC_DATEHIRED, "");
    SetDlgItemText(IDC_FIRSTNAME, "");
    SetDlgItemText(IDC_MI, "");
    SetDlgItemText(IDC_LASTNAME, "");
    SetDlgItemText(IDC_ADDRESS, "");
    SetDlgItemText(IDC_CITY, "");
    SetDlgItemText(IDC_STATE, "");
    SetDlgItemText(IDC_ZIPCODE, "");
    SetDlgItemText(IDC_COUNTRY, "");
    SetDlgItemText(IDC_SALARY, "");
    SetDlgItemText(IDC_EMAILADDRESS, "");
    SetDlgItemText(IDC_HOMEPHONE, "");
    SetDlgItemText(IDC_NOTES, "");

    reinterpret_cast<CButton *>(GetDlgItem(IDC_MARITALSTATUS))-
    >SetCheck(FALSE);
```

```
        m_pSet->AddNew();
        m_pSet->Update();
        m_pSet->Requery();
    }
```

9. Test the application and return to MSVC.

8.1.5. Deleting a Record

1. Add a Delete menu item :

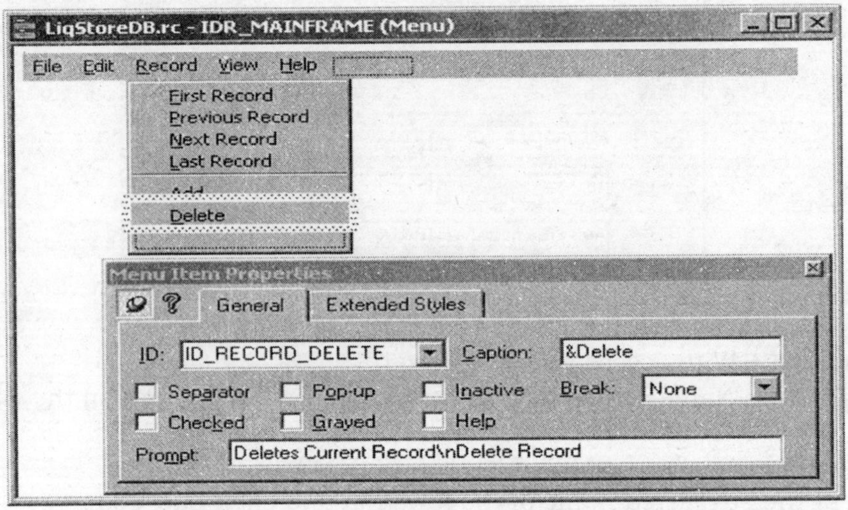

Fig. 8.18.

2. On the toolbar, add a button that has the same identifier as the Delete menu item :

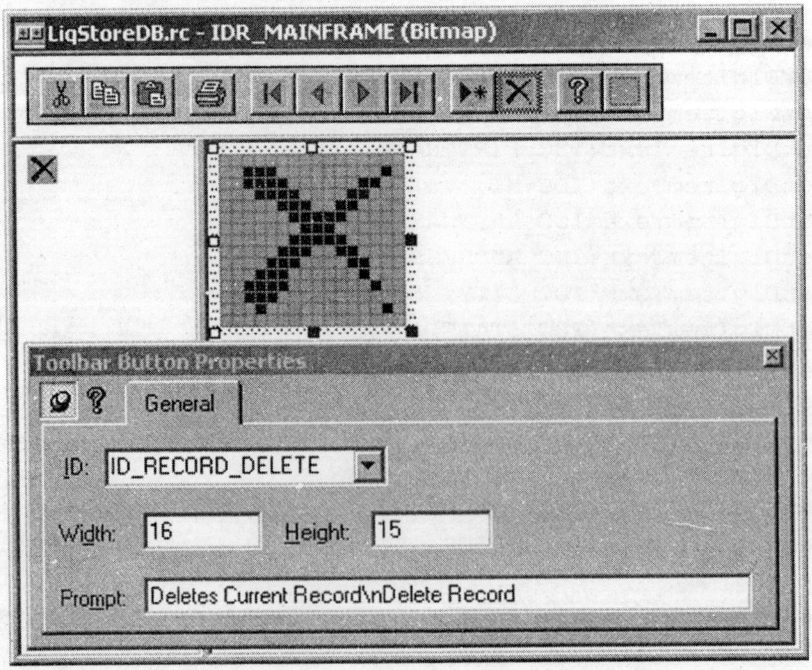

Fig. 8.19.

3. Implement its COMMAND function as follows :

```
void CLiqStoreDBView::OnRecordDelete()
{
    m_pSet->Delete();
    m_pSet->MoveNext();
    if( m_pSet->IsEOF() )
        m_pSet->MoveLast();
}
```

4. Test the application.

8.2. USING VISUAL BASIC

The latest release of Visual Basic has eased the Database applications development. Visual Basic 6.0 supports many data access methods. Besides using the traditional DAO (Data Access Objects) you can also use Remote Data Object (RDO) and ActiveX Data Object (ADO). It provides a built in access to Jet Database engine which is used in Microsoft Access. You can connect your databases to ISAM databases and even Oracle databases using ODBC. This unit covers the DAO. At the end of this unit you will be able to create database applications using data-aware controls Data controls and even using Data access objects.

8.2.1. Database Management **(V.Imp)**

A database can be viewed as a set of interrelated data. Database management deals with the storage, retrieval and editing of the data within the database. One of the most important functions of Database management is the Database security and Database validations.

Visual Basic makes it simple by supplying a variety of tools and approaches for data management. For significant database applications, utilising database functionality that already exists becomes necessary. Visual Basic 6.0 provides full-fledged database engine called Jet. This engine contains virtually all the database functions that applications may need. The simplest and easiest way to access the Jet database engine is through bound data control.

Data Access Object (DAO) has been the standard database access method for Visual Basic programmers. DAO was the first database interface included with Visual Basic in its third version.

Remote Data Object (RDO) was then introduced to make data-access more flexible and functional in VB 4. The most recent release by Microsoft with Visual Basic is the ActiveX data object, which is in a short while going to take over all the other methods of data access.

8.2.2. Data Access Object **(MDU 2008-09)**

Data Access Object enables you to use a programming language to access and manipulate data in local or remote databases and to manage databases, their objects and their structure. DAO utilises Microsoft's Jet database engine and can be used to access a wide variety of database formats, including, ISAM (Indexed Sequential Access Method) and Text.

In addition to this, DAO can also make use of Open Database Connectivity (ODBC) databases through its ODBC Direct Object Model.

Generally ODBC is a special technique to access a database residing on a separate Server such as Informix, Oracle etc.

The DAO object model has 17 types of objects like Databases, RecordSets, Fields etc. Databases are accessed by opening an existing Database object or creating a new one. The individual fields in a table make up a Recordset's Field object and table of a database is held in a Recordset Object.

Data Access object is implemented through an ActiveX control called the Data Control. This allows a large amount of DAO's power to be utilised without having to add a single line of code. When used in conjunction with Data aware controls such as TextBox, DBList, DbCombo, DbGrid etc, the Data control becomes very powerful.

Information from a database table is fetched by the DAO and then supplied to various Data aware controls like TextBox, DBListBox etc.

Databases accessible through the DB Engine are FoxPro 2.0 to 3.0 and VFP, dBase III, IV and 5.0, Paradox 3x, 4x and 5x, Microsoft Access 1x to 3x, ODBC databases, Microsoft Excel 3.0 to 8.0, Lotus 1-2-3, WKS, WK1, WK3 and WK4, ASCII tabular Format, HTML Tabular Data and Microsoft Exchange 4.0 The DBEngine consists of several Dynamic Link Libraries that are linked to your Visual Basic program at runtime. DBEngine is controlled by our application through DBEngine's object classes, their methods and properties.

8.2.3. Data Binding **(MDU 2007-08)**

In order to display the values of the columns accessed by the Data Control, some controls are required which support the data control. These control are called Data Bound controls and are simpler to use e.g., TextBox, Label DBGrid, DBList etc.

These controls have basically two standard properties other than their regular properties. The DataSource property enables you to specify the name of the Data Control to which the client control will bind. The name of the Data Control(s) available on your form is already present in the dropped down Combo Box of this property in the property window.

The Datafield property, a string property of the Client Control, specifies which of the Database field names available from the data control the client will bind.

After the DataSource and Datafield properties are set, the Data Control is properly initialized, the Client Control automatically displays the contents of the bound field as the data control is used to navigate though the recordset. You can then use an appropriate editable control, such as the textox, to change the Data contained in the Bound field. DBCombo and DBlist enable you to use the data control to provide a list of valid choices for a field, as well as to bind the client control to a specific field.

8.2.4. Data Control and The Data Bound Controls

Visual Basic provides many sources to connect to a data source. Data Control is one of them. Data Control was the first medium to connect the applications to a database in the third version of Visual Basic.

The Data control is an excellent tool for rapidly developing a database information retrieval system. With very little code you can get a useful view into stored information. This is practically a drop-in-place process where by you can place a data control, A DBGrid and a Command Button on a form and be able to access your database in a matter of minutes. After you set properties your application is set to go.

8.2.5. Data Control **(MDU 2007)**

Data Control is one of the easiest modes to access databases. You can perform most data access operations using the Data control without writing any code at all. Data-aware controls bound to a Data control automatically display data from one or more fields for the current record or, in some cases, for a set of records on either side of the current record. The Data control performs all operations on the current record.

If the Data control is instructed to move to a different record, all bound controls automatically pass any changes to the Data control to be saved in the database. The Data control then moves to the requested record and passes back data from the current record to the bound controls when it's displayed.

The Data control is itself smart enough to handle a number of contingencies including empty recordsets, adding new records, editing and updating existing records, and handling some types of errors. However, in more sophisticated applications, you need to trap some error conditions that the Data control can't handle. For example, if the Microsoft Jet database engine has a problem accessing the database file, doesn't have permission, or can't execute the query as coded, a trappable error results. If the error occurs before your application procedures start or due to some internal errors, the Errors event is triggered.

8.2.6. Data Control in Action

Using the Data Control

The data control is an intrinsic control used to bind - that is, create a connection- between a database and control such as text boxes. A database is sent to data control *via* its properties. Then individual control is bound to data control.

To add a data control to a form :

1. With a form open, double-click the data control in the toolbox (Figure 9.1). The data control is placed on the form (Fig. 8.20.).

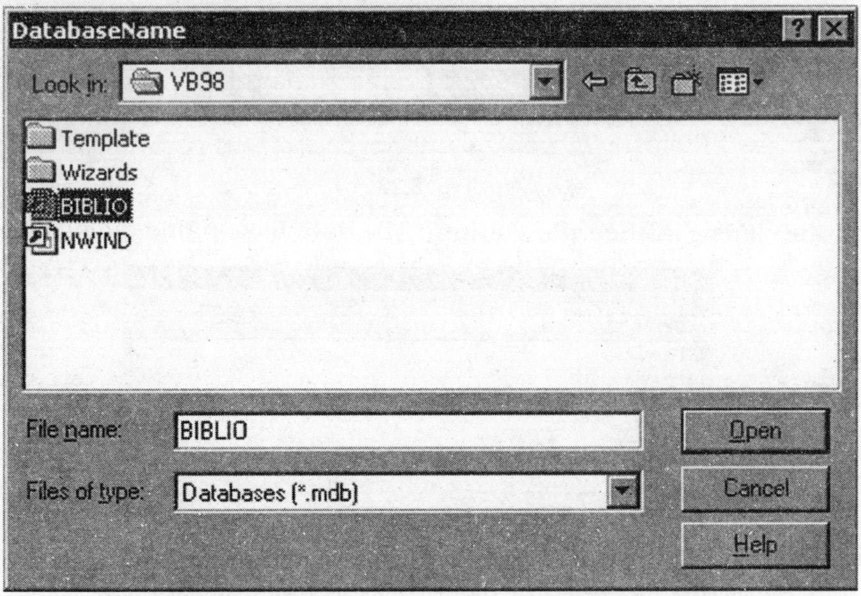

Fig. 8.20.

Fig. 8.21.

2. Using the properties window, change the caption of the data control to something descriptive, such as click here to control data display (Fig. 8.22).

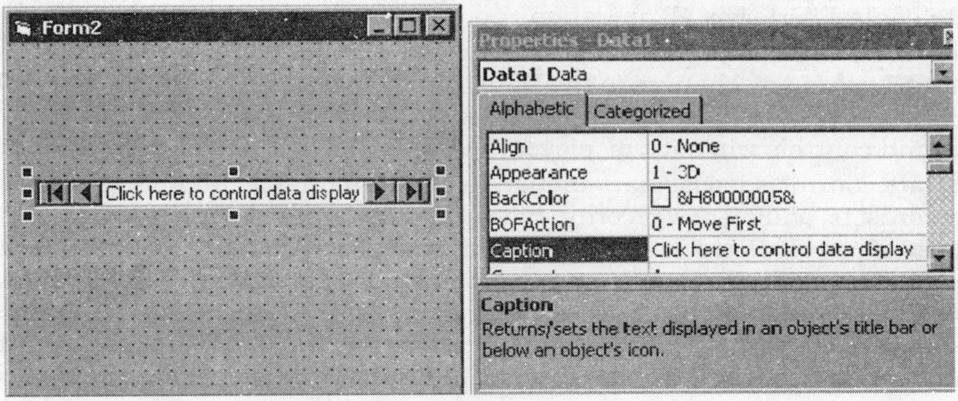

Fig. 8.22.

Tip :

Figure shows navigation buttons that are build into the data control.

To connect the data control to a base :

1. With the data control open in the properties windows, select the data base name properties in the left column (Fig. 8.23.)

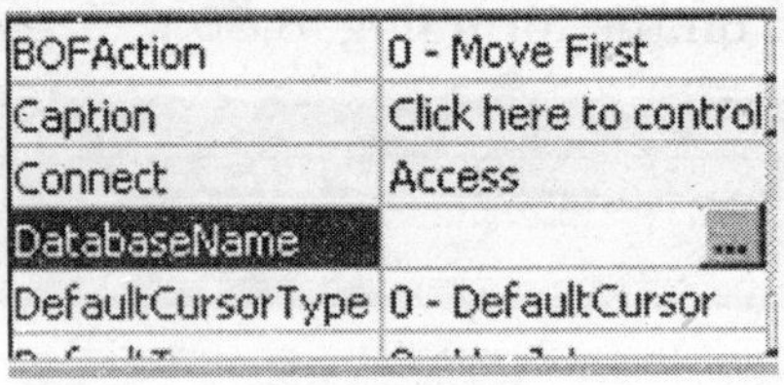

Fig. 8.23.

2. Click the ellipsis in the right column. The data base name dialog open (Fig. 8.24.)

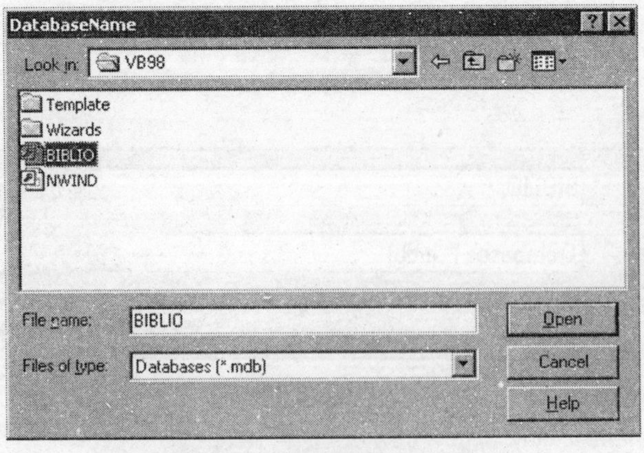

Fig. 8.24.

3. From your file system, select the database.
4. Click open.

BOFAction	0 - Move First
Caption	Click here to control data display
Connect	Access
DatabaseName	C:\Program Files\Microsoft Visual Studio\VB98\BIBLIO.MDB
DefaultCursorType	0 - DefaultCursor

Fig. 8.25.

The data base file now appears in the right column of the properties windows (Fig. 8.25.). The sample biblio.mdb database is shown in the Fig.

Tips :

- Visual basic is often used with Microsoft access databases. Access databases file have the .mdb suffix.
- Visual basic ships two sample access databases- biblio.mdb and Nwind.mdb- for you to experiment with. You can find these databases in your visual basic home directory.
- Data control properties assignments can also be made in code. Biblio consists of publisher, author, and book table. Nwind is an inventory database of fictional food-trading company.

End of File and Beginning of File Actions

The Data control can also manage what happens when you encounter a Recordset with no records. By changing the EOFAction property, you can program the Data control to enter Add new mode automatically. These properties have been discussed later.

Alignment

You can program the Data control to automatically snap to the top or bottom of its parent form by using the Align property. In either case, the Data control is resized horizontally to fill the width of its parent form whenever the parent form is resized. This property allows a Data control to be placed to an MDI form without requiring an enclosing Picture control.

Data Binding

In order to display the values of the columns accessed by the Data control, some controls are required which support the data control. These controls are called Data Bound controls and are simpler to use *x* TextBox, Label, DBGrid, DBList etc.

These controls have basically two standard properties apart from their regular properties. The DataSource property enables you to specify the name of the Data Control to which the client control will bind. The name of the Data Control(s) available on your form is already present in the drop down Combo Box of this property in the property window.

The Datafield property, a string property of the Client Control, specifies which of the Database field names available from the data control, the client will bind.

After the DataSource and Datafield properties are set, the Data Control is properly initialised, the Client Control automatically displays the contents of the bound fields as the data control is used to navigate though the recordset. You can use an appropriate editable control, such as the TextBox, to change the data contained in the bound field. DBCombo and DBList enables you to use the data control to provide a list of valid choices for a field, as well as to bind the client control to a specific field.

8.2.7. Properties of The Data Control

BOFAction and EOFAction Properties

As discussed earlier the BOFAction and EOFAction properties return or set a value indicating what action the data control takes when the BOF or EOF is encountered by Data control.

Syntax :

```
DataControlName.Recordset.BOFAction = value
DataControlName.Recordset.EOFAction = value
```

Different settings for the properties are :

Constant	*Value*	*Description*
VbBOFActionMoveFirst	0	Moves the record pointer back to the first valid record.
VbBOFActionBOF	1	Causes a validation event on the first record, repositions to the first record.

Above given values are for the BOFAction property, values for EOFAction property are:

Constant	*Value*	*Description*
VbEOFActionMoveLast	0	Moves the record pointer back to the last record.
VbEOFActionEOF	1	Causes a validation event on the last record, repositions to the last record.
VbEOFActionAddNew	2	After moving past the last record, triggers an AddNew event after Validation event.

Connect Property

The Connect property sets or returns a value that provides information about the source of an open connection, an open database, a database used in a pass through query, or a linked table. This property is read/write for database objects, new connection objects linked tables and tabledef objects. This property is read only for Querydef objects and basic tables.

Syntax :

```
Object.Connect = databasetype; parameters
```

Database type specifies the type of database. This parameter can be excluded for Microsoft Jet Databases.

Database Property **(V.Imp)**

This property returns a reference to a database connected through a data control.

Syntax :

```
Object.Database
```

The Database object created by the Data control is based on the control's Database Name, Exclusive Read Only and Connect properties. Database objects have properties and methods you can use to manage your data. You can use any method of a Database object with the Database property of a Data control such as close and execute. You can also examine the interval structure of database by using its TableDefs collection and in turn, the Fields, Indexes Collections of individual TableDef Objects.

DatabaseName Property

The DatabaseName Property reads or sets the name and location of the database. Setting the value in the data control's Database Name Property while your program is running, changes the database to which the control is connected. You must use the Refresh method to re open the control after changing databases for a data control.

Syntax :

```
DataControlName.DatabaseName = PathName
```

The following example closes a database referred by DBName variable after confirming from the user.

```
Private Sub OpenDatabaseNew (DBName as String)
    ans = Msgbox("Close" + Data1.DatabaseName+"Now", VBYesNo,"Confirm")
    If ans = VBYes Then Data1.DatabaseName = DBName
    Data1.Refresh
    End If
End Sub
    Exclusive Property
```

The Exclusive property reads or sets whether a database is opened for single user or multiuser access. The data control exclusively reads or writes the database at both design time and run-time, if set to True.

Syntax :

```
Object.Exclusive = True | False
    ReadOnly Property
```

Read Only property returns or sets a value that determines whether the control's Database is opened for read-only access.

Syntax :

```
Object.ReadOnly [ =boolean]
```

The object expression evaluates to a data object and a Boolean expression determines read/write access.

The control's Database object is opened with read-only access when set to True. Changes to data aren't allowed. The control's Database is opened with read/write access to when set False.

You can use the ReadOnly property with a Data control to specify whether data in the underlying Database can be changed. For example, you might create an application that only displays data. Accessing a Database using read-only results in faster access.

For a Data control, this property is used only the first time a database is opened by your application. If your application subsequently opens other instances of the database, the property is ignored. For a change in this property to take effect, you must close all instances of the databases and then use the Refresh method.

Recordset Property

The Recordset property sets or returns a Recordset object defined by a Data control's properties or by an existing Recordset object.

You can use the Recordset property like any other Recordset object. For example, you can use any of the Recordset methods or properties to examine the structure of the Recordset object's underlying schema.

You can also request the type of Recordset to be created by setting the Data control's Recordset Type property. If you don't request a specific type, a dynaset-type Recordset is created. Using the RecordsetType property, you can request to create either a Table -, Snapshot or dynaset type Recordset.

The syntax for the Recordset property is as follows :

```
Set Object.RecordSet = value
```

Object refers to a valid Data control name and value is any object variable containing a reference to a RecordSet object.

RecordSource Property

The RecordSource Property specifies the source of the records accessible through bound controls on your Form.

Syntax :

```
Object.RecordSource = [SQL Query | Table Name | Procedure Name]
```

The object expression evaluates to a Data control. Values can be the name of any table in the database specified by the DatabaseName property of the Data control or any table defined in the database object's TableDef's collection. Value can be also a string variable or a string containing a SQL.query. It can also be the name of any stored procedure in the Database.

If you set the RecordSource property to the name of an existing table in the database, all of the fields in that table are visible to the bound controls attached to the Data control. For table-type recordsets (RecordsetType = vbRSTypeTable), the order of the records retrieved is set by the Index object that you select using the Index property of the Recordset. For dynaset-type and snapshot-type Recordset objects, you can order the records by using an SQL statement with an Order By clause in the RecordSource property of the Data control. Otherwise, the data is returned in no particular order.

If you set the RecordSource property to the name of an existing QueryDef in the database, all fields returned by the QueryDef are visible to the bound controls attached to the Data control. The order of the records retrieved is set by the QueryDef object's query. For example, the QueryDef may include an ORDER BY clause to change the order of the records returned by the Recordset created by the Data control or a WHERE clause to filter the records.

You must use the Refresh method on a Data control to reopen the database if the RecordSource, property settings have changed and rebuild the dynaset in the control's Recordset property.

Example 1

```
Data1.RecordSource = "Select * from Emp where EMPID >=1001"
Data1.Refresh
```

The Refresh Method

The refresh method when applied to a data control, opens or reopens a database specified in the DatabaseName property of Data control.

Syntax :

```
Object.Refresh
```

This refresh method also resets the current record back to the first record or (row) in the table.

Important Events Of The Data Control

Error Event

This event occurs only as the result of a data access error that takes place when no Visual Basic code is being executed. The general form of this event procedure is :

```
Private Sub object_Error([Index As Integer,] dataerr As Integer,
response As Integer)
End Sub
```

Index identifies the control if it's in a control array, contains the index number of the Data control. Dataerr returns the error number and response is a number corresponding to the response you want to take, as described below :

Constant	***Value***	***Description***
DataErrContinue	0	Ignores the error and continues the execution of the program.
VbDataErrDisplay	1	Displays the Error message. This is by default.

Run-time errors can occur when none of your code is running, as when :

- A user clicks a Data control button.
- The Data control automatically opens a database and loads a Recordset object after the Form_Load event.
- A custom control performs an operation such as the MoveNext method, and AddNew method or the Delete method.

If an error results from one of these actions, the Error event occurs.

If an event procedure has not been written for the Error event, Visual Basic itself displays the message associated with the error.

Reposition Event

Reposition event occurs after a record becomes the current record. When a Data control is loaded, the first record in the Recordset object becomes the current record. Whenever a user clicks any button on the Data control, moving from record to record, or if you use one of the Move methods, such as MoveNext, the Find methods, such as FindFirst, or any other property or method that changes the current record, the Reposition event occurs after, when each record becomes current. Delete, Unload, Update are few actions causing this event. The general form of this event is :

```
Private Sub Object.Reposition([Index As Integer])
End Sub
```

Index specifies the index number of control if it is a part of control array. In contrast, the Validate event occurs before moving to a different record.You can use this event to perform calculations based on data in the current record.

Validate Event

The Validate event occurs before the reposition event (or a different record becomes the current record). Validate event procedure has two arguments Index number as integer for the Data control, if it is a part of a Data control array, Save a Boolean expresssion specifying whether bound Data has changed or not and Action an integer specifying operation that causes this event. Values for action are :

Constant	***Value***	***Description***
VbDataActionCancel	0	Cancels the operation when the sub-exits.
VbDataActionMoveNext	3	Caused due to MoveNext method.
VbDataActionMoveLast	4	Caused due to MoveLast method.
VbDataActionAddNew	5	Caused due to AddNew method.

VbDataActionUpdate	6	Caused due to Update method.
VbDataActionDelete	7	Caused due to Delete method.
VbDataActionFind	8	Caused due to Find method.
VbDataActionUnload	11	Caused due to Form being unloaded.

This event occurs even if no changes have been made to data in the bound controls and even if no bound controls exist. You can use this event to change values and update data. You can also choose to save data to stop whatever action is causing the event to occur and substitute a different action.

You can change the action argument to convert one action into another. You can change the various Move methods and the Add new method, which can be freely exchanged. Any action can be stopped by setting action to 0.

You can check the data in each bound control where DataChanged is True. You can then set DataChanged to False to avoid saving that data in the database. If it is invalid or corrupted, you can't use any methods such as MoveNext on the underlying Recordset object during this event.

8.2.8. Using Objects

Visual Basic allows you to create objects to represent a database, Recordsets, Fields etc. This makes the job easier in programming a Database application. You can then easilty manipulate the objects or change the DataSource according to your requirement in run-time.

Data Access Object (DAO) has been the simple Database access methods for Visual Basic programmers. You can use the Database and RecordSet objects. The RecordSet and Database objects, each has properties and methods of their own. You can either use a Database, RecordSet object created by a Data control or create an instance of you own, to manipulate the data.

8.2.9. Database Object **(MDU 2008)**

A Database object represents a Database. A Database object contains Recordsets, QueryDefs, and Procedures. A Database object is declared within a Workspace object, which is a part of Workspace collection.

A database object can be declared using the following code :

```
Dim MyDb As Database
Dim MyWrk As WorkSpace
Set MyWrk = WorkSpaces(0)
Set MyDB = MyWrk. OpenDatabase
("C:\ProgramFiles\MicrosoftOffice\Office\Samples\_ Northwind.MDB)
```

The OpenDatabase method of the Workspace object opens a database and returns its reference to a valid Database type object variable.

Important Methods Of A Database Object Are

Close	Closes an open Database Access Object.
Execute	Runs an Action Query or executes an SQL statement.
ExecuteSQL	Executes an Action Query as an SQL statement on an ODBC database.
OpenRecordset	Creates a new Recordset object and applies it to the Recordset collection.
Refresh	Refresh method closes or rebuilds the Recordset object.

Important Properties Of the Database Object

Connect	Specifies the source of an open database. The default source for ODBC and certain ISAM drivers in Jet databases.
Count	Returns the number of objects in a collection.
Name	Returns or sets a user defined name for data access objects.

8.2.10. Recordset Object

A Recordset object represents the entire set of records from a base table or the set of records resulted by an executed query. At any time, Recordset object refers only to a single record within the set as the current record.

A Record Set object can be used to manipulate, and access the data. In order to navigate records in the Database you can use various methods of the Record Set object.

The number of records returned by the Recordset can be determined by moving to the last record in the Recordset or by examining the Recordset object's RecordCount property. Before you move to the last record, the value returned by the property to display the number of records in a Data control's Recordset :

```
Data1.Recordset.MoveLast
MsgBox "Records" & Data1.Recordset.RecordCount
```

If you create a Recordset object using either code for another Data Control, you can set the Recordset property of the Data control to this new Recordset. Any existing Recordset in the Data control, and the Database object associated with it are released when a new Recordset is assigned to the Recordset property.

The five types of RecordSet objects are :

- Table type Recordset
- Dynaset type Recordset
- Snapshot type Recordset
- Forward only type RecordSet
- Dynamic type RecordSet

All the above mentioned RecordSet objects represent a set of records. Let us discuss these different types of RecordSets.

Table Type RecordSet Object

The first type of RecordSet object is the **Table-type** RecordSet. A table type object can be created from a table in a Microsoft Access database, but not from an Open Database Connectivity (ODBC) or a linked table. When you create a table-type the Jet database engine opens the actual table and your subsequent data manipulations operate directly on base-table data. A table-type can be opened on only one table; it cannot be opened on a union query or a select query with a join.

The best part of this type of object is that you can use index created for the underlying table. This allows much faster sorting and searching than the other types.

DynaSet Type RecordSet Object

Dynaset-type Recordset object represents a result of a query that can have updatable records. A dynaset-type object can be created from either a local or a linked table, or with a row-returning query that joins tables. It's actually a set of references to records in one or more tables. With a dynaset you can extract and update data from more than one table, including linked tables from other databases. Heterogeneous updatable joints are a unique feature of dynaset-they enable you to use updatable queries against tables in different types of databases. You can use the Seek methods rather than Find methods for a Dynaset-type Recordset object for better performance.

SnapShot Type Recordset Object

A Snapshot-type object is a static copy of a set of records as it exists at the time the snapshot is created. A snapshot-type object can contain fields from one or more tables in a database. You can't update a snapshot.

The main advantage of a snapshot is that it creates less processing overheads than the other types, so it can run queries and return data faster, especially when working with ODBC data sources.

Forward Only

A Forward-only type object is identical to a SnapShot type object. A forward-only-type object, sometimes referred to as a forward-scrolling snapshot or a forward-only snapshot, provides a sub-set of the capabilities of a snapshot. With forward-only snapshots, you can move only in a forward direction thourgh the records. Objects of this type cannot be cloned and only support the Move and MoveNext methods. Like snapshots, you can't update a forward-only-type object.

Dynamic Type Recordset Object

Dynamic type RecordSet object is a result of query from one or more base tables. You can add, change or delete records from a row-returning query. Besides this, records other users add, delete or edit in the base table are also present in the RecordSet.

Although you can work with any of the above RecordSet object, you must always choose the correct Record Set type, that suits or satisfies your desires. For example, you would like to work with the sorted data or a data where searching operations like FindFirst, Seek could be performed in a short span of time, then you will go for indexed table type Recordset. Similarly, in order to scan all the records you must use a Forward only type Recordset for better performance.

You can use the type argument of the Recordset object to choose the RecordSet type.

Important Methods of the RecordSet Object

AddNew

The Ms Access engine maintains a copy buffer where it keeps the pending data that it will be writing to the database.

The addnew method clears the copy buffer and moves the current record to the end. This is like adding a new row.

Syntax :

```
Datacontrolname.Recordset.AddNew
```

The Addnew method does not actually add the information to the database. This is done by the UPDATE method.

Update

This method sends the contents of the copy buffer to the table or dynaset.

Syntax :

```
Datacontrolname.Recordset.Update
```

Example 2

```
Data1.Recordset.Addnew
Data1.Recordset.Fields("Author") = "Homes"
```

```
Data1.Recordset.Fields("Titles") = Iliad"
Data1.Recordset.Update
```

This code adds a record to a table which has two fields Authors and Title.

Update Controls

This method can be used to retain the original data in the bound controls and ignore the changes made to the data by a user.

For *e.g.,* this method can be used in the Cancel button to reset the contents of bound controls to their original value.

Edit

This method copies the current record into the copy buffer for editing.

Example :

```
Data1.Recordset.Edit
Authorname = Data1.Recordset.Fields("Author")
Call Changefield(Authorname)
Data1.Recordset.Fields("Author") = Authorname
Data1.Recordset.Update
```

This code uses a user defined procedure called Changefield, which changes the value of the field.

Delete

This method deletes the current record in the recordset.

Syntax :

```
Datacontrolname.Recordset.Delete
```

After delete you must move the record pointer away from the deleted record by a move method.

Example 3

```
Data1.Recordset.Delete
Data1.Recordset.Moveprevious
```

Close

To close an open database attached to a data control or to close the specific recordset currently attached to the control, use the following :

Syntax :

```
ObjectName.Close
```

ObjectName can be an open database, recordset, dynaset, snapshot or a table object.

You must use the update method before you close the database to save all the pending changes made to the database.

UpdateRecord Method

Saves the current values of bound controls. Doesn't support named arguments.

Syntax :

```
Object.UpdateRecord
```

The Object represents a RecordSet object.

Remarks :

Use this method to save the current contents of bound controls to the database during the Validate event without triggering the Validate event again. Using this method avoids creating a cascading event. The UpdateRecord method has the same effect as executing the Edit method, changing a field, and then executing the Update method, except that no events occur. You can use this method to avoid triggering the Validate event.

Summary of the Methods of RecordSet Object

Name	***Description***
Properties	Contains all the property objects for a specific instance of the RecordSet object.
Fields	Contains all stored Field objects of the Recordset object.
AddNew	Creates a new record for an updatable Recordset object.
CancelUpdate	Cancels any changes made to the current record or to a new record prior to calling the Update method.
Clone	Creates a duplicate Recordset object from an existing Recordset object.
Close	Closes the open Recordset object and any dependent objects.
Delete	Delets the current record or a group of records.
GetRows	Retrieves multiple records of a Recordset object and makes that record the current record.
Move	Moves the position of the current record in a Recordset object.
MoveFirst	Moves to the first record in the Recordset object and makes that record the current record.
MoveLast	Moves to the last record in the Recordset object and makes that record the current record.
MoveNext	Moves to the next record in the Recordset object and makes that record the current record.
MovePrevious	Moves to the previous record in the Recordset object and makes that record the current record.
NextRecordset	Advances through a series of commands by clearing the current Recordset object and returns the next recordset.
Open	Opens a cursor.
Requery	Updates the data in a Recordset object by re-executing the query that is the basis of the object.
Resync	Re-freshes the data in the current Recordset object from the underlying database.
Supports	Determines whether a specified Recordset object supports a particular type of functionality.
Update	Saves any changes you make to the current record of a Recordset object.
UpdateBatch	Writes all pending batch updates to disk.

Properties of the Recordset Object

Name	***Description***
AbsolutePage	Sets the "page" number on which the current record is located. Uses the PageSize property to logically divide the Recordset object into a series of pages, each of which has the number of records equal to PageSize. The provider must support the appropriate functionality for this property to be available.
AbsolutePosition	Sets the Ordinal position of a Recordset object's current record.
ActiveConnection	Shows to which Connection object the specified Recordset object currently belongs.
BOF	Shows that whether the current record position is before the first record in a Recordset object, or not. Returns True or False.
BookMark	Returns a bookmark that uniquely identifies the current record in a

	Recordset object. Sets the current record in a Recordset object or the record identified to be a valued bookmark.
CacheSize	Shows the number of records from a Recordset object that are cached locally in memory.
CursorLocation	Sets or returns the location of the cursor engine.
CursorType	Shows the type of cursor used in a Recordset object-for example, Read Only or Updatable.
EditMode	Shows the editing status of the current record.
EOF	Shows that whether the current record position is after the last record in a Recordset object, or not. Returns True or False.
Filter	Shows that the current record position is after the last record in a Recordset object.
LockType	Shows the type of locks placed on records during editing.
MarshalOption	Indicates which records are to be marshaled back to the server.
MaxRecords	Shows the maximum number of records to return to a Recordset from a query.
PageCount	Shows how many pages of data the Recordset object contains.
PageSize	Shows how many records-constitute one "page" in the Recordset.
RecordCount	Shows the current number of records in a Recordset object.
Source	Shows the source for the data in a Recordset object (command object, SQL statement, table name, or stored procedure) or the name of the object or application that originally generated an error.
State	Describes the current state of an object.
Status	Shows the status of the current record with respect to batch updates or other bulk operations.

You can also declare a RecordSet type object variable using the following code :

```
Dim MyTable as Recordset
```

A reference to a RecordSet can be given to MyTable using the following line.

```
Set MyTable = DB.OpenRecordSet("Authors")
```

or your new RecordSet can be result of a new query.

```
Qry = "Select * from Authors where Su_ID >= 10"
Set MyTable = DB.OpenRecordSet(Qry, dbOpenDynaset)
```

SUMMARY

1. Visual C++ provides you with four different technologies for using and accessing databases in your applications, Data Access Objects (DAO), ODBC, OLE DB, and ActiveX Data Objects (ADO).
2. Visual Basic 6.0 supports many data access methods. Besides using the traditional DAO (Data Access Objects) you can also use Remote Data Object (RDO) and ActiveX Data Object (ADO).
3. A database can be viewed as a set of interrelated data. Database management deals with the storage, retrieval and editing of the data within the database. One of the most important functions of Database management is the Database security and Database validations.
4. In order to display the values of the columns accessed by the Data Control, some controls are required which support the data control. These control are called Data Bound controls and are simpler to use *e.g.,* TextBox, Label DBGrid, DBList etc.
5. Data Control is one of the easiest modes to access databases. You can perform most data access operations using the Data control without writing any code at all. Data-aware controls bound to a Data control automatically display data from one or more fields for the current record or, in some cases, for a set of records on either side of the current record.

6. In order to display the values of the columns accessed by the Data control, some controls are required which support the data control. These controls are called Data Bound controls and are simpler to use *e.g.*, TextBox, Label, DBGrid, DBList etc
7. The DatabaseName Property reads or sets the name and location of the database. Setting the value in the data control's Database Name Property while your program is running, changes the database to which the control is connected.
8. Reposition event occurs after a record becomes the current record. When a Data control is loaded, the first record in the Recordset object becomes the current record. Whenever a user clicks any button on the Data control, moving from record to record, or if you use one of the Move methods, such as MoveNext, the Find methods, such as FindFirst, or any other property or method that changes the current record, the Reposition event occurs after, when each record becomes current.
9. Data Access Object (DAO) has been the simple Database access methods for Visual Basic programmers. You can use the Database and RecordSet objects. The RecordSet and Database objects, each has properties and methods of their own. You can either use a Database, RecordSet object created by a Data control or create an instance of you own, to manipulate the data.
10. ObjectName can be an open database, recordset, dynaset, snapshot or a table object.You must use the update method before you close the database to save all the pending changes made to the database.

SHORT QUESTIONS WITH ANSWER

Q.1. How would Delete a Record?

Ans.

1. Add a Delete menu item :

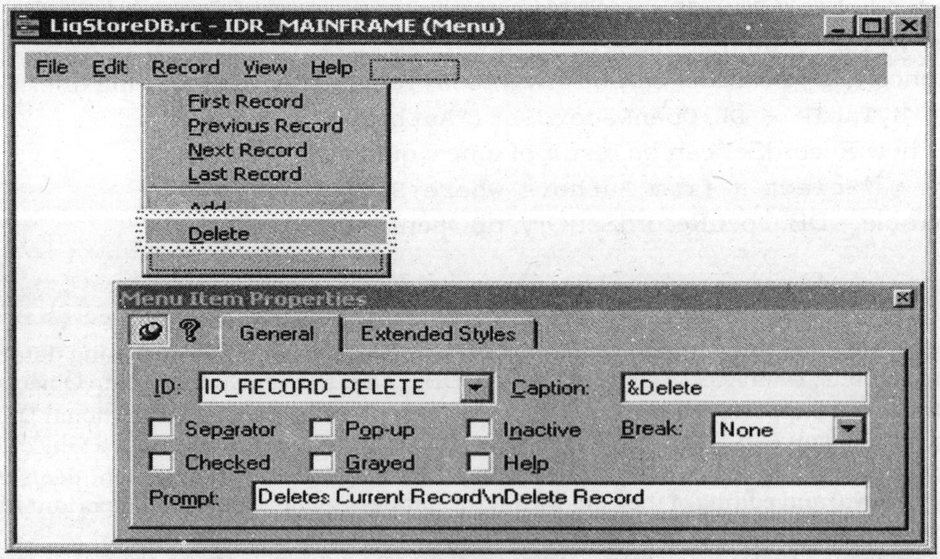

Fig. 8.26.

2. On the toolbar, add a button that has the same identifier as the Delete menu item :

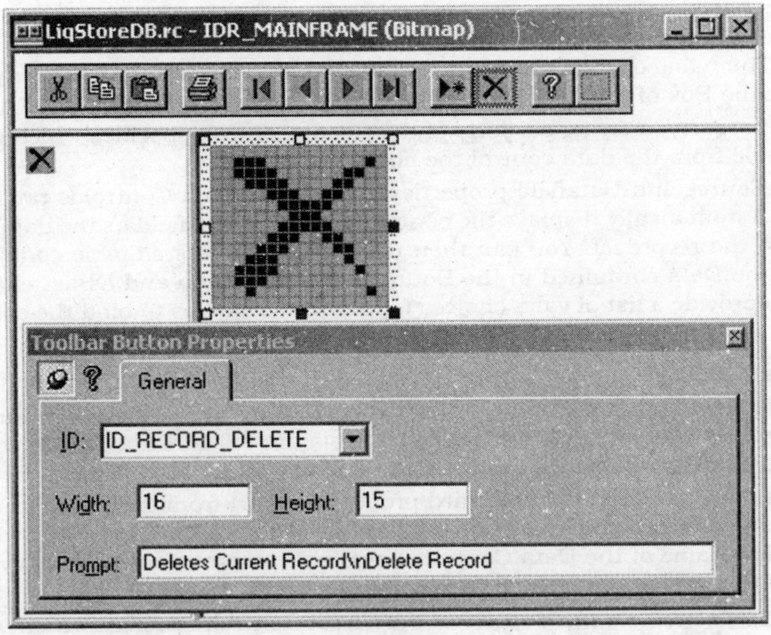

Fig. 8.27.

3. Implement its COMMAND function as follows :

```
void CLiqStoreDBView::OnRecordDelete()
{
    m_pSet->Delete();
    m_pSet->MoveNext();
    if( m_pSet->IsEOF())
        m_pSet->MoveLast();
}
```

4. Test the application.

Q.2. What is Database Management?

Ans. A database can be viewed as a set of interrelated data. Database management deals with the storage, retrieval and editing of the data within the database. One of the most important functions of Database management is the Database security and Database validations.

Visual Basic makes it simple by supplying a variety of tools and approaches for data management. For significant database applications, utilising database functionality that already exists becomes necessary. Visual Basic 6.0 provides full-fledged database engine called Jet. This engine contains virtually all the database functions that applications may need. The simplest and easiest way to access the Jet database engine is through bound data control.

Data Access Object (DAO) has been the standard database access method for Visual Basic programmers. DAO was the first database interface included with Visual Basic in its third version.

Remote Data Object (RDO) was then introduced to make data-access more flexible and functional in VB 4. The most recent release by Microsoft with Visual Basic is the ActiveX data object, which is in a short while going to take over all the other methods of data access.

Q.3. What is Data Binding? **(MDU 2008)**

Ans. In order to display the values of the columns accessed by the Data Control, some controls are required which support the data control. These control are called Data Bound controls and are simpler to use *e.g.*, TextBox, Label DBGrid, DBList etc.

These controls have basically two standard properties other than their regular properties. The DataSource property enables you to specify the name of the Data Control to which the client control will bind. The name of the Data Control(s) available on your form is already present in the dropped down Combo Box of this property in the property window.

The Datafield property, a string property of the Client Control, specifies which of the Database field names available from the data control the client will bind.

After the DataSource and Datafield properties are set, the Data Control is properly initialized, the Client Control automatically displays the contents of the bound field as the data control is used to navigate though the recordset. You can then use an appropriate editable control, such as the textox, to change the Data contained in the Bound field. DBCombo and DBlist enable you to use the data control to provide a list of valid choices for a field, as well as to bind the client control to a specific field.

Q.4. What is Data Binding?

Ans. In order to display the values of the columns accessed by the Data control, some controls are required which support the data control. These controls are called Data Bound controls and are simpler to use e.g., TextBox, Label, DBGrid, DBList etc.

These controls have basically two standard properties apart from their regular properties. The DataSource property enables you to specify the name of the Data Control to which the client control will bind. The name of the Data Control(s) available on your form is already present in the drop down Combo Box of this property in the property window.

The Datafield property, a string property of the Client Control, specifies which of the Database field names available from the data control, the client will bind.

After the DataSource and Datafield properties are set, the Data Control is properly initialized, the Client Control automatically displays the contents of the bound fields as the data control is used to navigate though the recordset. You can use an appropriate editable control, such as the TextBox, to change the data contained in the bound field. DBCombo and DBList enables you to use the data control to provide a list of valid choices for a field, as well as to bind the client control to a specific field.

Q.5. What is Recordset Property? **(V.Imp)**

Ans. The Recordset property sets or returns a Recordset object defined by a Data control's properties or by an existing Recordset object.

You can use the Recordset property like any other Recordset object. For example, you can use any of the Recordset methods or properties to examine the structure of the Recordset object's underlying schema.

You can also request the type of Recordset to be created by setting the Data control's Recordset Type property. If you don't request a specific type, a dynaset-type Recordset is created. Using the RecordsetType property, you can request to create either a Table -, Snapshot or dynaset type Recordset.

The syntax for the Recordset property is as follows :

```
Set Object.RecordSet = value
```

Object refers to a valid Data control name and value is any object variable containing a reference to a RecordSet object.

Q.6. X-window:

Ans. X-Window: The **X Window System** (commonly **XII** or **X**) is a system which implement the X display protocol and provides windowing on bitmap displays. It provides the standard took it and protocol with which to build graphical user interfaces (GUIs) on most Unix—like operating systems and Open VMS, and has been ported to many other contemporary general purpose operating systems. X provides the basic framework, or primitives, for building GUI environments: drawing and moving windows on the screen and interacting with a mouse and/or keyboard. X does not mandate the user interface—individual client programs handle this. As such, the visual styling of X-based environments varies greatly; different programs may present radically different interfaces. X is not an integral part of the operating system kernel. Unlike previous display protocols, X was specifically designed to be used over network connections rather than on an integral or attached display device. X features network transparency: the machine where an application program (the client application) runs an differ from the user's local machine (the display server).

Q.7. Client-area and non-client area messages.

Ans. Client Area and Non-Client Area Messages: The system sends a **WM_NCPAINT** message to the window whenever apart of the non-client area of the window, such as the title bar, menu bar, or window frame, must be updated. The system may also send other messages to direct a window to update a portion of its client area; for example, when a window becomes active or inactive, it sends the **WM_NCACTIVATE** message to update its title bar. In general, processing these messages to **DelWindowproc** for default processing.

An application that creates custom non-client areas for its windows must process these messages. When doing so, the application must use a window device context to carry out drawing in the window. The window device context enables the application to draw in all portions of the window, including the nonclt area. An application retrieves, a window device context by using the **GetWindowDC** or **GetDCEx** function and, when drawing is complete, must release the window device context y using the **ReleaseDC** function.

The system maintains an update region for the non-client area. When an application receives a **WM_NCPAINT** message, the wParam parameter contains a handle to a region defining the dimensions of the update region. The application can use the handle to combing the update region with the clipping region for the window device context. The system does not automatically combine the update region when retrieving the window device context unless the application uses **GetDCEx** and specifies both the region handle and the **DCX_INTERSECTRGN** flag. If the application does not combine the update region, only drawing operations that would otherwise extend outside the window are clipped. The application is not responsible for clearing the update region, regardless of whether it uses the region.

If an application processes the **WM_NCACTIVATE** message, after processing it must return TRUE to direct the system to complete the change of active window. If the window is minimized when the application receives the **WM_NCACTIVATE** message, it should pass the message to **DefWindowProc**. In such cases, the default function redraws the label for the icon.

Q.8. (*a*) Is it possible to add a table from another data base in VB ? State reasons.

(*b*) Differentiate between DDE and OLE.

Ans. OLE and DDE allow you to perform similar actions. Both enable you to send commands to another application, perform actions in that application, and return data to your Power Builder application. There are however, some fundamental differences between OLE and DOE. OLE is currently implemented using the DOE protocol. Application using OLE are not aware that DOE is being used, nor should they rely on thin, because the implementation mechanism will not be there in future releases of the OLE components.

Q.9. Differentiate between linked objects and embedded objects.

Ans. Linked object v/s embedded object: Embedding objects: When you embedded an object, a copy of the object is inserted into the destination document. There's no limit to the original file. When you change information in the source document, no changes will be reflected in the destination document. The actual data for the object is stored within the destination file. To make changes to the embedded object, double click it and it will launch the original application the source file was in.

Linking objects: Information is updated when you modify the original source file when you use a linked object. This dynamic updating is very handly for things such as the aforementioned monthly report. You can open up the Excel spreadsheet that is referenced within your Word document. Make changes to the spreadsheet, close Excel, and when you open your Word document... viola! The changes are already there. If the object is linked to ten other. Word files, the changes are already in those ten files, too! actually linking a embedding an object is fast and easy.

Q.10. How you can create a dialog box in VC++? How controls can be added to it?

Ans. Creating dialog box in VC++:

1. To create a class for the dialog box and display the dialog box in the application, change the file as follows:

```
#includs <afxwin.h>
#include "Resource.h"
class CExerciseApp: public CWinApp
{public: BOOL Initlnstance( );};
class CExoDialog: public CDialog{
public: enum{lDD = IDDJ_DIALOG1};
CExoDialog();}CExoDialog::CExoDialog():
CDialog(CExoDialog:: IDD, NULL) {} BOOL CExerciseAppr. instance( ){
CExoDialog Dig;
m_pMain Wnd = &DLg; Dlg.domodal ( ); return TRUE;) CEexercise App
the App;
```

```
At run time, to change the caption, call the CWnd::SetWindowTextO method and
provide th necessary string as argument. Here is an example:
BOOLCExoDialog2Dlg:: OnlnitDialog()
CDialog::OnlnitDialog( );
.............
//TODO: Add extra initialization here
SetWindowText("Go Slow, Dont Rush It!!!");
return TRUE; II return TRUE unless you set the focus to a control}
```

Q.11. How a multiple document program is different than a single document program? Explain with example

Ans. Multiple document interface: Unlike an SDI applied an MDI application can display multiple forms or documents at the same time. For example, in MS-Office applications such as MS-Word or MS-Excel, you can open more that one document at the same time. You can also navigate between the two documents.

In an MDI applications, the main window is called the parent window. The rest of the windows, which are displayed in the main window, are called child windows.

Q.12. How to use the visual C++ debugger?

Ans. Debugger is used to fix the bugs in a particular program. Break prints are used in a program and program is exerted upto that point. When a breakpoint is hit, program execution stops and the code that is debugged will appear so that what's going on can be examined.

To start debugging at a particular line of code, a break print is set there. To set the breakpoint at a particular line, go to the line and press F9. This will place a Red flower bubble in front of the line. Now run the program with the debugger. For that go to Build → Start Debug → Go. After reaching to the line the program will stop and display the code in V C++. Tostop into the code, F11 key is pressed: If a block of code is not to debug, code can be step out by pressing shift + Fl 1 key. Ctrl + F10 executes all the code upto the cursor. F10 key is used to single step through code.

Variable and its value can be viewed into the Auto window. This window holds the most recent variables that V C++ encountered. In the locals tab, the variables defined in the current method or code block can be viewed.

To keep track of a specific variable at all times when debugging a program, type it into the watch window 1 at lower right in V C++. By simply dry and drop-variables into the watch window.

This is all about VC++ debugger.

LONG QUESTIONS WITH ANSWER

Q.1. What is the procedure for Creating an ODBC Data Source? **(V.Imp)**

Ans. Here are the steps to create the needed data source :

1. Start or open Control Panel and, in Control Panel, double-click Administrative Tools.
2. In the Administrative Tools window, double-click Data Sources (ODBC). Alternatively, from the Taskbar, you could have clicked Start -> Programs -> Administrative Tools -> Data Sources (ODBC).
3. In the ODBC Data Source Administrator property sheet, click the Add button :
4. In the **Create New Data Source** wizard, scroll down in the list box and click **SQL Server** :
5. Click Finish.
6. In the Name edit box of the Create A New Data Source To SQL Server wizard, type **LiqStore** and **press Tab**.
7. In the Description edit box, type **The Liquor Store Application**.
8. Click the arrow of the Server combo box and select the server where the above (LiquorStore) database resides :

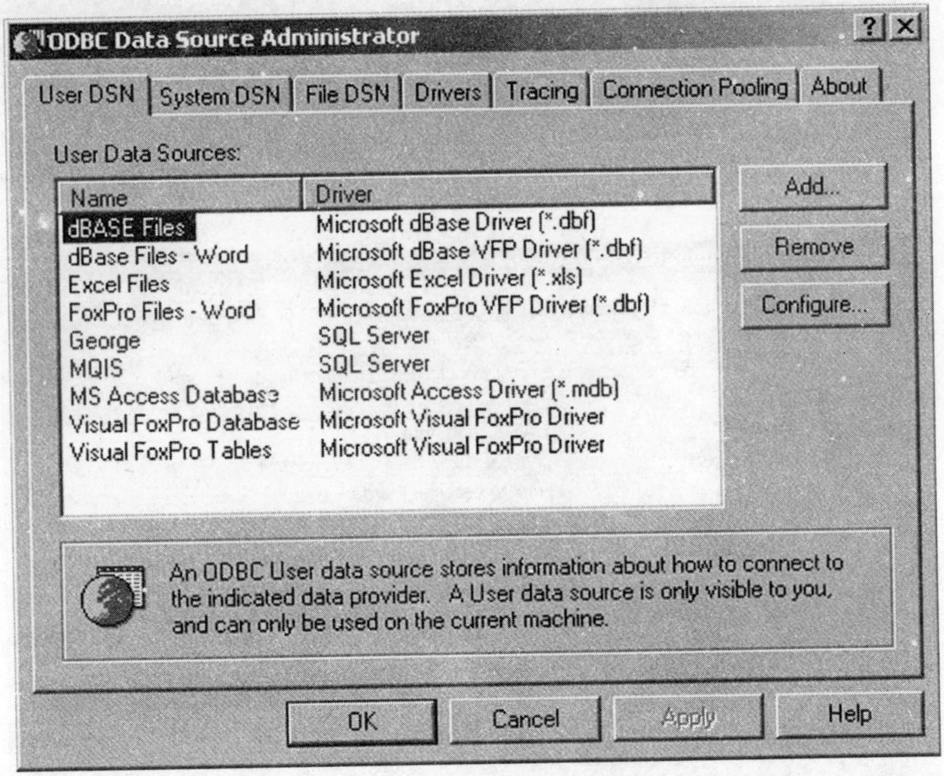

Fig. 8.28.

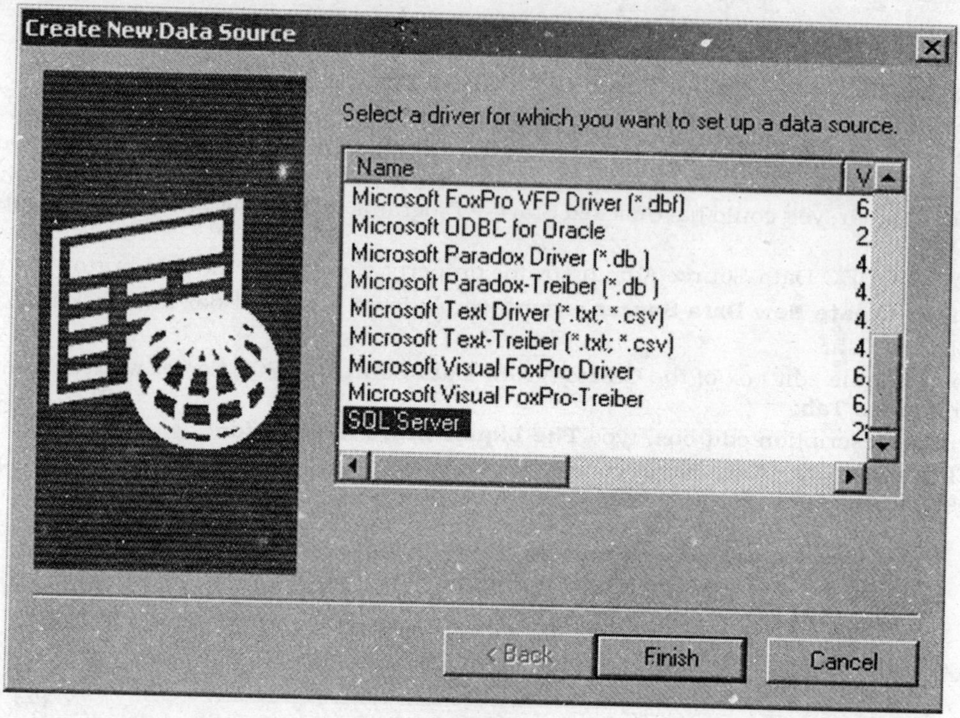

Fig. 8.29.

Fig. 8.30.

9. Click Next.
10. Select the desired authentication. For this example, I accept the **Windows NT Authentication.** I also accept to connect to SQL Server with Default Settings by leaving the check box untouched :

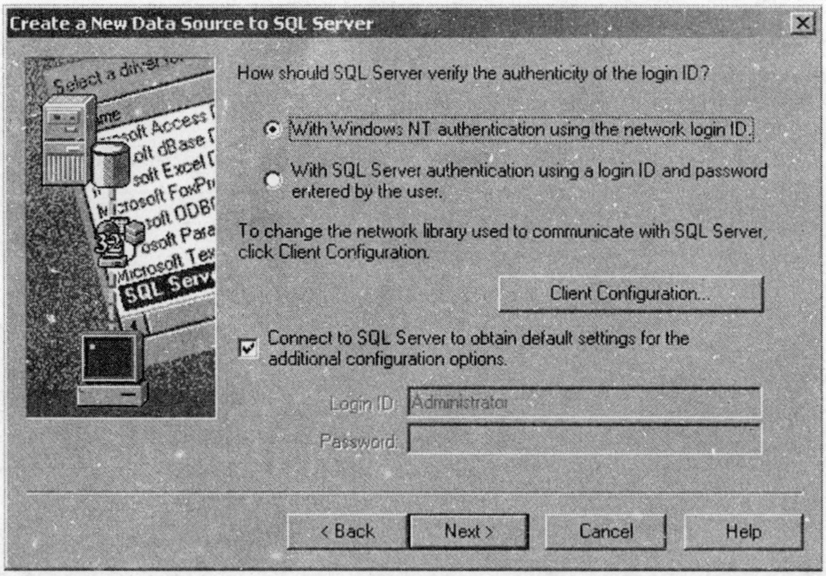

Fig. 8.31.

11. Click Next.
12. In the new page of the Create A New Data Source To SQL Server wizard, click the **Change The Default Database To** check box.
13. Once the combo box is enabled, click its arrow and select LiquorStore :

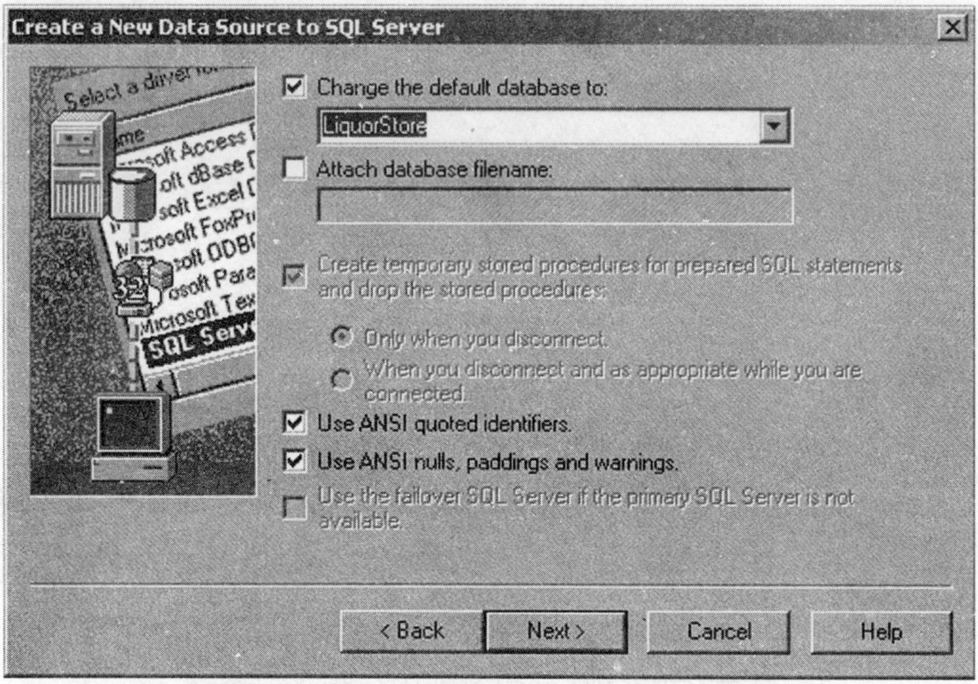

Fig. 8.32.

14. **For this example, leave the Attach Database FileName check box unchecked and accept the other defaults. Click Next.**

15. Unless you have any other reason, accept the default (language, encryption, etc.) settings of the new page :

Fig. 8.33.

16. Click Finish.
17. On the ODBC Microsoft SQL Server Setup dialog box, click the Test Data Source button :

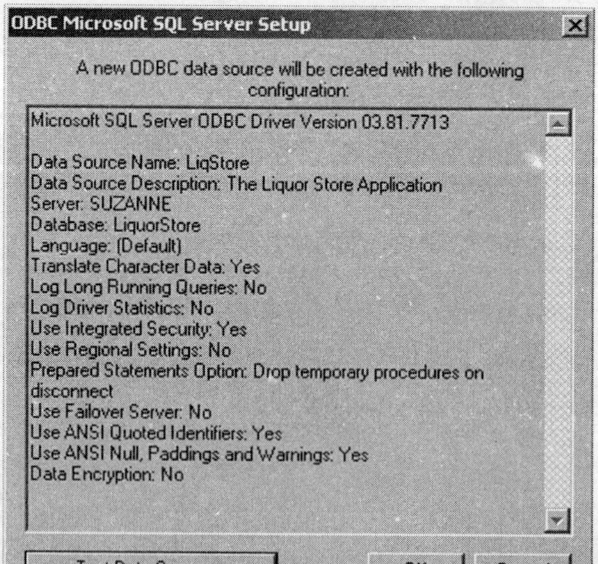

Fig. 8.34.

18. When the SQL Server ODBC Data Source Test dialog box confirms that the TESTS COMPLETED SUCCESSFULLY!, click OK :

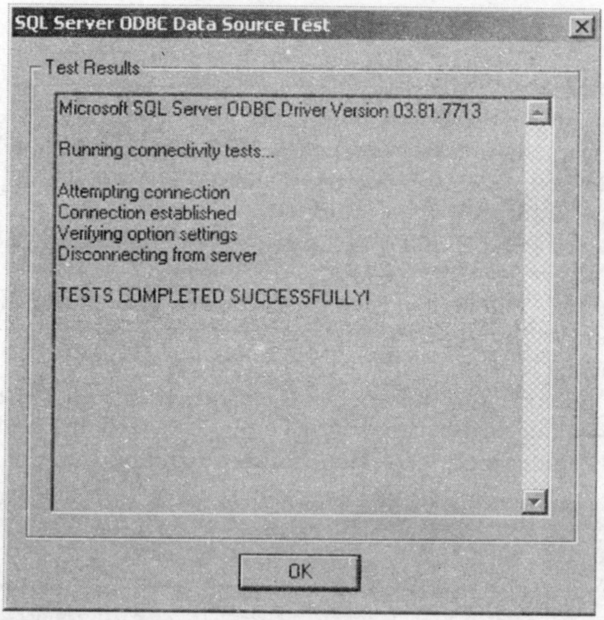

Fig. 8.35.

19. On the ODBC Microsoft SQL Server Setup dialog box, click OK.
20. In the User DSN property page of the ODBC Data Source Administrator dialog box, make sure that the LiqStore data store is listed in the User Data Sources list box :

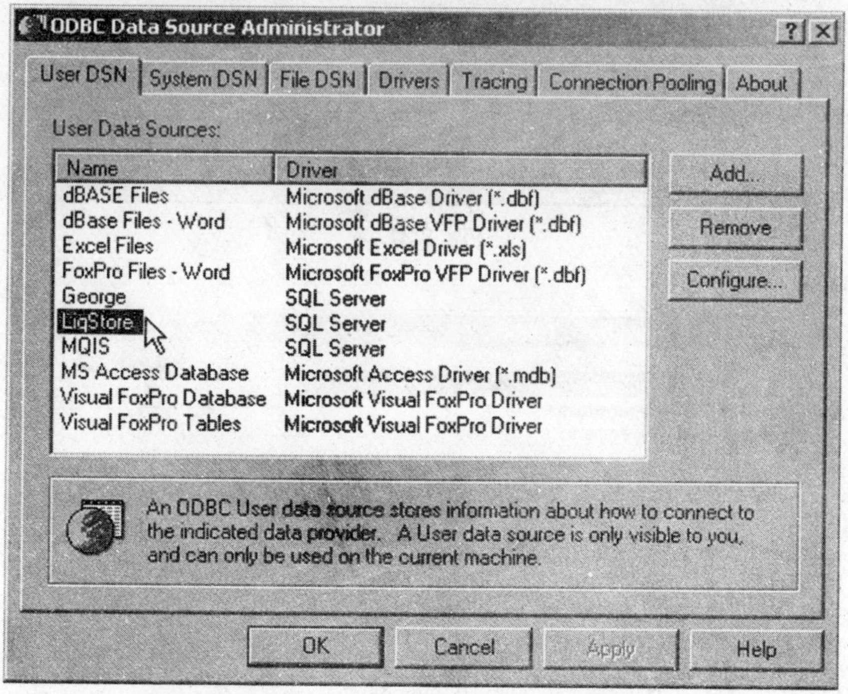

Fig. 8.36.

21. Click OK

Q.2. Write procedure for Creating the Application in VC++.

Ans.

1. Start Microsoft Visual C++.
2. On the main menu, click File -> New.
3. In the New dialog box, click Projects and click **MFC AppWizard (exe)**.
4. In the Location edit box, select or type the desired folder for the application. In the Project Name edit box, type **LiqStoreDB** and **click OK**.
5. In the MFC AppWizard - Step 1 dialog box, click the **Single Document** radio button and click Next.
6. In the MFC AppWizard - Step 2 of 6, click the **Database View Without File Support** (because we will not need and will not use file support for this exercise).
7. Click the **Data Source** button.
8. In the Database Options dialog box, in the **Datasource** section, make sure the **ODBC** radio button is selected and click the arrow of its combo box. **Select LiqStore**.
9. In the **Recordset Type** section, click the **Dynaset** radio button.

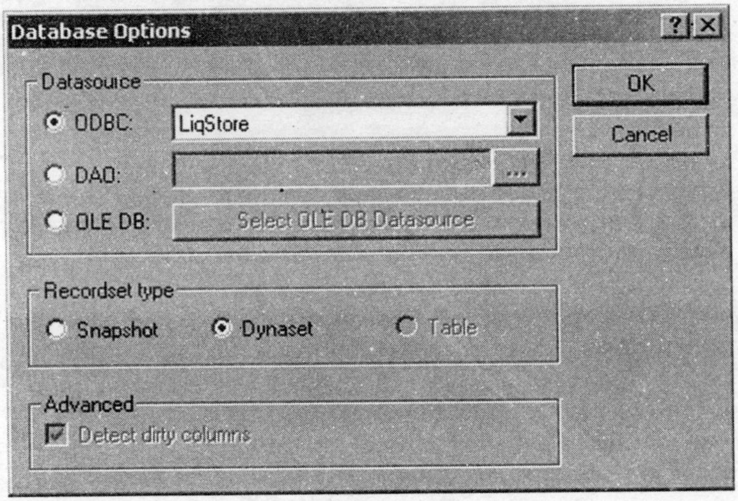

Fig. 8.37.

10. Click OK.

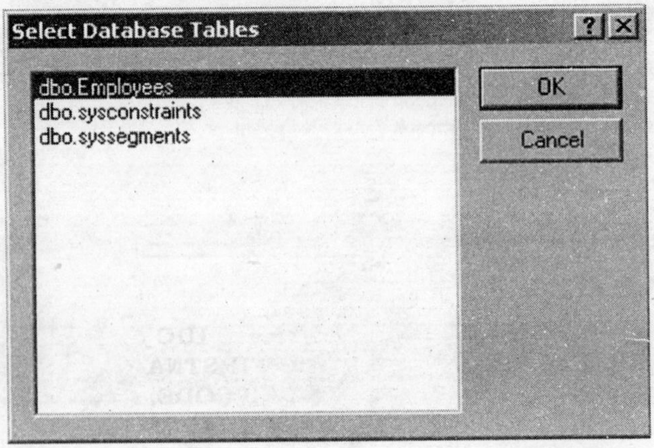

Fig. 8.38.

11. In the Select Database Tables, click dbo.Employees and click OK.
12. In the MFC AppWizard - Step 2 of 6, click Next.
13. In the MFC AppWizard - Step 3 of 6, accept all defaults and click Next.
14. In the MFC AppWizard - Step 4 of 6, uncheck the **Printing And Print Preview** check box.
15. Click the **Advanced** button and click the **Window** Styles property page. Uncheck the Maximise Box check box and click Close.
16. Click Next.
17. In the MFC AppWizard - Step 5 of 6, accept all defaults and click Next.
18. In the MFC AppWizard - Step 6 of 6, accept all defaults and click Finish.
19. In the New Project Information dialog box, click OK.
20. In the Workspace, click the ResourceView tab, expand the String Table, and double-click String Table. Double-click IDR_MAINFRAME and replace its entire Caption with Four-Corner Liquor Store - Employees Records and close the floating window. Also close the String Table window.
21. Using the Controls window, design form as follows :

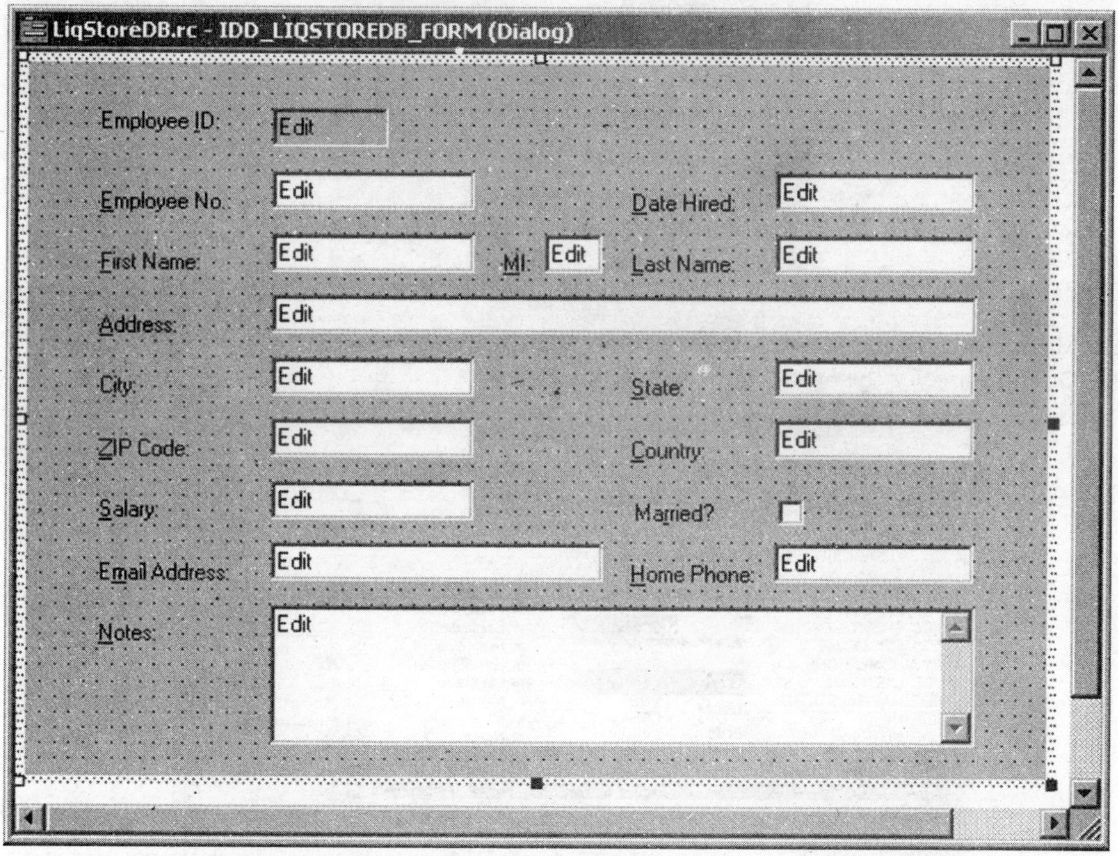

Fig. 8.39.

22. Set the IDs of the edit controls as follows : **IDC_EMPLOYEEID (Read-Only), IDC_EMPLOYEENBR, IDC_DATEHIRED, IDC_FIRSTNAME, IDC_MI, IDC_LASTNAME, IDC_ADDRESS, IDC_CITY, IDC_STATE, IDC_ZIPCODE, IDC_COUNTRY, IDC_SALARY, IDC_EMAILADDRESS, IDC_HOMEPHONE,** and **IDC_NOTES** (Multiline, Vertical Scroll, Want Return). Change the ID of the check box to **IDC_MARITALSTATUS**.
23. Save everything.

24. In the Workspace, click ClassView tab. Expand the LiqStoreDB classes node and double-click CLiqStoreDBSet. Notice that the CRecordset class already has variables declared for each control of our form.
25. To associate each control with the appropriate variable, on the main menu, click View -> ClassWizard.
26. In the MFC ClassWizard dialog box, click the Member Variables property sheet. In the Class Name combo box, select CLiqStoreDBView.
27. In the list box, double-click IDC_ADDRESS.
28. In the Add Member Variable dialog box, click the arrow of the Member Variable Name combo box and select m_pSet->m_Address :

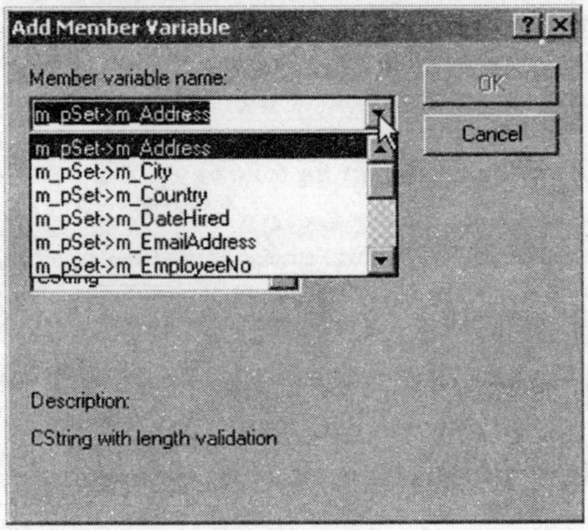

Fig. 8.40.

29. Make sure that the **Variable Type** is set to **CString** and click OK.
30. Do the same for the other controls :

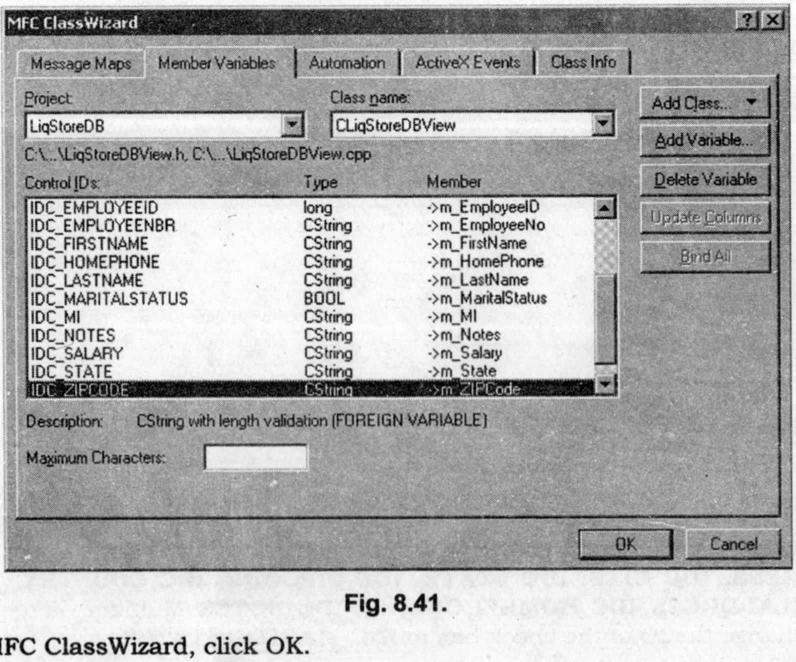

Fig. 8.41.

31. **On the MFC ClassWizard**, click OK.

32. Test the application.

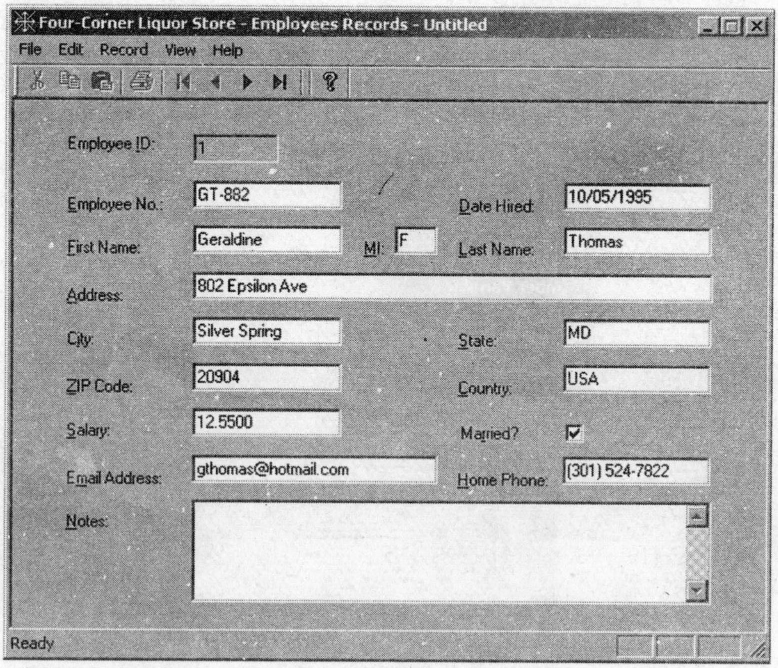

Fig. 8.42.

33. Close the applicationAs you can see, this database was created without a single line of code.

Q.3. How would you Add a Record?

Ans.

1. In the Workspace, click the ResourceView tab. Expand the LiqStoreDB Resource tree and expand the Menu node.
2. Double-click IDR_MAINFRAME and add an Add menu item :

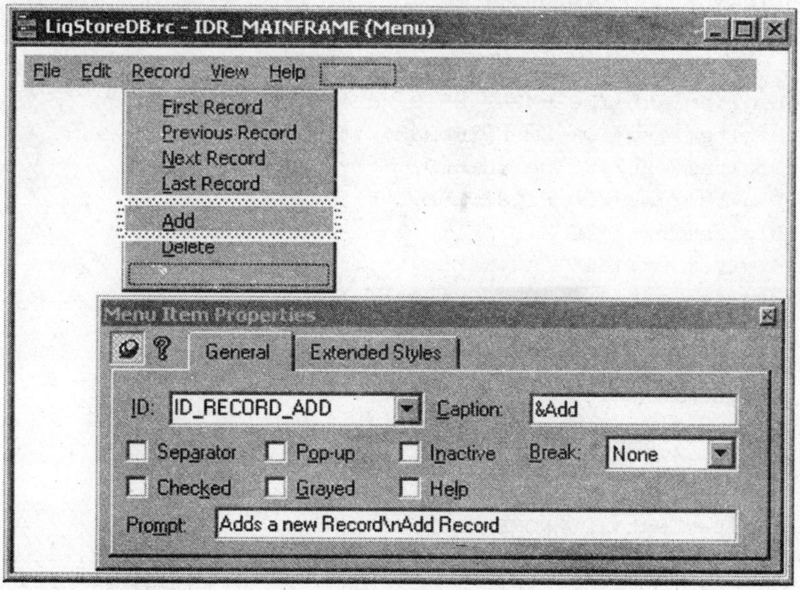

Fig. 8.43.

3. Add a button on the toolbar and associate to it the same identifier as the Add menu item :

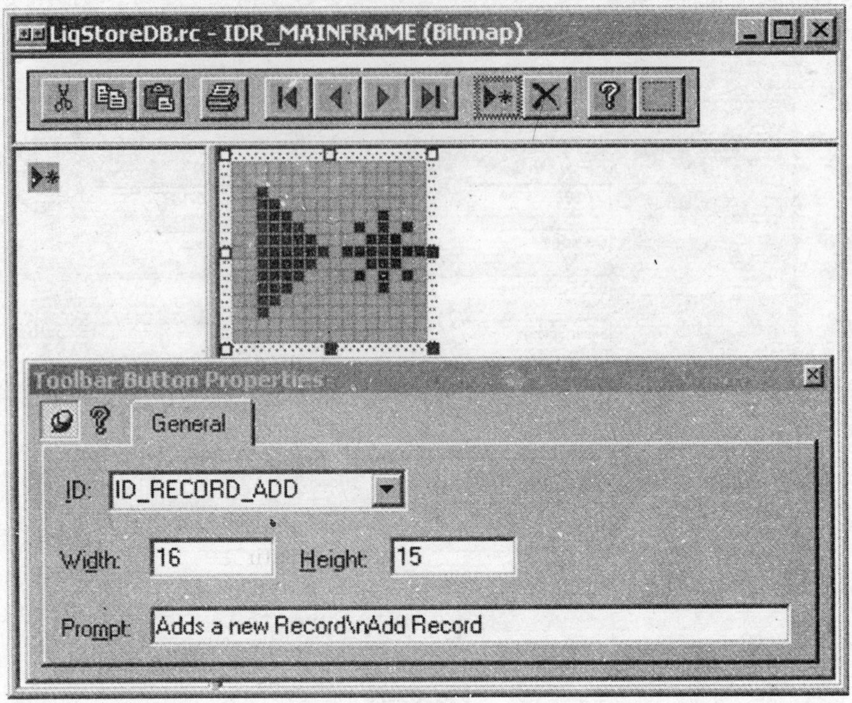

Fig. 8.44.

4. Press Ctrl + W to access the ClassWizard.
5. In the Class Name combo box, select CLiqStoreDBView and, in the Object IDs list box, select ID_RECORD_ADD.
6. In the Messages list box, double-click COMMAND.
7. Accept the suggested name of the function and click Edit Code.
8. Implement the function as follows :

```
void CLiqStoreDBView::OnRecordAdd()
{
    SetDlgItemText(IDC_EMPLOYEEID, "");
    SetDlgItemText(IDC_EMPLOYEENBR, "");
    SetDlgItemText(IDC_DATEHIRED, "");
    SetDlgItemText(IDC_FIRSTNAME, "");
    SetDlgItemText(IDC_MI, "");
    SetDlgItemText(IDC_LASTNAME, "");
    SetDlgItemText(IDC_ADDRESS, "");
    SetDlgItemText(IDC_CITY, "");
    SetDlgItemText(IDC_STATE, "");
    SetDlgItemText(IDC_ZIPCODE, "");
    SetDlgItemText(IDC_COUNTRY, "");
    SetDlgItemText(IDC_SALARY, "");
    SetDlgItemText(IDC_EMAILADDRESS, "");
    SetDlgItemText(IDC_HOMEPHONE, "");
    SetDlgItemText(IDC_NOTES, "");
```

```
        reinterpret_cast<CButton *>(GetDlgItem(IDC_MARITALSTATUS))-
>SetCheck(FALSE);

        m_pSet->AddNew();
        m_pSet->Update();
        m_pSet->Requery();
    }
```

9. Test the application and return to MSVC.

Q.4. What is Data Access Object? **(V.Imp)**

Ans. Data Access Object enables you to use a programming language to access and manipulate data in local or remote databases and to manage databases, their objects and their structure. DAO utilizes Microsoft's Jet database engine and can be used to access a wide variety of database formats, including, ISAM (Indexed Sequential Access Method) and Text.

In addition to this, DAO can also make use of Open Database Connectivity (ODBC) databases through its ODBC Direct Object Model.

Generally ODBC is a special technique to access a database residing on a separate Server such as Informix, Oracle etc.

The DAO object model has 17 types of objects like Databases, RecordSets, Fields etc. Databases are accessed by opening an existing Database object or creating a new one. The individual fields in a table make up a Recordset's Field object and table of a database is held in a Recordset Object.

Data Access object is implemented through an ActiveX control called the Data Control. This allows a large amount of DAO's power to be utilised without having to add a single line of code. When used in conjunction with Data aware controls such as TextBox, DBList, DbCombo, DbGrid etc, the Data control becomes very powerful.

Information from a database table is fetched by the DAO and then supplied to various Data aware controls like TextBox, DBListBox etc.

Databases accessible through the DB Engine are FoxPro 2.0 to 3.0 and VFP, dBase III, IV and 5.0, Paradox 3x, 4x and 5x, Microsoft Access 1x to 3x, ODBC databases, Microsoft Excel 3.0 to 8.0, Lotus 1-2-3, WKS, WK1, WK3 and WK4, ASCII tabular Format, HTML Tabular Data and Microsoft Exchange 4.0 The DBEngine consists of several Dynamic Link Libraries that are linked to your Visual Basic program at runtime. DBEngine is controlled by our application through DBEngine's object classes, their methods and properties.

Q.5. What is Data Control ? **(V.Imp)**

Ans. Visual Basic provides many sources to connect to a data source. Data Control is one of them. Data Control was the first medium to connect the applications to a database in the third version of Visual Basic.

The Data control is an excellent tool for rapidly developing a database information retrieval system. With very little code you can get a useful view into stored information. This is practically a drop-in-place process where by you can place a data control, A DBGrid and a Command Button on a form and be able to access your database in a matter of minutes. After you set properties your application is set to go.

Data Control: Data Control is one of the easiest modes to access databases. You can perform most data access operations using the Data control without writing any code at all. Data-aware controls bound to a Data control automatically display data from one or more fields for the current record or, in some cases, for a set of records on either side of the current record. The Data control performs all operations on the current record.

If the Data control is instructed to move to a different record, all bound controls automatically pass any changes to the Data control to be saved in the database. The Data control then moves to the requested record and passes back data from the current record to the bound controls when it's displayed.

The Data control is itself smart enough to handle a number of contingencies including empty recordsets, adding new records, editing and updating existing records, and handling some types of errors. However, in more sophisticated applications, you need to trap some error conditions that the Data control can't handle. For example, if the Microsoft Jet database engine has a problem accessing the database file, doesn't have permission, or can't execute the query as coded, a trappable error results. If the error occurs before your application procedures start or due to some internal errors, the Errors event is triggered.

Q.6. Write short notes on following:

(a) Queued and non-queued messages

Ans. Queued and Non Queued Messages: The system can display any number of windows at a time. To route mouse and keyboard input of the appropriate window, the system uses messages queues. The system maintains a single system message queue and one thread-specific message queue for each graphical user interface (GUI) thread. To avoid the overhead of creating a message queue for non-GUI threads, all threads are created initially without a message queue. The system creates a thread-specific message queue only when the thread makes its first call to one of the Use or Windows Graphics Device Interface (GDI) functions. Whenever the user moves the mouse, clicks the mouse buttons, or types on the keyboard, the device driver for the mouse or keyboard converts the input into messages and places them in the system message queue. The system removes the messages, one at a time, from the system message queue, examines them to determine the destination window, and then posts them to the message queue of the thread that created the destination windows. A thread's message queue receives all mouse and keyboard messages for the windows created by the thread. The thread removes messages from its queue and directs the system to send them to the appropriate window procedure for processing with the exception of the **WM_PAINT** message, the **WM_TIMER** message, and the **WM_QUIT** message, the system always posts messages at the end of a message queue. The ensures that a window receives its input messages in the proper first in, first out (FIFO) sequence. The **WM_PAINT** message, the **WM_TIMER** message, and the **WM_QUIT** message, however, are kept in the queue and are forwarded to the window procedure only when the queue contains no other messages. In addition, multiple **WM_PAINT** message for the same window are combined into a single **WM_PAINT** message, consolidating all invalid parts of the client area into a single area. Combining **WM_PAINT** messages reduces the number of times a window must redraw the contents of its client area.

The system posts a message to a thread's message queue by filling an MSG structure and then copying it to the message queue. Information in **MSG** includes; the handle of the window for which the messages is intended, the message identifier, the two messages parameters, the time the message was posted, and the mouse cursor position. A thread can post a message to its own message queue or to the queue of another thread by using the postMessage or postThreadMessage function. A thread can post a message to its own message queue or to the queue of another thread by using the postMessage or postThreadMessage function. An application can remove a message from its queue by using the GetMessage function. To examine a message without removing it from its queue, an application can use the PeekMessage function. This function fills **MSG** with information about the message.

After removing a message from its queue, an application can use the DispatchMessage function to direct the system to send the message to a window procedure for processing. **DispatchMessage** takes a pointer to **MSG** that was filled by a previous call to the **GetMessage** or **peekMessage** function. **DispatchMessage** passes the window handle, the message identifier, and the two message parameters to the window procedure, but it does not pass the time the message was posted or mouse cursor position. An application can retrieve this information by calling the GetMessage Time and GetMessagePos functions while processing a message. A thread can use the WaitMessage function to yield control to other threads when it has no messages in its message queue. The function suspends the thread and does not return until a new message is placed in the thread's message queue. You can call the SetMessageExtraInfo function to associate a value with the current thread's message queue. Then call the GetMessage ExtraInfo function to get the value associated with the last message retrieved by the **GetMessage** or **peekMessage** function.

Non-queued Messages: Non-queued messages are sent immediately to the destination window procedure, bypassing the system message queue and thread message queue. The system typically sends non-queued messages to notify a window of events that affect it. For example, when the user activates a new application windows, the system sends the window a series of messages, including WM_ACTIVATE, WM_SETFOCUS, and WM_SETCURSOR. These messages notify the window that it has been activated, that keyboard input is being directed to the window, and that the mouse cursor has been moved within the borders of the window. Non-queued messages can also result when an application calls certain system functions. For example, the system sends the WM_WINDOWPOSCHANGED message after an application uses the SetWindowPos function to move a window.

Some functions that send non-queued messages are BroadcastSysteMessage, BroadcastSystemMessageEx. SendMessage, SendMessageTimeout and SendNotify Message.

Q.7. Code convention and Calling convention:

Ans. Code Convention and Calling Convention: Code convention: Member functions are functions that are associated with classes

- Data members are variables that are part of classes.

Identifier Names:

- All variable and constant identifiers begin with a lowercase letter.
- Global variables begin with the (for example, the Windows).
- Constants begin with lowercase c (for example, cMaxWindows).
- Class data members begin with its (for example, its Canvas).
- Static data members begin with their (for example, their Total).
- Both member and non-member functions begin with an uppercase letter (for example, TextStyle).
- Names of classes begin with an uppercase letter (for example, Canvas).

Control Structures: The control structures are as follows:

```
if (expression) {
        statements; }  else { statements;)
    for (expression; expression; expression) {statements;}
    do                          { statements;
                                } while (expression);
while (expression)  { statements;}
switch (expression) { caseconstant;
    statements;
    break;
    default;
    statements;
    break; }
```

Calling Convention: This is something that happens at a very low level, say at the time your C or C++ code is compiled to native CPU instructions and comes into the picture then. A good understanding of calling conventions can help you write better code, debug the code better, and finally enable you to use a sources level debugger like **Softies**. When a function is invoked, whether it resides in a DLL or in your own source files, it receives its arguments/parameters through the stack. These arguments/parameters are set up for the called function by the calling function. We can use these parameters, but how does this frame of stack values get clean up ? Of course it'll be done by either the caller or the called. And this, finally, is specified by the specific calling convention used. VC++ compiler always decorates function names that use the_stdcall calling convention. In specific cases where a module definition file has explicitly been used while building the DLL, the functions with_stdcall calling convention are not decorated or mangled. Essentially, one has less work todo if a def file is used while building the DLL.

Q.8. How to create an active X control?

Ans. Creating active X controls means placing the controls in other programs, *i.e.*, create files Boxer Ctl.h and BoxerCtl.cpp connect the WM_LBUTTON DOWN message to control to handle mouse clicks.

```
In CBoxerCtl header:
Class CBoxer Ctrl: public Cole Control
    {
    :
    //Implementation
    protected:
        -CBoxer Ctrl();
        CRect box 1;
        CRect box 2;
        CRect box 3;
        CRect box 4;
        boolean fill I;
        boolean fill 2;
        boolean fill 3;
        boolean fill 4;
        :
        }
        CBoxer Ctrl:: CBoxer Ctrl()
        {
            Initialize IIDs (&IiD_DBoxer, & IIDDBoder Events); Fill 1-
    Fill 2 = fill 3 = fill 4 = false;
        }
        Void CBOxer Ctrl:: OnLButtonDown UINTnFlags, Cpoint point)
        {
            fill 1 = box 1. PtlnRect (point);
            fill 2 ? box 2. PtlnRsct (point);
            fill 3 = box 3. PtlnRect(point);
            fill 4 = box 4. PtlnRect (point);
            Invalidate();
            Coletwatrol:: OriLButtonDown (nFlags, point);
        }
Void CBoxerCtr!:: On Draw (CDC * pDC, const CRect * & rcBound, const CRect & rc
Invalid)
{
        pDc → Fill Rect(rcBouns, CBrush:: From Handle ((HBRUSH)
    Get Stockobject (WHITE-BRUSH)));
    box 1 = CRect (rcBousids. left. rcBounds.trp, rcBounds.right/2,
rcBounds.bottom/2);
    box 2 = CRect (rcBounds. left, rcBounds.button/2, rcBounds.right/2,
rcBounds.bottom);
    box 3 = CRect (rcBounds. right/2, rcBounds.trp, rcBounds.right, r
cBounds.bottom);
    box 3 = CRect (rcBounds.right/'2, rcBounds.bottom/2, rcBounds.right,
rcBounds.bottom);
    pDc → Rectangle (& box 1);
```

```
pDc → Rectangle (& box 2);
pDc → Rectangle (& box 3);
pDc → Rectangle (& box 4);
if (fill I) pDc → Fill Solid Rect (& box 1, RGB (0,0,0));
if (fill 2) pDc → Fill Solid Rect (& box 2, RGB (0,0,0));
if (fill 3) pDc → Fill Solid Rect (& box 3, RGB (0,0,0));
if (fill 4) pDc -* Fill Solid Rect (& box 4, RGB (0,0,0));
}
```

Q.9. Write short notes on the following:

(a) OLE object technologies

(b) Combo Box

(c) Keystokes

(d) API's.

Ans. (a) OLE Object Technologies: Object linking and embedding allows users to create documents that contain embedded sections of documents from other applications and law heat those embedded document sections as through they were a ratural part of current document. In OLE, we can embed items from other application into our program. Ms ignificance can be explained with an example of embedding excel in a program. Embedding enables the user to credit the cells in our program.

When we double-clicks the embedded item, level in automatically open and we can change the data there. When user is done, he can choose the update item in Excel's file menu to update date in our program. The user can also add text to the same document that contains the excel cells. In this way.user law embed and coordinate tem from many different applications into one. Here, the editor program is called container, because can colour our items. The program is responsible for actually maintaining those items. Excel here is called server. All Oil server are regulation with windows in a special file that we can edit with window ability Reg Edit.

(b) Combo Box: Combo box are very popular windows controls. These controls are combination of textbox, drop down list box and a button, the user can press to open the list box.

In this, e.g., we include a combo box and a text box in which to display the users selection.

Steps:

1. Start a program and select dialog based app. wizard program.
2. Initialize combo box by using class wizard to connect to member variable to the combo box control and name thus as m-combo.

Method:

1. **Add String**: Add a string to the end of the list in the list box of combo box.
2. **List Cun Sel**: Return the insides of currently selected item in the list box of combo box.
3. **Get LB Test:** Gets a string from the list box of a combobox. List Add String() method to add the strings.

(c) Keystokes: It is used for getting input from the user. Use VC++ class urgent to convert window message sent to us to a method called event handler. We handle WM-CHAR msg, connecting it to the event handler On Char ( ) which mean when the user strikes a key, the program calls Our OnChar ( ) method.

Divide the program into two parts:

1. Setting up storage for the keystroke.
2. Reading the keystroke.

1. **Setting up Storage for the Keystroke:** Program's document is to hold the data that is display in program's view. An user types key, add it to a string of text, which we store in an MFC string object. In view object, we display this text string. Call Text String as String Date and declare it is the document, i.e., document header full. Next initialize the string to an empty string.
2. **Reading KeyStrokes:** When user types a key, windows sends us a WM_CHAR msg and we connect that to a method in our view class named On Char( ). For then,

Start class wizard, select class wizard item from view menu. This opens class wound. Class wizard connects window message to our program through message map.

***(d)* API's (Application Programming Interface):** The windows API, win API is a microsoft's corset of application programming Interface (APIs) available in the microsoft window operating system. All windows program interact with the window APJ, the windows API is generally meant for the users so they can make the projects or program that is user friendly or we can say that project will be GUI. With GUI, we can also develop the event driven programming also. The microsoft windows application programming interface (API) provides serious used by all window based applications. We can provide our application with GUI; access system resources suqh as memory and derive; display graphics and formatted text; audio, video, networking or security.

EXERCISES

1. Add a button on the custom dialog that calls the MessageBox function with a Yes or No selection. Pass the result back to the main application dialog.
2. How can you specify that the text is to be underlined?
3. How can you print your text upside down?
4. How many times is the EnumFontFamProc callback function called by the operatingsystem?
5. What are the three values that are combined to specify a color?
6. What do you use to draw on windows without needing to know what graphics card the user has?
7. What size bitmap can you use to make a brush from it?
8. What event message is sent to a window to tell it to redraw itself?
9. How can you cause a window to repaint itself?
10. Make the second dialog window resizable, and make it adjust the figures drawn on it whenever it's resized.
11. Add a bitmap brush to the set of brushes used to create the rectangles and ellipses.

❑❑❑

APPENDICES

APPENDIX-A

GLOSSARY

Check Box : A square button in a GUI that can be checked or unchecked by clicking the left mouse button.

Child Window : Technically, and window contained in another window, but more commonly, the document windows of an MDI application.

Class : A class is a data type of objects. Class is synonymous with struct in C++.

Combo Box : A combination list control and edit control. Users may select an item on a drop down list, or may type in an item.

Configuration : A collection of settings maintained by a project.

Console Application : An application that must be run from a DOS console.

Control : Any element that appears on a GUI that accepts user input. Menus, toolbars, dialog boxes, buttons, list controls, etc. are all examples.

Desktop Window : A window that appears on the Windows desktop. Not a document window confined to the inside of another window.

Dialog Box : A control panel window that pops-up to gather more information in response to certain commands.

Docking Window : A window that, in its docked state, sticks to the inner border of a desktop window.

Document : A C++ object that encapsulates the application's data and logic.

Document File : A file that holds the document's data. When such a file is opened, its data is loaded into a document object. Saving a document writes its data to a document file.

Document Window : A window that displays the data contained in a document object.

DOS Console : Also called the Command Prompt, this is an interpreter that accepts DOS commands typed to its prompt.

Edit Control : A GUI control that accepts a typed string..

GUI : Graphical User Interface.

InfoViewer : The MSDS online help system. It consists of a web browser, a search engine, and a database of IVTs.

InfoViewer Title : An infoviewer page containing documentation on a particular topic.

InfoViewer Topic : Same as InfoViewer Title.

IV : InfoViewer

IVT : InfoViewer Title

List Box : A GUI control containing a list of strings the user can select from.

List Control : Like a list box, but it may also contain graphical entities.

MDI : Multiple Document Interface: A Windows application that can have several documents opened at the same time.

Menu : A list of selections that is normally hidden. There are drop down menus (the menu bar) and pop up windows (*e.g.,* shortcut menus).

Menu Bar : The main menu of an application (File, Edit, View, etc.)

MFC : Microsoft Foundation Classes: a framework for creating Windows applications.

Modal Dialog : A dialog box that blocks further work being done in the application until it is dismissed.

Modeless Dialog : A dialog box that allows further user interaction with the application while it remains visible.

MSDS : Microsoft Developers Studio.

Object : The basic component of an object oriented program. An object encapsulates functions and variables. Each object is an instance of a class.

Parent Window : Technically, any window that contains other windows, but more commonly, this is the main window of an MDI application.

Project : A collection of files and configurations.

Push Button : A button on a GUI that can be pushed by clicking on it with the left mouse button.

Radio Button : A circular button that can be turned on and off by clicking on it with the left mouse button. In its on state, a dark, solid circle appears in the center of the circle.

Scroll Bar : The bars on the righ side and bottom of most document and desktop windows. These bars allow users to see all of the data contained in a window.

SDI : Single Document Interface: A Windows application that can only have one document opened at a time.

Setting : The selected options for the compiler, debugger, linker, and other MSDS components. These selections are stored in a configuration.

Slider : A sliding control that appears on a GUI.

Title Bar : The strip across the top of document and desktop windows that displays the window's title.

Tree Control : A list control consisting of leafs items and parent items. A parent item can be expanded to show the items it contains or collapsed to hide its contents. The left half of Windows NT explorer is a list control as is the workspace window in MSDS.

Toolbar : A docking window normally containing push buttons that correspond to menu selections.

UI : User interface.

View : An object that displays the data in a document. Windows are one type of view, although menus, dialogs, and graphs can be views, too.

Windows : Either the Windows 3.1, Windows 95, Windows NT, or Windows 98 operating systems.

windows : Technically, all user interface elements are called windows. A frame is a window that may contain other windows, such as a desktop, document, or docking window.

Windows Application : An application with its own GUI.

Workspace : A collection of projects.

❑❑❑

APPENDIX-B

LAB MANUAL

PROGRAM #1 Study Window's API and Their Relationship with MFC classes.

API is an acronym for Application Programming Interface. It is simply a set of functions that are part of Windows OS. Programs can be created by calling the functions present in the API. The Programmer doesn't have to bother about the internal working of the functions. By just knowing the function prototype and return value he can invoke the API Functions.

A good understanding of Windows API would help you to become a good Windows programmer. Windows itself uses the API to perform its amazing GUI magic.The Windows APIs are of two basic varieties:

- API for 16-bit Windows(Win16 API)
- API for 32-bit Windows(Win 32 API)

Each of these have sub-APIs within it. If you are to program Windows 3.1 then you have to use Win16 API, whereas for programming Windows95 and WindowsNT you have to use Win32 API.

Win16 is a 16-bit API that was created for 16-bit processors, and relies on 16-bit values.

Win32 is a 32-bit API created generation of 32-bit CPUs and it relies on 32-bit values.

Win16 API	Win32 API	Description
USER.EXE	USER32.DLL	The USER components is responsible for window management, including messages, menus, cursors, communications, timer etc.
GDI.EXE	GDI32.DLL	The GDI management is the Graphics Device Interface; it takes care of the user interface And graphics drawing, including Windows metafiles, bitmaps, device contexts, and fonts.
KRNL386.EXE	KERNEL32.DLL	The KERNEL component handles the low level functions of memory, task, and resource Management that are the heart of Windows.

The Win16 and Win32 APIs ere similar in most respects, but the Win16 API can be considered as a subset of Win32 API.Win32 API contains almost everything that the Win16 API has, and much more.

At its core each relies on three main components to provide most of the functionality of Windows. These core components along with their purpose are shown in the table given above.

Although the Win16 versions of these componets have .EXE extensions, they are actually all DLLs and cannot execute on their own.

The Win32 API has many advantages, some obvious and others that are not so obvious. The following lists major advantages that applications have when developed with the Win32 API and a 32-bit compiler (such as Microsoft's Visual C++) :

- **True multithreaded applications:** Win32 applications support true preemptive multitasking when running on Windows 95 and Windows NT.
- **32-bit linear memory:** Applications no longer have limits imposed by segmented memory. All memory pointers are based on the applications virtual address and are represented as a 32-bit integer.
- **No memory model:** The memory models (small, medium, large, etc.) have no meaning in the 32-bit environment. This means there is no need for near and far pointers, as all pointers can be thought of as far.
- **Faster:** A well-designed Win32 application is generally faster. Win32 applications execute more efficiently than 16-bit applications.
- **Common API for all platforms:** The Win32 API is supported on Windows 95, Windows NT on all supported hardware, Windows CE, and the Apple Macintosh.

MFC (Microsoft Foundation's Class)

The C++ class library that Microsoft provides with its C++ compiler to assist programmers in creating Windows-based applications. MFC hides the fundamental Windows API in class hierarchies, so that, programmers can write Windows-based applications without needing to know the details of the native Windows API. The MFC classes are coded in C++. It provides the necessary code for managing windows, menus and dialog-boxes. It also helps in carrying out basic tasks like performing basic input-output, storing the collection of data objects etc. It provides the basic framework for the application on which the programmer can build the rest of the customized code.

MFC Library is a collection of classes in hierarchical form, where classes are derived from some base class. For most of the classes provided by MFC, base class is CObject from which most of the classes are inherited. The rest few classes are independent classes. The classes in MFC Library can be categorized as:

- Root Class: CObject
- MFC Application Architecture Classes
- Window, Dialog, and Control Classes
- Drawing and Printing Classes
- Simple Data Type Classes
- Array, List, and Map Classes
- File and Database Classes
- Internet and Networking Classes
- OLE Classes
- Debugging and Exception Classes

In addition to these classes, the Microsoft Foundation Class Library contains a number of global functions, global variables, and macros.

The Microsoft Foundation Class Library (MFC) supplies full source code. The full source code is supplied in the form of Header files (.H) that are in the MFC\Include directory and implementation files (.Cpp) that are in the MFC\Src directory.

MFC Library was especially designed to provide an object oriented interface to Windows, simultaneously providing the compatibility with the windows programs written in C language. The compatibility with C language was required as the Windows was developed long before MFC came into existence and hence all the applications for Windows were initially written in C, a large number of which is still in use.

MFC Library provides the following features:

- Significant reduction in the effort to write an application for Windows.
- Execution speed comparable to that of the C-language API.
- Minimum code size overhead.
- Ability to call any Windows C function directly.
- Easier conversion of existing C applications to C++.
- Ability to leverage from the existing base of C-language Windows programming experience.
- Easier use of the Windows API with C++ than with C.
- Easier-to-use yet powerful abstractions of complicated features such as ActiveX, database support, printing, toolbars, and status bars.
- True Windows API for C++ that effectively uses C++ language features.

Advantages of MFC: Microsoft Foundation Class Library greatly reduces the development time, makes the code more portable, provides support without reducing the programming freedom and flexibility and provides easy access to user interface elements and technologies like ActiveX and Internet programming which are otherwise very hard to program.

By using MFC, developers can add many capabilities to their applications in an easy, object-oriented manner. MFC simplifies database programming by providing Data Access Objects (DAO) and Open Database Connectivity (ODBC) and network programming through Windows Sockets.

MFC offers many advantages like:

- Application Framework
- Largest base of reusable C++ source code
- Integration with Visual C++
- Flexible and fast database access
- Support for Internet and ActiveX technology
- Support for messaging API
- Support for multithreading

MFC is not only library of classes. MFC is also an application framework.MFC helps to define the structure of an application and handles many routine chores on the application's behalf.

Starting with CWinApp, the class that represents the application itself, MFC encapsulates virtually every aspect of a program's operation. The framework supplies the WinMain() function, and WinMain() in turn calls the application object's member functions to make the program go. One of the CWinApp member functions called by WinMain()- Run() – encapsulates the message loop that literally runs the program.

PROGRAM #2 Study essential classes in Document View Architecture and Their Relationship with each other.

Parts of Application: The application source code generated for you by AppWizard actually consists of 4 major classes. All these classes can be seen by expanding the class listing in the Class View tab. For each of these classes corresponding header file (.h) and an implementation file (.cpp) is generated.

The MainFrameWindow Class: This class is named CMainFrameand is derived from the MFC class CFrameWnd. This class manages the main window of the application that contains the window frame, Menu bar, Toolbar, Status bar, System menu and Minimise, Maximise and Close boxes and also contains the view window of the application.

The Application Class: This class named C ProjectNameAppis derived from the MFC class CWinApp. It manages the application as a whole by managing the tasks, like initialising the application instance, managing the message loop and performing final cleanup of the program.

The Document Class: This class named C ProjectNameDoc is derived from the MFC class CDocument. It is responsible for storing the program data as a single string and reading and writing this data to disk files.

The View Class: The class named C projectNameView is derived from CView class from the MFC classes.It manages the view window, which is used to display program data on the screen and also used in processing the input from the user.

MFC and AppWizard use the Document/View architecture to organize programs written for Windows. Document/View separates the program into four main classes:

- A document class derived from CDocument
- A view class derived from CView
- A frame class derived from CFrameWnd
- An application class derived from CWinApp

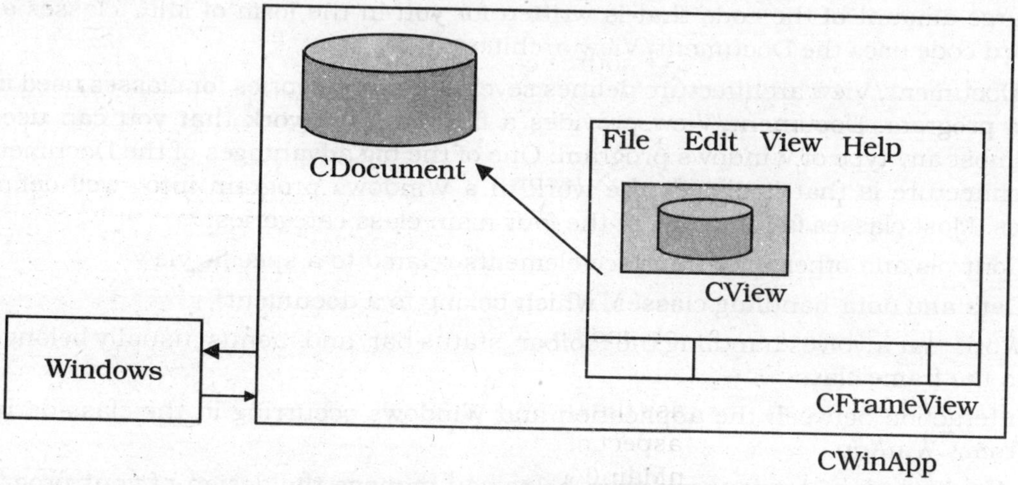

Fig. 1. The Document/View architecture

Each of these classes has a specific role to play in an MFC Document/View application. The document class is responsible for the program's data. The view class handles interaction between the document and the user. The frame class contains the view and other user interface elements, such as the menu and toolbars. The application class is responsible for actually starting the program and handling some general-purpose interaction with Windows. Figure shows the four main parts of a Document/View.

Although the name "Document/View" might seem to limit you to only word-processing applications, the architecture can be used in a wide variety of program types. There is no limitation as to the data managed by CDocument; it can be a word processing file, a spreadsheet, or a server at the other end of a network connection providing information to your program. Likewise, there are many types of views. A view can be a simple window, as used in the simple SDI applications presented so far, or it can be derived from CFormView, with all the capabilities of a dialog box. You will learn about form views in Section 23, "Advanced Views."

SDI and MDI Applications

There are two basic types of Document/View programs:

- SDI, or Single Document Interface
- MDI, or Multiple Document Interface

An SDI program supports a single type of document and almost always supports only a single view. Only one document can be open at a time. An SDI application focuses on a particular task and usually is fairly straightforward.

Several different types of documents can be used in an MDI program, with each document having one or more views. Several documents can be open at a time, and the open document often uses a customized toolbar and menus that fit the needs of that particular document.

Why Use Document/View?

The first reason to use Document/View is that it provides a large amount of application code for free. You should always try to write as little new source code as possible, and that means using MFC classes and letting AppWizard and ClassWizard do a lot of the work for you. A large amount of the code that is written for you in the form of MFC classes and AppWizard code uses the Document/View architecture.

The Document/View architecture defines several main categories for classes used in a Windows program. Document/View provides a flexible framework that you can use to create almost any type of Windows program. One of the big advantages of the Document/View architecture is that it divides the work in a Windows program into well-defined categories. Most classes fall into one of the four main class categories:

- Controls and other user-interface elements related to a specific view
- Data and data-handling classes, which belong to a document
- Work that involves handling the toolbar, status bar, and menus, usually belonging to the frame class
- Interaction between the application and Windows occurring in the class derived from CWinApp

Dividing work done by your program helps you manage the design of your program more effectively. Extending programs that use the Document/View architecture is fairly simple because the four main Document/View classes communicate with each other through well-defined interfaces. For example, to change an SDI program to an MDI program, you must write little new code. Changing the user interface for a Document/View program impacts only the view class or classes; no changes are needed for the document, frame, or application classes.

Documents

Documents are the basic elements that are created and manipulated by the application. Windows provides a graphic interface that gives a user a natural way to use the application. To implement this interface, the developer has to provide ways to see and interact with the information that the application creates and uses.

A document is simply a place to collect common data elements that form the processing unit for the application.

Documents and the Application Relationship

The application class uses documents as a way to organize and present information to the user. Each application derived from the MFC defines at least one type of document that is a part of the application. The type of document and the number of documents that the application uses are defined in the code of the application class.

As we have already discussed, MFC supports two types of applications – MDI and SDI. MDI - In applications that support Multiple Document Interface, a number of text files can be opened for editing at once; each in a different window. Each of the open files has a corresponding document. Also, in MDI, the same document can have multiple views, where a window is split. Each pane of the window can show a different portion of the data whereas this data is coming from a common source, the document.

SDI - In applications with Single Document Interface only one document is open at a time. A SDI application does not have a Window menu and the File menu does not have a Close option because only one document can be open at a time, opening a document automatically closes the current document. These are two common characteristics of a SDI application.

Why Documents

The fundamental responsibility of the document is to store all the elements that constitute an application unit. An application may support more than one type of data like numbers, text, drawings etc.

The document also controls all the views associated with it. As the user opens and manipulates windows on the screen, views are created and the document is associated with each view. The document is responsible for controlling and updating the elements within a given view.

The view may either request the document to draw its components or directly request the components to draw themselves. It must provide a device context where the drawing occurs. All elements of a drawing must be able to display themselves correctly upon request.

Views

The view is the user's window to the document. The user interacts with a document using the view. Each active document will have one or more active views available on the display. Generally, each view is displayed in a single window.

Why Views

The view gives the document a place to display information. It is the intermediary between the document, which contains the information and the user.

The view organises and displays the document information onto the screen or printer and takes in the user input as information or operation on the document.

The view is the active area of the document. It has two functions:

- Acts as the display area for the document data and
- Acts as the input area where the user interacts with the document, normally providing additional commands or data for the document to process.

All the information passes through the view before reaching the document. Therefore, if an application handles mouse messages in functions like OnLButtonDown(), it first receives the message, translates the information into an appropriate form and then calls the required document function to process the mouse command.

Although view is responsible for displaying the document and its information, there are two possibilities:

- Let the view directly access the document's data elements
- Create a document function that handles access to the appropriate data member

The choice of writing directly in the view has two advantages:

- It is the responsibility of the view to handle document display, so having the code there is more appropriate.
- Placing the code in the view keeps all the code regarding the device context in the view where it seems natural.

A view is always tied to a single document.

One document can have several views.

Opening another window on the document can create two views of the same document. Alternative ways can be presented to look at the same information by implementing different types of views for the document. For example, as we discussed in the beginning of the chapter, in MS Excel same data can be viewed either in the form of a table or in the form of different graphs.

Gaining Access to Document Data from the View

The view accesses its document's data either with the GetDocument()function, which returns a pointer to the document or by making the view class a C++ friend of the document class.

The view then uses its access to the data to obtain the data when it is ready to draw or otherwise manipulate it.

For example, from the view's OnDraw() member function, the view uses GetDocument() to obtain a document pointer. Then it uses that pointer to access a CString data member in the document. The view passes the string to the TextOut() function.

PROGRAM #3 Draw a free hand drawing as Mouse is Drag.

```
#include<windows.h>


long _stdcall func(HWND,UINT,UINT,long);
WNDCLASS a;
int flag=0,x1,y1,x2,y2;
```

```
int _stdcall WinMain(HINSTANCE i,HINSTANCE j,char *k,int l)
{
    HWND h;
       MSG m;

       a.hInstance=i;             //instance handle
    a.lpszClassName="preeti";  //long pointer to class name
       a.lpfnWndProc=func;        //long pointer to Window procedure
    a.hbrBackground=(HBRUSH)GetStockObject(WHITE_BRUSH);
                               //set background color of Window Client area

    RegisterClass(&a);         // Register the Window Class before use
    h=CreateWindow("preeti","Title",WS_OVERLAPPEDWINDOW,20,20,300,200,
0,0,i,0);
    ShowWindow(h,3);           // Create Window Application

    while(GetMessage(&m,0,0,0)) // extracts message from Message Queue
       DispatchMessage(&m);    // dispatches the message received
    return 0;
}

long _stdcall func(HWND w,UINT x,UINT y,long z)
{
    HDC d;
    switch(x)
    {
    case WM_LBUTTONDOWN:
       if(flag==0)
       {
          x1=LOWORD(z);
          y1=HIWORD(z);
          flag=1;
       }
          break;

    case WM_MOUSEMOVE:
       if(flag==1)
       {
          x2=LOWORD(z);
```

```
        y2=HIWORD(z);
        d=GetDC(w);
            MoveToEx(d,x1,y1,0);// current position is updated to (x1,y1)
        LineTo(d,x2,y2);       // draw line from current pos to (x2,y2)
        ReleaseDC(w,d);
        x1=x2;
        y1=y2;
        }
        break;

    case WM_LBUTTONUP:
        flag=0;
        break;

    case WM_DESTROY:
        PostQuitMessage(0);  // used to terminate while loop in WinMain()
        break;

    default:
        return DefWindowProc(w,x,y,z);
    }
    return 0L;
}
```

/* Output*/

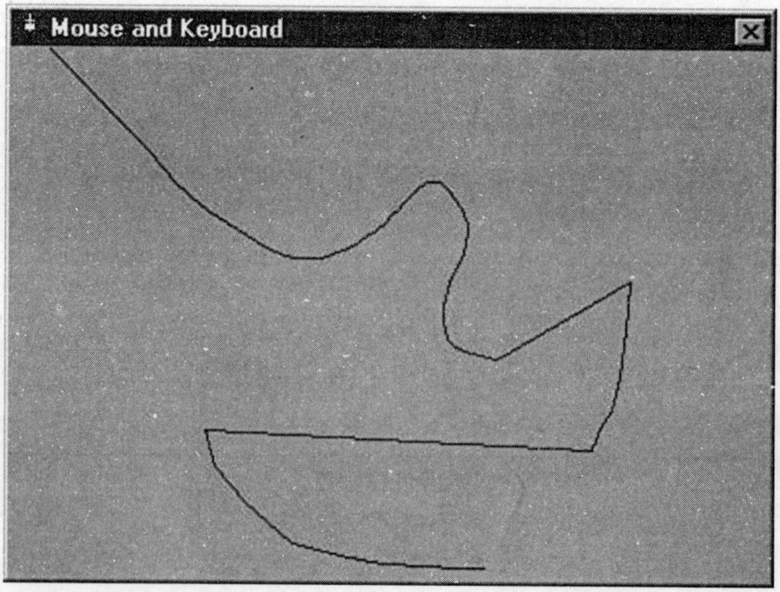

PROGRAM # 4 Create a simple model dialog box to read the information about a student i.e. name,rollno.,class using simple fields.

Difference between Modal and Modless Dialog

In some discussion boards i saw lot of questions regarding what is the difference between the modal dialog and modeless dialog boxes. as to my opinion modal dialog if once invoked it will not allow to access the parent window. but modeless dialog is will allow the user to access the parent window.

How to create a modal dialog

A modal dialog can easily created using the following steps.

Start your visual studio Editor 6

Step 1

Select File->New

in the project tab select MFC appWizard(exe)

Give "DialogModalDemo" in the project name text box

Give OK

Step 2

Select single document

uncheck document/architecture support

click next

Step 3

Click Finish

Now you created the farameword for implementing or Modal and Modeless dialog

Step 4

Now go to the Resource view and select the menu item on there.

Add a New Menu item as Dialog Demo to the main menu and also add tw2o sub menu item as "Modal Dialog" and "Modeless Dialog"

Step 5

Now press the Ctrl+R and add a new dialog Templete to the project

Rename the template id from IDD_DIALOG1 to IDD_MODAL_DIALOG

Now go to to the CMainFrame class and press Ctrl + W

add a handler for the ID_DIALOGDEMO_MODALDIALOG item

Now go to the CFrameWnd eader implementation file and add #include "ModalDialog.h" on there now go to the handler CMainFrame::OnDialogdemoModaldialog() that we added and add the code on there as following looks like

```
ModalDialog dlg;
dlg.DoModal();
```

Now run the application and Go to the Modal Dialog Menu and click it. Check weather it is working file. If not Please let me know.

How to Create Modless Dialog

Now i will explain how you can create a modeless dialog.

Go to the CmainFrame class and add a handler for the ID_DIALOGDEMO_MODELESSDIALOG menu item to the CMainFrame class. Next go to the CMainFrame window class definition header file and decalre a variable of ModalDialog in the CMainFrame class as like following

ModalDialog modelessdialog;

Now go to the handler void CMainFrame::OnDialogdemoModelessdialog() that we have created befor and add the following code on there.

```
modelessdialog.Create(IDD_MODAL_DIALOG,this);
modelessdialog.ShowWindow(SW_SHOW);
```

Now run the application and check weather it is working for you.

PROGRAM#5 Write a simple console application to create a archive class object from file class that reads and stores a simple structure.

Creating an MFC Application

1. The gdcs1 project should still be opened.

   On the main menu, click Project -> gdcs1 Properties...

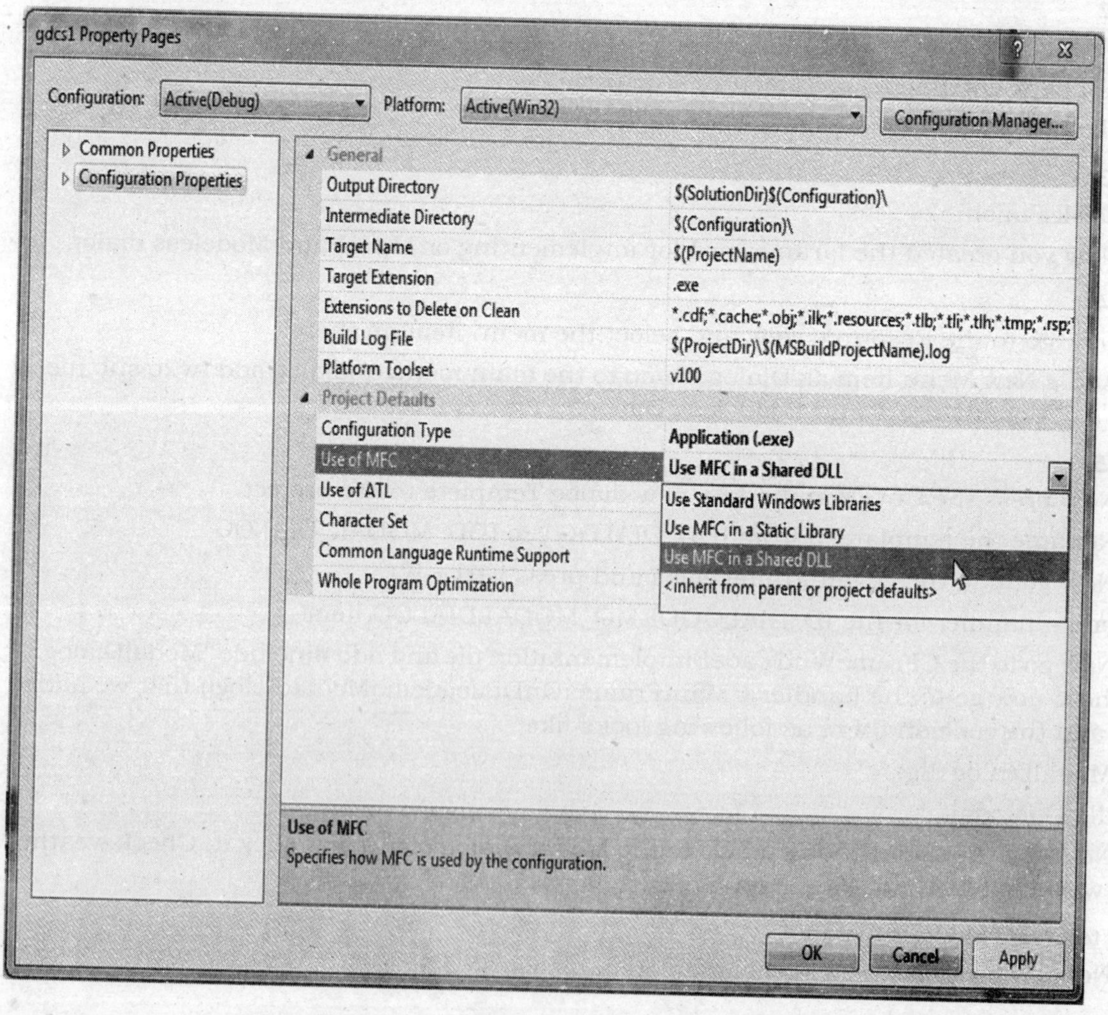

2. In the left section, click Configuration Properties (or double-click it to expand; then, under Configuration Properties, click General).
   In the right list, click Use of MFC to reveal its combo box
3. Click the arrow of the Use of MFC combo box and select Use MFC in a Shared DLL.
4. Click OK
5. Change the file as follows:
6. #include <iostream>
7. #include <Windows.h>
8. using namespace std;
9.
10. void main()
11. {
12.   printf_s("Georgetown Dry Cleaning Services");
13.
14.   getchar();
    }
15. To execute the application, on the main menu, click Debug -> Start Debugging
16. Press Enter to close the DOS window and return to your programming environment

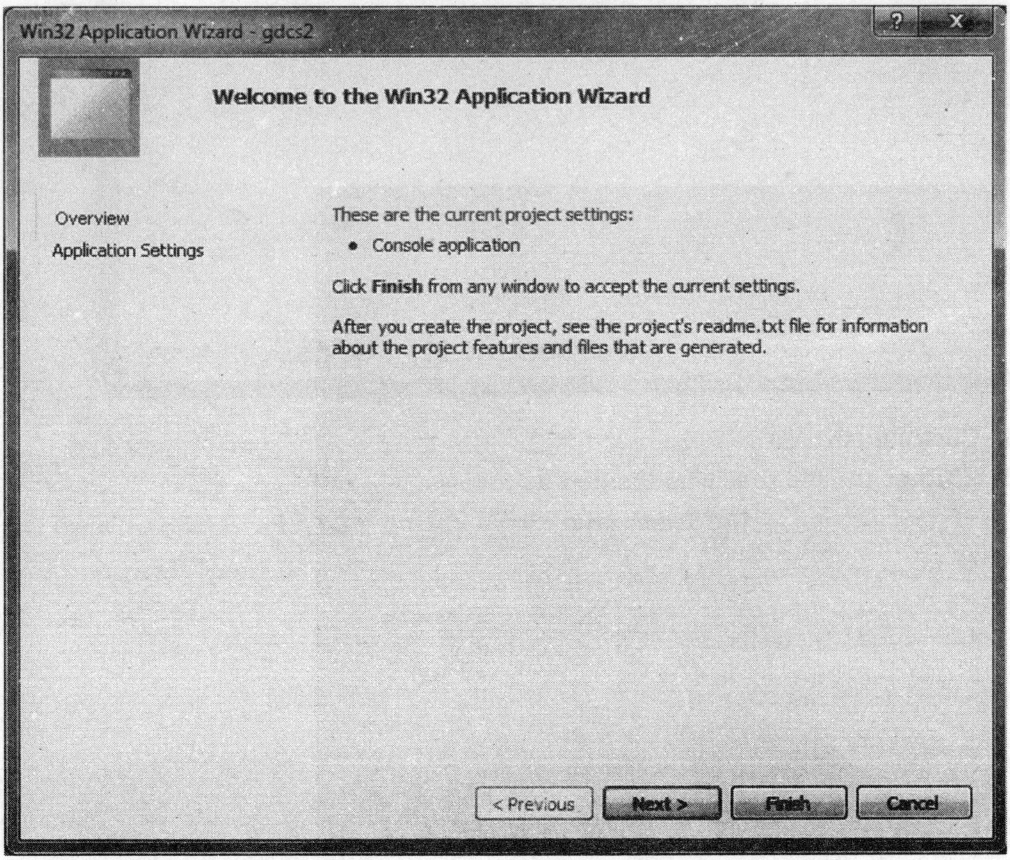

17. To start a new application, on the main menu, click File -> New Project...
18. In the middle list, click Win32 Console Application
19. Change the name to **gdcs2** and click OK
20. In the first page of the wizard, click Next
21. In the second page of the Win32 Application Wizard - gdcs2, make sure Console Application is selected. Click Precompiled Header (or make sure it is clicked). Click the MFC check box.

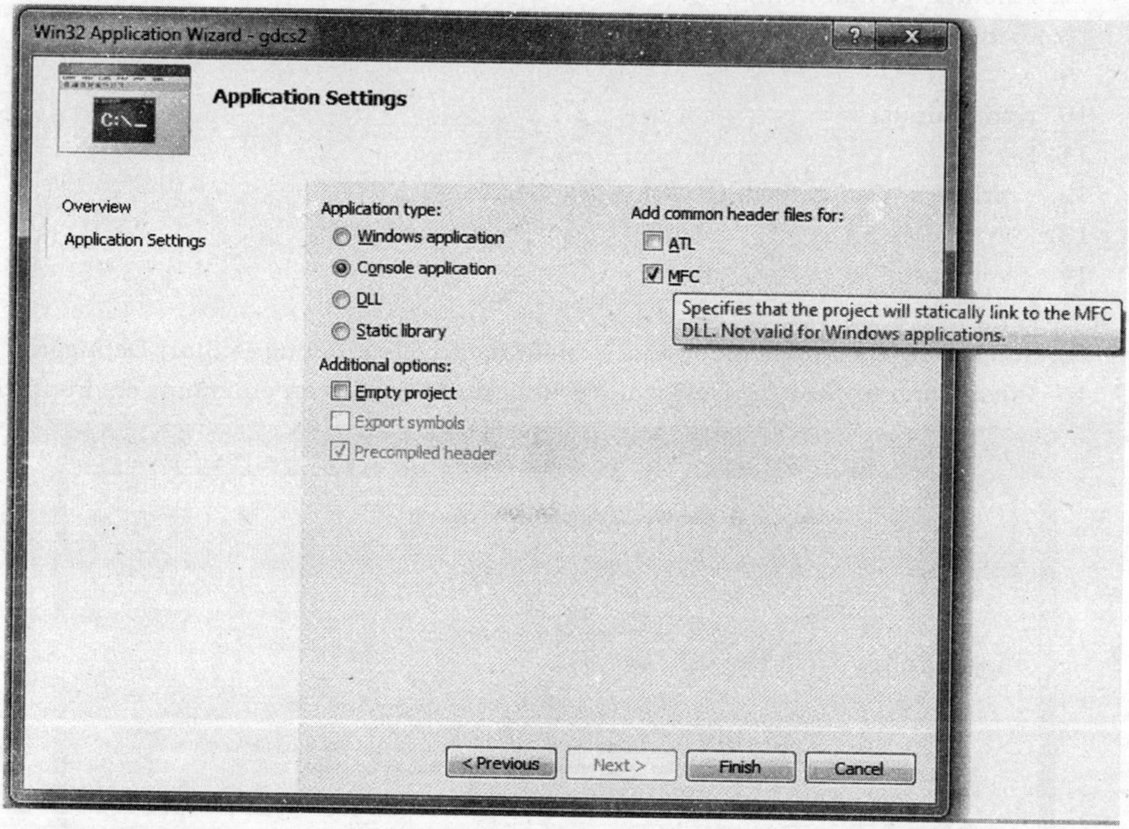

22. Click Finish
23. Change the file that was created as follows:

```
// gdcs2.cpp : Defines the entry point for the console application.
//

#include "stdafx.h"
#include "gdcs2.h"

#ifdef _DEBUG
#define new DEBUG_NEW
#endif
```

```
// The one and only application object

CWinApp theApp;

using namespace std;
int _tmain(int argc, TCHAR* argv[], TCHAR* envp[])
{
    int nRetCode = 0;

    HMODULE hModule = ::GetModuleHandle(NULL);

    if (hModule != NULL)
    {
        // initialize MFC and print and error on failure
        if (!AfxWinInit(hModule, NULL, ::GetCommandLine(), 0))
        {
            // TODO: change error code to suit your needs
            _tprintf(_T("Fatal Error: MFC initialization failed\n"));
            nRetCode = 1;
        }
        else
        {
            // TODO: code your application's behavior here.
            printf_s("Georgetown Dry Cleaning Services - ");
            cout << "Processing Customers Orders - ";
        }
    }
    else
    {
        // TODO: change error code to suit your needs
        _tprintf(_T("Fatal Error: GetModuleHandle failed\n"));
        nRetCode = 1;
    }

    return nRetCode;
}
```

24. To execute the application, on the main menu, click Debug -> Start Debugging Press Enter to close the DOS window and return to your programming environment.

PROGRAM#6 Create a simple database in MS Access Database and connect it to VB using DAO or ADO.

To use the dao library, go to the menu Project->References... This brings up a list of the libraries that your project is currently referencing and can hence use, along with a list of the other available libraries on your computer. Scroll down to "Microsoft DAO 3.6 Object Library" and click in the check box to reference it. If it does not exist in the list, you may not have this library installed. If there is a similar item, use that. Otherwise try clicking on "Browse..." and going to "C:\Program Files\Common Files\Microsoft Shared\Dao\" This directory should contain the DAO libraries.

Click OK to close the references window and then go to "View->Object Browser" or press F2 to open the Object Browser window. This window can be used to view the current libraries that you are referencing - it is a very handy tool. Select "DAO" from the drop down list at the top left and you will be able to see everything in the DAO library. For instance, if you click on "Database" in the list on the left, this will then list all of the properties, events and methods that are available for the Database object.

Creating a Database

Creating a database is very easy. In the "globals" list for the DAO library, there is a method "CreateDatabase".

Function CreateDatabase(Name As String, Locale As String, [Option]) As Database

Name is the full path to the database that you want to create.

Locale is one of the constants defined in LanguageConstants in the Object Browser.

We are not considering the Option argument.

Place the following code into a Form.

```
Private Sub CreateDB(Path As String)
' Define a Database variable
Dim OurDB As Database

' Create our database
Set OurDB = CreateDatabase(Path, dbLangGeneral)

' OurDB now refers to the database we just created. We don't want to do anything else,
so close it.
OurDB.Close

' Destroy the database object to free memory
Set OurDB = Nothing
End Sub

Private Sub Form_Load()
CreateDB "C:\OurDB.mdb"
End Sub
```

Running this program will create a database at C:\OurDB.mdb and then close it. If you have MS Access installed, you can open this file - some versions of Access will try to persuade you to convert the database to a newer version. Don't do this or you won't be able to access it with your program again.

You will notice that the database is a bit useless as it stands - there is nothing whatsoever in it. The next step is to add a table so that we can store some information. Creating a table is done with the CreateTableDef function which is a member of the Database object.

Function CreateTableDef([Name], [Attributes], [SourceTableName], [Connect]) As TableDef

Add the following code to CreateDB() after the line "Set OurDB = CreateDatabase(...)"

```
Dim OurTable As TableDef

' Create a table
Set OurTable = OurDB.CreateTableDef("OurTable")

' Add some fields to the table
OurTable.Fields.Append OurTable.CreateField("Name", dbText)
OurTable.Fields.Append OurTable.CreateField("Age", dbInteger)

' Add the table to our database
OurDB.TableDefs.Append OurTable

Set OurTable = Nothing
```

This requires some further explanation. One of the properties of a Table is the Fields property. This contains a collection of Field objects that define all of the information for the Table fields. We want to add some fields to our table, so we use the Fields.Append method which takes a Field as an argument. To create this field we have used the Table method CreateField.

Function CreateField([Name], [Type], [Size]) As Field

The Field types are defined in DataTypeEnum which can be found in the Object Browser.

In the same way that a Table has a collection of Fields, a Database has a collection of Tables called TableDefs. Our Table can be added to the collection using the TableDefs.Append method.

Before you try and run this code you will need to delete the file C:\OurDB.mdb.

If you open C:\OurDB.mdb in Access again, you will now be able to see a Table called "OurTable". Opening the table will show the two fields that we added.

Adding Data

Adding data is a little more complicated. First let us assume that we have already got a database which of course we do. We will now look at opening an existing database.

Add the following code to the form

```
Private Sub AddRecord(Path As String)
Dim OurDB As Database
Set OurDB = OpenDatabase(Path)

OurDB.Close
Set OurDB = Nothing
End Sub
```

This is the code to open and close an existing database. To add data we need to open a record set. Add the next piece of code before the "OurDB.Close" line.

```
Dim OurRS As RecordSet
' Open the table as a recordset
Set OurRS = OurDB.OpenRecordset("OurTable", dbOpenTable)
' Add a new record
OurRS.AddNew
' Set the values of the new record
OurRS.Fields(0).Value = InputBox$("Input name")
OurRS.Fields(1).Value = CInt(InputBox$("Input age"))
OurRS.Update
OurRS.Close
```

Add a command button called cmdAdd and add this code to the form:

```
Private Sub cmdAdd_Click()
AddRecord "C:\OurDB.mdb"
End Sub
```

Now you can add a new record to the table by clicking the cmdAddButton. Open the database in Access to view the changes.

Reading Data

After having added data, reading it back again is not so difficult.

```
Private Sub ReadAllData(Path As String)
Dim OurDB As Database
Dim OurRS As Recordset
Dim i As Integer

Set OurDB = OpenDatabase(Path)
Set OurRS = OurDB.OpenRecordset("OurTable", dbOpenTable)
If OurRS.RecordCount > 0 Then
' Go to the first record in the list
OurRS.MoveFirst
```

```
For i = 0 To OurRS.RecordCount - 1
Debug.Print OurRS.Fields(0).Value & " " & OurRS.Fields(1).Value
OurRS.MoveNext
Next
End If
OurRS.Close
OurDB.Close
Set OurRS = Nothing
Set OurDB = Nothing
End Sub
```

Deleting Data

To delete a record, we first have to find it. When we have an open recordset the simplest way to do this is in a similar way to our ReadAllData method.

```
Private Sub DeleteRecord(Path As String)
Dim OurDB As Database
Dim OurRS As Recordset
Dim i As Integer
Dim OurAge As Integer
Dim OurName As String

OurName = InputBox$("Input name to delete")
OurAge = CInt(InputBox$("Input age"))

Set OurDB = OpenDatabase(Path)
Set OurRS = OurDB.OpenRecordset("OurTable", dbOpenTable)
If OurRS.RecordCount > 0 Then
OurRS.MoveFirst
For i = 0 To OurRS.RecordCount - 1
If LCase$(OurRS.Fields(0).Value) = LCase$(OurName) And OurRS.Fields(1).Value =
OurAge Then
' Delete the currently selected record
OurRS.Delete
Exit For
End If
OurRS.MoveNext
Next
End If
OurRS.Close
OurDB.Close
```

```
Set OurRS = Nothing
Set OurDB = Nothing
End Sub
```

This assumes that we want to delete the first matching record and that the name is not case sensitive.

Editing Data

My record editing example is identical to the record deleting example but substituting the "OurRS.Delete" line with:

```
OurRS.Edit
OurRS.Fields(1).Value = CInt(InputBox$("Input new age"))
OurRS.Update
```

This allows you to change the age of the people in the database.

Data Bound Controls

A data bound control is a control that connects to a data source - a database. Many controls can be data bound, we are going to look at just two though; the Microsoft ADO Data Control and the Microsoft Hierarchical FlexGrid Control. To add these controls to your project, go to Project->Components... and select them from the list.

Add one each of the two new controls to your form. For simplicity we shall keep the default names of MSHFlexGrid1 and Adodc.

If you now select MSHFlexGrid1 and look at the DataSource property, you should see that there is an entry for Adodc1 - this means that the Adodc is a compatible data source for the HFlexGrid. Our Adodc1 control will be bound to our database and it will provide the data to MSHFlexGrid1 for displaying.

Edit the Form_Load event so that it appears as follows:

```
Private Sub Form_Load()
' Connect to our database
Adodc1.ConnectionString = "Provider=Microsoft.Jet.OLEDB.4.0;Persist Security
Info=False;Data Source=C:\OurDB.mdb"

' Select the records that we want to view
Adodc1.RecordSource = "select Name,Age from OurTable"

' Refresh Adodc1 and any connected controls (ie. MSHFlexGrid1)
Adodc1.Refresh
End Sub
```

If you are using a different version of the DAO library, the provider string may need to be altered slightly. For dao350.dll for instance, use

```
Adodc1.ConnectionString = "Provider=Microsoft.Jet.OLEDB.3.51;Persist Security
Info=False;Data Source=C:\OurDB.mdb"
```

Other than that, the only important part for us is "...Source=C:\OurDB.mdb". This defines which database to use and could of course be specified with a variable:

```
Adodc1.ConnectionString = "Provider=Microsoft.Jet.OLEDB.3.51;Persist Security
Info=False;Data Source=" & Path
```

The RecordSource statement uses SQL. SQL stands for Structured Query Language and is a standard way of selecting data from a database. This is not meant to be a tutorial on SQL so I shall not be offering many different commands to put in this string. In the form shown, we are selecting the fields Name and Age of all records from the table OurTable. If we only wanted to look at the Name field we could use the following line:

```
Adodc1.RecordSource = "select Name from OurTable"
```

Add a command button called cmdAdd to the form along with this code:

```
Private Sub cmdAdd_Click()
AddRecord "C:\OurDB.mdb"
Adodc1.Refresh
End Sub
```

Then run the program and click on the button to add a new record. After you have finished the HFlexGrid should show your new record.

Searching for Data

A very important part of databases is the ability to search for data. Add a text box called txtSearch and two command buttons called cmdSubSearch and cmdWordSearch to the form. We are going to use an extended version of the SQL select command to do our searching:

```
Private Sub cmdSubSearch_Click()
Adodc1.RecordSource = "select Name,Age from OurTable where instr(Name,"" &
txtSearch.Text & "")>0"
Adodc1.Refresh
End Sub

Private Sub cmdWordSearch_Click()
If Len(txtSearch) > 0 Then
Adodc1.RecordSource = "select Name,Age from OurTable where Name= "" &
txtSearch.Text & """
Else
Adodc1.RecordSource = "select Name,Age from OurTable"
End If
Adodc1.Refresh
End Sub
```

cmdSubSearch searches Name for the substring specified in txtSearch whereas cmdWordSearch searches for the exact name.

An alternative way of doing this would be to put the code from cmdSubSearch in the txtSearch_Change event - the search would then occur as you type.

PROGRAM#7 Write a simple program that reads a text and changes its font, font size as selected by the user and display it in a list box.

1. Open Visual C++ and click the New item in the File menu, opening the New dialog box.
2. Now select the MFC AppWizard(exe) entry in the New dialog box.
3. Give the new program the name Keystrokes in the Project name box.
4. Click OK to starts the Visual C++ AppWizard.
5. Click the option marked Single Document in the AppWizard, click Finish.
6. Here, the AppWizard indicates four classes will be created : CKeystrokesApp, CMainFrame, CKeystrokesDoc, CKeystrokesView .
7. Declare the StringData variable on the document's header file, `KeystrokesDoc.h, in the protected part.`

```
class CKeystrokesDoc : public CDocument
{
protected: // create from serialization only
        CKeystrokesDoc();
          DECLARE_DYNCREATE(CKeystrokesDoc)
        CString StringData;
        .
        .
};
```

8. Initialize that string to an empty string -" "- in the document's constructor, which we find in KeystrokesDoc.cpp.

```
CKeystrokesDoc::CKeystrokesDoc()
{
  StringData=" ";
  // TODO: add one-time construction code here
}
```

9. Add a new event handler –OnChar( )- to our view class which is called every time the user types the character.
10. The character the user typed is now in the nChar parameter, which we store in the our data string object, StringData. The object in our document, so we need a pointer(pDoc) to our document object.
11. Add the character nChar to the string StringData.

```
void CKeystrokesView::OnChar(UINT nChar, UINT nRepCnt,UINT nFlags)
{
```

```
// TODO: Add your message handler code here and/or call default
        CKeystrokesDoc* pDoc=GetDocument();
    ASSERT_VALID(pDoc);
        pDoc->StringData+=nChar;
    Invalidate();

        CView::OnChar(nChar, nRepCnt, nFlags);
}
```

12. We'll handle the display of our data in the view's OnDraw( ). We need to draw the text string, which we do with TextOut( ).

```
void CKeystrokesView::OnDraw(CDC* pDC)
{
        CKeystrokesDoc* pDoc = GetDocument();
    ASSERT_VALID(pDoc);

        pDC->TextOut(0,0,pDoc->StringData);
    // TODO: add draw code for native data here
}
```

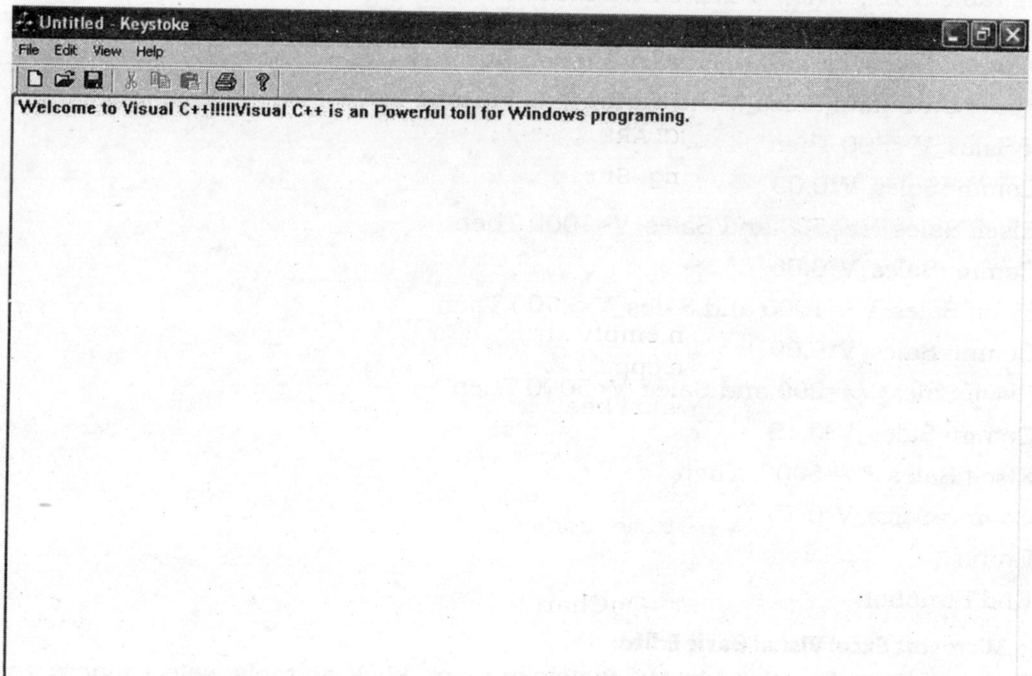

PROGRAM#8 With the help of VB create an object of excel application and implement any function of it.

The Needs to Create VBA Functions in MS-Excel

You can create your own functions to supplement the built-in functions in Microsoft Excel spreadsheet, which are quite limited in some aspects. These user-defined functions are also called Visual Basic for Applications functions, or simply VBA functions. They are very useful and powerful if you know how to program them properly. One main reason we need to create user defined functions is to enable us to customize our spreadsheet environment for individual needs. For example, we might need a function that could calculate commissions payment based on the sales volume, which is quite difficult if not impossible by using the built-in functions alone. The code for VBA is illustrated on the right.

Table : Commissions Payment TableSales

Volume($)	Commissons
<500	3%
<1000	6%
<2000	9%
<5000	12%
>5000	15%

In table 15.1, when a salesman attain a sale volume of $6000, he will be pjaid $6000x15%=$720.00. A visual basic function to calculate the commissions can be written as follows:

```
Function Comm(Sales_V  As Variant) as Variant
If Sales_V <500 Then
Comm=Sales_V*0.03
Elseif Sales_V>=500 and Sales_V<1000 Then
Comm=Sales_V*0.06
Elseif Sales_V>=1000 and Sales_V<2000 Then
Comm=Sales_V*0.09
Elseif Sales_V>=200 and Sales_V<5000 Then
Comm=Sales_V*0.12
Elseif Sales_V>=5000  Then
Comm=Sales_V*0.15
End If
End Function
```

Using Microsoft Excel Visual Basic Editor

To create VBA functions in MS Excel, you can click on tools, select macro and then click on Visual Basic Editor as shown in Figure 15.1.

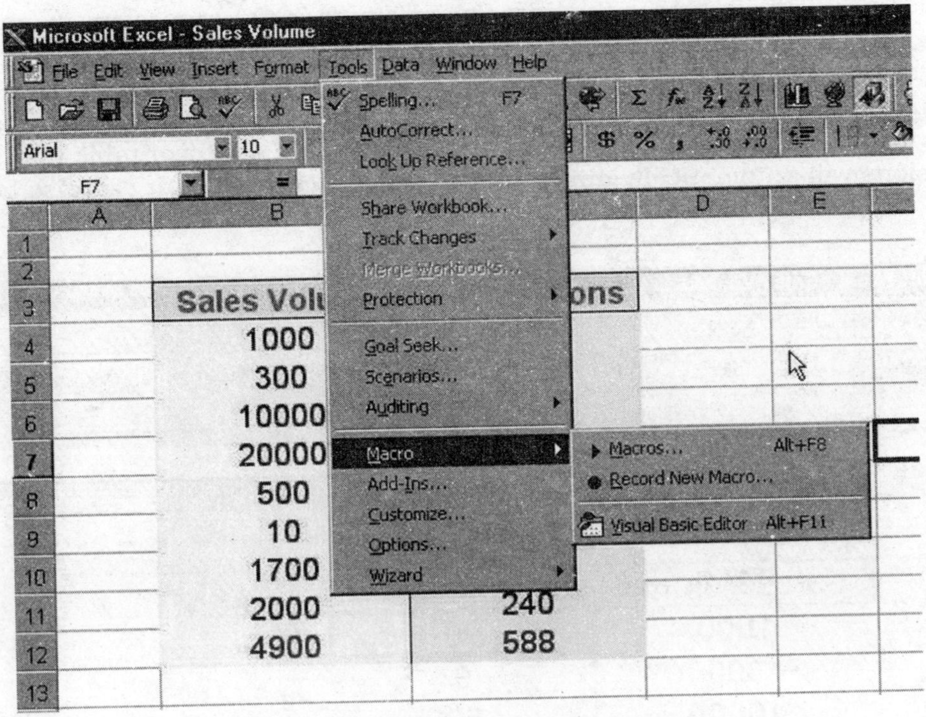

Fig. 1. Inserting MS_Excel Visual Basic Editor

Upon clicking the Visual Basic Editor, the VB Editor windows will appear as shown in figure 15.2. To create a function, type in the function as illustrated in section 15.1 above After typing, save the file and then return to the Excel windows.

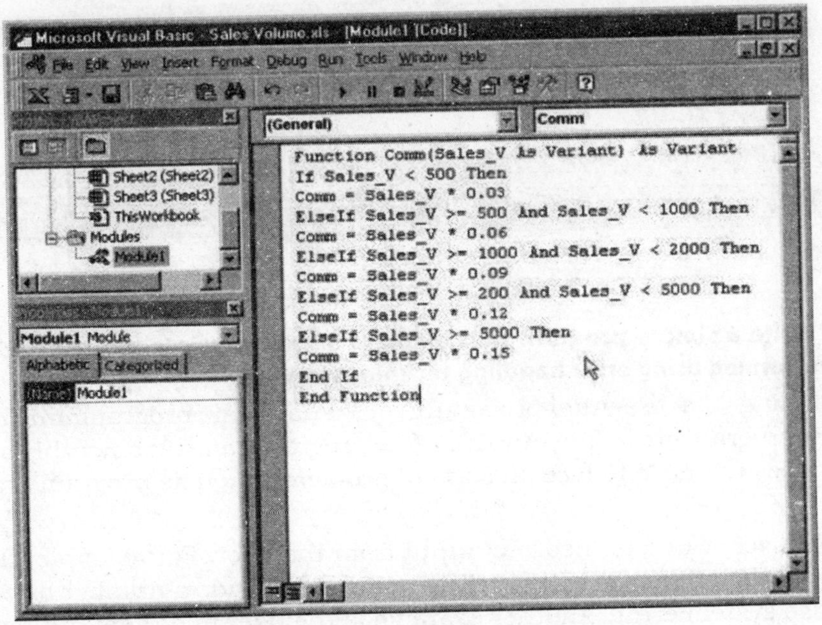

Fig. 2. The VB Editor

In the Excel window, type in the titles Sales Volume and Commissions in any two cells. By referring to Fig. 3.key-in the Comm function at cell C4 and by referencing the value in cell B4, using the format Comm(B4). Any value appear in cell B4 will pass the value to the Comm function in cell C4. For the rest of the rows, just copy the formula by dragging the bottom right corner of cell C4 to the required cells, a nice and neat table that shows the commissions will automatically appear (as shown in figure 15.3). It can also be updated anytime.

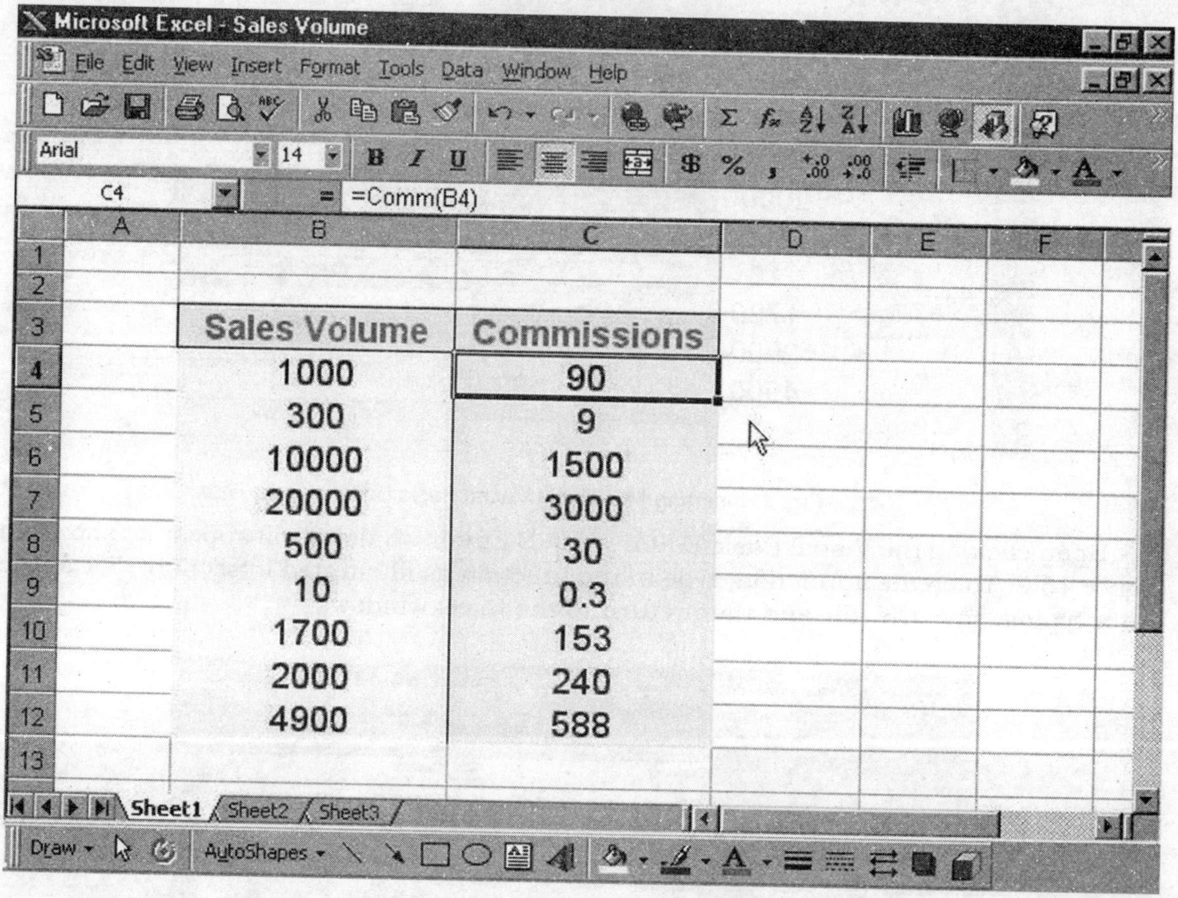

Fig. 3. MS Excel Windows – Sales Volume

PROGRAM#9 Write a simple program that displays an appropriate message when the illegal operation is performed using error handling technique in VB.

Error handling is an essential procedure in Visual Basic programming because it can help make the program error-free. An error-free program can run smoothly and efficiently, and the user does not have to face all sorts of problems such as program crash or system hang.

Errors often occur due to incorrect input from the user. For example, the user might make the mistake of attempting to ask the computer to divide a number by zero which will definitely cause system error. Another example is the user might enter a text (string) to a box that is designed to handle only numeric values such as the weight of a person, the computer will not be able to perform arithmetic calculation for text therefore will create an error. These errors are known as synchronous errors.Therefore a good programmer should be more alert to the parts of program that could trigger errors and should write errors handling code to help the user in managing the errors. Writing errors handling code should be considered a good practice for Visual Basic programmers, so do try to finish a program fast by omitting the errors handling code. However, there should not be too many errors handling code in the program as it create problems for the programmer to maintain and troubleshoot the program later.

Writing the Errors Handling Code

We shall now learn how to write errors handling code in Visual Basic. The syntax for errors handling is

```
On Error GoTo    program_label
```

where ***program_label*** is the section of code that is designed by the programmer to handle the error committed by the user. Once an error is detected, the program will jump to the ***program_label*** section for error handling.

Example : Division by Zero

```
Private Sub
CmdCalculate_Click()

Dim firstNum, secondNum As
Double
firstNum =
Txt_FirstNumber.Text
secondNum =
Txt_SecondNumber.Text
On Error GoTo error_handler
Lbl_Answer.Caption = firstNum
/ secondNum
Exit Sub   'To prevent error handling even the inputs are valid
error_handler:
Lbl_Answer.Caption = "Error"
Lbl_ErrorMsg.Visible = True
Lbl_ErrorMsg.Caption = "You attempt to divide a number by zero!Try again!"

End Sub

Private Sub
Txt_FirstNumber_GotFocus()

Lbl_ErrorMsg.Visible = False

End Sub

Private Sub
```

```
Txt_SecondNumber_GotFocus()

Lbl_ErrorMsg.Visible = False

End Sub
```

The Output Window

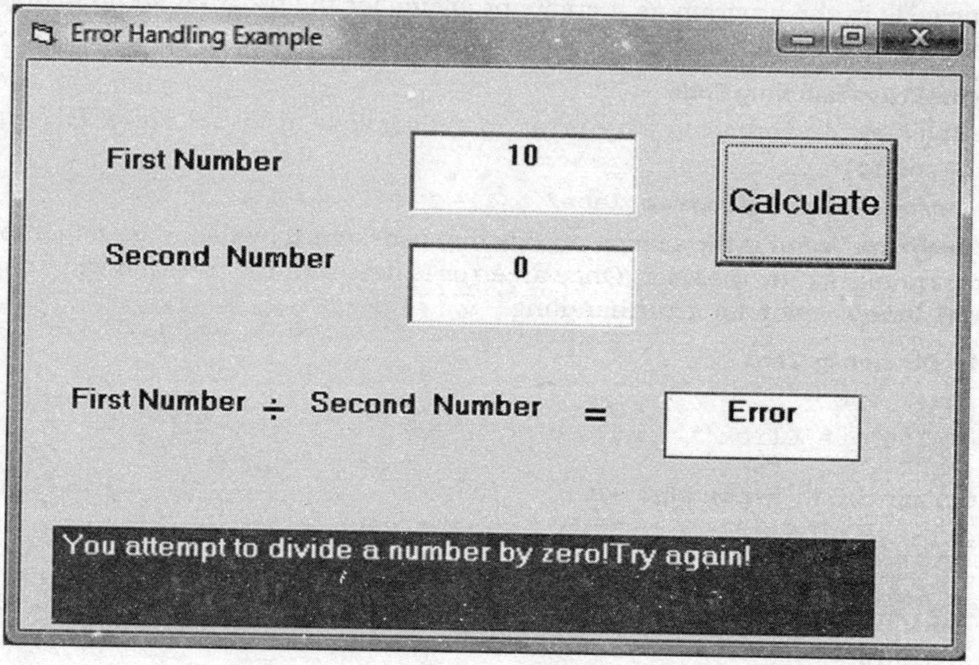

Explanation

In this example, you design the interface as above. Name the first textbox as Txt_FirstNumber and the second textbox as Txt_SecondNumber. Insert one command button as label it as Calculate. Insert one label and name it as Lbl_Answer to display the answers. If the user enter 0 in the second textbox as shown above, the program will jump to the label **error_handler**, and the procedure that is executed. It will show an error in the Txt_Answer label and an error message in the Lbl_ErrorMsg label.

Notice that Exit sub after the division. It is important because it can prevent the program to execute the error_handler code even though the user does not enter zero in the second textbox.

Lastly, after the error message appeared, the user will click on the textboxes again. When this occur, the error message will disappear both from the answer label and error message label. This is achieved by using he event procedure GotFocus, as shown in the code.

Example : Nested Error Handling Procedure

We need to consider other errors probably will be made by the user, such as entering non-numeric inputs like letters. Therefore, we need to write error handling code for this error too. It should be put in the first place as soon as the user input something in the textboxes. And the error handler label ***error_handler1*** for this error should be put after the ***error_handler2*** label. This means the second error handling procedure is **nested** within the first error handling procedure. Notice that you have to put an Exit Sub for the second error handling procedure to prevent to execute the first error handling procedure again. The code is as follow:

```
Private Sub CmdCalculate_Click()

Dim firstNum, secondNum As Double
On Error GoTo error_handler1
firstNum = Txt_FirstNumber.Text
secondNum = Txt_SecondNumber.Text
On Error GoTo error_handler2
Lbl_Answer.Caption = firstNum / secondNum
Exit Sub 'To prevent errror handling even the inputs are valid

error_handler2:
Lbl_Answer.Caption = "Error"
Lbl_ErrorMsg.Visible = True
Lbl_ErrorMsg.Caption = " You attempt to divide a number by zero!Try again!"
Exit Sub
error_handler1:
Lbl_Answer.Caption = "Error"
Lbl_ErrorMsg.Visible = True
Lbl_ErrorMsg.Caption = " You are not entering a number! Try again!"

End Sub

Private Sub Txt_FirstNumber_GotFocus()
Lbl_ErrorMsg.Visible = False
End Sub

Private Sub Txt_SecondNumber_GotFocus()
Lbl_ErrorMsg.Visible = False
End Sub
```

The Output window

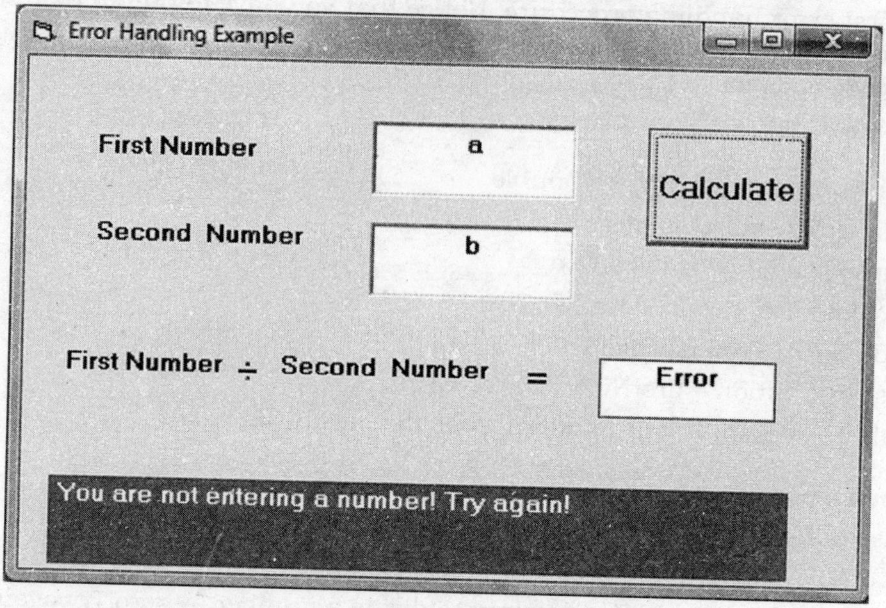

PROGRAM#10 Make an Active X control of your own using VB.

In this lab you will learn how to do the following:

- Create an ActiveX control that can be used in InfoPath
- Add the ActiveX control to the **Controls** task pane
- Insert the ActiveX control into your form
- Install and register the ActiveX controls on users' computers

Setup

Before beginning this lab, install Microsoft Visual Basic 6.0 Professional for Windows® or Microsoft Visual Studio® 6.0. In addition, you will need to obtain a code-signing certificate from a trusted certificate authority (CA).

Exercises

Exercise 1: Create an ActiveX control

Contoso uses a combination of list and text box controls on a frequent basis. As a member of the IT department, you need to write an ActiveX control containing both of these controls so form designers can simply insert this one custom control into their forms. You'll integrate the logic related to retrieving the current price for medications into the ActiveX control so that form designers don't have to worry about writing this code in the future.

Create a new ActiveX control

1. Start Visual Basic.
2. When the **New Project** dialog box appears, click **ActiveX Control**, and then click **Open**.
3. Visual Basic creates a skeleton project with an empty ActiveX control called UserControl1.

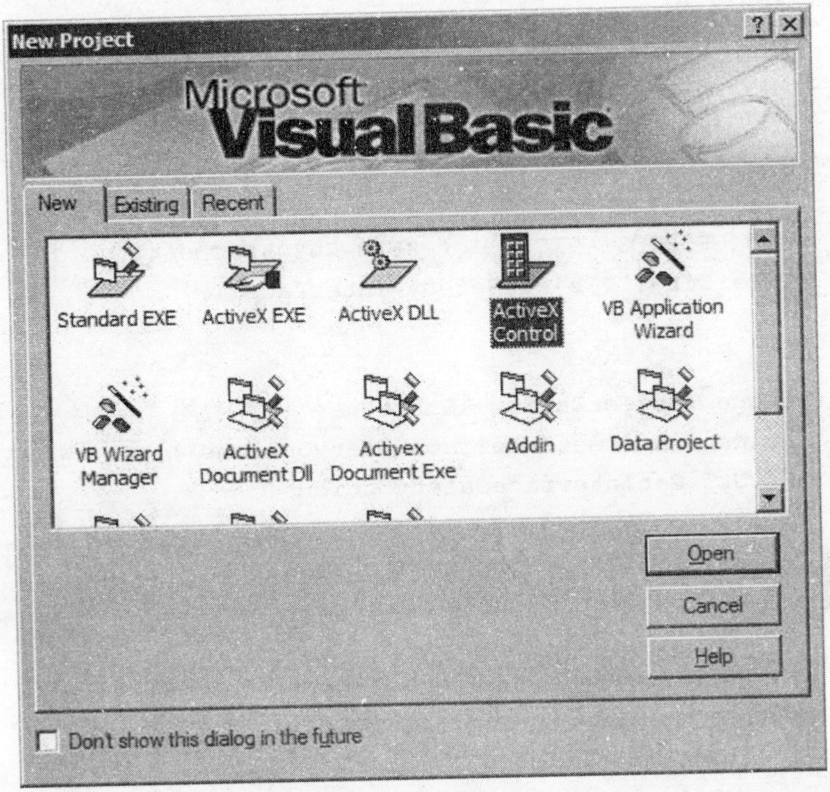

In the next step, you'll add two controls to this ActiveX control, and then add code to retrieve price information. ActiveX controls used in InfoPath have restrictions that are more strict than ActiveX controls used in Microsoft Internet Explorer. For example, InfoPath requires that ActiveX controls be marked as both safe for scripting and safe for initialization. If you write custom ActiveX controls for use in forms, you must implement the **IObjectSafety** interface so that InfoPath knows that a particular control is marked safe for scripting and safe for initialization. Because the **IObjectSafety** interface is not included in the Visual Basic references, you will need to build a type library (.tlb) file that Visual Basic can reference. The process for doing so is described in the following Knowledge Base article:

http://support.microsoft.com/support/kb/articles/Q182/5/98.asp

The first step is to import the following code and reference it in your project.

Mark the control as safe for scripting and initialization

1. Get the OLE Automation Type Library Generator from the Visual Basic CD-ROM. To do this, copy all four files from the \Common\Tools\VB\Unsupprt\Typlib\ folder to your project folder (If you have not saved your VB project, then you should create a folder to store your project and these files in).
2. Copy the following code into a text editor, such as Microsoft Notepad, and then save the file in the project folder as **Objsafe.odl:**

```
[
    uuid(C67830E0-D11D-11cf-BD80-00AA00575603),
    helpstring("VB IObjectSafety Interface"),
```

```
    version(1.0)
  ]
  library IObjectSafetyTLB
  {
    importlib("stdole2.tlb");
    [
      uuid(CB5BDC81-93C1-11cf-8F20-00805F2CD064),
      helpstring("IObjectSafety Interface"),
      odl
    ]
    interface IObjectSafety:IUnknown {
      [helpstring("GetInterfaceSafetyOptions")]
      HRESULT GetInterfaceSafetyOptions(
        [in] long riid,
        [in] long *pdwSupportedOptions,
        [in] long *pdwEnabledOptions);

      [helpstring("SetInterfaceSafetyOptions")]
      HRESULT SetInterfaceSafetyOptions(
        [in] long riid,
        [in] long dwOptionsSetMask,
        [in] long dwEnabledOptions);
      }
  }
```

3. To reference the file in Visual Basic, you must compile it into a .tlb file. Open a command prompt window, and then switch to the folder containing your project. This is the folder where you saved the **Objsafe.odl** file.
4. At the command prompt, type the following command:
   MKTYPLIB objsafe.odl /tlb objsafe.tlb
   After creating the .tlb file, you can reuse it whenever you create ActiveX controls in the future.
5. On the **Project** menu, click **References.**
6. In the **References – Project1** dialog box, click **Browse**, and then locate and open the **Objsafe.tlb** file in your project folder.
   The reference, called **VB IObjectSafety Interface**, now appears in the **Available References** list.
7. Add a new module named **basSafeCTL** to your project.
8. Insert the following code inside the module:
   Option Explicit:

```
Public Const IID_IDispatch = "{00020400-0000-0000-C000-000000000046}"
Public Const IID_IPersistStorage = "{0000010A-0000-0000-C000-
```

```
000000000046}"
    Public Const IID_IPersistStream = "{00000109-0000-0000-C000-
000000000046}"
    Public Const IID_IPersistPropertyBag = "{37D84F60-42CB-11CE-8135-
00AA004BB851}"

    Public Const INTERFACESAFE_FOR_UNTRUSTED_CALLER = &H1
    Public Const INTERFACESAFE_FOR_UNTRUSTED_DATA = &H2
    Public Const E_NOINTERFACE = &H80004002
    Public Const E_FAIL = &H80004005
    Public Const MAX_GUIDLEN = 40

    Public Declare Sub CopyMemory Lib "kernel32" Alias "RtlMoveMemory"
       (pDest As Any, pSource As Any, ByVal ByteLen As Long)
    Public Declare Function StringFromGUID2 Lib "ole32.dll" (rguid As
       Any, ByVal lpstrClsId As Long, ByVal cbMax As Integer) As Long

    Public Type udtGUID
        Data1 As Long
        Data2 As Integer
        Data3 As Integer
        Data4(7) As Byte
    End Type

    Public m_fSafeForScripting As Boolean
    Public m_fSafeForInitializing As Boolean

    Sub Main()
        m_fSafeForScripting = True
        m_fSafeForInitializing = True
    End Sub
```

9. On the **Project** menu, click **Project1 Properties**.
10. On the **General** tab in the **Project1 Properties** dialog box, click **Sub Main** in the **Startup Object** list.

    The **m_fSafeForScripting** and **m_fSafeForInitializing** variables are used to specify the values of the safe for scripting and initialization variables. For the ActiveX control to work in InfoPath, both values must be set to **True**.
11. Open the code window for your ActiveX control.

    Tip You should see UserControl1 in the title bar.
12. Add the following line of code at the top of the code window:

Implements IObjectSafety: This indicates that your ActiveX control implements the necessary **IObjectSafety** interface.

13. Now that you've referenced the **IObjectSafety** interface in your ActiveX control, you must implement the methods in **IObjectSafety**. Because this implementation is the same for any ActiveX control, you can copy the two methods below and paste them into your control's code:

```
Private Sub IObjectSafety_GetInterfaceSafetyOptions(ByVal riid As
Long, pdwSupportedOptions As Long, pdwEnabledOptions As Long)

    Dim Rc       As Long
    Dim rClsId   As udtGUID
    Dim IID      As String
    Dim bIID()   As Byte

    pdwSupportedOptions = INTERFACESAFE_FOR_UNTRUSTED_CALLER Or _
                                     INTERFACESAFE_FOR_UNTRUSTED_DATA

    If (riid <> 0) Then
       CopyMemory rClsId, ByVal riid, Len(rClsId)

       bIID = String$(MAX_GUIDLEN, 0)
       Rc = StringFromGUID2(rClsId, VarPtr(bIID(0)), MAX_GUIDLEN)
       Rc = InStr(1, bIID, vbNullChar) - 1
       IID = Left$(UCase(bIID), Rc)

       Select Case IID
          Case IID_IDispatch
             pdwEnabledOptions = IIf(m_fSafeForScripting, _
          INTERFACESAFE_FOR_UNTRUSTED_CALLER, 0)
             Exit Sub
          Case IID_IPersistStorage, IID_IPersistStream, _
             IID_IPersistPropertyBag
             pdwEnabledOptions = IIf(m_fSafeForInitializing, _
          INTERFACESAFE_FOR_UNTRUSTED_DATA, 0)
             Exit Sub
          Case Else
             Err.Raise E_NOINTERFACE
             Exit Sub
        End Select
     End If
 End Sub
```

```
    Private Sub IObjectSafety_SetInterfaceSafetyOptions ByVal riid As
Long, ByVal dwOptionsSetMask As Long, ByVal dwEnabledOptions As Long)
        Dim Rc        As Long
        Dim rClsId    As udtGUID
        Dim IID       As String
        Dim bIID()    As Byte

        If (riid <> 0) Then
            CopyMemory rClsId, ByVal riid, Len(rClsId)

            bIID = String$(MAX_GUIDLEN, 0)
            Rc = StringFromGUID2(rClsId, VarPtr(bIID(0)), MAX_GUIDLEN)
            Rc = InStr(1, bIID, vbNullChar) - 1
            IID = Left$(UCase(bIID), Rc)

            Select Case IID
                Case IID_IDispatch
                    If ((dwEnabledOptions And dwOptionsSetMask) <> _
                INTERFACESAFE_FOR_UNTRUSTED_CALLER) Then
                        Err.Raise E_FAIL
                        Exit Sub
                Else
                    If Not m_fSafeForScripting Then
                        Err.Raise E_FAIL
                    End If
                    Exit Sub
                End If

                Case IID_IPersistStorage, IID_IPersistStream, _
            IID_IPersistPropertyBag
                    If ((dwEnabledOptions And dwOptionsSetMask) <> _
                INTERFACESAFE_FOR_UNTRUSTED_DATA) Then
                    Err.Raise E_FAIL
                    Exit Sub
                Else
                    If Not m_fSafeForInitializing Then
                        Err.Raise E_FAIL
                    End If
                    Exit Sub
            End If
```

```
            Case Else
                Err.Raise E_NOINTERFACE
                Exit Sub
            End Select
        End If
    End Sub
```

To make the ActiveX control useful, you must add the controls you want it to include and also specify properties for InfoPath to bind to. In this first exercise, you'll add two controls to the ActiveX control.

Add controls to the ActiveX control

1. On the Toolbox, double-click the combo box and text box controls to add them to the design surface, and then position them next to one another, as shown in the following illustration.

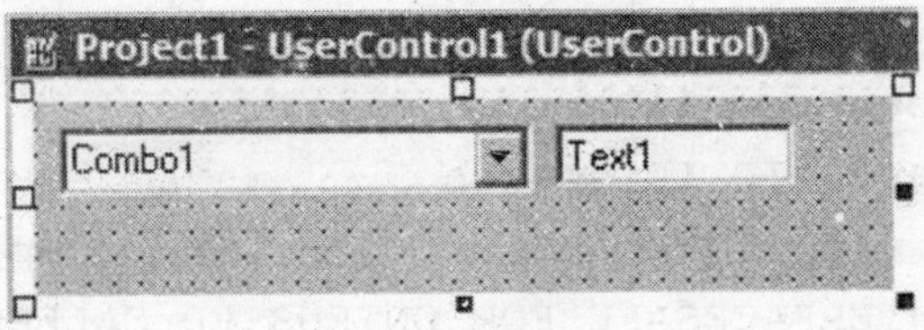

In your InfoPath form, the combo box will show a list of medications from which sales representatives can choose. The text box will show prices for the selected medication.

InfoPath interacts with ActiveX controls in a form through the use of Binding and Enabled properties. InfoPath uses the Binding property to store and retrieve data, and the Enabled property to either enable or disable a control. Visual Basic comes with a wizard that lets you quickly specify properties for an ActiveX control.

Specify properties for an ActiveX control

1. On the **Add-Ins** menu, click **ActiveX Control Interface Wizard**.

   Note If you do not see this command on the **Add-Ins** menu, you can add it by clicking **Add-In Manager** on the Add-Ins menu.
2. On the first page of the wizard, click **Next**.
3. In the **Selected names** list on the **Select Interface Members** page, remove everything except the **Enabled** property, as shown in the illustration below. Although you can use as many standard properties as you would like for your control, InfoPath only uses two—the Binding and Enabled properties.
4. To continue, click **Next**.
5. To create a Value property, on the **Create Custom Interface Members** page, click **New**.

   Note You don't have to name the property "Value," but it's helpful to do so because InfoPath automatically recognizes properties with this name as the binding property for the ActiveX control.
6. To continute, click **Next**.

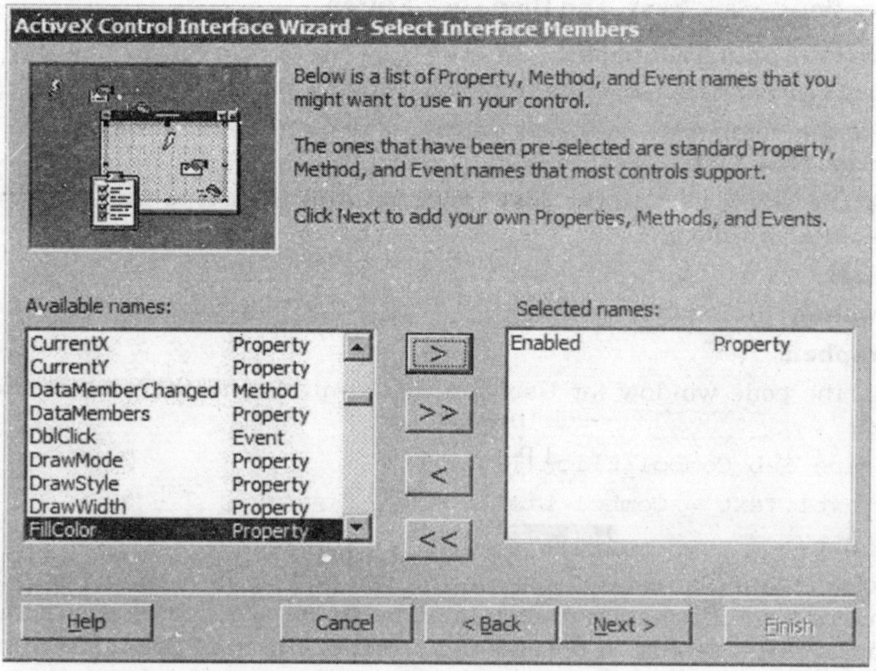

7. In the **Public Name** list on the **Set Mapping** page, click **Enabled**. In the **Control** list under **Maps to**, click **UserControl**, and then click **Enabled** in the **Member** list.

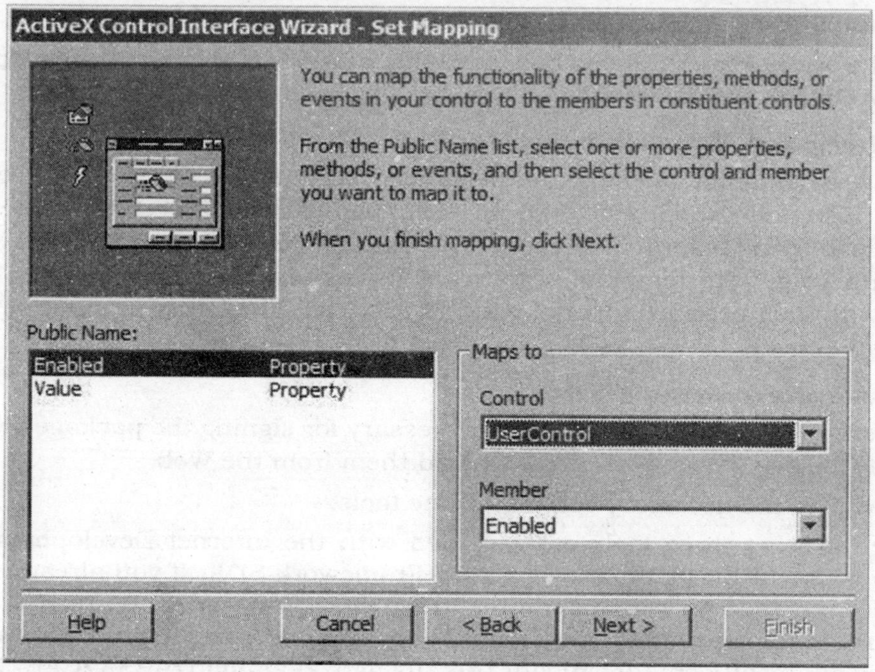

8. Click **Value** in the **Public Name** list.
9. In the **Control** list under **Maps to**, click **Text1**, and then click **Text** in the **Member** list.

10. To continue, click **Next**, and then click **Finish**.

Add code to retrieve pricing information

1. If it isn't already visible, open the form designer for UserControl1.
2. Select the combo box you just added, and then in the **Properties – Combo 1** window, click **List**.
3. Click the arrow next to the **List** property, and then type the following Contoso medicines as values:

   Petosel

   Accuphen

   Ciprophen
4. Open the code window for UserControl1 control, and then insert the following code:

```
Private Sub Combo1_Click()
   Text1.Text = Combo1.ListIndex
End Sub
```

If you were creating a real ActiveX control, you'd likely have more complicated code that retrieved pricing information from a database or Web service. However, making calls to a database or Web service is beyond the scope of this lab. For the purposes of this exercise, the code uses hard-coded values for the Contoso medicine.

At this point, you can compile your project.

Compile the ActiveX control

1. On the **File** menu, click Make **Project1.ocx**.
2. In the **Make Project** dialog box, save the .ocx file in your project folder, and then click **OK** to start compiling.

Exercise 2: Making ActiveX controls available to users

If you use a custom ActiveX control in a form, you must take measures to ensure that the control is installed and registered on users' computers. If you don't do this, users won't be able to open the form to fill it out. You can use InfoPath to package an installation .cab file with your form template. This .cab file installs the ActiveX control on users' computer (with their permission) and registers it for use. The following exercise shows you how to create this file.

Obtain the necessary command-line tools

Note Certain command-line tools are necessary for signing the package file. If you do not have the following files, you can download them from the Web.

1. Download the necessary command-line tools.
2. The following three tools are included with the Internet Development Software Development Kit (SDK) and the .NET Framework SDK. If you already have these SDKs installed on your computer, then there is no need to download anything extra. **cert2spc.exe, makecert.exe, setreg.exe, signcode.exe** http://www.microsoft.com/downloads/details.aspx?FamilyId=9B3A2CA6-3647-4070-9F41-A333C6B9181D&displaylang=en

   or

http://www.microsoft.com/msdownload/platformsdk/sdkupdate/

cabarc.exe

http://support.microsoft.com/default.aspx?scid=KB;en-us;310618

If you do not already have a code signing certificate from a Certificate Authority, you will need to get a test certificate to use on your computer.

Obtain a test certificate

1. Open a command prompt window.
2. The makecert.exe command-line tool can generate a test certificate for testing purposes. At the command prompt, type **makecert newCert.cer –sv privatekey.pvk**.
3. Because this certificate is only good for testing, your computer will not have the necessary certificate authority (CA) certificate root on your computer. To turn this on, at the command prompt, type **setreg 1 true**.

Now that you have a certificate installed on your computer, you will need to prepare it for signing.

Prepare your certificate for signing

If your certificate file is a .cer file, you will need to generate a .spc file.

1. Open a command prompt window.
2. An .spc file can be created by running the cert2spc.exe command-line tool on your .cer file. At the command prompt, type Cert2Spc newCert.cer output.spc.

You are now ready to sign the ActiveX control.

Sign the ActiveX control

1. Copy the following files into one directory:
   - Your ActiveX control (.dll or .ocx) and any supplementary files that the control needs. (For the purposes of this lab, there aren't any extra files for the control you created.)
   - Your certificates and private keys (.spc and .pvk).
2. Open a command prompt window.
3. Package all your files up into a .cab archive. At the command prompt, type: Usage: **cabarc N cabfile.cab**.

   Note If you have an input-mask project, you can type **cabarc N inputmask.cab inputmask.ocx** at the command prompt. If you have multiple files, you can type **cabarc N inputmask.cab inputmask.ocx setupfile.inf otherfiles.txt** at the command prompt.
4. Use the signcode tool to digitally sign the .cab file. At the command prompt, type signcode -spc output.spc -v privatekey.pvk -t http://timestamp.verisign.com/scripts/timstamp.dll inputmask.cab.

   Tip Instead of following the preceding step, you can choose to run signcode.exe, and a wizard will guide you through the necessary steps.

Exercise 3: Add your ActiveX control to InfoPath

Now that the ActiveX control is complete and you have packaged it, it is ready to be used within InfoPath. In the following exercise, you'll learn how to add the control to the **Controls** task pane.

Add the ActiveX control to the Controls task pane

1. In design mode, click **More Controls** on the **Insert** menu.
2. At the bottom of the **Controls** task pane, click **Add or Remove Custom Controls**.
3. In the **Add or Remove Custom Controls** dialog box, click **Add**.
4. On the first page of the Add Custom Control Wizard, click the ActiveX control that you just created in the **Select a control** list, and then click **Next**.

   By default, the ActiveX control is called Project1.UserControl1.

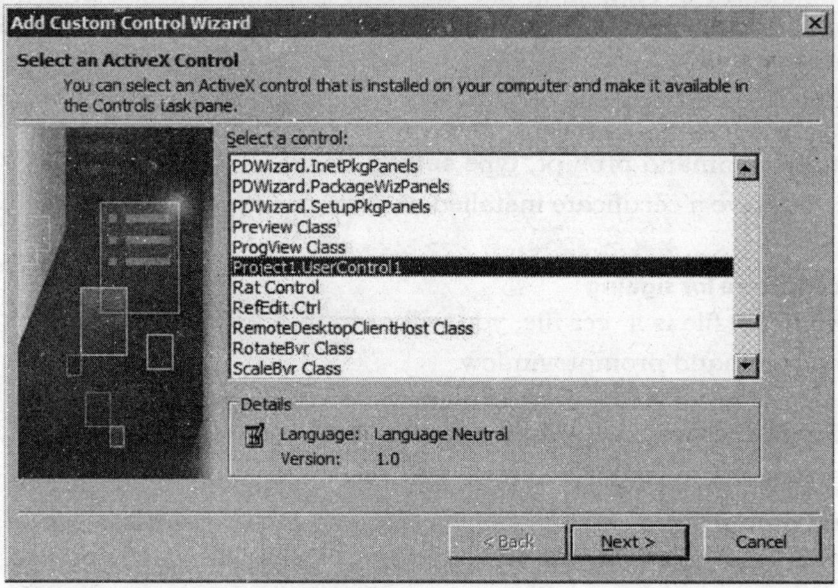

5. On the next page of the wizard, click **Include a .cab file**, and then click **Browse** to search for the .cab file you created in Exercise 2.

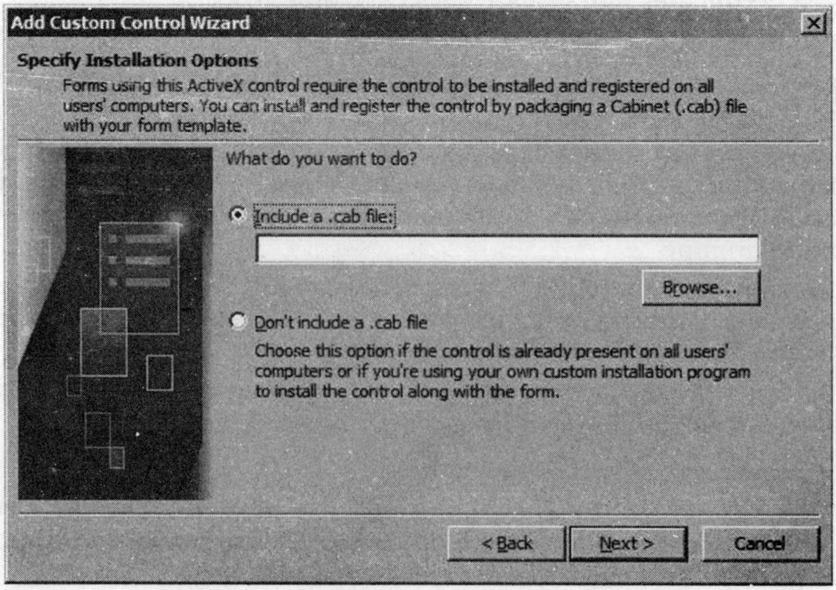

6. To continue, click **Next**.
7. In the **Binding property** list on the next page of the wizard, click a property to bind to. The ActiveX uses this binding property to receive and store XML data.

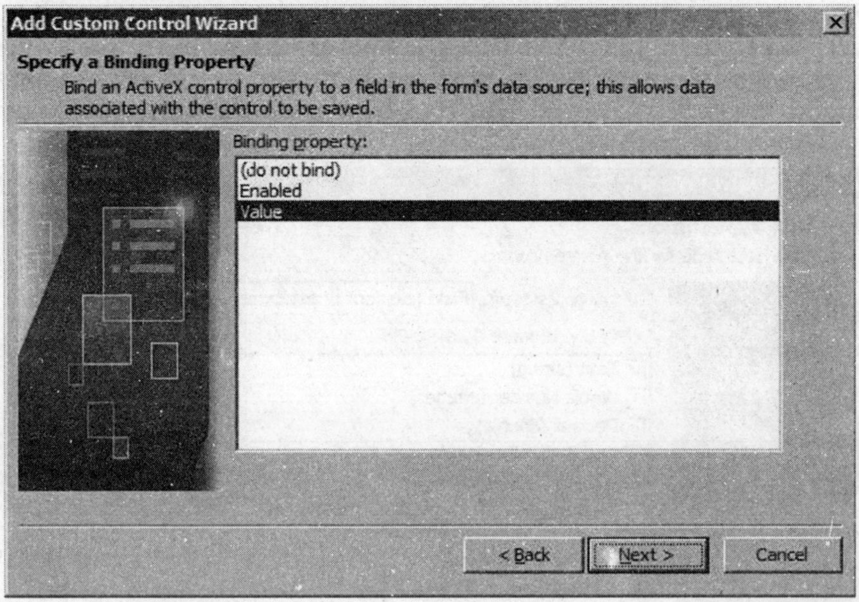

8. To continue, click **Next**.
9. In the **Enable or Disable property** list on the next page of the wizard, click **Enabled**, and then make sure that the value is set to **true**. This step is necessary for InfoPath forms that use digital signatures, conditional formatting, or Reduced Functionality Mode.

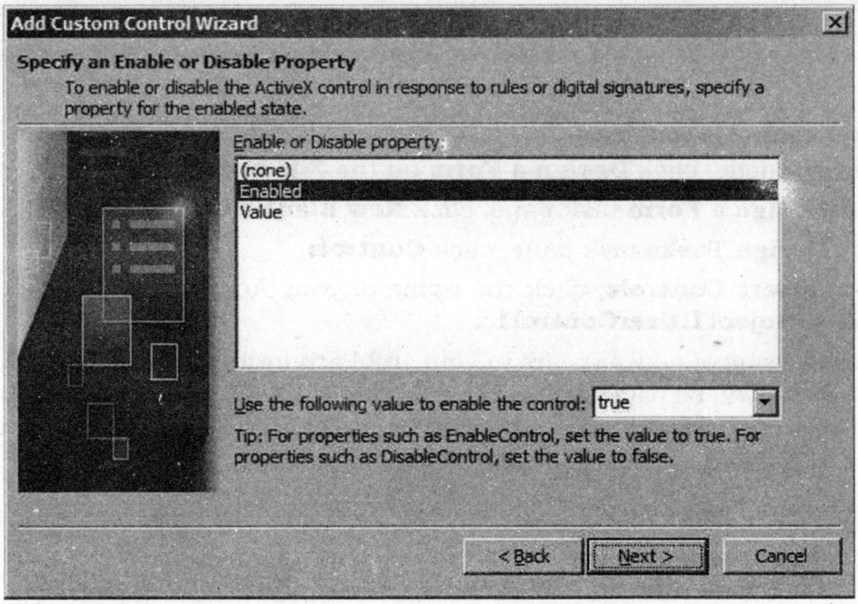

10. To continue, click **Next**.
11. If you specified a binding property in step 7, you must now specify binding options for the ActiveX control. Although there are three types of binding options, for the purposes of this lab, you'll bind the control to a field with a simple data type. This data type is appropriate when the ActiveX control deals with single values. By default, **Text (string)** is selected in the **Select one or more data types** list, and also appears as the default data type. InfoPath will use the default data type when you insert the ActiveX control into your form from the **Controls** task pane.

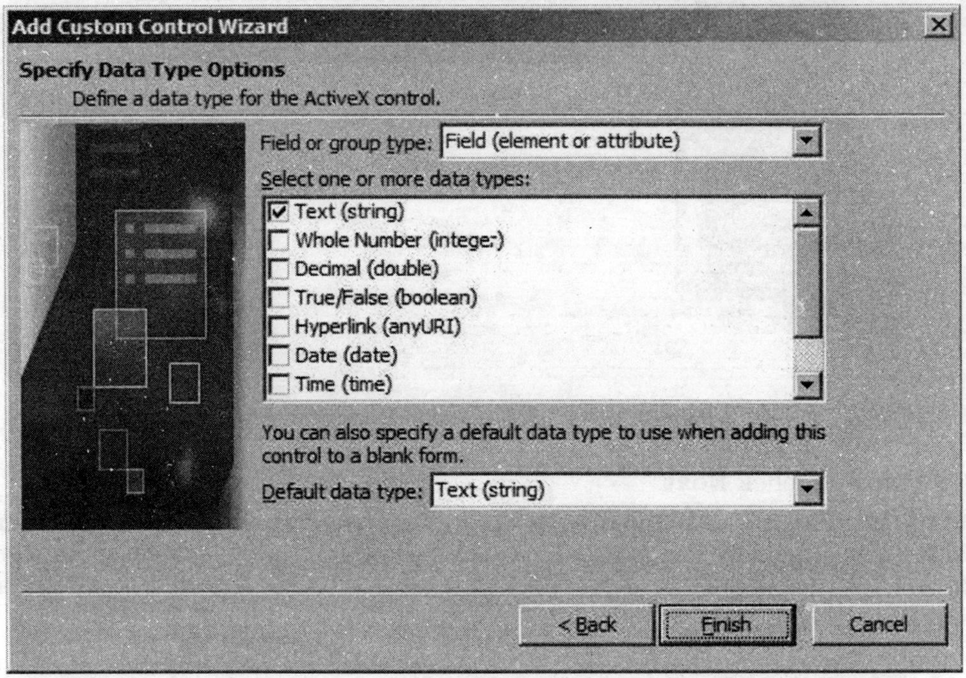

Now that you have added the ActiveX control the **Controls** task pane, you can add it to your form.

Add the ActiveX control to your form

1. In design mode, click **Design a Form** on the **File** menu.
2. In the **Design a Form** task pane, click **New Blank Form**.
3. In the **Design Tasks** task pane, click **Controls**.
4. Under **Insert Controls**, click the name of your ActiveX control. By default, the name is **Project1.UserControl1**.

Your ActiveX control now appears in your InfoPath form. When you save your form, a copy of the .cab file will be packaged along with the form template (.xsn) file. InfoPath uses this .cab file to install and register the ActiveX control for users who don't already have it installed and registered on their computers.

APPENDIX-C

INSTALLING VISUAL STUDIO

You must have administrator credentials to install Visual Studio. However, you do not have to have these credentials to use Visual Studio after it is installed.

If an earlier version of Visual Studio is installed on the computer, read about how to run multiple versions.

Installing By Using Installation Media

Start installing Visual Studio by running Setup.exe, which is located in the root directory on the Visual Studio installation media.

Installing By Downloading from the Product Web Site

Start installing Visual Studio by going to the Visual Studio 2010, download page on the MSDN web site.

Installing Visual Studio in Virtualised Environments

If you run Windows Server 2008 or Windows Server 2008 R2 with Hyper-V enabled and an accelerated graphics adapter, you may experience system slow downs that are caused by an interaction of Hyper-V, CPU features, and accelerated graphics adapters (see KB 961661). This may affect customers who want to use the graphics-rich Visual Studio 2010 environment while they use Hyper-V for testing. To avoid these slowdowns, reconfigure the system as follows:

- On Windows Server 2008, you can use a non-accelerated graphics card, a basic graphics driver (VGA.sys), or an XP-based graphics driver. Any of these options will cause reduced graphics performance and functionality, for example, loss of multi-monitor support.
- On Windows Server 2008 R2, you can use any of the options for Windows Server 2008, or you can ensure that the system supports a feature known as "Nested Page Tables" (AMD), "Rapid Virtualisation Indexing" (AMD), or "Extended Page Tables" (Intel). This feature is present on AMD Phenom and Opteron processors, and on Intel Nehalem processors (Core i7 series, Xeon 5500 series). The processor feature is the preferred option because it enables full graphics performance and functionality without system slow downs.

USING COMMAND-LINE PARAMETERS

The following table lists command-line parameters that are available when you run Set up.

Parameter	Description
/?	Display command-line parameters.
/q	User interface (UI) is suppressed.
/remove	Uninstall Visual Studio.

/f	Repair installed Visual Studio components.
/full	Install all features (only works in quiet mode).
/norestart	Never restart. The installer never restarts the computer during or after installation. Set up will still return 3010, if a restart is required.
/CreateUnattend filename.ini	Create a settings file that is named filename.ini for running unattended installations.
/UnattendFile filename.ini	Run in unattended mode by using the settings in filename.ini.
/NoEventLog	Turn off event logging.
/NoErrorLog	Turn off error logging.
/CreateTransform="filename.mst"	Configure the installation and save the configuration in the specified transform file.

TROUBLESHOOTING INSTALLATION PROBLEMS

For more information about possible installation issues, see Additional Information About Set up and Installation Errors.

Enterprise Network Deployment

For information about how to deploy Visual Studio over a network, see Visual Studio Administrator Guide.

After Visual Studio Is Installed

After you have installed a Visual Studio edition, we recommend that you register your copy of the product and check for updates.

Registering Visual Studio

To register Visual Studio

- On the Help menu, click Register Product.

> **Note :** Microsoft Passport Network credentials (the e-mail address and password on the Microsoft Passport network) and the Product Identification Number (PID) are required to register. The PID is displayed in the About dialog box on the Help menu.

- Or, use the registration card in the product box.

Repairing Visual Studio

To repair Visual Studio on Windows XP or earlier

1. In Control Panel, on the Add or Remove Programs page, select the product edition you want to repair and then click Change/Remove.
2. In the Set up wizard, click Next.
3. Click Repair or Reinstall.

To repair Visual Studio on Windows Vista

1. In Control Panel, on the Programs and Features page, select the product edition you want to repair and then click Uninstall/Change.
2. In the Set up wizard, select Repair or Reinstall and then click Next.
3. Follow the remaining instructions.

Installing Optional Components

To install optional components on Windows 7 or Vista

1. In Control Panel, on the Programs and Features page, select the product edition you want to add to and then click Uninstall/Change.
2. In the Set up wizard, select Add Optional Products and then click Next.
3. Select the components you want to install and then click Next.
4. Follow the remaining instructions.

To Install Optional Components on Windows XP or Earlier

1. In Control Panel, on the Add or Remove Programs page, select the product edition you want to add to and then click Change/Remove.
2. In the Set up wizard, click Next.
3. Click Add or Remove Features.
4. In the Select features to install list, select the items that you want to install and then click Update.

Checking for Service Releases and Product Updates

To Check for Service Releases

- On the Help menu, click Check for Updates.
  This opens the Microsoft Update Web site where service releases are available for download.

> **Note :** If you have changed the settings for Security and Privacy in the Internet Options dialog box in Internet Explorer, you might receive an error message when you try to check for updates. Reset the Security and Privacy settings to the default values and then check for updates again.

To Check for Other Updates

1. Go to the Visual Studio Shipped Software Updates Website.
2. Browse for updates that may be available for your version.

UNINSTALLING VISUAL STUDIO

To uninstall Visual Studio

- To uninstall Visual Studio, see the readme file in the root directory on the installation media, or see Readme on the Microsoft Download Center Website.

INSTALLING VISUAL C++ 6.0 STANDARD EDITION

This paragraph describes the installation of Visual C++ 6.0 standard edition. The installation of other versions or editions might be different; just follow the prompts on the screen. The installation of Visual C++ 6.0 involves two phases, the installation of an Internet Explorer component and the actual installation of Visual C++. If a recent Internet Explorer already exists, the first phase (step 6 to 15) is automatically skipped.

1. Log in as Administrator
2. Insert the CDROM '*Visual C++*'
3. A window '*Visual C++ 6.0 Standard Edition*' will appear:
   - Click **Next**
4. A window '*End User license agreement*' will appear:
   - Click **I accept the agreement**, click **Next**
5. A window '*Product number and User ID*' will appear :
   (a) Enter the number which is on the back of the CDROM cover
   (b) Enter your name or accept the default (if it is correct)
   (c) Enter your company name or accept the default (if it is correct)
   (d) Click **Next**

   Step 6 to 15 are automatically skipped, if a recent version of Internet Explorer is already installed.
6. A window '*Internet Explorer x.yz*' will appear:
   - Click **Next**
7. A window '*License agreement*' will appear:
   (a) Click **I accept the agreement**
   (b) Click **Next**
8. A window '*Installation Option*' will appear:
   - Click **Next**
9. A window '*Windows Desktop Update*' will appear:
   (a) Click **No** (unless you want the Windows Desktop)
   (b) Click **Next**
10. A window '*Active Channel Selection*' will appear:
    (a) Select your country
    (b) Click **Next**
11. A window '*Destination Folder*' will appear:
    - Click **Next** (to accept the default destination) and wait.
12. A window '*Setup has finished installing components*' will appear:
    - Click **OK** and wait...
13. A window '*Set up must restart your computer*' will appear:
    - Click **OK** (to reboot)
14. Log in as Administrator
15. A window '*Install Internet Explorer 4*' will appear:
    - Click **Next**
16. A window '*Choose a common Install Folder*' will appear:
    - Click **Next** (to accepts the default directory)
17. A window '*Visual C++ Standard Setup*' will appear:

(a) Click **Continue**, click **OK**

(b) Click **Typical** (for a typical installation)

18. A window '*Set up Environment Variables*' will appear:

    (a) Unselect **Register Environment Variables** (if it is selected)

    (b) Click **OK** and wait...

19. A window '*Windows NT Debug Symbols*' will appear:

    - Click **OK**

20. A window '*Visual C++ 6.0 Standard Set up*' will appear

    - Click **OK**

21. A window '*Install MSDN*' will appear:

    (a) Unselect **Install MSDN** (You can select MSDN online documentation but it is not required for the NMR Suite.)

    (b) Click **Next**, click **Yes**

22. A window '*Register over the Web Now*' will appear:

    - Unselect **Register** now, click **Finish**

23. If a window '*Restart Windows*' will appear:

    Click **Restart Windows** (to reboot the computer).

□□□

APPENDIX-D

SOLVED QUIZES

How to Read a Text File Line by Line?

```
void CClassName::ReadTextFile(LPCTSTR pszFileName, CStringArray& arrLines)
{
arrLines.RemoveAll();
CString strLine;
TRY
{
CStdioFile file(pszFileName, CFile::modeRead);
while(file.ReadString(strLine)) arrLines.Add(strLine);
}
CATCH_ALL(e)
{
e->ReportError();
// shows what's going wrong
} END_CATCH_ALL}
```

Above code segment will read file line by line and add the lines into arrLines CstringArray type of variable .Later on you can get these lines from addLines Array.

How to Get the Application Name?

In case you have MFC Application you can use . AfxGetApp()->m_pszExeName;

In all other cases the function 'GetModuleFileName()' can be used..

Using STL string

```
#include <string>
char szAppPath[MAX_PATH] = "";
std::string strAppName;
::GetModuleFileName(0, szAppPath, MAX_PATH);
// Extract name
strAppName = szAppPath;
strAppName = strAppName.substr(strAppName.rfind("\\") + 1);
CString str =strAppName.c_str();
MessageBox(str);
```

Using CString (MFC)

```
char szAppPath[MAX_PATH] = " ";
```

```
CString strAppName;
::GetModuleFileName(0, szAppPath, MAX_PATH);
// Extract name
strAppName = szAppPath;
strAppName = strAppName.Right(strAppName.ReverseFind('\\') + 1);
MessageBox(strAppName);
```

How to Convert String to CString (LPSTR)

```
string str ("Hello");
CString str1=str.c_str();
```

HOW TO CONVERT FROM CSTRING TO STD::STRING?

```
CString strSomeCstring ("Hello");

// Use ANSI variant CStringA to convert to char*
std::string strStdString (CStringA (strSomeCstring));
```

What Types of Strings are There?

- **'char*'** -> this is also called a C-style string, or an ANSI string.
- **'wchar_t*'** -> this is wide character string, *i.e.*, an **'unsigned short*'** .
- **CString'** -> this is a string wrapper class, which is part of the Microsoft Foundation Classes (MFC).
- **'std::string'** -> this is a Standard C++ Class wrapping a char string. It is part of the Standard Template Library, or STL.
- **'std::wstring'** -> this is a Standard C++ Class wrapping a wchar_t string. It is part of the Standard Template Library, or STL.
- **'BSTR'** -> this known as basic string or binary string, and is a pointer to a wide character string used by Automation data manipulation functions.

C++ String: How to Convert a Numeric Type to a String?

```
First Method char c[10];
int i = 1234;
sprintf(c, "%d", i); Second Method int i = 1234;
CString cs;
cs.Format("%d", i);
```

How to Convert Between a 'CString' and a 'BSTR'?

```
CString to BSTR
CString str("Pakistan");
BSTR bstr =cs.AllocSysString();
::SysFreeString(bstr);
```

How to Convert String to Hexadecimal

```
sscanf(string, %04X, &your_word16);
```

How to Convert Hexdecimal to String

```
CString strvalue;
unsigned char Buff[1];
Buff[0] = 0x01;
strvalue.Format("0x0%x",Buff[0])
```

How to Convert COleVariant into CString

```
CString strvalue;
COleVariant olevar;
olevar = "Lahore";
strvalue = olevar.bstrVal;
```

How to Convert CString to Char * (char pointer)

```
char *charBuffer = new char[1024];
CString strvalue = "Pakistan";
charBuffer = strvalue.GetBuffer(sizeof(charBuffer));
```

How to Convert char Pointer to CString

```
char * charvalue ="Lahore";CString strvalue =charvalue;
```

How to Append Text to an Edit Control?

```
void CClassName::AppendTextToEdit(CEdit& edit, LPCTSTR pszText)
{
int Length = edit.GetWindowTextLength();
edit.SetSel(Length,Length);
edit.ReplaceSel(pszText);
edit.LineScroll(edit.GetLineCount());
}
//In case you are using multiline Edit Control. If you want to append
text in multiline edit control . set its property multiline =True; and
append \r\n in the text to whom you want to add like "abc\r\n"
```

How to Browse Folder in MFC (Visual C++)

```
BROWSEINFO bi;
memset(&bi, 0, sizeof(bi));
bi.hwndOwner = this->GetSafeHwnd();
bi.lpszTitle = _T("Choose Folder");
LPITEMIDLIST pIDL =::SHBrowseForFolder(&bi);
if(pIDL != NULL)
{
// Create a buffer to store the path, then get the path.
char buffer[_MAX_PATH] = {'\0'};
if(::SHGetPathFromIDList(pIDL, buffer) != 0)
{
```

```
this->txtPath=buffer;
UpdateData(FALSE);
}
// free the item id list CoTaskMemFree(pIDL);
```

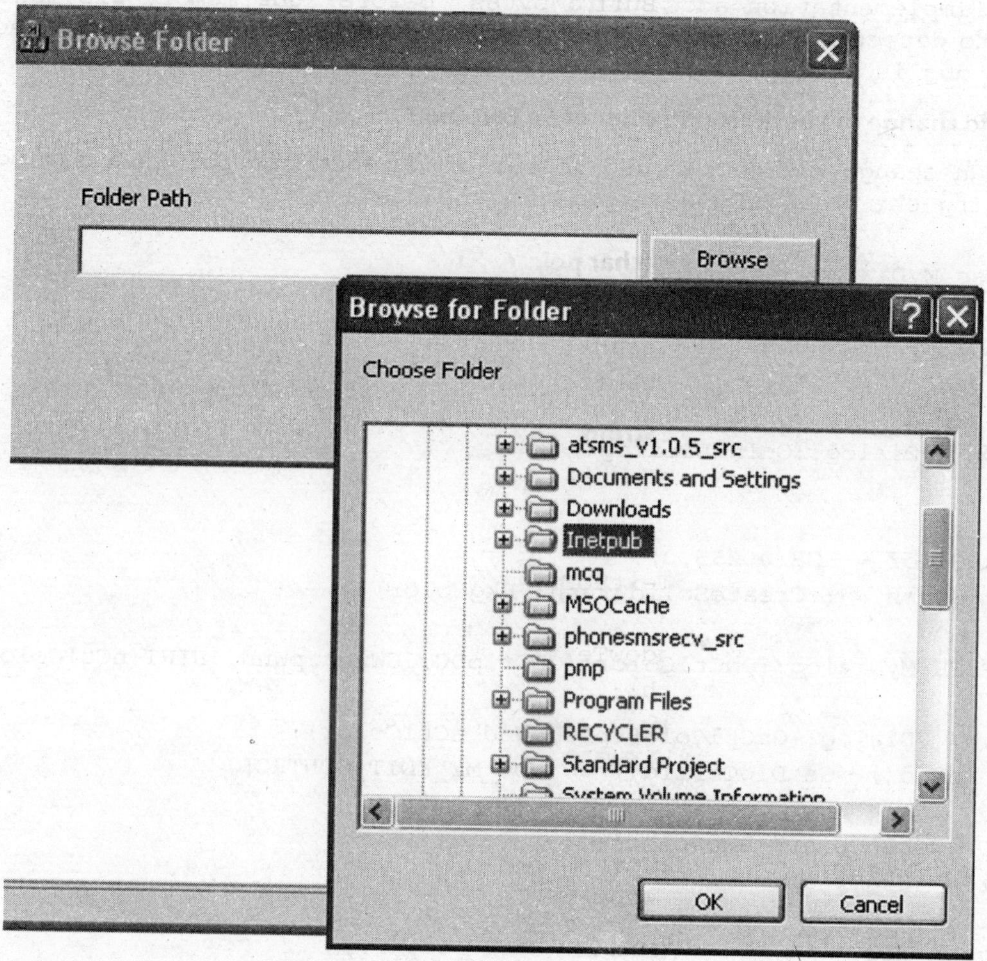


```
}
```

Download Source Code

Is There any Function to minimise the Window?

If you are using MFC: this->ShowWindow(SW_MINIMIZE);

If you are using Win32 API based application: ShowWindow(hWnd,SW_MINIMIZE);

What are Guard bytes ?

malloc_dbg calls malloc, but requests a few more bytes. These bytes are filled with a predefined value. free_dbg then checks if these extra bytes still hold this value. If they did, you probably wrote beyond the allocated memory (the VC Runtime prints a diagnostic message to the debug output). Typical implementations add 4 bytes at the end.

In this case, using free instead of _free_dbg is often not a problem, but depending on the implementation, it might leak the guard bytes. But if an implementation adds guard bytes *before* the memory provided to you (to detect underflows, which are less common), using free instead of -free_dbg is likely to corrupt the heap.

How to change the background color of an Edit Box?

You can change the background colour of an edit control in a dialog by handling the WM_CTLCOLOR message in the dialog:

```
class MyDialog : public CDialog {
//...
COLORREF _BkgColor;
HBRUSH _BkgBrush;
};
BOOL MyDialog::OnInitDialog()
{
//...
_BkgColor = RGB(0,255,0);
_BkgBrush = ::CreateSolidBrush(_BkgColor);
}
HBRUSH MyDialog::OnCtlColor( CDC* pDC, CWnd* pWnd, UINT nCtlColor )
{
hbr = CDialog::OnCtlColor(pDC,pWnd,nCtlColor);
if (pWnd()->GetDlgCtrlId() == IDC_MY_EDIT_CONTROL)
{
pDC->SetBkColor(_BkgColor);
hbr = _BkgBrush;
}
return hbr;
}
```

In the OnInitDialog() handler for the dialog, we initialise the background color value and create a brush in that color. The WM_CTLCOLOR handler is called OnCtlColor. The edit control for which we want to change the color has the resource ID IDC_MY_EDIT_CONTROL. We set the text background to our background color using SetBkColor(), and the overall background for the control by returning the brush we reated.

How to Change Attributes for a Directory?

```
if (! SetFileAttributes("c:\\ResOrg",FILE_ATTRIBUTE_READONLY))
{
DWORD dwError = GetLastError();
}
```

How to Set root Folder in CFileDialog?

CFileDialog has a member m_ofn, which is an OPENFILENAME structure. In that structure, you have to set the lpstrInitialDir variable to point to a string containg the folder in question.

```
CFileDialog FileDialog(TRUE);
FileDialog.m_ofn.lpstrInitialDir ="C:\\MyDirectory";
FileDialog.DoModal();
```

How Do I convert an Integer or Short to a String?

Using a character array

ASCII

itoa(integer,chararray,10); //integer to string array, 10 = radix, returns the string array

TCHAR

_itot(integer,chararray,10);

Unicode

_itow(integer,chararray,10);

Using wsprintf

wsprintf(buff,"Value: %d", integer)

params are TCHAR *, so Unicode or ASCII may be used returns "number of characters stored in the output buffer, not counting the terminating null character."

%f =float

%s=string

%c=char

%d=integer

Using the MFC

CString: CString MFCString;

MFCString.Format("Value: %d", integer);

%f =float

%s=string

%c=char

%d=integer

How do I Convert a String to an Integer?

Using the Run-time Library int _ttoi(TCHAR* str)

returns 0 when the input could not be deciphered

returns the int when the input was deciphered correctly

How do I Write ANSI or Unicode to Console?

```
int puts( const char *string );
int _putws( const wchar_t *string );
```

puts also replaces the '\0' with a '\n' for a new line in the console.

Does the Length of Strlen and Wcslen Include the Terminating Null Character?

No.

How do I remove, for instance, a Password from the RAM while my Program is Running?

```
SecureZeroMemory(szPassword, sizeof(szPassword));
memset( szPassword, '*', sizeof(szPassword) ); //will replace the password, starting at szPassword, with the '*'
character, for a length szPassword
```

How do I Move Memory to a Different Location While My Program is Running?

Copy Memory

Standard

```
void *memcpy( void *dest, const void *src, size_t count );
returns the value of dest
```

Windows API

```
CopyMemory,
MoveMemory
```

How do I Create a Data Segment where I can Store Public Variables that can be Accessed by any Application?

```
#pragma data_seg(".SHARED")
char sharedvar[256] = {0}; // variables MUST be initialized,
#pragma data_seg()
#pragma comment(linker, "/section:.SHARED,RWS") //if you don't put this in the DLL, the data can't even be shared from function to function!
```

How can I make a Region of Data that can be Accessed by Other Applications?

```
GlobalAlloc() (slower than heap) (the clipboard and DDE uses this)
heap function
smore...
```

Is my Application Data Safe in the RAM?

Yes (mostly). All threads of a process can access its virtual address space. However, threads cannot access memory that belongs to another process, which protects a process from being corrupted by another process (unless the program crashes).

Explain the "Virtual Addresses" that are Used by all of my Programs.

Memory management is no longer a problem because the system is free to manage memory by moving pages of physical memory without affecting the virtual addresses. If the threads of a process attempt to use more physical memory than is currently available, the system pages some the memory contents to disk. The total amount of virtual address space available to a process is limited by physical memory and the free space on disk available for the paging file. Physical storage and the virtual address space of each process are organized into pages, units of memory, whose size depends on the host computer. For example, on x86 computers the host page size is 4 kilobytes. When a page is moved in physical memory, the system updates the page maps of the affected processes. When the system needs space in physical memory, it moves the least recently used pages of physical memory to the paging file.

How can I Make a Char Array the Exact Size to hold Two strings?

```
char gray[] = "Gray";
char cat[] = "Cat";
char * graycat = malloc(strlen(gray) + strlen(cat) + 1);
strcpy(graycat, gray);
strcat(graycat, cat);
```

How do I know if a file exists?

```
/**
* Returns true if the file or directory exists and false if neither
exists.
*
* @param filepath file path, can be relative or absolute
* @return true if the file or directory exists
*/
bool FileExists(const TCHAR* filepath)
{
   struct _stat buf;
   return _tstat(filepath, &buf)==0;
}
```

How do I know if a file can be Read?

```
MFC
CFile f;
int success=f.Open(fileLocation,CFile::modeRead);
if(success)
{
...

If you're not using the MFC, you can use this instead:

ASCII:

#include <stdio.h>
FILE *file;
file = fopen( fileLocation, "r" );
if( file == NULL )
break;
```

```
fclose(file);
(or _wfopen for wide characters)
```

Unicode:

```
#include <stdio.h>
FILE *file;
file = _wfopen( fileLocation, "r" );
if( file == NULL )
   break;
fclose(file);
```

How can I get Information about a file or Directory?

```
DWORD dwAttr = GetFileAttributes("C:\\Windows");
if(dwAttr == 0xffffffff)
{
DWORD dwError = GetLastError();
if(dwError == ERROR_FILE_NOT_FOUND)
{
// file not found
}
else if(dwError == ERROR_PATH_NOT_FOUND)
{
// path not found
}
else if(dwError == ERROR_ACCESS_DENIED)
{
// file or directory exists, but access is denied
}
else
{
// some other error has occured
}
}
else
{
if(dwAttr & FILE_ATTRIBUTE_DIRECTORY)
{
// this is a directory
if(dwAttr & FILE_ATTRIBUTE_ARCHIVE)
// Directory is archive file
if(dwAttr & FILE_ATTRIBUTE_COMPRESSED)
// Directory is compressed
if(dwAttr & FILE_ATTRIBUTE_ENCRYPTED)
```

```
// Directory is encrypted
if(dwAttr & FILE_ATTRIBUTE_HIDDEN)
// Directory is hidden
if(dwAttr & FILE_ATTRIBUTE_READONLY)
// Directory is read-only
if(dwAttr & FILE_ATTRIBUTE_REPARSE_POINT)
// Directory has an associated reparse point
if(dwAttr & FILE_ATTRIBUTE_SYSTEM)
// Directory is part or used exclusively by the operating system
}
else
{
// this is an ordinary file
if(dwAttr & FILE_ATTRIBUTE_ARCHIVE)
// File is archive file
if(dwAttr & FILE_ATTRIBUTE_COMPRESSED)
// File is compressed
if(dwAttr & FILE_ATTRIBUTE_ENCRYPTED)
// File is encrypted
if(dwAttr & FILE_ATTRIBUTE_HIDDEN)
// File is hidden
if(dwAttr & FILE_ATTRIBUTE_NOT_CONTENT_INDEXED)
// File will not be indexed
if(dwAttr & FILE_ATTRIBUTE_OFFLINE)
// Data of file is not immediately available
if(dwAttr & FILE_ATTRIBUTE_READONLY)
// File is read-only
if(dwAttr & FILE_ATTRIBUTE_REPARSE_POINT)
// File has an associated reparse point
if(dwAttr & FILE_ATTRIBUTE_SPARSE_FILE)
// File is a sparse file
if(dwAttr & FILE_ATTRIBUTE_SYSTEM)
// File is part or used exclusively by the operating system
if(dwAttr & FILE_ATTRIBUTE_TEMPORARY)
// File is being used for temporary storage
}
}
```

How do I Delete a file that is not Read-Only?

```
Windows API function
BOOL DeleteFile("myfile.txt")

Standard Library
#include <stdio.h>
```

```
if( remove( "myfile.txt" ) == -1 )
   "Error deleting file"
else
   "File successfully deleted
```

How do I Change a file from Read-Only to Writable?

```
#include <sys\stat.h>

if(FindFileData.dwFileAttributes & FILE_ATTRIBUTE_READONLY)
_wchmod(FileName, _S_IWRITE); // change read-only file mode

_S_IWRITE = Writing permitted
_S_IREAD | _S_IWRITE = Reading and writing permitted
```

How do I Extract a Resource While my Program is Running?

```
//e.g. ExtractBinResource(TEXT("BIN"), IDR_BIN1, TEXT("E:\\file.exe"));
void ExtractBinResource( LPCTSTR ResName, int nResourceId, LPCTSTR
strOutputName )
{
   HGLOBAL hResourceLoaded; // handle to loaded resource
   HRSRC hRes; // handle/ptr. to res. info.
   char *lpResLock; // pointer to resource data
   DWORD dwSizeRes;
   // find location of the resource and get handle to it
   hRes = FindResource( NULL, MAKEINTRESOURCE(nResourceId), ResName);
   // loads the specified resource into global memory
   hResourceLoaded = LoadResource( NULL, hRes );
   // get a pointer to the loaded resource!
   lpResLock = (char*)LockResource( hResourceLoaded );
   // determine the size of the resource, so we know how much to write
   out to file!
   dwSizeRes = SizeofResource( NULL, hRes );
   /*
   LPCTSTR lpFileName,
   DWORD dwDesiredAccess,
   DWORD dwShareMode, 0 = do not allow other processes to read / write
   until we're finished
   LPSECURITY_ATTRIBUTES lpSecurityAttributes,
   DWORD dwCreationDisposition,
   DWORD dwFlagsAndAttributes,
   HANDLE hTemplateFile
   */
   HANDLE file =
CreateFile(strOutputName,GENERIC_WRITE,0,NULL,CREATE_ALWAYS,FILE_ATTRIBUTE_NORMAL,NULL);
```

```
if(file!=INVALID_HANDLE_VALUE)
{
    printf("File opened.");
    /*
    HANDLE hFile,
    LPCVOID lpBuffer,
    DWORD nNumberOfBytesToWrite,
    LPDWORD lpNumberOfBytesWritten,
    LPOVERLAPPED lpOverlapped
    */
    DWORD bytesWritten;
    WriteFile(file,(LPCVOID)lpResLock,dwSizeRes,&bytesWritten,NULL);
    CloseHandle(file);
    }
}
```

How do I Write to a File using the Platform SDK?

```
HANDLE file =
CreateFile(strOutputName,GENERIC_WRITE,0,NULL,CREATE_ALWAYS,FILE_AT
    TRIBUTE_NORMAL,NULL);
if(file!=INVALID_HANDLE_VALUE)
{
printf("File opened.");
/*
HANDLE hFile,
LPCVOID lpBuffer,
DWORD nNumberOfBytesToWrite,
LPDWORD lpNumberOfBytesWritten,
LPOVERLAPPED lpOverlapped
*/
DWORD bytesWritten;
WriteFile(file,(LPCVOID)lpResLock,dwSizeRes,&bytesWritten,NULL);
CloseHandle(file);
}
```

How do I Read from a file using the Platform SDK?

```
HANDLE file =
CreateFile(InfoClient,GENERIC_READ,FILE_SHARE_READ,NULL,OPEN_EXISTING,
FILE_ATTRIBUTE_NORMAL,NULL);
if(file!=INVALID_HANDLE_VALUE)
{
    CloseHandle(file);
}
```

How do I get a file's size using the Platform SDK?

```
DWORD GetFileSize(
HANDLE hFile,
LPDWORD lpFileSizeHigh (this can equal NULL)
) != INVALID_FILE_SIZE
```

How do I Get the Working Directory or folder that my Program is Running in?

```
Using the standard libraries
#include <direct.h>
_getcwd(buff,sizeof(buff));

Using the MFC
//sets the working directory, such as C:, E:\my folder, etc.
void SetWorkingDirectory()
{
   TCHAR str[MAX_PATH];
   GetModuleFileName(NULL, str, MAX_PATH);
   CString strDir(str);
   workingDir = strDir.Left(strDir.ReverseFind(_T('\\')));
}
```

How do I delete a Directory and all of the files inside of it, including the read-only files?

```
Uses the Windows API:

BOOL DeleteDirectory(const TCHAR* sPath) {
   HANDLE hFind; // file handle
   WIN32_FIND_DATA FindFileData;
   TCHAR DirPath[MAX_PATH];
   TCHAR FileName[MAX_PATH];

   _tcscpy(DirPath,sPath);
   _tcscat(DirPath,"\\*"); // searching all files
   _tcscpy(FileName,sPath);
   _tcscat(FileName,"\\");

   hFind=FindFirstFile(DirPath,&FindFileData); // find the first file
   if(hFind == INVALID_HANDLE_VALUE) return FALSE;
   _tcscpy(DirPath,FileName);

   bool bSearch = true;
   while(bSearch) { // until we finds an entry
      if(FindNextFile(hFind,&FindFileData)) {
         if(IsDots(FindFileData.cFileName)) continue;
         _tcscat(FileName,FindFileData.cFileName);
         if((FindFileData.dwFileAttributes & FILE_ATTRIBUTE_DIRECTORY)){
```

```
            // we have found a directory, recurse
            if(!DeleteDirectory(FileName)){
               FindClose(hFind);
               return FALSE; // directory couldn't be deleted
            }
            RemoveDirectory(FileName); // remove the empty directory
            _tcscpy(FileName,DirPath);
         }
         else {
            if(FindFileData.dwFileAttributes & FILE_ATTRIBUTE_READONLY)
            _chmod(FileName, _S_IWRITE); // change read-only file mode
            if(!DeleteFile(FileName)) { // delete the file
            FindClose(hFind);
            return FALSE;
         }

            _tcscpy(FileName,DirPath);
         }
      }
      else {
        if(GetLastError() == ERROR_NO_MORE_FILES) // no more files there
        bSearch = false;
        else {
           // some error occured, close the handle and return FALSE
           FindClose(hFind);
           return FALSE;
           }

      }
    }
    FindClose(hFind); // closing file handle

    return RemoveDirectory(sPath); // remove the empty directory

  }

  BOOL IsDots(const TCHAR* str) {
     if(_tcscmp(str,".") && _tcscmp(str,"..")) return FALSE;
     return TRUE;
  }
```

How do I Enable Unicode in my MFC Program?

MFC

If the symbol _UNICODE is defined for a build of your program, TCHAR is defined as type wchar_t, a 16-bit character encoding type; otherwise, it is defined as char, the normal 8-bit character encoding. Under Unicode, then, CStrings are composed of 16-bit characters. Without Unicode, they are composed of characters of type char.

Windows API

If the symbol UNICODE is defined before including <windows.h>, the windows functions will be set for Unicode. The standard libraries use TCHAR, so define both _UNICODE and UNICODE.

Placing L before a string converts it to Unicode, but this does not work correctly for Windows functions for some reason.

Use the _T( ) macro to acquire Unicode strings for a MFC application, and use the TEXT( ) macro to acquire Unicode strings for a Windows application.

Make these substitutions for char

TCHAR Where you would use char.

LPTSTR Where you would use char* or TCHAR*

LPCTSTR Where you would use const char*. CString provides the operator LPCTSTR to convert between CString and LPCTSTR, which is highly convenient.

I've defined _UNICODE but now my MFC app says: unresolved external symbol "_WinMain@16". Help!

In the Output category of the Link tab in the Project Settings dialog box, set the Entry Point Symbol to wWinMainCRTStartup

How do I Convert char* to wchar_t* (or ASCII to Unicode)?

Here's a function I wrote, but it can be improved if you don't like it:

```
//after calling this function make sure to free() the result when done
wchar_t* ASCIIToUnicode(const char *ascii, const int length )
{
    //first it would be best to discover where the '\0' is located
    int i;
    for(i=0;i<length;++i)
    {
        if(ascii[i]=='\0')
            break;
    }
    ++i;
    //now we can make the actual Unicode array
    char *unicode = (char*)calloc(i,sizeof(wchar_t));
    if(unicode==NULL)
        exit(1);
```

```
     for(int j=0;j<i;++j)
     {
        unicode[j*2]=ascii[j];
        unicode[j*2+1]='\0';
     }
        return (wchar_t*)unicode;
  }
  Or try this function (I couldn't get it to work however)
  mbstowcs( server, argv[1], MAXLEN ); (server = w_char with length
MAXLEN, argv[1] = constant char)
```

How do I get the Handle (HWND) of my Console Program?

```
  HWND GetConsoleWindow();
```

How do I get the HWND of the Desktop?

```
  GetDesktopWindow();
```

How do I read or write to the Registry?

```
  HKEY hKey; //remember to call RegCloseKey when finished
  DWORD exists; //will become either REG_CREATED_NEW_KEY or
REG_OPENED_EXISTING_KEY
  /*
  HKEY hKey,
  LPCTSTR lpSubKey,
  DWORD Reserved,
  LPTSTR lpClass,
  DWORD dwOptions,
  REGSAM samDesired, = needs to have KEY_QUERY_VALUE so we can query the
value (KEY_ALL_ACCESS includes that)
  LPSECURITY_ATTRIBUTES lpSecurityAttributes,
  PHKEY phkResult,
  LPDWORD lpdwDisposition
  */
  if(::RegCreateKeyEx(HKEY_CURRENT_USER,TEXT("Software\\My
Folder\\myvalues"),NULL,NULL,REG_OPTION_NON_VOLATILE,KEY_ALL_ACCESS,NULL,
&hKey,&exists)==ERROR_SUCCESS)
  {
     //a new key was created
     if(exists==REG_CREATED_NEW_KEY)
     {
        printf("New key\n");
        //write a value to the registry
        /*
        HKEY hKey,
        LPCTSTR lpValueName,
```

```
        DWORD Reserved,
        DWORD dwType,
        const BYTE* lpData,
        DWORD cbData
        */
        DWORD type = REG_DWORD;
        DWORD longValue = 40;
        DWORD size = sizeof(DWORD);
        if(::RegSetValueEx(hKey,TEXT("Version"),NULL,type,(BYTE*)
(&longValue),size)==ERROR_SUCCESS)
        {
           printf("Wrote a value of 40 to the key.\n");
        }
        else
        {
           printf("Failed to write a value of 40 to the key.\n");
        }
     }
     //the original key was opened
     else
     {
        printf("Opened old key\n");
        //get a value from the key
        /*
        HKEY hKey,
        LPCTSTR lpValueName,
        LPDWORD lpReserved,
        LPDWORD lpType, = REG_BINARY (anything), REG_DWORD (long), REG_SZ
(string)
        LPBYTE lpData,
        LPDWORD pcbData = First you input the size of the pvData buffer.
After the function completes, it stores the amount of bytes that were
copied to pvData (useful for strings), otherwise we get an error
        */
        DWORD type = REG_DWORD;
        DWORD longValue;
        DWORD size = sizeof(DWORD);
        if(::RegQueryValueEx(hKey,TEXT("Version"),NULL,&type,(LPBYTE)
&longValue,&size)==ERROR_SUCCESS)
        {
           printf("Acquired the value.");
        }
        else
        {
```

```
        printf("Failed to acquire the value.");
    }
  }
  RegCloseKey(hKey);
  }
  else
  {
     printf("Could not open key.\n");
  }
```

How do I Delete a value in the Registry?

```
if(RegDeleteValue(hKey,TEXT("Version"))==ERROR_SUCCESS)
printf("Deleted the value.");
else
printf("Failed to delete the value.");
```

How do I delete a key (and all of its subkeys) in the registry?

```
//import shlwapi.lib

#include <shlwapi.h>

if(SHDeleteKey(HKEY_CURRENT_USER,TEXT("Software\\My
Folder"))==ERROR_SUCCESS)
    printf("Deleted the key.");
else
    printf("Failed to delete the key.");
```

How do I stop my Dialog box from Closing when the user presses Enter (Return) or Escape?

Place this function in the dialog's class and add it to the header as well.

```
// stops the default behavior of the Return and Escape keys
BOOL CCreateDirDlg::PreTranslateMessage(MSG* pMsg)
{
    // TODO: Add your specialized code here and/or call the base class
    if(pMsg->message==WM_KEYDOWN)
    {
       if(pMsg->wParam==VK_RETURN || pMsg->wParam==VK_ESCAPE)
       pMsg->wParam=NULL ;
    }
    return CDialog::PreTranslateMessage(pMsg);
}
```

How do I Create a Modeless Dialog?

```
NewDialog *dlg;
dlg = new NewDialog;
```

```
dlg->Create(IDR_MYDIALOG);
dlg->ShowWindow(1);
```

How do I set a hotkey for my Application?

When the program starts...

```
if( RegisterHotKey(m_hWnd, 200, MOD_ALT | MOD_SHIFT | MOD_WIN,
'R')==FALSE ) //200 should be a unique id
    AfxMessageBox("Failed to set the hotkey.");
```

To handle the hotkey in BEGIN_MESSAGE_MAP(CHotKeyTestDlg, CDialog)...

```
ON_MESSAGE(WM_HOTKEY,OnHotKey)
```

Subsequently you define the function...

```
LRESULT OnHotKey(WPARAM wParam, LPARAM lParam);
```

You need to check wParam in the function to see if it contains the id of your hotkey

```
if( wParam == 200)
```

When the window is closed (OnStop() and OnDestroy() to be extra sure)...

```
UnregisterHotKey(m_hWnd,200);
```

How do I get whether a key or Mouse Button is pushed without using Hooks?

```
GetAsyncKeyState();
```

How do I get the position of the Mouse Cursor?

Definition

```
BOOL GetCursorPos(LPPOINT lpPoint);
```

MFC Example

```
POINT point;
GetCursorPos(&point);
AfxMessageBox(_itow(point.x,textArray,10));
```

How do I Simulate a Mouse press or keyboard Press?

```
//mouse input
INPUT input;
input.type=INPUT_MOUSE;
MOUSEINPUT mouseInput;
mouseInput.dx=0;
mouseInput.dy=0;
mouseInput.mouseData=NULL;
mouseInput.dwFlags=MOUSEEVENTF_LEFTDOWN;
mouseInput.time=0;
mouseInput.dwExtraInfo=0;
input.mi=mouseInput;
if(!::SendInput(1,&input,sizeof(INPUT)))
{
    AfxMessageBox(_T("Error sending input."));
```

```
}
//(Unicode) keyboard input
INPUT inputKey;
inputKey.type=INPUT_KEYBOARD;
KEYBDINPUT keyInput;
keyInput.wVk=0;
keyInput.wScan=_T('a');
keyInput.dwFlags=KEYEVENTF_UNICODE;
keyInput.time=0;
keyInput.dwExtraInfo=0;
inputKey.ki=keyInput;
if(!::SendInput(1,&inputKey,sizeof(INPUT)))
{
   AfxMessageBox(_T("Error sending input."));
}
```

How do I let a Program run in Invisible Mode where no Windows or Console Boxes come up?

```
#pragma comment(linker, "/subsystem:\"windows\" /
entry:\"mainCRTStartup\"")
```

How do the Arguments of a Console Program Work?

You start off with:

```
int main( int argc, char *argv[] )
```

argv[0] should be the complete path of the executable

argc is the number of arguments provided (arguments with spaces in between), and should always be greater than one, since argv[0] is the path.

If you want to make a Windows console that features Unicode arguments, use:

```
void wmain( int argc, TCHAR *lpszArgv[ ])
```

or you could use yet another startup:

```
int _tmain(int argc, TCHAR* argv[], TCHAR* envp[])
```

How do I log in with LogonUser?

```
HANDLE hToken;
if(LogonUser(TEXT("User"),TEXT("."),TEXT("password"),
LOGON32_LOGON_NETWORK,LOGON32_PROVIDER_DEFAULT,&hToken))
   printf("Logged in");
else
   printf("Login failed");
```

Why put :: before function names?

You may put :: before a C functions name, like the Windows Software Development Kit's function ::PostMessage(). The Windows SDK ::PostMessage has four paramters. On the other hand, the CWnd class has a PostMessage function too, but this one only has three parameters. If you are writing a CWnd class function in which you wish to call a function which happens to have the same name, you place :: before it.

❑❑❑

INDEX

Win Prizes

ATTENTION READERS

Win Prizes

We request for your frank assessment regarding various aspects of the book as given below :

RAPID APPLICATION DEVELOPMENT USING VISUAL BASIC & VISUAL C++

By: Pankaj Sharma

Please Fill up the Sheet in Neat and Clean Writing

(i) What topics of your syllabus, important for your examination point of view, are not covered in this book ?

....

....

....

....

(ii) What are the chapters wherein the treatment of the subject matter is not systematic or organised or updated ?

....

....

....

....

(iii) Have you come across misprints/mistakes/factual inaccuracies in the book? Please specify the chapters and the page numbers.

....

....

....

....

(iv) What other books on the same subject have you found/heard better than the present book? Please specify in terms of price and quality

....

....

....

(v) Further suggestions and comments for the improvement of this book.

....

....

....

....

(PS: If need be please attach a separate sheet to write your views)

OTHER DETAILS

(i) Who recommended you this book ? (Please tick (✓) in the box near the option relevant to you.

☐ Teacher ☐ Friends ☐ Bookseller

....

....

....

....

(ii) Name of the recommending teacher, his designation and address :

....

....

....

....

(iii) Name and address of the bookseller you purchased the book from :

....

....

....

....

(iv) Name and address of your institution (Please mention the University or Board as the case may be)

....

....

....

(v) Your name and complete postal address :

....

....

....

....

(vi) Course you are enrolled in :

....

....

....

....

The best assessment will be awarded each month: The award will be in the form of our publications as decided by the Editorial Board.

Please mail the filled up coupon at your earliest to :

S.K.KATARIA & SONS®
4760-61/23, Ansari Road,
Darya Ganj, New Delhi-110002
☎: 23243489, 23269384
Email : katariabooks@yahoo.com
katson_sanjeev@yahoo.co.in